FOOD POLYMERS, GELS, AND COLLOIDS

Special Publication No. 82

Food Polymers, Gels, and Colloids

Based on the Proceedings of an International Symposium
organized by the Food Chemistry Group of
The Royal Society of Chemistry at Norwich
from 28th–30th March 1990

Edited by
Eric Dickinson
Procter Department of Food Science, University of Leeds,
England

British Library Cataloguing in Publication Data
Food polymers, gels and colloids.
  1. Food. Constituents: Colloids
I. Dickinson, Eric   II. Royal Society of Chemistry   III. Series
  641.1

  ISBN 0-85186-657-3

Published by The Royal Society of Chemistry,
Thomas Graham House, Science Park, Cambridge CB4 4WF

Typeset by Paston Press, Loddon, Norfolk
and printed by Henry Ling Ltd., at the Dorset Press, Dorchester

Preface

Manufactured foodstuffs typically exist in the form of complex, multi-phase, multi-component, colloidal systems. The two main classes of structural entities in food colloids are polymer molecules (proteins and polysaccharides) and dispersed particles (droplets, crystals, globules, granules, bubbles, *etc.*). The constituent phases in food colloids may exist in various states of matter—gas, liquid, solution, crystal, glass, gel, liquid crystal, microemulsion, and so on. In bulk aqueous phases, interactions between high molecular weight polysaccharides controls biopolymer association and gelation behaviour, which is an important factor influencing the structure, texture, and rheology of many foodstuffs. An important role of proteins is in adsorption at oil–water and air–water interfaces during the formation and subsequent stabilization of emulsions and foams. Overall stability, texture, and microstructure of these food colloids depend on the state of aggregation of the dispersed particles. This in turn depends on interactions between the food polymer molecules (protein–protein, protein–polysaccharide, *etc.*), and on the influence of other food components such as lipids, simple salts, and low molecular weight sugars. One way to try to make sense of this chemical and structural complexity is to study simple model systems in which the nature and properties of the polymer molecules and dispersed particles are relatively well known.

This volume records the proceedings of an International Symposium on 'Food Polymers, Gels, and Colloids' held at the University of East Anglia, Norwich, England, on 28th–30th March 1990. The meeting was the third in a series of biennial Spring Symposia on food colloids to be organized by the Food Chemistry Group of The Royal Society of Chemistry, the first ('Food Emulsions and Foams') having been held in 1986 at the University of Leeds, and the second ('Food Colloids') in 1988 at Unilever, Colworth House. The main theme of the Norwich conference was the role of food macromolecules in determining the stability, structure, texture, and rheology of food colloids, with particular reference to gelling behaviour and interactions between macromolecules and interfaces. The Symposium was attended by over 190 people, of whom about a third were from overseas. This book collects together most of the invited lectures, contributed papers, and poster presentations. Many of the lectures attracted lively discussion, some of which is also reproduced here. A notable feature of the collection of papers assembled in this book is the wide range of physico-chemical techniques which are now being used to address the challenging problems in this field. It is a personal pleasure for me to have the opportunity also to include in this volume an article by my ex-colleague and research collaborator, Dr George Stainsby. The lecture entitled 'Food polymers, gels, and colloids—a teacher's view' was delivered at the Norwich conference to coincide with his being awarded the Food Chemistry Group Senior Medal for 1989 in recognition of his outstanding contribution to the field.

I am very pleased to acknowledge the enthusiastic assistance of Dr Rod Bee, Dr Alex Lips, and Professor Peter Richmond in setting up the scientific programme, and all those at IFRN who helped towards the smooth running of the local arrangements. I am grateful to contributors for submitting manuscripts on time, and especially to those authors and questioners who took the trouble to send me a written version of their discussion remarks.

E. Dickinson
May 1990

Contents

Aggregation Mechanisms in Food Colloids and the Role of Biopolymers

By A. Lips, I. J. Campbell, and E. G. Pelan

UNILEVER RESEARCH, COLWORTH LABORATORY, COLWORTH HOUSE, SHARNBROOK, BEDFORD MK44 1LQ

1 Introduction

Biopolymers play a central role in the stability of oil-in-water food emulsions. More commonly they act by positive adsorption at interfaces, which, as is now appreciated in some theoretical detail,[1] can lead to both stability (*steric stabilization*) and instability (*bridging flocculation*). At a qualitative level, these phenomena are well documented for milk proteins adsorbed on model dispersions[2] and food emulsions.[3–5] Such studies have also addressed the relative effects of milk protein components and their competitive action[6–9] including mixtures of proteins and emulsifiers.[10] The evidence for negative adsorption effects (*depletion flocculation*)[1,11] in food emulsions is less definitive.[12] However, by analogy with model studies on concentrated dispersions,[13,14] depletion flocculation can be expected in concentrated food emulsions in the presence of non-adsorbing polysaccharide thickeners at solution concentrations close to or greater than the overlap value c^*.

In recent years there have been significant advances in statistical *ab initio* theories of steric interactions of polymers between surfaces. An important development has been the elaboration of self-consistent treatments for non-anchored homopolymers[1,15] to complement earlier advances in the description of terminally anchored chains.[16–19] The lattice-based theory of Scheutjens and Fleer[2] (S–F theory) provides a comprehensive framework for representing adsorption and depletion effects for homopolymers between parallel plates for both *full thermodynamic equilibrium* (chains free to leave the gap) and *restricted equilibrium* (chains trapped but subject to local thermodynamic equilibrium). Under the *full equilibrium* approach the interaction between the plates is predicted to be always attractive. Such a situation is relevant for non-adsorbing or very weakly adsorbing polymers and possibly also for flexible surfaces, *e.g.* liquid emulsion films. The model of *restricted equilibrium* is held to be the more appropriate for typical kinetic conditions of approach of coated surfaces. In restricted equilibrium, the free energy will always be greater than that at full equilibrium. For large and intermediate plate separations, the models do not deviate substantially in their prediction of the attractive well due to bridging. At shorter range, however, the restricted equilibrium model always indicates a steric barrier. The parameters of lattice treatments are the number of segments of the polymer chain N, the familiar Flory–Huggins solvency parameter χ, and an adsorption energy per segment χ_S usually measured in units of kT. For adsorbing polymers under conditions of restricted equilibrium, the free energy of the polymer interaction between approaching plates is a subtle interplay of attractive contributions from bridging and repulsive contributions from segmental overlap and loss of conformational entropy. In general, the attractive minimum shifts to larger separations with increasing surface coverage and occurs at a separation d comparable to the radius of gyration R_G of the isolated polymer chain. The depth of the bridging minimum increases with surface coverage to a maximum at an intermediate coverage. For high adsorbed amounts in good solvents, the minimum can disappear altogether and only repulsion is then predicted. An interesting aspect of the theory is that the interaction is expected to have little dependence on the molecular weight of the polymer provided that the adsorbed amount is the same. However, since, at a given solution concentration, the surface coverage increases with molecular weight, shorter chains can yield deeper minima; complete steric stabilization is possible only with polymers of very high molecular weight.

Direct measurements of interaction forces between polymer covered mica surfaces[20–22] lend qualitative support to the S–F theory for restricted equilibrium. At short range, strong repulsion is seen, and bridging minima are observed even in good solvents at distances comparable to R_G. The minimum disappears for high adsorbed amounts in good solvents. Colloid stability studies on dispersions with polymers adsorbed from good solvents generally support the theoretical conclusion that bridging can be a dominant mechanism at low to intermediate surface coverage and that therefore it is not necessary to invoke poor solvent conditions ($\chi \geq \frac{1}{2}$) to explain instability. Quantitative comparison between theory and experiment, however, is still difficult as the level of adsorption of polymer in the model interaction studies is not easily determined. Also, the steric effects can be strongly dependent on heterodispersity, and in practice it is difficult to obtain a polymer fraction of adequate monodispersity to test theories. Another problem is that typical experimental situations, for example in emulsification or whipping, could imply an approach of surfaces at greater rates than those of local equilibration of tightly confined polymer. Even the restricted equilibrium model may then be inappropriate.

With concentrated dispersions subject to depletion flocculation, the link between theory and experiment is more fully established. It is assumed then that $\chi_S \approx 0$.[1,11] Statistical mechanical predictions of the phase behaviour of concentrated dispersions,[23] with interparticle potentials modelled on the volume depletion argument of Asakura and Oosawa[24] and Vrij,[25] are in reasonable agreement with experiment. However, quantitative discrepancies[14] are apparent when particle and polymer are of comparable size.

Regarding more complicated macromolecular behaviour, a successful model[26–28] has been elaborated for the adsorption of heterodisperse polymers. Advanced statistical treatments for polyelectrolyte adsorption[29–31] are also now available. Self-consistent lattice treatments have recently been developed to represent the adsorption of copolymers[32–33] as well as interactions between adsorbed layers of block copolymers.[33] These studies include the complications from bulk association of the polymers. Non-random block copolymers with strongly adsorbing blocks and non-adsorbing protruding moieties are not expected to yield bridging minima in good solvents even under conditions of full equilibrium. The behaviour predicted for restricted equilibrium can approach that for terminally anchored chains. The adsorption of a random copolymer can be modelled in terms of an equivalent homopolymer with a suitably weighted average segment adsorption energy.

Pragmatic theories of steric stabilization are useful for simple baseline predictions. The earliest is that of Fischer[34] which models the overlap of the steric layers attached to two spheres on the basis of changes in mixing free energy for an assumed constant density of polymer segments in the gap. This approach can be criticized in that the model implies terminally anchored as opposed to volume-restricted chains and, more seriously, in that, even in tightly confined situations, the segment density is non-uniform. Also the mixing term is an incomplete representation of the free energy in the interpenetrational regime $(2R_G > d > R_G)$ and certainly in the interpenetrational/compressional domain $(d < R_G)$ where elastic repulsion due to volume restriction can be predominant.[11] Napper has considered more realistic segment distributions.[11]

Measurements of the osmotic disjoining pressure of concentrated dispersions,[35,36] and the compression of particles located at surface monolayers,[37] indicate that the pragmatic theories substantially overestimate the magnitude of the steric repulsion at short range. This may be due to more complicated solution thermodynamics (χ dependent on concentration), but, more probably, it is due to an inadequate representation of local (restricted) equilibrium in the pragmatic models. Another factor could be desorption of polymer or partial redistribution of segments away from the region of closest contact between the curved surfaces.[37] To date no *ab initio, self-consistent* theory for steric interactions between curved surfaces has been advanced; the existing treatments are all for the flat plate geometry and implicit in their extension to spherical geometry has been the use of the Derjaguin approximation.[38] Such an approach may be questioned, even for large particles and small gaps, since polymer segments remain correlated on length scales larger than the gap width. One of the assumptions implicit in the Derjaguin approximation is that local properties are averaged on length scales which are small compared with the separation between particles.

A major goal in understanding the behaviour of food emulsions is to relate the magnitude of steric interactions to electrostatic and van der Waals forces, external shear forces, and other possible mechanisms of interaction such as crystal bridging[39] and protein micellization. Factors associated with quiescent storage (sedimentation, creaming, yield stress, stability) are governed by relatively weak polymer interactions at longer range. These can frequently be net attractive owing to bridging. On the other hand, food macroemulsions, even in gentle shear, experience a local Stokes force between particles which is greatly in excess of thermal forces. From a technological standpoint the behaviour of emulsions under shear (*e.g.*, during emulsification, whipping, packing, *etc.*) is of major concern. With increasing shear the magnitude of steric repulsion is tested at progressively shorter separations—well into the interpenetrational/compressional domain ($d < R_G$). As stated above, it is in this region where our understanding of polymer forces is least certain and where the geometry of the contact region may be a controlling factor in a way not yet fully understood. Relevant here is the fact that liquid droplets can undergo contact flattening under shear;[5,40–42] food emulsion droplets, often semi-solid, can therefore display a range of elasto-hydrodynamic responses and contact geometries. While it is well established that oil-in-water emulsions derive their stability to coalescence from adsorbed milk proteins through steric barriers at short range, the magnitude of these forces is not adequately elucidated. A greater insight, including the role of curved contact geometry, would help to clarify some important mechanistic questions: *e.g.*, whether sheared fat emulsions are destabilized by crystal bridging[39] or by an inadequacy of steric forces.

The aim of the present study is to consider the extent to which the above theories can serve as a framework for understanding, at a more informed and less qualitative level, the steric effects of food biopolymers between surfaces. Such an endeavour may be too ambitious since biopolymers display a much greater complexity than is currently within the scope of theory. Milk protein emulsifiers are mixtures of proteins with intricate secondary and tertiary structure, they are polyelectrolytes ('copolymeric'), and they are subject to involved, ion-mediated, 'micellar' mass action equilibria. Polysaccharides can be highly polydisperse and branched, strongly heteropolymolecular and 'copolymeric' (*e.g.*, gum arabic), polyelectrolytes, and extremely stiff (*e.g.*, xanthan). A striking feature can be their ability to form gels by conformational transitions involving regions of alignment on more than one chain. There is considerable evidence for two-dimensional network formation and denaturation in adsorbed protein layers. Despite these complexities, some general conformity to the above theories might still be expected. Indeed, the case for exploring this is stronger now as the solution behaviour of food macromolecules is becoming better characterized in terms both of macromolecular structure and thermodynamic solution behaviour.

Our experimental approach is based on comparative studies of the colloid stability of dilute dispersions of model polystyrene latex particles in the presence of a range of food polymers. Time-resolved light scattering has been utilized following a previously established methodology.[43–47] We believe that this technique can yield sensitive information on polymer interactions between particles and that such an approach has not yet been fully exploited with food polymers. Dynamic light scattering investigations have also been carried out in preliminary characterization of micellar interactions in caseinate solutions and measurement of adsorbed layer thickness. Kinetic studies of colloid stability offer advantages of relative simplicity and the ability to study polymer effects on non-polar substrates. The information is necessarily less direct than that obtained by a force balance approach. Whether the behaviour of model colloids is adequately correlated with the stability of typical food emulsions remains to be fully confirmed. We will, however, show an example where such a connection can be made.

2 Experimental

Materials—Sodium caseinate (commercial grade 'Spraybland') was supplied by DMV, and samples of dextran T500, T110, and T40 (molecular weights respectively 500 000, 110 000, and 40 000) by Pharmacia. A commercial grade of guar (TH/225 Hercules) was used. Keltrol F (Kelco) was used as a source of xanthan. Gum arabic (4282) was supplied by Merck, acid gelatin (Bloom Strength 250) by Croda. Stock solutions of the caseinate, dextrans, and gum arabic were prepared by dissolution in cold water with mild stirring, gelatin was dissolved in warm water, and guar and xanthan were stirred into hot (95 °C) water. A range of conditions of electrolyte, pH, and temperature was examined.

Three monodisperse polystyrene latices were used: latex A (diameter 207 nm), prepared by the method of Kotera *et al.*[48] with persulphate as initiator; latex B (diameter 88 nm), prepared with SDS and persulphate initiator followed by removal of the SDS by passage through an ion retardation resin; and latex C (diameter 48 nm), supplied by Polysciences (Cat. 0891).

Methods—Several time-resolved light-scattering methods were employed to study coagulation kinetics.

The stability of latex A was assessed using a Sofica wide-angle scattering instrument with incoherent source and detection at a scattering angle of 30° and wavelength $\lambda = 546$ nm. For the chosen range of number concentration, $5 \times 10^8 – 5 \times 10^9$ cm^{-3}, which corresponds to a diffusion-limited coagulation half time in the range 40–400 s, the initial slopes of intensity *versus* time curves can be reliably measured and, using a model previously established,[43,44] estimates can be obtained of absolute rate constants of coagulation. Latex A was free from pre-aggregation as ascertained from the good agreement between the measured angular distribution of the scattered light and that calculated from Mie theory.[43] The stability of latex A was studied in the presence of salt and dextran flocculant; the stability factor W (ratio of the most rapid diffusion-limited rate measured in excess electrolyte to the rate under more stable conditions) could be determined in the range $0 < \log W < 2.5$.

The stability of latex B was examined in the presence of different levels of electrolyte and a range of food biopolymers. A turbidimetric technique was employed at $\lambda = 500$ nm. The minimum measurement time of this technique proved to be 5 s. This is several times the Smoluchowski diffusion half-time at the higher particle number concentration, 4×10^{11} cm^{-3}, chosen for latex B to achieve an adequate level of turbidity. It was not therefore possible to measure initial rapid rates in excess electrolyte. However, using a random polycondensation model[45] to represent the structure factor of a coagulating dispersion of spheres and integrating over all angles to represent turbidity,[49] it was possible to model the long-time behaviour of the turbidity change and to extract a reasonable estimate, $(1.7 \pm 0.7) \times 10^{-18}$ m^3 s^{-1} at 28 °C, for the rapid rate constant in high electrolyte. Details will be presented elsewhere of the model fit which also indicated a significant degree (*ca.* 0.5 of a half-life) of pre-aggregation in latex B, an inference which was consistent with the difference between the measured initial turbidity and that calculated from Mie theory[49] assuming the electron microscope diameter of latex B and complete dispersion. While rapid rates are less accurately defined, dispersions of smaller particles with consequent higher number densities offer access to a wider range of stability factor W; in the case of latex B, this range proved to be $0 < \log W < 4.5$.

Simultaneous dynamic and static light scattering studies, using an instrument developed by ALV-Langen, were carried out on latex C. The particle number density was 8×10^{12} cm^{-3}; other conditions were: 0.01 mol dm^{-3} sodium borate, pH 8, and temperature in the range 5–50 °C. An attempt was made at measuring the adsorbed layer thickness of sodium caseinate on the polystyrene particles at a protein dose of 3 mg m^{-2} and as a function of temperature. A difference method between coated and uncoated particles was employed: hydrodynamic radii were inferred from a cumulant analysis of the autocorrelation function; changes in polydispersity on addition of polymer were assessed by a Laplace transform method to yield the z-distribution, and more directly from the measured time averaged absolute intensity. Polymer adsorption can change the stability of the latex and therefore difficulties can arise in the interpretation of dynamic measurements.[50] The selection of small particles is preferred. The measured hydrodynamic radius of the bare latex particles C was 24 ± 0.5 nm, which is in good agreement with the electron microscope diameter, suggesting only minor complications from pre-aggregation. The protein was first adsorbed on the particles, and the coated particles were then filtered through 0.22 μm Millipore filters. The scattering measurements were performed at angles in the range 30–120°.

Model concentrated emulsions (45 wt%) based on sunflower oil and 1 wt% caseinate as the emulsifier were prepared on a pilot plant at an homogenization pressure of 100 bar. The storage stability of these emulsions in the presence of 0.1 mol dm^{-3} sodium chloride was studied at 5 and 25 °C. Concentrated emulsions (45 wt%) based on pure tripalmitin and 1 wt% caseinate were prepared under similar processing conditions. Low-resolution NMR indicated that the tripalmitin particles were 100% solid fat globules in the temperature range 5–35 °C. The whip times of the tripalmitin emulsion in this temperature range were measured by the Mohr method. Individual whip experiments were carried out under isothermal conditions.

3 Results and Discussion

Interactions between Uncoated Particles—Initial studies of the stability of the latices to electrolyte alone indicated reasonable conformity to DLVO theory. Figure 1 illustrates a $\log W$ *versus* $\log C$ stability plot for latex B, where C is the concentration of sodium chloride. The measured slope and intercept interpreted by the classical method of Reerink and Overbeek[46,47,51] yielded a realistic Hamaker constant of 1×10^{-20} J for the polystyrene particles which was therefore used to characterize the electrostatic barrier between the charged latex particles. The difficulties of measurement and interpretation of colloid stability plots have been discussed elsewhere;[46,47] the simplest analysis which has been adopted here is, we believe, sufficient to gain a first-order appreciation of the

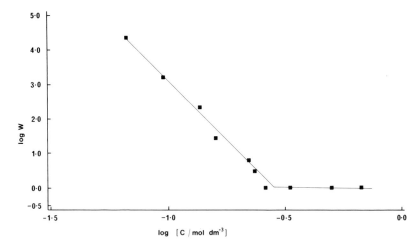

Figure 1 *Log* W versus *log* C *colloid stability plot for latex B.* W *is the stability factor and* C *is the concentration of sodium chloride* (*latex diameter* 88 nm, *particle density* 4.5×10^{11} cm^{-3}, 28 °C, *pH* 5.9)

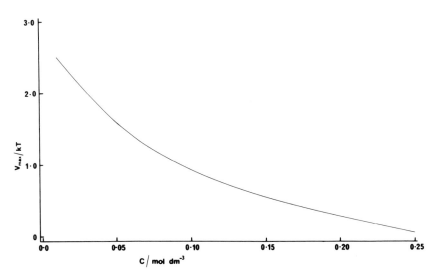

Figure 2 *Predicted DLVO maximum* V_{max}/kT *for uncoated polystyrene particles.* C *is the concentration of sodium chloride* (*latex diameter* 88 nm, *pH* 5.9, 28 °C)

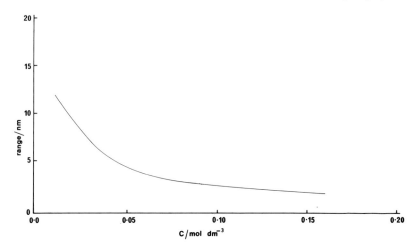

Figure 3 *Predicted maximum range of DVLO barrier for uncoated polystyrene particles. C is the concentration of sodium chloride (latex diameter 88 nm, pH 5.9, 28 °C)*

maximum magnitude (Figure 2) and span (Figure 3) of the electrostatic barrier of the uncoated particles.

Effects of Dextran—Figure 4 shows the dependence of the stability factor W for latex A on the level of dextran addition expressed in mg added per m^2 of available polystyrene latex surface. The concentration of sodium chloride was held constant at $0.095 \, \text{mol dm}^{-3}$. This corresponds to a relatively weak electrostatic barrier of *ca.* $4 \, kT$ with a narrow span of 2.5 nm which is less than the radius of gyration of the smallest chain (dextran T40). In this case bridging is highly efficient. At low to intermediate coverage, both the high and low molecular weight polymers induce flocculation and completely counteract the electrostatic barrier ($\log W = 0$). At high polymer addition, restabilization can be observed for dextran T500 but not for the smaller chains. This behaviour is consistent with the predictions of the S–F theory for restricted equilibrium in that steric effects are controlled by the level of adsorbed polymer, the pseudo-plateau limit of which is expected to increase with molecular weight.[52] The observed greater bridging efficiency of the high molecular weight dextran *at low coverage* is less easily explained. It is remarkable that effects of bridging at levels as low as $0.001 \, \text{mg m}^{-2}$ (corresponding to less than one polymer chain for every ten polystyrene particles) can be detected with dextran T500. Figure 5 also illustrates that bridging, at a fixed low polymer dose of $0.0537 \, \text{mg m}^{-2}$, increases with increasing molecular weight. The concentration of electrolyte required to induce significant instability increases with decreasing molecular weight. On the basis of DLVO calculations, this would imply that the range of the electrostatic barrier required to limit bridging is *ca.* 17 nm for dextran T500, 7.5 nm for T110, and

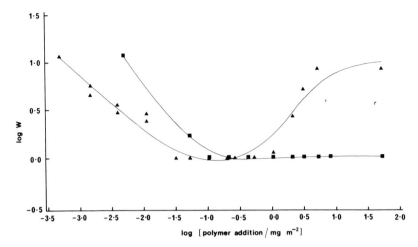

Figure 4 *Bridging flocculation with dextrans. W is the stability factor, and the polymer addition is scaled to available latex area (latex diameter 207 nm, particle density 4×10^9 cm^{-3}, 25 °C, pH 5.8, sodium chloride concentration 0.095 mol dm^{-3}): ■, dextran T40; ▲, dextran T500*

4.5 nm in the case of T40. The reasonable correspondence of these interaction ranges with the radii of gyration of the chains is noteworthy.

High molecular weight polymers in general show high affinity adsorption with pseudo-plateau levels at very low equilibrium concentrations of polymer in solution.[52] In most cases, it can be reasonably assumed that, at levels of polymer

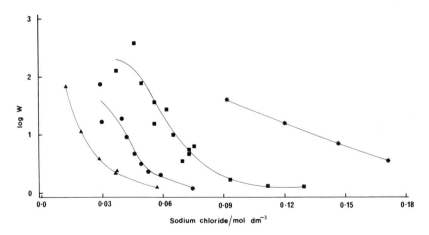

Figure 5 *Bridging flocculation with dextrans. Effect of molecular weight at low coverage (polymer addition 0.0537 mg m^{-2}, latex diameter 207 nm, 25 °C, pH 5.8): ▲, dextran T500; ●, dextran T110; ■, dextran T40; ✸, stability of bare latex particles*

addition below the saturation capacity of the pseudo-plateau, the large majority of any added polymer molecules are adsorbed. If this condition did apply, then the levels of adsorption below the pseudo-plateau (in the Henry region) would be independent of molecular weight. According to S–F theory, bridging flocculation, irrespective of whether under full or restricted equilibrium control, should not then be strongly dependent on molecular weight. In the present case, however, the maximum possible equilibrium phase volume of dextran, at the levels at which bridging is first observed, is typically 10^{-9}–10^{-7}. Since dextran adsorbs only weakly on polystyrene,[14] the observed dependence of bridging on molecular weight is perhaps attributable to relatively incomplete adsorption of the low molecular weight dextran. On the basis of calculations and theory reported in the literature,[52–54] this would imply that the value of χ_S (*i.e.* the difference in adsorption energy per segment–solvent exchange) only slightly exceeds the critical value for adsorption,[53,54] by at most $0.1\,kT$. While such an argument could reconcile an apparent contradiction between theory and experiment, it remains unclear why, at very low coverage, more strongly adsorbing polymers of comparable molecular weight (*e.g.* PVP) are far less efficient than dextran at bridging flocculation. One explanation could be that, by virtue of weak adsorption, macromolecules such as dextran are less restricted between surfaces; at low coverage this may result in deeper bridging minima close to those expected for full equilibrium.

Recently, Pelssers *et al.*[55] have proposed that reconformation rates of attached polymer can be of the order of the times of collisions between particles in typical colloid stability experiments. This has led to a distinction between the concepts of 'equilibrium' and 'non-equilibrium' bridging flocculation. The former is intended to describe fully relaxed polymer on the surface, and the latter the case of polymer temporarily extended and more actively bridging. The evidence for this is an observed breakdown in second-order coagulation kinetics: at higher concentrations 'bridging particles' are unexpectedly reactive. In the case of bridging with dextran, we observe a similar deviation from second-order kinetics. However, we attribute this to a different phenomenon. Figure 6 shows the dependence of the structure factor $S(\tau)$ (= ratio of the intensity of light scattered at an angle of 30° to the corresponding intensity of the unaggregated dispersion) on time of flocculation. The horizontal axis represents a reduced time $\tau = K^{\dagger} t n_0$, where n_0 is the initial particle number concentration of latex A, t the actual time of flocculation, and $K^{\dagger}$ a constant which has the significance of a second-order forward rate constant. The curves relate to different number concentrations each at a constant added concentration, $0.057\,\mathrm{mg\,m^{-2}}$, of dextran T500 and at a sodium chloride concentration of $0.029\,\mathrm{mol\,dm^{-3}}$. If strict second-order kinetics applied, a single curve should represent all the data points (with constant $K^{\dagger}$ independent of n_0). As can be seen, however, the curves are coincident only at short reduced times. This suggests that, while the *initial* kinetics are second order, there is also a second kinetic process, a fragmentation process, which limits the extent of aggregation. The deviations increase with decreasing n_0, indicating a progressively lower degree of aggregation. Reversibility can be clearly demonstrated. The flocculation of the most concentrated dispersion, $n_0 = 4.3 \times 10^9\,\mathrm{cm^{-3}}$ (curve A), was allowed to proceed to completion until there was no further change in light

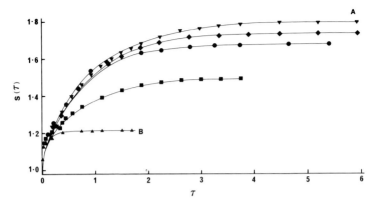

Figure 6 *Bridging flocculation with dextran T500; reversible flocculation equilibria. S(τ) is the structure factor (reduced scattering intensity) and τ is the reduced time of flocculation (see text) (polymer addition 0.0537 mg m^{-2}, latex diameter 207 nm, sodium chloride concentration 0.029 mol dm^{-3}, pH 5.8, 25 °C). Curves relate to different number concentrations n$_0$ ($\times 10^9$ cm^{-3}): ▲, 0.43 (B); ■, 0.86; ●, 2.15; ◆, 2.87; ▼, 4.3 (A)*

scattering intensity with time. An aliquot of this flocculated latex was then diluted ten-fold under iso-molar conditions of sodium chloride. Immediately upon dilution, the scattering intensity was monitored and found to decrease over a period of 10 minutes. The new equilibrium level was identical to the plateau structure factor (curve B) of the corresponding dilute dispersion. We will show elsewhere that the data in Figure 6 can be modelled on random polycondensation theory with reversible kinetics (second-order forward and first-order fragmentation)[47,56] and a cascade description for light scattering from aggregating spheres.[45] The main conclusion is that bridging minima of finite depth can give rise to fully reversible flocculation. Reconformation kinetics are then not relevant.

Bridging Efficacy of Food Biopolymers—Figure 7 illustrates the relative flocculating powers of a range of food biopolymers. The measurements relate to latex B at $n_0 = 4.5 \times 10^{11}$ cm^{-3}, pH 5.9, 28 °C, and sodium chloride at 0.067 mol dm^{-3}. This corresponds to an electrostatic barrier of *ca.* 13 kT and range 4 nm. Bridging flocculation followed by restabilization can be observed for caseinate, gelatin, gum arabic, and dextran. Guar and xanthan show some instability, but at much higher polymer additions, and there is no evidence of restabilization in the accessible range of solution concentrations. The widest window of instability occurs with dextran T500, which can be partly attributed to its weak binding affinity reducing restabilization and its unusual bridging efficacy at low coverage (see above). Caseinate and gelatin display narrow regions of instability. Gum arabic has the narrowest region of instability; moreover, unlike caseinate or gelatin, it is unable fully to counteract the electrostatic barrier (log $W > 0$). These

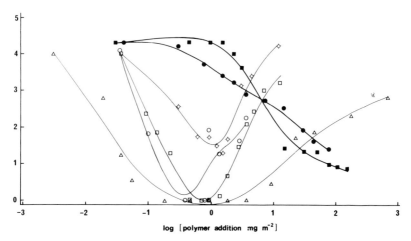

Figure 7   *Comparative flocculation efficiency of food biopolymers (latex diameter*
88 nm, *particle density* 4.5×10^{11} cm^{-3}, *sodium chloride concentration*
6.7×10^{-2} mol dm^{-3}, 28 °C, *pH* 5.9): △, *dextran T500;* ○, *gelatin;* □,
caseinate; ◇, *gum arabic;* ●, *xanthan;* ■, *guar*

differences probably reflect differences in level and type of adsorption. Figure 8
displays the stability diagrams for caseinate and dextran T500 together with
published adsorption isotherms for caseinate[2] and dextran T500[14] on polystyrene. The isotherms were obtained under somewhat different conditions from
those employed in the present stability studies. Nonetheless the relative tendency

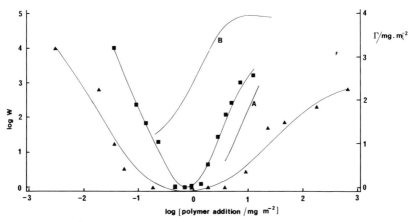

Figure 8   *Bridging flocculation and adsorption behaviour (latex diameter* 88 nm,
particle   density 4.5×10^{11} cm^{-3},   *sodium   chloride* 6.7×10^{-2}
mol dm^{-3}; 28 °C, *pH* 5.9). *Stability factors log* W: ▲, *dextran T500;* ■,
caseinate. Amounts adsorbed Γ *(right hand vertical axis)*: A, *dextran*
T500 *(ref. 14)*; B, *caseinate (ref. 2)*

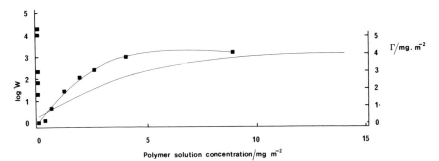

Figure 9   *Bridging flocculation and adsorption behaviour of sodium caseinate on polystyrene latex (latex diameter 88 nm, particle density 4.5 × 10¹¹ cm⁻³, sodium chloride 6.7 × 10⁻² mol dm⁻³, 28 °C, pH 5.9). The polymer solution concentrations were estimated from levels of addition on the basis of the isotherm published in ref. 2 and assuming scaling appropriate to that for adsorption of polydisperse polymer (ref. 26). ━■━, dependence of log W; ——, amounts adsorbed Γ (right hand vertical axis)*

to restabilize appears to be regulated by adsorption affinity as would be expected on the basis of S–F theory. It should be noted that in scaling the isotherms to the stability studies it was assumed that the adsorption conforms to polydisperse behaviour, *i.e.* adsorbed levels per unit area scale with the amount of free polymer expressed per unit area of adsorbent rather than directly with the bulk solution concentration of free polymer.[26-28] Indeed, the plots in Figure 9 of adsorption of caseinate per unit area and log W against the amount of free polymer in solution (also scaled to unit area of latex surface) display a close correspondence in the region of restabilization. Gum arabic has polyelectrolyte character and can adsorb strongly on polystyrene (>5 mg m⁻²) from high ionic strength media.[57,58] It is heteropolymolecular and adopts highly branched block-type structures. In addition, it contains fractions of arabinogalactan–protein complex in which carbohydrate blocks are covalently linked to a central polypeptide chain. The relatively weak flocculation power of gum arabic (Figure 7) is possibly attributable to this non-random copolymeric nature.

   The stability diagrams for latex B in Figures 10 and 11 illustrate the effect of electrolyte on the bridging action of caseinate. As expected, the region of instability widens with increasing electrolyte. At the lowest salt level, 0.0067 mol dm⁻³, corresponding to an electrostatic interaction range of *ca.* 17 nm (Figure 3), the caseinate is unable fully to counteract the electrostatic barrier between the latex particles at any coverage. Adsorbed layer measurements on latex C, by dynamic light scattering, suggest a layer thickness of *ca.* 7 nm at a protein loading of 3 mg m⁻². It is not surprising therefore that bridging is substantially limited at the lowest electrolyte concentration. Restabilization above a loading of *ca.* 1 mg m⁻² is generally observed. It is important to note, however, that the recovery of stability is limited (Figure 11): at low electrolyte,

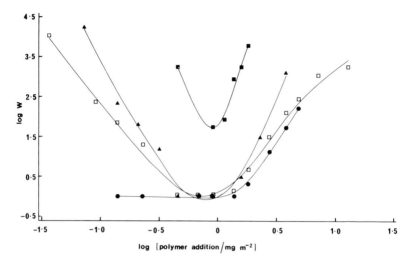

Figure 10 *Bridging flocculation with caseinate: dependence on electrolyte (latex diameter* 88 nm, *particle density* 4.5×10^{11} cm^{-3}, 28 °C, *pH* 5.9). *Concentrations of sodium chloride* (mol dm^{-3}): ■, 0.0067; ▲, 0.0267; □, 0.067; ●, 0.261*

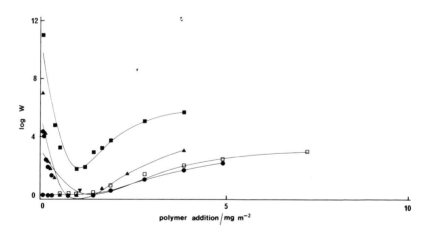

Figure 11 *Bridging flocculation with caseinate: dependence on electrolyte. Symbols and conditions as in Figure* 10. *The displayed trends of log* W *at low polymer addition are estimates based on a linear extrapolation of the stability plot in Figure* 1 *for log* W > 5. *It can be seen that the restabilization at high polymer addition does not always enhance the stability of the bare particles*

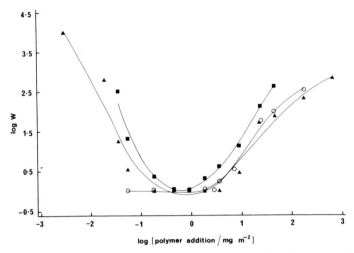

Figure 12 *Bridging flocculation with dextran T*500: *dependence on electrolyte (latex diameter* 88 nm, *particle density* 4.5×10^{11} cm^{-3}, 28 °C, *pH* 5.9). *Concentrations of sodium chloride* (mol dm^{-3}): ■, 0.0067; ▲, 0.067; ○, 0.261

the limiting stability at high protein dose is generally *less* than that of the corresponding uncoated particles. The caseinate is therefore strictly acting as a flocculant rather than as a stabilizer even at high polymer dose. A net stabilizing effect is observed only in excess electrolyte (0.261 mol dm^{-3}) when the electrostatic interaction between the uncoated particles is screened. This behaviour is not specific to caseinate and can be observed with simple homopolymers, *e.g.*, polyethylene oxide[55] and dextran (Figure 12). It can be understood within the framework of S–F theory on the basis that the saturation levels of adsorbed polymer in the pseudo-plateau limit can be insufficient to prevent the presence of residual bridging minima at long range.

It is of interest to explore whether the flocculation effects seen with xanthan and guar (Figure 7) can be attributed to negative adsorption, *i.e.* *depletion* flocculation. Neglecting short-range hydrodynamic effects,[59] the stability factor W is related to the interparticle potential $V(r)$ by[60]

$$W = 2a \int_{2a}^{\infty} \exp\{V(r)/kT\}/r^2 \, \mathrm{d}r, \tag{1}$$

where r is the centre-to-centre distance between the colloidal particles, and a is the particle radius. Equation (1) reduces to[51]

$$W \approx (2\kappa a)^{-1} \exp(V_{\text{max}}/kT) \tag{2}$$

for DLVO potentials, where κ is the inverse Debye–Hückel screening length, and V_{max} is the maximum height of the electrostatic barrier. The contribution to the interaction potential from depletion flocculation can be modelled by[24,25]

$$V_{dep}(r) = -(4\pi/3)(a + R_G)^3[1 - (3r/4)/(a + R_G) + (1/16)(r/(a + R_G))^3]P_{osm},$$
(3)

where R_G is the radius of gyration of the chain and P_{osm} is the osmotic pressure of the polymer solution given by

$$P_{osm} = RT(c/M + A_2c^2 + \cdots)$$
(4)

with R as the gas constant, M the polymer molecular weight, c the polymer concentration, and A_2 the second virial coefficient. Using realistic values for these parameters, the effect of the depletion potential on W has been estimated. Figure 13 shows the predicted and measured change in log W as a function of polymer concentration for dextran T500. Similar comparisons for xanthan and guar are depicted in Figures 14 and 15. For dextran T500 the macromolecular parameters[14] used in the depletion calculation were $R_G = 16$ nm, $M = 5 \times 10^5$ dalton and $A_2 = 4 \times 10^{-4}$ mol cm^3 g^{-2}. Since there is now unambiguous evidence[14] that dextran adsorbs positively, albeit weakly, on polystyrene, it is not surprising that depletion cannot account for the observed behaviour. Reasonable parameters for native xanthan[61] are $R_G = 300$ nm, $M = 3 \times 10^6$ dalton, and $A_2 = 5 \times 10^{-4}$ mol cm^3 g^{-2}. In this case, the depletion calculation accounts rather well for the measured change in log W (Figure 14). The comparison for guar is less definitive, and in any case is more difficult as R_G, estimated as 106 nm, cannot be measured directly. Our estimate is based on the

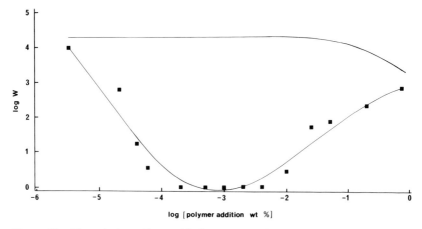

Figure 13 *Flocculation of latex with dextran T500 (latex diameter 88 nm, particle density 4.5 × 10^{11} cm^{-3}, 28 °C, pH 5.9, sodium chloride concentration 6.7 × 10^{-2} mol dm^{-3}): —■—, experimental data; —, depletion calculation on the basis of ref. 25*

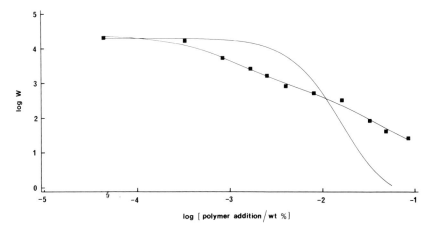

Figure 14 *Flocculation of latex with xanthan. Conditions as in Figure* 13: —■—, *experimental data;* —, *depletion calculation on the basis of ref.* 25

value 12.6 reported for the macromolecular characteristic ratio C_∞ of a typical guar chain[62] and a calculation based on blob theory (the sliding rod model)[63] assuming good solvent conditions, $M = 1.7 \times 10^6$ dalton, a segment length of 0.54 nm, and $A_2 = 10^{-4}$ mol cm^3 g^{-2}. In the present studies we have not taken the special precautions (*e.g.* initial dissolution in urea) which are necessary for the characterization of these polymers by light scattering or intrinsic viscosity. The effective radii of gyration may therefore be larger than those quoted in the literature and this could account for the greater than expected depletion effect observed with guar.

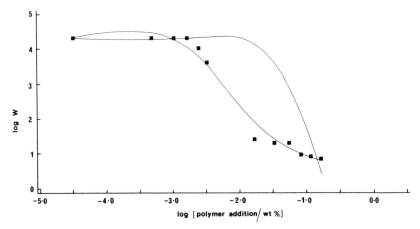

Figure 15 *Flocculation of latex with guar. Conditions as in Figure* 13: —■—, *experimental data;* —, *depletion calculation on basis of ref.* 25

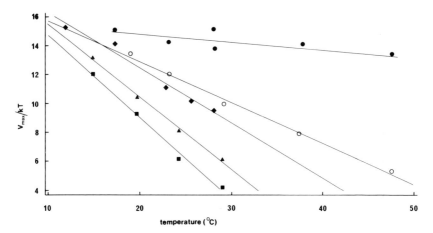

Figure 16   *Effect of temperature on flocculation with caseinate (latex diameter 88 nm, pH 5.9, sodium chloride 6.7×10^{-2} mol dm^{-3}). Barrier heights V_{max}/kT are inferred from measured stability factors on basis of equation 2. Levels of addition (mg m^{-2}) of caseinate: ■, 1.84; ▲, 2.10; ◆, 2.76; ○, 3.84; ● (bare particles)*

Figure 16 illustrates that the bridging efficacy of caseinate increases with temperature. The barrier height V_{max}/kT (inferred from W by equation (2)) decreases with temperature in the presence of caseinate; with the uncoated particles there is no appreciable change. This temperature dependence could

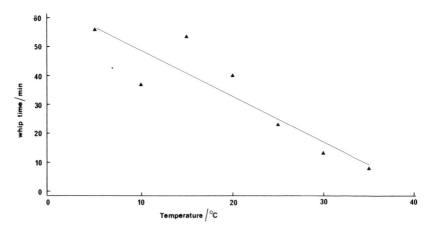

Figure 17   *Effect of temperature on Mohr whip time for model emulsion containing 100% solid fat globules. Oil-in-water emulsion: 45 wt% pure tripalmitin, 1 wt% caseinate, pH 6.6*

offer an explanation for the stability loss with temperature sometimes observed in protein stabilized oil-in-water emulsions. We find that concentrated emulsions based on liquid sunflower oil (45 wt%) in the presence of 1 wt% caseinate and sodium chloride at 0.1 mol dm^{-3} thicken substantially by forming macroscopic networks of aggregated droplets on prolonged storage at room temperature. The same system stored at 5 °C remains fluid and unaggregated. The whip behaviour of corresponding model emulsions based on pure tripalmitin (which provides for globules which are 100% fat below 35 °C), with 1% caseinate but without added salt, shows a similar dependence on temperature (Figure 17). The decrease in Mohr whip times over the range 5–35 °C cannot be attributed to a change in fat crystal content and may reflect a loss in the stabilizing capacity of the caseinate coat adsorbed on the fat particles.

Dynamic light scattering measurements of the adsorbed layer thickness of caseinate on latex C, at pH 8 and 0.01 mol dm^{-3} sodium borate and at a protein loading of 3 mg m^{-2} (corresponding to 75% of the pseudo-plateau value), are shown in Figure 18. The interpretation of dynamic light scattering measurements of adsorbed layer thickness is not completely unambiguous and we will present a full analysis elsewhere. Nonetheless, a decrease in layer thickness with temperature is strongly indicated. This would be consistent with the predictions of S–F theory, suggesting that water becomes a poorer solvent for caseinate with increasing temperature, the decreased solvency (increasing χ) resulting in a progressive collapse of the chains on the surface and a commensurate loss in steric stabilization. It appears then that the temperature effects can be rationalized within the framework of S–F theory.

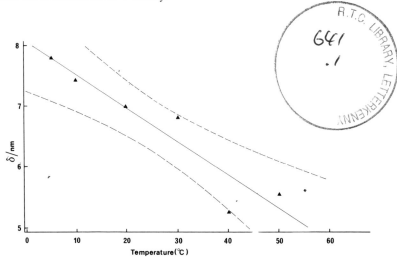

Figure 18 *Effect of temperature on adsorbed layer thickness, δ, of caseinate on polystyrene (latex diameter 48 nm, pH 8, sodium borate 0.01 mol dm^{-3}, caseinate addition 3 mg m^{-2}): ———, 95% confidence limits for linear regression*

4. Conclusions

The Scheutjens–Fleer theory provides a useful framework for discussing the relative interaction effects of food polymers. All the adsorbing food polymers studied have the capacity to induce bridging minima when adsorbed from good solvents. Of these gum arabic shows the least flocculating power, which may be consistent with its relatively pronounced non-random block copolymeric nature. Dextran is an excellent flocculant possibly because of its weak adsorption, enabling stronger interaction minima to be realized more in line with full rather than restricted equilibrium theory. The observed flocculation behaviour of xanthan and guar is not inconsistent at a quantitative level with the predictions of theories of depletion flocculation for non-adsorbing polymers. The adsorbed layer thickness and the interaction between adsorbed layers of caseinate display a consistent temperature dependence expected within the framework of Scheutjens–Fleer theory in the direction of water becoming a poorer solvent for caseinate with increasing temperature. This finding has relevance for understanding the storage and whip stability of food emulsions.

References

1. J. M. H. Scheutjens and G. J. Fleer, *Macromolecules*, 1983, **18**, 1882.
2. E. Dickinson, E. W. Robson, and G. Stainsby, *J. Chem. Soc., Faraday Trans. 1*, 1983, **79**, 2937.
3. H. Mulder and P. Walstra, 'The Milk Fat Globule', Pudoc, Wageningen, 1974.
4. E. Dickinson and G. Stainsby, 'Colloids in Food', Applied Science, London, 1982.
5. P. Walstra, in 'Encyclopedia of Emulsion Technology', ed. P. Becher, Marcel Dekker, New York, 1983, Vol. 1, p. 57.
6. E. Dickinson, *Food Hydrocolloids*, 1986, **1**, 3.
7. E. Dickinson, S. E. Rolfe, and D. G. Dalgleish, in 'Food Colloids', ed. R. D. Bee, P. Richmond, and J. Mingins, Royal Society of Chemistry, Cambridge, 1989, p. 377.
8. E. Dickinson, F. O. Flint, and J. A. Hunt, *Food Hydrocolloids*, 1989, **3**, 389.
9. E. Dickinson, S. E. Rolfe, and D. G. Dalgleish, *Food Hydrocolloids*, 1989, **3**, 193.
10. E. Dickinson and C. M. Woskett, in 'Food Colloids', ed. R. D. Bee, P. Richmond, and J. Mingins, Royal Society of Chemistry, Cambridge, 1989, p. 74.
11. D. H. Napper, 'Polymeric Stabilization of Colloidal Dispersions', Academic Press, London, 1983.
12. Workshop on Emulsion Stabilization, chaired by D. F. Darling, in 'Gums and Stabilisers for the Food Industry', ed. G. O. Phillips, D. J. Wedlock, and P. A. Williams, IRL Press, Oxford, 1988, Vol. 4, p. 507.
13. P. R. Sperry, *J. Colloid Interface Sci.*, 1981, **84**, 409.
14. P. D. Patel and W. B. Russel, *J. Colloid Interface Sci.*, 1989, **131**, 192.
15. P. G. de Gennes, *Macromolecules*, 1982, **15**, 492.
16. D. J. Meier, *J. Phys. Chem.*, 1967, **71**, 1981.
17. F. Th. Hesselink, A. Vrij, and J. Th. G. Overbeek, *J. Phys. Chem.*, 1971, **75**, 2094.
18. A. K. Dolan and S. F. Edwards, *Proc. Roy. Soc. (London)*, 1974, **A337**, 509.
19. S. Levine, M. M. Tomlinson, and K. Robinson, *Faraday Discuss. Chem. Soc.*, 1978, **65**, 202.
20. J. Klein, *Nature (London)*, 1980, **288**, 248.
21. J. Israelachvili, M. Tirrell, J. Klein, and Y. Almog, *Macromolecules*, 1984, **17**, 204.
22. J. Klein and P. Luckham, *Macromolecules*, 1984, **17**, 1041.
23. A. P. Gast, C. K. Hall, and W. B. Russel, *J. Colloid Interface Sci.*, 1983, **96**, 251.
24. S. Asakura and F. Oosawa, *J. Chem. Phys.*, 1954, **22**, 1255.
25. A. Vrij, *Pure Appl. Chem.*, 1976, **48**, 471.

26. M. A. Cohen Stuart, J. M. H. Scheutjens, and G. J. Fleer, *J. Polym. Sci., Polym. Phys. Ed.*, 1980, **18**, 559.
27. L. K. Koopal, *J. Colloid Interface Sci.*, 1981, **83**, 116.
28. V. Hlady, J. Lyklema, and G. J. Fleer, *J. Colloid Interface Sci.*, 1982, **87**, 395.
29. F. Th. Hesselink, *J. Colloid Interface Sci.*, 1977, **60**, 448.
30. H. A. Van der Schee and J. Lyklema, *J. Phys. Chem.*, 1984, **88**, 6661.
31. J. Papenhuijzen, H. A. Van der Schee, and G. J. Fleer, *J. Colloid Interface Sci.*, 1985, **104**, 540.
32. B. van Lent, 'Molecular structure and interfacial behaviour of polymers', Ph.D. Thesis, University of Wageningen, 1989.
33. O. A. Evers, 'Statistical thermodynamics of block copolymer adsorption', Ph.D. Thesis, University of Wageningen, 1990.
34. E. W. Fischer, *Kolloid Z.*, 1958, **160**, 120.
35. A. M. Homola and A. A. Robertson, *J. Colloid Interface Sci.*, 1976, **54**, 286.
36. F. W. Cain, R. H. Ottewill, and J. B. Smitham, *Faraday Discuss. Chem. Soc.*, 1978, **65**, 33.
37. A. Doroszkowski and R. Lambourne, *J. Colloid Interface Sci.*, 1973, **43**, 97.
38. B. V. Derjaguin, *Kolloid Z.*, 1934, **39**, 155.
39. M. A. J. S. van Boekel, 'The influence of fat crystals in the oil phase on stability of oil-in-water emulsions, Ph.D. Thesis, University of Wageningen, 1980.
40. R. S. Allan, G. E. Charles, and S. G. Mason, *J. Colloid Sci.*, 1961, **16**, 150.
41. D. S. Dimitrov and I. B. Ivanov, *J. Colloid Interface Sci.*, 1978, **64**, 97.
42. D. S. Dimitrov, N. Stoicheva, and D. Stefanova, *J. Colloid Interface Sci.*, 1984, **98**, 269.
43. A. Lips, C. Smart, and E. Willis, *Trans. Faraday Soc.*, 1971, **67**, 2979.
44. D. Giles and A. Lips, *J. Chem. Soc., Faraday Trans. 1*, 1974, **74**, 733.
45. A. Lips, *J. Chem. Soc., Faraday Trans. 2*, 1987, **83**, 221.
46. R. M. Duckworth and A. Lips, *J. Colloid Interface Sci.*, 1978, **64**, 311.
47. A. Lips and R. M. Duckworth, *J. Chem. Soc., Faraday Trans. 1*, 1988, **84**, 1223.
48. A. Kotera, F. Furusawa, and Y. Takeda, *Kolloid Z.*, 1970, **239**, 677.
49. M. Kerker, 'The Scattering of Light and Other Electromagnetic Radiation', Academic Press, New York, 1969.
50. D. S. Duckworth, A. Lips, and E. J. Staples, *Faraday Discuss. Chem. Soc.*, 1978, **65**, 288.
51. H. Reerink and J. Th. G. Overbeek, *Faraday Discuss. Chem. Soc.*, 1954, **18**, 74.
52. G. J. Fleer and J. M. H. M. Scheutjens, *Adv. Colloid Interface Sci.*, 1982, **16**, 341.
53. M. A. Cohen Stuart, T. Cosgrove, and B. Vincent, *Adv. Colloid Interface Sci.*, 1986, **24**, 143.
54. R. J. Roe, *J. Phys. Chem.*, 1974, **42**, 2101.
55. E. G. M. Pelssers, M. A. Cohen Stuart, and G. J. Fleer, *Colloids Surf.*, 1989, **38**, 15.
56. P. van Dongen and M. H. Ernst, *J. Phys. A*, 1983, **16**, L327.
57. M. J. Snowden, G. O. Phillips, and P. A. Williams, *Food Hydrocolloids*, 1987, **1**, 291.
58. M. J. Snowden, G. O. Phillips, and P. A. Williams, in 'Gums and Stabilisers for the Food Industry', ed. G. O. Phillips, D. J. Wedlock, and P. A. Williams, IRL Press, Oxford, 1988, Vol. 4, p. 489.
59. L. A. Spielman, *J. Colloid Interface Sci.*, 1970, **33**, 562.
60. N. Fuchs, *Z. Phys.*, 1934, **89**, 736.
61. T. Coviello, K. Kajiwara, W. Burchard, M. Dentini, and V. Crescenzi, *Macromolecules*, 1986, **19**, 2826.
62. G. Robinson, S. B. Ross-Murphy, and E. R. Morris, *Carbohyd. Res.*, 1982, **107**, 17.
63. M. Bemouna, A. Z. Akcasu, and M. Daoud, *Macromolecules*, 1980, **13**, 1703.

Conformation and Physical Properties of the Bacterial Polysaccharides Gellan, Welan, and Rhamsan

By Geoffrey Robinson,* Charles E. Manning,† and Edwin R. Morris

DEPARTMENT OF FOOD RESEARCH AND TECHNOLOGY, CRANFIELD INSTITUTE OF TECHNOLOGY, SILSOE COLLEGE, SILSOE, BEDFORD MK45 4DT

1 Introduction

Following the spectacular commercial success of their first bacterial poly-saccharide xanthan, Kelco (now a Division of Merck) have developed three new commercial polymers from microbial fermentation: gellan, welan, and rhamsan. In the present work we have explored the relationships between the primary structure, conformation, and physical properties of these materials.

Gellan—Gellan gum is a deacylated form of the extracellular bacterial poly-saccharide from *Auromonas elodea* (ATCC 31461) and has a tetrasaccharide repeating sequence[1,2] of: →3)-β-D-Glcp-(1 → 4)-β-D-GlcpA-(1 → 4)-β-D-Glcp-(1 → 4)-α-L-Rhap-(1→. In the native form it has an L-glyceryl substituent on O-2 of the 3-linked glucose and an acetyl group on approximately half of the O-6 atoms of the same residue,[3] but both these groups are lost during normal commercial extraction.

Both the acylated and deacylated polysaccharides form gels in the presence of moderate concentrations (≥60 mM) of monovalent ions (typically Na+) or with much lower concentrations of divalent ions.[4] In comparison with other uronic acid containing polysaccharides such as alginate or pectin, the selectivity for different divalent cations is very small,[5] particularly within the alkaline earths. The gels vary in mechanical properties from soft and elastic for the acylated form to hard and brittle for the fully deacylated polysaccharide.

In the solid state, gellan adopts a double helix structure,[6] with two left-handed, three-fold helical strands organized in a parallel fashion in an intertwined duplex

---

* Present address: Kruss (UK) Ltd., Carrington House, 37 Upper King Street, Royston, Herts SG8 9AZ.
† Present address: Department of Biotechnology, South Bank Polytechnic, 103 Borough Road, London, SE1 0AA.

in which each chain is translated by half a pitch ($p = 5.64$ nm) with respect to the other, an arrangement closely similar to the double helix structure of iota-carrageenan.[7]

Welan—Welan, from *Alcaligenes* ATCC 31555 (formerly known as S-130), has[8] the same backbone sequence as gellan, but with a monosaccharide sidechain at O-3 of the 4-linked glucose; this can be either α-L-rhamnose or α-L-mannose, in the approximate ratio 2:1.

Welan does not gel, but forms viscous solutions which can suspend particles or stabilize emulsions in a manner similar to xanthan 'weak gels'.[9] These properties are stable to high temperatures and the polymer is intended primarily for commercial use in oil recovery (hence the name).

Preliminary X-ray results from fibre diffraction of welan[10] suggested a pitch of ~3.7 nm. However, more recent work (R. Chandrasekaran, personal communication) indicates a three-fold structure similar to that of gellan ($p \approx 5.6$ nm).

Rhamsan—Rhamsan, the exopolysaccharide from *Alcaligenes* ATCC 31961 (formerly known as S-194), is also a branched variant of gellan. In this case the side-chain is a disaccharide, β-D-Glc$p$-$(1 \rightarrow 6)$-α-D-Glc$p$, at O-6 of the 3-linked glucose residue in the tetrasaccharide repeating sequence of the polymer backbone[11] (the same position as the acetyl substituents in native gellan). Rhamsan solutions are more viscous than those of welan, but rhamsan is less thermally stable.

2 Materials and Methods

Samples of gellan (deacylated form), welan, and rhamsan were kindly supplied by Kelco Division of Merck & Co. Inc. The polymers were exchanged into appropriate ionic forms (Na^+; Me_4N^+) on an Amberlite IR 120 column. Xanthan (Kelco Inc., Keltrol F) was used as received.

Intrinsic viscosities were determined on a Contraves Low Shear 30 viscometer with concentric cylinder geometry over a maximum shear-rate range of 0.05–$100\,s^{-1}$. For solutions that showed detectable shear-thinning, the maximum, constant viscosity at low shear rates was used in calculation of intrinsic viscosity values. In all cases a stock solution was prepared to a relative viscosity slightly above 2.0 and dialysed to equilibrium against the appropriate salt solution. The dialysate was then used for all subsequent dilutions to maintain constant chemical potential. Extrapolation to obtain the intrinsic viscosity was by a combined Huggins and Kraemer treatment.[12]

A few solutions were also measured in a U-tube capillary viscometer. The peak shear rate $\dot{\gamma}_{max}$ generated at the wall of the capillary was estimated from the standard relationship

$$\dot{\gamma}_{max} = \rho g r / 2 \eta_s \eta_{rel}, \qquad (1)$$

where ρ is solution density, g is acceleration due to gravity, η_s is solvent viscosity

(0.89 mPa s for water at 25 °C), η_{rel} is the viscosity of the solution relative to that of the solvent, and r is the radius of the capillary (determined from the weight of a column of mercury filling a measured length). For the viscometer used, we have $\dot{\gamma}_{max}/s^{-1} = 1927/\eta_{rel}$.

Rheological measurements at higher concentrations were performed on a Sangamo Viscoelastic Analyser, using a 5 cm cone and plate configuration with a cone angle of 2°. Differential scanning calorimetry (DSC) measurements were made on a Seteram biocalorimeter. Optical rotation studies were carried out on a Perkin Elmer 241 polarimeter, using a jacketed 10 cm cell with temperature control by a Haake circulating water bath. High-resolution proton NMR spectra were recorded at 200 MHz on a Brucker AM 200, using a pyrazine standard as reference for signal intensity.

3 Conformational Ordering in Gellan Solutions

As described above, gellan forms gels with a variety of monovalent and divalent metal ions.[4,5] As in the case of certain other gelling polyelectrolytes, such as carrageenan,[13] however, gelation can be inhibited by using large organic cations (*e.g.* tetramethylammonium). Under these non-gelling conditions, the tempera-

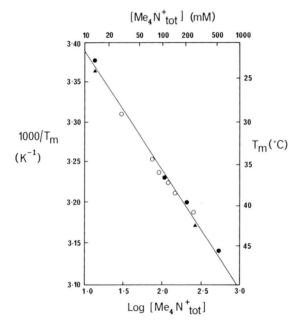

Figure 1 *Variation in midpoint temperature* T_m *for the conformational transition of* Me_4N^+ *gellan in solution on varying cation concentration* $[Me_4N^+_{tot}]$ *(counter-ions to the polymer plus added* Me_4NCl*). Results obtained by Crescenzi et al.[14] using optical rotation at 302 nm with a gellan concentration of 0.08 wt% (○) are compared with those obtained in the present work at a gellan concentration of 1.0 wt%, using optical rotation at 436 nm (●) and DSC (▲)*

ture dependence of optical rotation shows a sharp sigmoidal change indicative of a conformational transition in solution.[14] The transition is fully reversible, with no thermal hysteresis between results obtained on heating and on cooling.

DSC measurements on the same solutions show a well defined endotherm on heating and a corresponding exotherm on cooling, with the same absolute value of transition enthalpy in both directions ($5.6 \pm 0.2 \text{ kJ mol}^{-1}$). In this case there is an apparent displacement of the temperature at the peak maxima (T_{max}) between heating and cooling, but this is due simply to thermal lag in the instrument; the discrepancy decreases linearly with decreasing scan rate, and on extrapolation to zero rate the T_{max} values for heating and cooling coincide, as expected from the optical rotation studies.

Ionic-Strength Dependence—Addition of salt (Me_4NCl) raises the transition temperature. As in other polyelectrolyte systems, the reciprocal of the mid-point temperature T_m varies linearly (Figure 1) with the logarithm of total cation concentration (counter-ions to the polymer plus added salt). The values obtained in the present work by optical rotation and by DSC are in close agreement, indicating that both techniques are monitoring the same molecular process, and are also in excellent agreement with optical rotation results reported previously by Crescenzi and co-workers,[14] using a much lower polymer concentration (0.08 wt% in comparison with 1.0 wt% in the present work).

Rheological Changes—These non-gelling samples flow like normal solutions both above and below the conformational transition, but examination by mechanical spectroscopy (Figure 2) shows fundamental differences. At high temperatures,

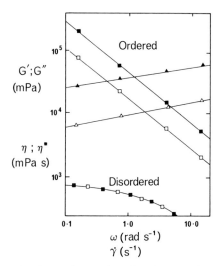

Figure 2   *Solution rheology of Me_4N^+ gellan (1.0 wt%) in the disordered coil form (0.1 M Me_4NCl; 30 °C) and in the ordered state (0.25 M Me_4NCl; 5 °C). Representative data points are shown for G' (▲), G" (△), η^* (■) and η (□)*

shear-thinning behaviour typical[15] of polysaccharide 'random coils' is observed, but at temperatures below the transition the solutions show 'weak gel' properties similar to those observed for the ordered form of xanthan.[9,12] Specifically, elastic response (characterized by the storage modulus, G') exceeds viscous response (characterized by the loss modulus, G''), with little frequency-dependence in either modulus; and 'small-deformation' dynamic viscosity η^*, measured under low-amplitude oscillation, is substantially greater, at equivalent values of frequency $\omega(\text{rad s}^{-1})$ and shear rate $\dot{\gamma}(\text{s}^{-1})$, than 'large-deformation' steady-shear viscosity η under rotation.

4 Solution Rheology of Welan and Rhamsan

The branched polysaccharides welan and rhamsan give solutions with 'weak gel' properties similar to those shown in Figure 2 for the ordered form of gellan. This gel-like character extends to a c^2 dependence of G', as seen[16,17] for true polysaccharide gels such as those of alginate, pectin, and agarose at concentrations well above the minimum critical gelling concentration, c_0. We also observed a similar dependence of modulus on the square of polymer concentration for xanthan weak gels. Thus, by the criterion of mechanical spectroscopy, the solution properties of welan and rhamsan are quite different from those of entangled 'random coils',[12,15] but show striking similarities to those of gellan and xanthan in their ordered conformation, suggesting that welan and rhamsan may also be conformationally ordered in solution.

5 Conformation of Welan and Rhamsan in Solution

Unlike gellan and xanthan, welan and rhamsan show no evidence of a conformational transition between 0 and 100 °C. One interpretation of this behaviour[14] is that both exist as disordered coils throughout the accessible temperature range. An alternative explanation, however, is that they are locked in stable, ordered structures which resist thermal denaturation to temperatures above 100 °C. As described below, evidence from NMR linewidths and the salt-dependence of macromolecular dimensions (as characterized by intrinsic viscosity) strongly favours the latter possibility.

NMR Evidence—High-resolution NMR gives a good, simple indication of the conformational state. For example, xanthan shows reasonably sharp NMR spectra at high temperature, but on cooling through the disorder–order transition the spectrum collapses into the baseline[18] due to extreme line broadening as the polymer loses its segmental mobility. Similar collapse has been observed for other polysaccharides and has been related to changes in spin–spin relaxation time.[19]

Gellan, under non-gelling conditions (Me_4N^+ salt form), displays the expected loss of high-resolution proton NMR spectrum on cooling through the conformational transition identified by optical rotation and DSC, fully consistent with conversion from a fluctuating coil conformation at high temperature to the ordered double helix structure at low temperature. In the case of welan and rhamsan, however, the high-resolution spectra remain featureless from 0 to 100 °C, indicating conformational rigidity throughout this entire temperature range.

Response to Salt—Conformational rigidity is also indicated by their response to changes in ionic strength. At low ionic strength I, disordered polyelectrolytes are expanded by electrostatic repulsions within the coil. As the ionic strength is increased, these forces are gradually screened out and the coil dimensions collapse. Quantitatively, the polyelectrolyte intrinsic viscosity $[\eta]$ increases linearly with $I^{-1/2}$, and the gradient S relative to the intrinsic viscosity $[\eta]_{0.1}$ at a fixed, reference ionic strength ($I = 0.1$ M) gives a flexibility parameter (or 'B value')[20] which is inversely proportional to the persistence length:

$$a/\text{nm} = 0.26/B = 0.26[\eta]_{0.1}^{1.3}/S. \tag{2}$$

B values for welan and rhamsan (Na$^+$ salt form), and the persistence lengths estimated from them, are listed in Table 1, along with corresponding values for other polyelectrolytes. In comparison with relatively inflexible disordered polysaccharides such as alginate, welan and rhamsan show little change in intrinsic viscosity with increasing ionic strength (Figure 3) indicating rigid, ordered structures in solution.

Molecular Size and Stiffness—The intrinsic viscosity of rhamsan (extrapolated to infinite ionic strength, *i.e.* $I^{-1/2} = 0$) is ~135 dl g^{-1}, the highest yet measured for a polysaccharide, with a B value of ~0.003 (corresponding to a persistence length of ~88 nm); thus the rhamsan molecule appears to be at least as stiff as the double-stranded helix of DNA, which also has a comparable intrinsic viscosity (>100 dl g^{-1}).

Table 1 *Polyelectrolyte* B *values and persistence lengths* a

Polyelectrolyte	B	a/nm
Welan	0.001[a]	250[b]
Rhamsan	0.003	88[b]
Xanthan	0.00525[c]	50[b]
DNA	0.0055	45[d]
Alginate (guluronate-rich)	0.031	7.8[d]
Alginate (mannuronate-rich)	0.040	6.5[d]
Carboxymethylcellulose	0.065	4.1[d]
Carboxymethylamylose	0.20	1.28[d]

[a] Due to the extremely low slope of $[\eta]$ *versus* $I^{-1/2}$ for welan, this value is subject to a large experimental error.
[b] Calculated from B by the relationship $a/\text{nm} = 0.26/B$.
[c] From B. Tinland and M. Rinaudo, *Macromolecules*, 1989, **22**, 1863.
[d] Derived by the relationship $b = 2a$ from values of statistical segment length b tabulated by O. Smidsrød and A. Haug, *Biopolymers*, 1971, **10**, 221.

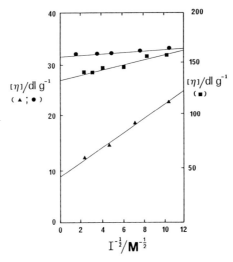

Figure 3 *Variation of intrinsic viscosity [η] with ionic strength I for alginate (▲),*
welan (●), and rhamsan (■). Values for alginate are taken from
Smidsrød and Haug.[20] Note that results for rhamsan (right-hand axis)
are much higher than for alginate and welan (left-hand axis)

The value of the intrinsic viscosity for welan is substantially smaller ($\sim$32 dl g^{-1} at $I^{-1/2} = 0$), but it is still much higher than for normal commercial 'random coil' polysaccharides (typically 5–20 dl g^{-1}) As shown in Figure 3, the plot of [η] *versus* $I^{-1/2}$ for welan is almost parallel to the $I^{-1/2}$ axis, so that the value of S, and those of B and a derived from it, are subject to large experimental error. By the criterion of the response of intrinsic viscosity to increasing ionic strength, however, welan appears to be at least comparable in stiffness to rhamsan and double-stranded DNA (Table 1).

6   Comparison with Previous Work

Previously published values of intrinsic viscosity for rhamsan and welan are much smaller than those obtained in the present work. For example, Crescenzi and co-workers[14] reported a values of $\sim$8 dl g^{-1} for the intrinsic viscosity of rhamsan (at $I^{-1/2} = 0$), with a B value of about 0.05, suggesting behaviour more like that of a random coil polymer. The same authors found virtually no ionic strength dependence of welan, obtaining values of [η] $\approx$ 20 dl g^{-1} throughout the concentration range 0.8–150 mM Me$_4$NCl. Over a wider range of salt concentrations (0.1–100 mM NaCl), however, Urbani and Brant[21] found a slope of $\sim$0.075 for [η] *versus* $I^{-1/2}$ with [η]$_{0.1}$ $\approx$ 17 dl g^{-1}, corresponding to a B value of 0.0019 and a persistence length of 138 nm.

Thus, for welan, the previous studies are in qualitative agreement with the present work in suggesting a highly persistent structure, whose dimensions are virtually independent of ionic strength, but the absolute values of intrinsic viscosity are almost a factor of 2 lower, while for rhamsan the discrepancy in intrinsic viscosities approaches a factor of 20, and there is qualitative disagreement on whether the molecule is ordered or disordered.

Consistency of Experimental Evidence—The two main differences between the present work and the earlier study of welan and rhamsan by Crescenzi and co-workers[14] are (i) the measurement here of viscosity under rotation at very low shear rates, rather than in a capillary viscometer, and (ii) the use of Na^+ as counter-ion here rather than Me_4N^+. The first of these differences, but not the second, also applies to the study of welan by Urbani and Brant.[21]

To verify that the large discrepancies in experimental evidence were not simply due to differences in, for example, molecular weight or sample preparation, we measured the intrinsic viscosity of Me_4N^+ rhamsan in 0.1 M Me_4NCl, using a capillary viscometer as in the earlier studies, and obtained a value in reasonable agreement with those reported by Crescenzi and co-workers (~ 11 dl g^{-1}).

Effect of Tetramethylammonium Ions—Similar measurements for Na^+ rhamsan in 0.1 M NaCl gave a much higher value of apparent intrinsic viscosity (~ 22 dl g^{-1}). During the course of these studies we observed the gradual appearance of a precipitate in solutions of rhamsan in Me_4NCl, indicating 'salting-out' of the polymer. No such precipitation was observed for solutions of Na^+ rhamsan in NaCl. Thus the differences (*ca.* 2-fold) in the viscosities obtained for these two different salt forms are probably due simply to reduction in the concentration of rhamsan chains in solution in the presence of Me_4NCl.

Shear-thinning in Dilute Solution—The above changes in apparent viscosity on changing the cation are minor, however, in comparison with the differences between values obtained from capillary measurements and those recorded under rotation at low shear rate. The shear rates generated in capillary viscometers are known to be high, typically of the order of 1000 s^{-1}. For intrinsic viscosity measurements on normal disordered polymers, this does not present a problem, since, at the very low concentrations required to give relative viscosities in the appropriate range (1.2–2.0), shear-thinning effects are usually negligible. For rhamsan, however, even at the lowest polymer concentrations at which reliable capillary measurements could be made, we observed very obvious shear-thinning. For example, Figure 4 shows the shear-rate dependence of viscosity for a 0.008 wt% solution of Na^+ rhamsan in 0.15 M NaCl. Capillary measurements gave a relative viscosity of 1.221 which, from equation (1), corresponds to a shear rate of 1578 s^{-1}. This value fits smoothly on to the shear thinning curve obtained for the same solution under rotation over the shear-rate range 0.05–70 s^{-1}. The 'zero shear' specific viscosity η_{sp} in the Newtonian plateau region at low shear rates is about 25 times higher than the corresponding capillary value (*ca.* 5.4 in comparison with *ca.* 0.22).

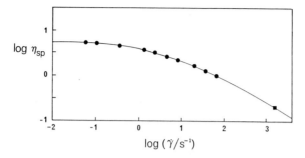

Figure 4 *Shear-thinning of Na^+ rhamsan in dilute solution (0.008 wt%; 0.15 M NaCl), showing values obtained under rotation in cone and plate geometry (●) and in a capillary viscometer (■)*

Origin of Conflicting Results—It seems clear from the above evidence that the major discrepancies between the intrinsic viscosity results obtained for rhamsan in the present work and those reported previously by Crescenzi and co-workers,[14] and the smaller discrepancies for welan, arise from very substantial shear-thinning at the shear rates generated in capillary viscometers, even at the extremely low concentrations appropriate to determination of intrinsic viscosity. The probable origin of this unusual behaviour is alignment of the stiff, ordered chains in the direction of flow, with consequent reduction in their resistance to movement through the solution. Since interpretation of intrinsic viscosity in terms of molecular size (hydrodynamic volume) requires measurements to be made under 'unperturbed' conditions (*i.e.* within the low-shear Newtonian plateau region, as in the present work), the values reported previously[14,21] for welan and rhamsan, and the conclusions drawn from them, must be regarded as invalid.

7   Salt-dependence of Molecular Interactions

In contrast to those of xanthan,[9] the 'weak gel' rheological properties of welan and rhamsan show little, if any, dependence on salt concentration. Gellan, however, forms 'true' gels with low concentrations of Group II cations and somewhat higher concentrations of Group I cations. The onset of gel formation is accompanied by dramatic changes in the thermal transition observed on heating (Figure 5).

Biphasic Melting of Gellan Gels—At concentrations of NaCl below those required for gel formation, a single DSC endotherm is observed at the same temperature (after compensation for thermal lag in the instrument) as the exotherm observed on cooling (as in solutions of the tetramethylammonium salt form). As the salt concentration is increased to give progressively stronger gels, the cooling exo-therm remains as a single peak, showing the expected increase in temperature illustrated in Figure 1 for non-gelling conditions. The heating endotherm,

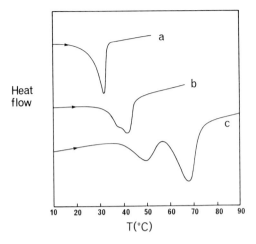

Figure 5 *DSC heating scans* (0.1 K min^{-1}) *for* Na^+ *gellan* $(1.0 \text{ wt\%})$, *showing the progressive development of biphasic melting behaviour with increasing concentration of* Na^+ *(counter-ions to the polymer,* 15 mM, *plus added NaCl):* (a) 25 mM, (b) 67 mM, (c) 115 mM. *The traces were recorded at different sensitivities; the overall enthalpy change (i.e. the peak area for the same sensitivity) shows only a moderate increase (from ca.* 6 kJ mol^{-1} *to ca.* 9 kJ mol^{-1})

however, gradually splits into two peaks, one at approximately the same temperature as the exotherm observed on cooling, and the other moving to progressively higher temperatures with increasing salt concentration. The relative magnitude of this peak also increases with increasing salt concentration. The combined enthalpy from the two endothermic transitions, however, shows only a small net increase with increasing salt, and remains numerically equal to that of the single exotherm observed in the cooling direction.

8 Conclusions

We propose the following 'minimum interpretation' of the above results, as outlined schematically in Figure 6.

At high temperature, gellan exists in solution as a disordered coil. On cooling under non-gelling conditions, it converts reversibly to the double-helical structure characterized by X-ray diffraction. Weak association of these helices by, for example, van der Waals attraction, gives rise to the 'weak gel' solution properties.

In the presence of appropriate cations, such as sodium or calcium, a proportion of the helices are able to associate into cation-mediated aggregates which cross-link the gel network, the proportion of aggregated helices increasing with increasing salt concentration. On heating, the non-aggregated helices melt out first at the same temperature as they formed on cooling; the aggregates then melt out at higher temperature. Acyl substituents present in native gellan interfere with the aggregation process, giving weaker gels.

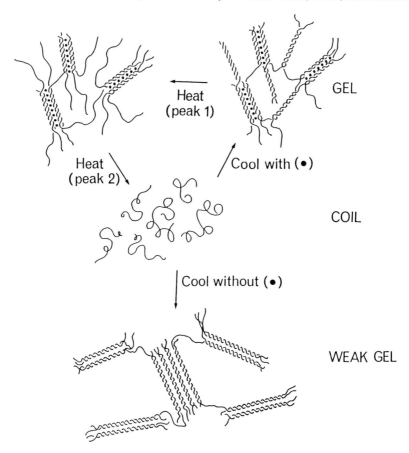

Figure 6   *Schematic representation of proposed model for the conformation and functional interactions of gellan, showing (top) the processes envisaged on cooling and reheating in the presence of cations (●) that promote gel formation (such as Na^+ and Ca^{2+}) and (bottom) the 'weak gel' networks developed when such cations are absent (or present in insufficient concentrations to stabilize association of helices)*

In the branched variants this process goes a stage further, and the side-chains totally abolish the cation-induced aggregation process, perhaps by screening the uronic acid groups and making them unavailable for cation binding. Thus, in welan and rhamsan, intermolecular association is limited to 'weak gel' formation only. The side-chains, however, appear to stabilize the ordered structure, perhaps by folding down and packing along the polymer backbone, so that the industrially useful weak gel properties persist to very high temperatures.

References

1. P.-E. Jansson, B. Lindberg, and P. A. Sandford, *Carbohydr. Res.*, 1983, **124**, 135.
2. M. A. O'Neil, R. R. Selvendran, and V. J. Morris, *Carbohydr. Res.*, 1983, **124**, 123.
3. M.-S. Kuo, A. J. Mort, and A. Dell, *Carbohydr. Res.*, 1986, **156**, 173.
4. G. R. Sanderson and R. C. Clark, in 'Gums and Stabilisers for the Food Industry', ed. G. O. Phillips, D. J. Wedlock, and P. A. Williams, Pergamon, Oxford, 1984, Vol. 2, p. 201.
5. H. Grasdalen and O. Smidsrød, *Carbohydr. Polym.*, 1987, **7**, 371.
6. R. Chandrasekaran, R. P. Millane, S. Arnott, and E. D. T. Atkins, *Carbohydr. Res.*, 1988, **175**, 1.
7. S. Arnott, W. E. Scott, D. A. Rees, and C. G. A. McNab, *J. Mol. Biol.*, 1974, **90**, 253.
8. M. A. O'Neil, R. R. Selvendran, V. J. Morris, and J. Eagles, *Carbohydr. Res.*, 1986, **147**, 295.
9. S. B. Ross-Murphy, V. J. Morris, and E. R. Morris, *Faraday Symp. Chem. Soc.*, 1983, **18**, 115.
10. P. T. Attwool, E. D. T. Atkins, M. J. Miles, and V. J. Morris, *Carbohydr. Res.*, 1986, **148**, C1.
11. P.-E. Jansson, B. Lindberg, J. Lindberg, and E. Maekawa, *Carbohydr. Res.*, 1986, **156**, 157.
12. E. R. Morris, in 'Gums and Stabilisers for the Food Industry', ed. G. O. Phillips, D. J. Wedlock, and P. A. Williams, Pergamon, Oxford, 1984, Vol. 2, p. 57.
13. E. R. Morris, D. A. Rees, and G. Robinson, *J. Mol. Biol.*, 1980, **138**, 349.
14. V. Crescenzi, M. Dentini, and I. C. M. Dea, *Carbohydr. Res.*, 1987, **160**, 283.
15. E. R. Morris, A. N. Cutler, S. B. Ross-Murphy, D. A. Rees, and J. Price, *Carbohydr. Polym.*, 1981, **1**, 5.
16. J. M. V. Blanshard and J. R. Mitchell, *J. Text. Studies*, 1980, **7**, 341.
17. A. H. Clark, R. K. Richardson, S. B. Ross-Murphy, and J. M. Stubbs, *Macromolecules*, 1983, **16**, 367.
18. E. R. Morris, D. A. Rees, G. Young, M. D. Walkinshaw, and A. Darke, *J. Mol. Biol.*, 1977, **110**, 1.
19. S. Ablett, A. H. Clark, and D. A. Rees, *Macromolecules*, 1982, **15**, 597.
20. O. Smidsrød and A. Haug, *Biopolymers*, 1971, **10**, 221.
21. R. Urbani and D. A. Brant, *Carbohydr. Polym.*, 1989, **11**, 169.

Recent Developments in Polyelectrolyte Adsorption: Theory and Experiment

By G. J. Fleer

DEPARTMENT OF PHYSICAL AND COLLOID CHEMISTRY, WAGENINGEN
AGRICULTURAL UNIVERSITY, DREIJENPLEIN 6, 6703 HB WAGENINGEN,
THE NETHERLANDS

1 Introduction

Polyelectrolytes are widely used to modify surfaces and to stabilize or destabilize colloidal dispersions.[1,2] Applications can be found in food technology, mineralogy, paint production, and in a wide variety of other industrial processes. Fundamental insight into the adsorption of polyelectrolytes is of crucial importance for these applications and for new developments.

In recent years, much progress has been made in understanding the behaviour of uncharged macromolecules near surfaces.[1–4] A theoretical model[5–7] has been developed that describes the main trends quite satisfactorily. Theory for polyelectrolyte adsorption is still in the early stages of development and experiments on well defined systems are scarce. Earlier attempts to incorporate the electrostatics in a polymer adsorption model have considered the polymer and the solvent as entities having a finite size, but treated the small ions as point charges without volume. These ions were then assumed to distribute themselves according to the Poisson–Boltzmann equation. In order to prevent unrealistically high ion concentrations near the surface, a Stern layer approach was incorporated in the model. This model has first been described for strong polyelectrolytes with a fixed charge[8,9] and compared with experiments on well-defined systems.[10,11] Evers et al.[12] applied the same model to weak polyelectrolytes with a degree of dissociation that depends on the local pH (i.e. on the distance from the surface). Owing to a lack of suitable data, they could not make a complete comparison with model experiments. Very recently, Böhmer et al.[13] have developed a general multi-Stern-layer model in which all the components, including the small ions, are treated as entities with volume. Simultaneously, a new set of experimental data has become available for polyacrylic acid on highly positively charged latex.[14]

In this paper, we review briefly the background of the new model of Böhmer et al.,[13] present a few illustrative results, and compare the predictions with the experimental data.[14] It will be shown that for polyelectrolyte adsorption a clear picture is now gradually emerging, in which the main trends are very well covered by recent theory.

2 Model

We do not present the model here in its most general form, but restrict ourselves to the simplest system of a surface (which may be charged), a weak polyanion in water, an indifferent electrolyte, and one type of potential determining (pd) ions. Even then, five components have to be distinguished, which we number $i = 1, 2, \ldots, 5$ as follows: 1 = solvent (H_2O), 2 = cation (*e.g.* Na^+), 3 = anion (*e.g.* Cl^-), 4 = pd-ion (*e.g.* H_3O^+), 5 = polyanion. Components 1–4 are monomeric and occupy one lattice site each and component 5 is polymeric and consists of r segments, occupying r consecutive sites in the lattice. Figure 1 gives a schematic representation of a solid adsorbent, three polyanion molecules in specified conformations (c, d, and e), solvent, and small ions. In the diagram, no distinction is made between Na^+ ions and H_3O^+ ions. The lattice layers are

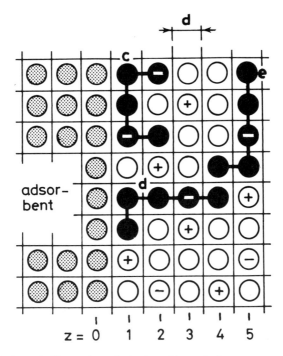

Figure 1 *Schematic illustration of a lattice with a solution next to a solid adsorbent. The solution sites are filled with solvent (open circles), small ions, and three polyelectrolyte chains (of five segments each) in specified conformations c, d, and e. Conformations c and d are adsorbed and have $r^c(1) = 3$ and $r^d(1) = 2$ segments, respectively, in the first layer. Conformation e is free (nonadsorbed), with $r^e(4) = 1$ and $r^e(5) = 4$*

numbered $z = 0, 1, 2, \ldots$, where $z = 0$ corresponds to the centre of the atoms in the surface layer of the adsorbent, and $z = 1$ to (the centre of) the first layer in the liquid medium.

We use a mean-field approximation, where each segment, ion, or solvent molecule in a given layer is considered to have the same energy: each species in layer z has properties which are equal to the average over the whole layer. Fluctuations within layer z are neglected.

Each monomeric component and each polyelectrolyte segment can then be assigned a weighting factor $G_i(z)$, defined such that $G_i(\infty) = 1$; that is, we have

$$G_i(z) = e^{-u_i(z)/kT}, \qquad (i = 1, 2, \ldots, 5) \tag{1}$$

where $u_i(z)$ is the potential of a segment, solvent molecule, or ion with respect to the bulk of the solution $[u_i(\infty) = 0]$.

For each of the monomeric components, $G_i(z)$ is directly proportional to its volume fraction $\varphi_i(z)$ in layer z,

$$\varphi_i(z) = \varphi_i^b G_i(z) \qquad (i = 1\text{–}4) \tag{2a}$$

where $\varphi_i^b = \varphi_i(\infty)$ is the bulk solution volume fraction of component i. For the polymeric component 5, the relation between $\varphi_5(z)$ and φ_5^b is more complicated because of the many different possible conformations of the chain molecules. Scheutjens and Fleer[5-7] have shown that $\varphi_5(z)$ is obtained by summation over all possible conformations c, d, . . .:

$$\varphi_5(z) = \frac{\varphi_5^b}{r} \sum_c r^c(z) G^c. \tag{2b}$$

In equation (2b), G^c is a multiple product of $r - 1$ bond weighting factors λ_1 or λ_0 and r segmental weighting factors $G_5(z)$. The parameters λ_1 and $\lambda_0 = 1 - 2\lambda_1$ depend only on the lattice and represent the *a priori* probabilities to make a step to a neighbouring layer or within the same layer, respectively. In a simple cube lattice, with co-ordination number 6, we have $\lambda_1 = 1/6$ and $\lambda_0 = 4/6$; in a hexagonal lattice, we have $\lambda_1 = 3/12$ and $\lambda_0 = 6/12$. The conformation statistical weight G^c can be illustrated by the configuration drawn in Figure 1; in this case $r = 5$, and for the two adsorbed conformations c and d we have $G^c = G_5(2)\lambda_1 G_5(1)\lambda_0 G_5(1)\lambda_0 G_5(1)\lambda_1 G_5(2)$ and $G^d = G_5(1)\lambda_0 G_5(1)\lambda_1 G_5(2)\lambda_1 - G_5(3)\lambda_1 G_5(4)$. Similarly, we have $G^e = \lambda_1 \lambda_0^3 G_5(4) G_5(5)^4$. The quantity $r^c(z)$ in equation (2b) represents the number of segments that conformation c has in layer z; in Figure 1, we have $r^c(1) = 3$, $r^c(2) = 2$, and $r^d(1) = 2$, $r^d(2) = 1$, $r^d(3) = 1$, $r^d(4) = 1$. The summation in equation (2b) extends over all possible conformations, and can be performed most easily using a suitable matrix procedure.[5]

In order to find a solution, we have to express the potentials $u_i(z)$ in the local volume fractions $\varphi_i(z)$. Generally, we may write

$$u_i(z) = u'(z) + u_i^{ct}(z) + u_i^{el}(z), \tag{3}$$

where $u'(z)$ is the local hard-core potential which is independent of the segment type and ensures full occupancy of the lattice, $u_i^{ct}(z)$ represents the nearest neighbour contact interactions (adsorption energy and solvency effects), and $u_i^{el}(z)$ is the electrostatic contribution. For convenience, we assume that the monomeric components are identical except for their charge, so that the small ions mix athermally with the solvent and do not adsorb specifically. Then u^{ct} is given by[5,15]

$$u_i^{ct}(z)/kT = \chi\{\langle\varphi_5(z)\rangle - \varphi_5^b\}, \qquad (i = 1\text{--}4) \tag{4a}$$

$$u_5^{ct}(z)/kT = \chi \sum_{i=1}^{4} \{\langle\varphi_i(z)\rangle - \varphi_i^b\} + \chi_s\delta(z-1), \tag{4b}$$

where the contact fraction $\langle\varphi_i(z)\rangle$ is a weighted average over three layers:

$$\varphi_i(z) = \lambda_1\varphi_i(z-1) + \lambda_0\varphi_i(z) + \lambda_1\varphi_i(z+1). \tag{5}$$

The Flory–Huggins parameter χ expresses the (non-electrostatic) polymer–solvent (or polymer–ion) interaction, and $-\chi_s kT$ is the adsorption energy difference between a polymer segment and a solvent molecule or small ion. The Kronecker delta $\delta(z-1)$ is unity for $z = 1$ and zero otherwise.

The electrostatic contribution u_i^{el} is a function of the valency v_i and the degree of dissociation α_i of species i, and of the electrostatic potential $\psi(z)$,

$$u_i^{el}(z) = v_i\alpha_i(z)e\psi(z), \qquad (i = 1, 2, \ldots, 5) \tag{6}$$

where e is the elementary charge. For our system, we have $v_1 = 0$, $v_2 = v_4 = +1$, and $v_3 = v_5 = -1$. The degree of dissociation α_i is unity for $i = 2, 3$, and 4, and variable for the polyanion segments. The quantity α_5 depends on the local H_3O^+ concentration (hence, on the layer number z) through the equilibrium

$$HP + H_2O \rightleftharpoons H_3O^+ + P^-.$$

The (dimensionless) equilibrium constant of this reaction can be written as $K = [\alpha_5/(1-\alpha_5)][H_3O^+]/[H_2O]$. In the gradient near the surface, the ratio $[H_3O^+]/[H_2O]$ should be replaced by the ratio of contact fractions $\langle\varphi_4(z)\rangle/\langle\varphi_1(z)\rangle$. Hence, we have

$$\alpha_5(z)^{-1} = 1 + K^{-1}\langle\varphi_4(z)\rangle/\langle\varphi_1(z)\rangle. \tag{7}$$

In this expression, K is written in terms of the volume fractions $\varphi_i(z) = c_i(z)/c^*$, where $c_i(z)$ is the local molar concentration of component i and c^* is the molar concentration of a lattice (layer) fully occupied by solvent: $c^* = (10^3 N_{Av} a_s d)^{-1}$, where d is the lattice spacing, and a_s is the area per site. The constant K is related to the usual dissociation constant K_d (in mol l^{-1}) by $K = K_d/c^*$.

The only remaining problem is to express $\psi(z)$ in the local volume fractions. To

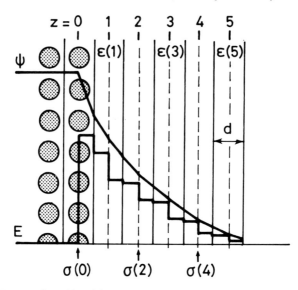

Figure 2   *Potential profile $\psi(z)$ and field strength profile $E(z)$ in the multi-Stern-layer model. The field strength changes at each charged plane $(z = 0, 1, \ldots)$ due to the plane charge density $\sigma(z)$, and at each boundary between the layers $(z = \frac{1}{2}, \frac{3}{2}, \ldots)$ due to a change in the dielectric constant $\varepsilon(z)$. The potential drops linearly in each half-layer because there is no space charge*

that end, we use a multi-Stern-layer model as illustrated in Figure 2. The charge in each layer is assumed to be concentrated at the mid-plane in each layer (*i.e.* $z = 0, 1, 2, \ldots$). The plane charge density is given by

$$\sigma(z) = \sum_i v_i \alpha_i(z) e \varphi_i(z)/a_s. \tag{8}$$

The surface charge density is denoted as $\sigma(0)$. It is also given by equation (8) if we consider the adsorbent S as the sixth component $i = S$. For $z > 0$, we have $\varphi_S(z) = 0$. In between the mid-planes, there is no charge and the electric displacement $D(z)$, which is the product of the dielectric permittivity $\varepsilon(z)$ and the field strength $E(z)$, is constant. At the charged planes, D changes discontinuously. According to standard electrostatics, we have

$$D(z) = \varepsilon(z)E(z) = \sum_{z'=0}^{z} \sigma(z'). \tag{9}$$

The field strength changes discontinuously at the mid-planes ($z = 0, 1, 2, \ldots$), but also at the layer boundaries ($z = \frac{1}{2}, \frac{3}{2}, \ldots$) because of the change in ε. For $\varepsilon(z)$

we take a linear combination of the permittivities ε_i of the pure components:

$$\varepsilon(z) = \sum_i \varepsilon_i \varphi_i(z). \tag{10}$$

From equations (8) to (10), $E(z)$ can be computed in each half-layer; a schematic picture is given in Figure 2. The electric potential $\psi(z)$ varies linearly in each half-layer. For equation (6), we need the values at $z = 1, 2, 3, \ldots$, given by

$$\psi(z + 1) = \psi(z) - \frac{d}{2}\{E(z) + E(z + \tfrac{1}{2})\}. \tag{11}$$

We now have, in equations (1) to (3), a set of five equations per layer in five unknown concentrations $\varphi_i(z)$ and one unknown parameter $u'(z)$ per layer. Closure of the equations is obtained by using as the sixth relation the full occupancy condition,

$$\sum_i \varphi_i(z) = 1, \tag{12}$$

for any z. For details about the numerical procedure, we refer to the original paper.[13] The problem may be solved for constant surface potential $\psi(0)$, for constant surface charge $\sigma(0)$, or even for a charge regulation model in which the quantities $\sigma(0)$ and $\psi(0)$ adjust themselves to local conditions. Extension towards two surfaces in interaction is straightforward.

3 Results and Discussion

We give only a few selected examples to illustrate the main points. First, we compare the outcome of the multi-Stern-layer model for a double-layer without polyelectrolyte with analytical results using the Poisson–Boltzmann equation. Then, we show some predictions of the model for strong and weak polyelectrolytes, and compare them with experimental data for well defined systems. More extensive results, both theoretical and experimental, have been published elsewhere.[10–14] For all the computations in this paper, a hexagonal lattice ($\lambda_0 = 1 - 2\lambda_1 = 1/2$) was used; the relative dielectric constants for the small ions and solvent were taken as 80, and those of the polymer and the adsorbent as 20.

Figure 3 shows the potential decay in the double-layer near a charged surface in the presence of simple electrolyte at three different salt concentrations c_s, for a relatively low surface potential $\psi(0) = 100$ mV. The distance D in Figure 3 equals zd, where z is the dimensionless layer number, and d is the lattice spacing (see Figure 2); c_s is identical to $\varphi_2^b c^* = \varphi_3^b c^*$. The curves in Figure 3 were calculated using the Goüy–Chapman theory in combination with a Stern layer of thickness d, and the points were obtained from the multi-Stern-layer model with lattice spacing d; four different values for d were chosen. The difference between both models is that in the latter case the volume of the ions (and, if desired, the

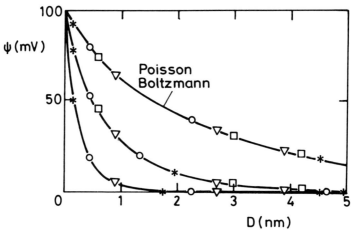

Figure 3   *Potential decay in the electrical double-layer in the absence of poly-electrolyte for three concentrations c_s of monovalent ions and $\psi(0) = 100$ mV. The full curves were computed from the Poisson–Boltzmann equation with the multi-Stern-layer model using various lattice spacings d: ∗, 0.1 nm; ○, 0.3 nm; ▽, 0.6 nm; □, 1.2 nm. The salt concentrations c_s are: 0.01 M (upper curve), 0.1 M (middle curve), and 1 M (lower curve)*

variation of the dielectric constant) is accounted for in all layers, and not only in the Stern layer. For a relatively low surface potential of 100 mV, the ion concentration outside the Stern layer is low everywhere and both models give virtually the same results. For higher surface potentials, the accumulation of ions is more pronounced, and the multi-Stern-layer model leads to a slightly slower decay of the potential, and, consequently, a lower surface charge than the Goüy–Chapman theory,[13] in full agreement with Monte Carlo simulations of Carney and Torrie.[16] An additional feature of the lattice model is that the specific adsorption of certain ions can be built in very easily by assigning these ions a (non-electrostatic) adsorption energy, adding a term $\chi_{s,i}$ to equation (4a).

In Figures 4 and 5, we show some results on the adsorption of strong polyelectrolytes ($\alpha_5 = 1$). Figure 4 gives semi-logarithmic concentration profiles near an uncharged surface, at various salt concentrations c_s, and at a constant polyelectrolyte bulk solution concentration $\varphi_5^b = 10^{-4}$. This figure was computed with the model of Evers *et al.*,[12] where the small ions have no volume, but are considered to be point charges distributing themselves according to the Poisson–Boltzmann equation. The results obtained with the multi-Stern-layer model are virtually the same; it is only at the higher salt concentrations that some minor differences occur. The most conspicuous feature of Figure 4 is the minimum at low ionic strength, which originates from the high potential generated by the adsorbing molecules, repelling non-adsorbing chains. Although the minimum is rather pronounced on the logarithmic scale used for $\varphi_5(z)$, it can probably not be detected experimentally, because of the low concentrations involved.

At low ionic strength the adsorbed amount is quite low [$\varphi_5(1)$ is only a few per

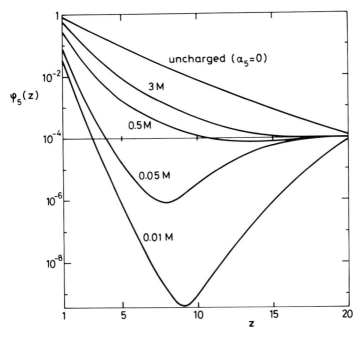

Figure 4 *Semi-logarithmic segment density profiles $\phi_5(z)$ of a strong polyelectro-lyte ($\alpha_5 = 1$) adsorbing on an uncharged surface at various salt concen-trations. The curves were computed with the model of Evers et al.,[12] using $\varphi_5^b = 10^{-4}$, r = 2000, d = 0.71 nm, $\chi_s = 1$, $\chi = 0.5$*

cent at $c_s < 0.1$ M] owing to the strong mutual repulsion between the highly charged segments. Under these conditions, the adsorbed chains lie essentially flat on the surface, with only a few short loops and hardly any tails. The chain-length dependence is very weak. With increasing c_s, the adsorbed amount (and also the chain-length dependence) increases, and the minimum in the profile becomes weaker. Obviously, this is due to a stronger screening of the intersegmental repulsion, allowing more loops and tails. However, even at $c_s = 3$ M, the screening is not yet complete and the adsorption remains below that of an uncharged polymer ($\alpha_5 = 0$).

Figure 5 shows the salt concentration dependence of the adsorbed amount θ_5^a, *i.e.* the number of segments belonging to adsorbing chains per surface site. The abscissa scale is linear in $\sqrt{c_s}$. Curves are given for three different surface charges. The three curves run almost parallel, and the adsorbed amount increases linearly with the surface charge. This is due to the electrostatic contribution to the adsorption energy. As a matter of fact, it can be shown[9] that each additional elementary charge on the surface leads to an extra adsorption of one segment.

At low c_s the dependence of θ_5^a on c_s is weak. In this region, the electrostatic attraction between surface and segments (incremented with χ_s) dominates, and virtually all adsorbed segments are in the surface layer. Recently, Blaakmeer *et*

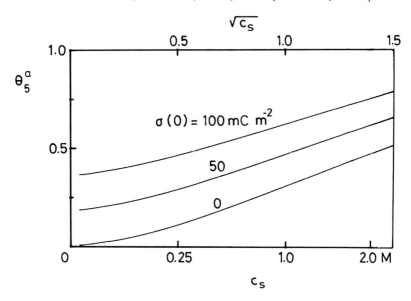

Figure 5   *The adsorbed amount θ_5^a of a strong polyelectrolyte ($v_5 = -1$) as a function of the salt concentration c_s at three surface charge densities $\sigma(0)$. The abscissa scale is linear in $\sqrt{c_s}$. Parameter values: $r = 500$, $\varphi_5^b = 10^{-4}$, $d = 0.6$ nm, $\chi_s = 1$, $\chi = 0.5$*

al.[14] corroborated this weak c_s dependence experimentally in the range $0.001 \text{ M} < c_s < 0.1 \text{ M}$. At higher c_s the stronger screening of the intersegmental repulsion leads to some loop and tail formation, and the adsorbed amount increases linearly with $\sqrt{c_s}$. This increase continues up to very high ionic strength, a feature which has also been demonstrated experimentally.[10,11,17,18]

Our last two examples deal with the adsorption of weak polyelectrolytes. Some years ago, Evers *et al.*[12] predicted that the amount of adsorption of a weak polyelectrolyte on to an oppositely charged surface should pass through a maximum as a function of the pH. The multi-Stern-layer model predicts an even more pronounced maximum. An example is given in Figure 6, where the full curve, corresponding to the right-hand-side ordinate axis, gives the theoretical dependence of θ_5^a on pH, for a polyelectrolyte ($pK_d = 4.25$) adsorbing on a highly charged surface. A pronounced maximum is found at a pH which is *ca.* 1.5 units below pK_d. The shape of the curve is explained as follows. At low pH, the polyelectrolyte is virtually uncharged and the adsorption behaviour is like that of an uncharged polymer, adsorbing with loops and tails. However, the adsorbed amount at high $\sigma(0)$ is lower than on an uncharged surface because the high charge leads to a strong accumulation of counter-ions near the surface, which compete for surface sites and displace polymer segments. With increasing pH, the polyelectrolyte segments acquire a negative charge and can compete more efficiently with the counter-ions, because they also have a non-electrostatic affinity as embodied in χ_s. Hence, θ_5^a increases with pH. In this pH region, the

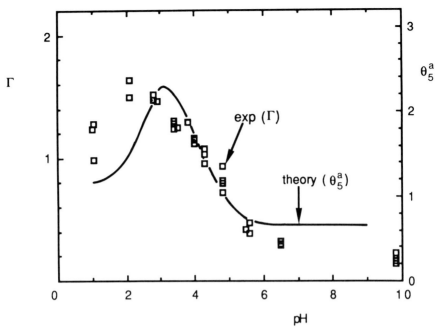

Figure 6 *Comparison of the experimental adsorbed amount* Γ *(in* mg m^{-2},
left-hand scale) for a weak polyacid with the theoretical θ_5^a *(in equivalent
monolayers, right-hand scale), as a function of pH. The experimental
points are for polyacrylic acid* ($M_w = 71\,000$ *daltons) onto a cationic
latex with* $\sigma(0) = 160$ mC m^{-2} *at an equilibrium concentration of*
100 *p.p.m. in* 10^{-1} M KNO$_3$. *The full curve was computed with the
model of Böhmer* et al.,[13] *using* $\varphi_5^b = 10^{-4}$, *p*K$_d = 4.25$,
$\sigma(0) = 160$ mC m^{-2}, $c_s = 0.1$ M, $d = 0.6$ nm, $r = 500$, $\chi_s = 2$, *and*
$\chi = 0.5$

degree of dissociation of adsorbed segments is higher than of segments in the bulk
solution or those in loops and tails (see also Figure 7), so that the intersegmental
repulsion is still weak. With increasing pH, the polyelectrolyte becomes progressively more dissociated and the mutual repulsion between the segments opposes
their accumulation, leading to lower adsorbed amounts. At very high pH, the
polymer is fully charged and the behaviour is the same as that of strong
polyelectrolytes, adsorbing in small amounts and forming thin layers.

Until very recently, such a maximum had not been measured experimentally
for well defined systems, although there is one indication of the phenomenon in
the literature for less characterized materials (modified starch on bleach kraft
pulp[19]). Blaakmeer *et al.*[14] carried out a detailed study of polyacrylic acid onto a
specially prepared cationic latex[20] of high surface charge; these data are also
plotted in Figure 6. The squares in this figure correspond to the left-hand ordinate
axis. Although the experimental maximum is situated at a somewhat lower pH
than the theoretical one, the qualitative agreement between theory and experi-

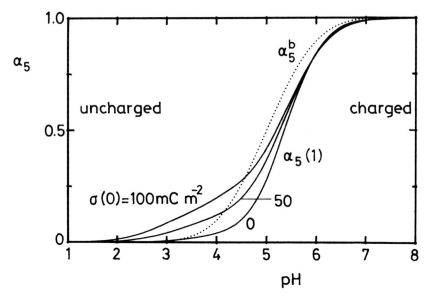

Figure 7 *Degree of dissociation $\alpha_5(1)$ on the surface and α_5^b in solution for a weak polyacid as a function of pH. The quantity $\alpha_5(1)$ was computed for three surface charge densities $\sigma(0)$. Parameter values: $r = 500$, $pK_d = 4$, $\varphi_5^b = 10^{-4}$, $c_s = 1$ M, $d = 0.6$ nm, $\chi_s = 0.5$, and $\chi = 0$*

ment is excellent. For a more quantitative comparison, one should convert the theoretical scale (segments per surface site) to weight per unit area, which is rather tricky. However, the order of magnitude is certainly reasonable. For example, if it is assumed that one segment corresponds to one monomer unit, and if its dimensions are estimated from a molecular model, a fully packed monolayer ($\theta_5^a = 1$) would correspond to about 0.4 mg m^{-2}. Full agreement in Figure 6 would be obtained for *ca.* 0.7 mg m^{-2} for a monolayer. Presumably, a better fit could be found by optimizing the theoretical parameters d, χ, χ_s, and r. In view of the rather crude geometry of the present model, in which polymer segments, water, and small ions are forced to occupy one lattice site each, such an optimization seems at present unwarranted. Moreover, water is here treated as a structureless solvent. Nevertheless, it is gratifying to note that even this rather simplified model describes the main trends excellently.

As discussed above, the degree of dissociation $\alpha_5(1)$ of adsorbed segments is different from that of segments in the bulk solution. Figure 7 gives some more detail of the dependence of $\alpha_5(1)$ at three surface charges (full curves) and of α_5^b. Owing to the accumulation of segment charges on an uncharged surface, the surface concentration of H$_3$O$^+$ is higher, and that of H$_2$O lower, than in the bulk solution, leading to a lower degree of dissociation according to the mass action law [equation (7)]. On a highly charged surface the same effect is found at high pH. However, at low pH the adsorbed segments adjust their degree of dissociation in order effectively to neutralize the surface charge. In terms of equation

(7), $[H_3O^+]$ on the positively charged surface is more strongly decreased than $[H_2O]$, resulting in a higher dissociation of the polyelectrolyte on the surface. Figure 7 illustrates the subtle balance of the various electrostatic and non-electrostatic interactions. Apparently, the theoretical model is able to describe this complicated interplay rather well.

An interesting feature of the present model is the prediction that the adsorption of an uncharged polymer on a highly charged surface is smaller than on an uncharged adsorbent. As far as we know, this effect has not yet been verified experimentally, although Figure 6 at low pH gives an indication of the effect. It would be interesting to devise suitable experiments to check this prediction more systematically.

4 Conclusions

Electrostatic interactions can be successfully incorporated in a lattice model for polymer adsorption. The main trends for the effect of parameters like the salt concentration, the surface charge, and, for weak polyelectrolytes, the pH, are fully corroborated by recent experiments.

Strong polyelectrolytes adsorb at low ionic strength in thin adsorbed layers, developing more loops and tails as the salt concentration increases. The adsorbed amount depends linearly on the surface charge. With weak polyelectrolytes, the degree of dissociation of the adsorbed segments adjusts itself so as to neutralize the surface charge effectively. The adsorbed amount on an oppositely charged surface passes through a maximum as a function of the pH. This maximum is situated at a pH where only a relatively small fraction, of the order of 10%, of the polyelectrolyte segments are dissociated. For weak polyelectrolytes, the effect of the ionic strength is relatively small below $c_s \approx 0.1$ M.

Acknowledgements. The author thanks Drs Marcel Böhmer and Jan Blaakmeer for helpful discussions and for their permission to use some of their theoretical and experimental results.

References

1. D. H. Napper, 'Polymeric Stabilization of Colloidal Dispersions', Academic Press, London, 1981.
2. T. Sato and R. Ruch, 'Stabilization of Colloidal Dispersions by Polymer Adsorption', Marcel Dekker, New York, 1980.
3. G. J. Fleer and J. Lyklema, in 'Adsorption from Solution at the Solid/Liquid Interface', ed. G. D. Parfitt and C. H. Rochester, Academic Press, London, 1983, p. 153.
4. M. A. Cohen Stuart, T. Cosgrove, and B. Vincent, *Adv. Colloid Interface Sci.*, 1986, **24**, 143.
5. J. M. H. M. Scheutjens and G. J. Fleer, *J. Phys. Chem.*, 1979, **83**, 1619.
6. J. M. H. M. Scheutjens and G. J. Fleer, *J. Phys. Chem.*, 1980, **84**, 178.
7. J. M. H. M. Scheutjens and G. J. Fleer, *Macromolecules*, 1985, **18**, 1882.
8. H. A. van der Schee and J. Lyklema, *J. Phys. Chem.*, 1984, **88**, 6661.
9. J. Papenhuijzen, H. A. van der Schee, and G. J. Fleer, *J. Colloid Interface Sci.*, 1985, **104**, 540.
10. J. Marra, H. A. van der Schee, G. J. Fleer, and J. Lyklema, in 'Adsorption from

Solution', ed. R. H. Ottewill, C. H. Rochester, and A. L. Smith, Academic Press, London, 1983, p. 245.

11. J. Papenhuijzen, G. J. Fleer, and B. H. Bijsterbosch, *J. Colloid Interface Sci.*, 1985, **104**, 530.

12. O. A. Evers, G. J. Fleer, J. M. H. M. Scheutjens, and J. Lyklema, *J. Colloid Interface Sci.*, 1986, **111**, 446.

13. M. R. Böhmer, O. A. Evers, and J. M. H. M. Scheutjens, *Macromolecules*, 1990, **23**, 2288.

14. J. Blaakmeer, M. R. Böhmer, M. A. Cohen Stuart, and G. J. Fleer, *Macromolecules*, 1990, **23**, 2301.

15. O. A. Evers and J. M. H. M. Scheutjens, *J. Chem. Soc., Faraday Trans.*, 1990, **86**, 1333.

16. S. L. Carney and G. M. Torrie, *Adv. Chem. Phys.*, 1984, **56**, 141.

17. B. C. Bonekamp, H. A. van der Schee, and J. Lyklema, *Croat. Chem. Acta*, 1983, **56**, 695.

18. T. Cosgrove, T. M. Obey, and B. Vincent, *J. Colloid Interface Sci.*, 1986, **111**, 409.

19. H. Tanaka, K. Tachiki, and M. Sumimoto, *Tappi*, 1979, **62**, 41.

20. J. Blaakmeer and G. J. Fleer, *Colloids Surf.*, 1989, **36**, 439.

Probe Studies of the Gelation of Gelatin using the Forced Rayleigh Scattering Technique

By William G. Griffin and Mary C. A. Griffin

AFRC INSTITUTE OF FOOD RESEARCH, READING LABORATORY, SHINFIELD, READING, BERKSHIRE RG2 9AT

1 Introduction

The mechanism of physical gelation and the structure of physical gels are matters of current theoretical[1,2] and experimental[3,4] interest. Questions about the degree of heterogeneity of physical gels (*e.g.* how it depends on the thermal history and gel composition) remain largely unanswered. The varied morphologies which can be created in certain types of physical gel have recently been analysed in terms of a theoretical model of the mechanism of gel formation by liquid–liquid phase separation arrested by a glass transition.[5–7] In gels which appear to develop as the result of junction-zone crystallite formation, subtle points about the relationship between polymer-chain topology and crystallization kinetics await detailed analysis. The non-ergodic nature of gels suggests connections with recent interesting discussions about the interpretation of light scattering experiments on colloidal glasses.[8,9] Finally, gels are of immense technological importance, not least in the food and food packaging industries, where the factors determining gel stability and transport phenomena in and through gels (*e.g.* certain types of food packaging material[10,11]) need to be thoroughly understood. A technique which can illuminate the time-dependent microscopic structure of gels and help in understanding the hydrodynamic aspects of diffusion of particles in gels and incipient gels (*e.g.* concentrated polymer networks) is therefore of considerable importance.

The forced Rayleigh scattering (FRS) technique was originally devised by Pohl and colleagues[12,13] for the measurement of thermal diffusion coefficients. Subsequently, the technique was extended by workers at the Collège de France[14–16] to enable measurements of molecular self-diffusion coefficients to be carried out. For molecular diffusion measurements the technique requires that the molecules under investigation contain a photochromic group. This form of FRS was first applied to the study of self-diffusion of an azobenzene derivative, Methyl Red, in liquid crystals.[14] The self-diffusion of synthetic polymers and other types of macromolecules has also been studied using FRS (see, for example, refs. 15–19). In order to do this the polymer must first be labelled with an appropriate photochromic group.

The principle of the FRS technique is as follows. A laser beam is split into components of equal intensity. These component beams (the 'writing' beams) are then recombined to produce a pattern of interference fringes which is allowed briefly to illuminate the dye-labelled polymer sample. This fringe pattern creates a transient diffraction grating in the sample by spatially modulating the optical properties of the photochromic molecules. A third laser beam is directed at the Bragg angle on to the decaying grating. The intensity of the diffracted laser beam is measured as a function of time. (Suitable types of photochromic dye include spiropyrans[15,16] and fluorescein,[17] as well as azo dyes.[14,18,19] The stable form of azobenzene compounds generally has a *trans* configuration. Photoisomerization of azo dyes involves rotation about the N=N bond giving rise to the higher energy *cis*-form;[20] this *trans–cis* isomerization of azobenzene and its derivatives is characterized by a change in the visible absorption spectrum.[20,21]) The grating decays at a rate determined by self-diffusion of the dye-labelled species. The technique measures the diffusion of the chromophoric groups, and can be used, therefore, to measure the diffusion of one species in a complicated mixture. In the work to be described here, we have measured the diffusion of a dye-labelled protein molecule, bovine serum albumin (BSA), in aqueous buffer solution and in an aqueous solution of gelatin. We have been able to probe the development of structure in the gelling mixture through changes in the diffusional characteristics of the BSA revealed by FRS.

Gelatin is prepared by either alkaline or acid treatment of collagenous tissue and is a form of hydrolysed, denatured collagen; it is well known for its ability to form food gels. Collagen has an unusual amino acid distribution, containing repeating triplets of -(Gly-X-Y)- where a large percentage of X and Y are proline or hydroxyproline residues. *In vivo* collagen self-assembles to give a triple helical structure; the structure is stabilized by the formation of chemical cross-links.[22] Above *ca.* 40 °C, non-cross-linked collagen unwinds and forms a more-or-less random, denatured configuration. On cooling, such denatured collagen gels. Gelation is thought to occur through development of junction zones with partial reformation of the collagen triple helix structure; such ideas have gained support from experiments on the annealing of denatured collagen chains,[23,24] and from spectroscopic investigations of gelatin solutions and gels.[25–27]

BSA is a large stable globular protein (66 267 daltons[28]) with seventeen disulphide bridges.[29] Its linear dimensions are sensitive to pH and ionic strength,[30,31] but at pH 7.0, at which our measurements have been made, it has a hydrodynamic diameter of *ca.* 8 nm. The choice of dye-reporter group for attachment to BSA was determined by the desirability of carrying out the coupling reaction under mild, non-denaturing conditions, such that changes in the protein structure do not affect its behaviour, *e.g.* its state of aggregation should be unaffected. We therefore chose a maleimide coupling group, as maleimides react primarily with cysteine thiols of proteins at pH 7.0, although they will also react with other nucleophilic groups; a detailed discussion of the reactions of proteins with maleimides is given elsewhere.[32] BSA, which has one free thiol,[29] was therefore modified with 4-dimethylaminophenyl-azophenyl-4'-

maleimide (DABMI) (whose synthesis was described originally by Chang *et al.*[33]).

Aqueous solutions of BSA are completely miscible with gelatin sols at 40–50 °C without any apparent aggregation of the BSA. It is therefore possible to mix dye-labelled BSA with gelatin solutions and carry out FRS measurements on the BSA in the gelatin + BSA mixture. In this paper, we report an FRS study of the self-diffusion of dye-labelled BSA in gelatin during the process of gelation.

2   Materials and Methods

Preparation of DABMI–BSA—BSA (Pentex® bovine albumin, monomer standard, supplied by Miles Ltd., Slough, UK) was dissolved at 4.6 mg ml^{-1} at 0 °C in 0.1 M sodium phosphate buffer pH 7.0 containing 0.02 wt% sodium azide. The solution was treated with 0.05 mM DABMI (supplied by Molecular Probes, Inc., Eugene, Oregon, USA), added, with rapid mixing, as 0.5 vol% of a 10 mM solution in acetone. After incubation for several hours at 0 °C, the reaction mixture was transferred to a controlled temperature room at 4 °C where it was left for 1–3 days. The reaction mixture was then extensively dialysed against the same phosphate buffer. During the course of the reaction, the mixture became slightly turbid, and so, before carrying out forced Rayleigh scattering measurements, the protein solution was filtered through a Millex-GV 0.22 μm filter (Millipore, Harrow, UK). It was then concentrated by centrifugation in a Centricon-10 apparatus. In a later development of the work-up procedure, the dialysis was carried out in the presence of another dialysis bag containing activated charcoal, to remove any free dye from solution. The extent of labelling was estimated as 0.5 moles DABMI per mole BSA from the molecular weights and the extinction coefficients of ε_{280}(BSA) = 0.667 1 g^{-1} cm^{-1},[34] and ε_{460}(DAB) = 24 800 M^{-1} cm^{-1}.[35] (A similar stoichiometry of reaction of BSA with N-ethylmaleimide has been reported.[36]) FRS measurements were carried out on DABMI–BSA in the concentration range 5–7 mg ml^{-1}.

Preparation of Gelatin Solutions and Gels—Gelatin solutions were prepared, following the method given by Djabourov,[26] by adding 10 ml 0.1 M sodium chloride to 1 g gelatin granules (photographic gelatin from Societé Rousselot, lime-processed and de-ionized). The mixture was left at 4 °C overnight, so that the gelatin became swollen, and was then warmed to 50 °C for a few hours until the gelatin had dissolved. A small volume of sodium hydroxide solution was added to bring the pH of the gelatin solution to 7.0.

In order to provide a standard FRS diffraction block, a small quantity of the same gelatin was modified by reaction with fluorescein isothiocyanate;[18] a 0.1 g ml^{-1} fluorescein-labelled gelatin solution in distilled water was prepared at above 50 °C and poured into a 5 mm fluorescence cuvette in which it formed a gel.

Photon Correlation Spectroscopy—Photon correlation spectroscopy (PCS) was carried out as described previously,[37] the normalized correlation functions being fitted by the cumulants method.[38–40]

Filtration—All buffers used were ultrafiltered using a stirred cell (model 52, Amicon Corp., Danvers, Mass., USA) fitted with a Diaflo YM10 membrane (Amicon Corp.) which has a nominal molecular weight cut-off of 10 000 daltons. A pressure of 276 kPa was applied to the cell, using a supply of nitrogen. Before making PCS or FRS measurements, BSA solutions (native or DABMI-modified) were filtered through Dynagard™ (Microgon, Inc., Laguna Hills, CA, USA) 0.2 μm syringe filters. The 4.7 wt% gelatin mixture was not filtered.

Forced Rayleigh Scattering: Description of Apparatus—The apparatus was designed and constructed to enable FRS measurements to be made over a wide range of decay times and scattering vectors. A schematic diagram of the apparatus is shown in Figure 1. Special attention was given to optimizing the stability of the mechanical and electronic components so that measurements at a given scattering angle could be carried out over long periods without significant drift. The temperature control of the sample was considered to be crucial, and several different devices for regulation and measurement of the sample temperature for different temperature ranges were constructed. The apparatus was set up to enable either analogue (DC) or pulse (photon counting) measurements of the diffracted light intensity. A more detailed description of the apparatus will appear elsewhere. What follows is a brief description of the apparatus used to gather the data described in this paper.

The writing beams were obtained from a Coherent Innova 90 argon ion laser fitted with etalon. Up to 500 mW of power at 488 nm was thus available. The blue beam was passed through a spatial filter (Newport Ltd.) and re-focused with a microscope objective so that the spot size at the sample was between 1.5 and 2.0 mm in diameter. A rigid beam-steering device was used to elevate the blue beam to the optical plane of the apparatus and point the beam along the centre line of a precision, low profile, optical rail on which were mounted a broad-band beam splitter and a 25 mm diameter, $\lambda/10$, front-aluminized plane mirror. Two blue beams of nearly equal intensity were thus created and these were directed on to a large, rectangular plane mirror ($\lambda/4$, front-aluminized) whence they converged to cross in the sample cell at a crossing angle θ. It was important to ensure that the scattering from the cell faces was minimal, and fused silica (Suprasil) fluorescence cuvettes (Hellma 101-QS) with 5 mm path length were used throughout the measurements reported here. These cells were found to resist surface scratching rather well. The sample cuvette was supported on a computer-controlled rotary table which in turn was mounted on a computer-controlled, motor-driven, three-axis adjustable platform. An electromechanical shutter (UniBlitz) was placed in front of the beam-steering optics so that the writing beams could be interrupted to produce short (~1 ms) pulses. A 5 mW helium–neon laser (Uniphase) was used to provide the reading beam. This beam ($\lambda = 632.8$ nm) was directed through a variable neutral density wedge and then, at grazing incidence, passed one edge of the large plane mirror. Coincidence of the three beams in the sample was ensured by means of a special alignment block machined out of brass with a 1 mm hole drilled at the crossing point such that the beams must cross in a plane midway between the two optical faces of the cuvette. The separation between the small mirror and the beam-splitter determined the

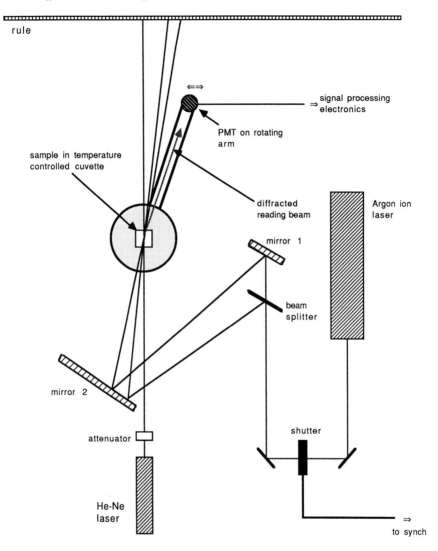

Figure 1 *Schematic diagram of forced Rayleigh scattering apparatus*

crossing angle of the two blue beams. For a given separation the blue beams were directed relative to the reading beam so that the Bragg scattering condition was satisfied. A microscope objective ($\times 100$, Vickers) could be positioned at the point where the writing beams crossed and precisely adjusted to project an image of the fringe pattern on to a screen. The fringe spacings and fringe profiles could then be directly measured as a check on the alignment. The scattering conditions were checked by writing a grating in the gelatin modified with fluorescein isothiocyanate. The diffracted reading beam spot appeared on the steel rule and the scattering angle could then be calculated. The diffracted light intensity was measured using a photomultiplier tube mounted on a swinging arm centred below the sample cell. The swinging arm was mounted centrally on a massive steel clutch plate which coupled the rotor to a motor-driven rotary table (Newport Ltd.). Alternatively a fibre optic coupler was mounted on the swinging arm and the diffracted light was passed via a single mode optical fibre to a photon counting PMT (EMI 9816B) enclosed in a water-cooled housing (Products for Research TE-104TSRF). In either case, the detector head was mounted on a precision screw jack (Newport Ltd.) and the diffracted beam was passed through a blackened brass tube containing a fixed aperture nearest the PMT and an iris diaphragm mounted on an XY-adjustment table at the farther end of the tube. This arrangement served to reduce stray light and ensure that the coherence conditions were satisfied.

In the analogue mode, the signal from the PMT (Hamamatsu R928) was presented to a current-to-voltage converter [EMI pre-amplifier (C632)]. The output from the pre-amplifier was taken directly to a digital oscilloscope (Nicolet 4094C with 4570 plug-in) with 12-bit vertical resolution and 5 MHz analogue bandwidth. Signal averaging was performed using the virtual averaging facility of the oscilloscope. Synchronization was achieved using a quad digital delay pulse generator (SRS DG535) which triggered the oscilloscope and the control elec-tronics (Uniblitz SD-1000) for the electromechanical shutter. The data was transferred to, and stored on, a Macintosh II microcomputer (Apple Computer Inc.) via an IEEE-488 interface (National Instruments). Preliminary data analysis was carried out using the microcomputer; non-linear regression analysis was carried out using a MicroVAX 3600 to which data was transferred via a serial link. The whole apparatus was constructed on a 30 cm thick steel honeycomb optical table (Newport Ltd.) supported on four gas-filled vibration de-coupling plinths. The optical table and electronics were set up in an air-conditioned room with air-temperature control to $\pm0.5\,^\circ$C. The air in the room was continuously filtered to remove dust particles. Temperature control of the sample was ensured by enclosing the side faces of the cuvettes within two copper blocks containing electrical heating elements and a PT100 platinum resistance element connected to a PID temperature control unit (CAL-9000). The room temperature was set to $17.5\,^\circ$C so that the sample temperature could be held at selected temperatures ($\pm0.2\,^\circ$C) above this.

Forced Rayleigh Scattering: Theory and Data Analysis—As discussed by Marmonier,[41] the signal level at the photocathode is proportional to $|E_S + E_{BC} + E_{BI}|^2$, where $E_S(t)$ is the time-dependent field diffracted by the transient grating, and E_{BC} and E_{BI} are the background scattered fields, which are, respectively, coherent and incoherent with respect to the diffracted signal. By definition E_{BI} satisfies the conditions:

$$|E_{BI} \cdot E_S| = 0, \qquad (1)$$

$$|E_{BI} \cdot E_{BC}| = 0. \qquad (2)$$

The photomultiplier response is therefore determined by the quantity

$$V = |E_S + E_{BC}|^2 + |E_{BI}|^2. \tag{3}$$

Where the FRS signal arises from a monodisperse population of dye-labelled species, the amplitude, E_S, is given by

$$E_S = A\,e^{-t/\tau}, \tag{4}$$

where

$$\frac{1}{\tau} = D_S q^2 + \frac{1}{\tau_{dye}}, \tag{5}$$

τ is the grating lifetime, q is the wave vector of the grating [$q = 2\pi/d$, where d is the fringe-spacing, given by the Bragg relation, $d = \lambda_{write}/(2 \sin(\theta/2))$], D_S is the self-diffusion coefficient of the dye-labelled species, and τ_{dye} is the lifetime of the excited dye species. The coefficient A depends on experimental conditions such as the dye concentration and its photochemistry, the change in refractive index and in absorption of the chromophores on photoexcitation, the power of both writing and reading laser beams, and the duration of the flash. Equations (3) and (4) yield the following expression, as shown by Marmonier,[41] for the time-dependence of the FRS signal level:

$$V(t) = (A\,e^{-t/\tau} + B)^2 + C^2. \tag{6}$$

In equation (6), B and C are proportional to the coherent and incoherent background levels, respectively; $(B^2 + C^2)$ is equal to the baseline, V_B, measured as the output voltage before the writing beam excitation. The ratio of the coherent background to the total background must be less than or equal to unity, and this inequality can be included as a constraint[41] if the data are fitted to a theoretical decay curve of the form

$$V(t) = A^2\,e^{-2t/\tau} + 2AV_B^{1/2} \tanh(\gamma)\,e^{-t/\tau} + V_B, \tag{7}$$

where $\gamma \equiv \tanh^{-1}(B/\sqrt{V_B})$. The experimental data were analysed by obtaining the best fit to equation (7). The sum of squares (over the N time points) of weighted residuals, χ^2, was minimized using a non-linear fitting routine.[42] χ^2 is defined by:

$$\chi^2 = \sum_{j=1}^{N} \frac{(V(t_j)^{data} - V(t_j))^2}{\sigma(t_j)^2} \tag{8}$$

Residuals were weighted by the inverse of the estimated error, $\sigma(t_j)$, given by $[V(t_j)]^{1/2}$, which corresponds to Poisson statistics for the noise. This choice was justified because the residuals thus computed for a good fit were distributed uniformly within a band of values over the whole decay timescale (see, for example, Figures 2b and 5b). Other weightings did not have this property. For example Gaussian weighting gave residuals which showed increasing scatter with time and uniform weighting gave residuals which showed decreasing scatter with time.

A quality estimator, Q, defined as

$$Q = 1 - \frac{\sum\limits_{n=1}^{N-1} \varepsilon_n \varepsilon_{n+1}}{\sum\limits_{n=1}^{N} \varepsilon_n^2}, \quad \text{where} \quad \varepsilon_n = \frac{V(t_n) - V(t_n)^{\text{data}}}{\sigma(t_n)}, \tag{9}$$

was used to define goodness of fit, and, except where otherwise stated, results presented here are only for $|Q - 1| < 0.1$.

As is described in the next section, some data did not give a good fit to equation (7). In order to allow for the effects of polydispersity, we used a cumulants method, similar to that sometimes used in photon correlation spectroscopy.[38-40] Where there is a distribution of sizes of dye-labelled particles, or a distribution of types of environment for the diffusers, there will be a distribution of diffusion coefficients, and the amplitude of the diffracted field can be represented by a sum of exponential decays. For a continuous, normalized distribution, $A(\Gamma)$, of decay rates Γ, we have:

$$E_S(t) \propto \int_0^\infty A(\Gamma)\, e^{-\Gamma t}\, d\Gamma. \tag{10}$$

The average decay rate, $\bar{\Gamma}$, is then given by

$$\bar{\Gamma} = \int_0^\infty d\Gamma\, \Gamma A(\Gamma), \tag{11}$$

and the moments of the distribution are given by

$$\mu_i = \int_0^\infty (\Gamma - \bar{\Gamma})^i A(\Gamma)\, d\Gamma. \tag{12}$$

Expanding the exponential function in equation (10) about $\bar{\Gamma}$, we obtain

$$E_S(t) \propto \int_0^\infty d\Gamma\, e^{-\bar{\Gamma} t} \left\{ 1 + \sum_{j=1}^\infty (-1)^j \frac{(\Gamma - \bar{\Gamma})^j t^j}{j!} \right\}$$

$$\propto e^{-\bar{\Gamma} t} \exp \log \left\{ 1 + \sum_{j=1}^\infty (-1)^j \frac{\mu_j t^j}{j!} \right\}$$

$$\cong A'\, e^{-\bar{\Gamma} t + \mu_2 t^2/2}, \tag{13}$$

where third and higher order moments have been neglected. Substituting for $|E_S(t)|$ in equation (3) we obtain

$$V(t) = \left[A' \exp\left(-\bar{\Gamma} t + \frac{\mu_2 t^2}{2} \right) + B' \right]^2 + C'^2. \tag{14}$$

A', B', and C' are now adjustable parameters with $B'^2 + C'^2 = V_B$, the measured background.

Data that could not be fitted well to equation (7) were fitted to equation (14) to give a mean value for $1/\tau$ ($=\bar{\Gamma}^{-1}$) and a normalized variance $\mu_2/\bar{\Gamma}^2$ of the distribution of decay rates. Where $1/\tau_{dye} \ll \bar{\Gamma}$, $\mu_2/\bar{\Gamma}^2$ may be taken as an approximate measure of the polydispersity in the self-diffusion coefficients. As described in the next section, in order to reduce the number of parameters, and hence improve accuracy of the fit, the ratio of B'^2 to V_B was estimated. Kim *et al.*[19] used a form of second-order cumulant model function to fit FRS data from dye-labelled polystyrene in toluene.

The form of FRS signal normally seen in our experiments is shown in Figures 2 and 5. Occasionally, anomalous decays were observed, especially when the writing beam power was high. In these cases, the signal decayed more slowly than the dominant decreasing exponential curve during the first few milliseconds after the excitation pulse. These anomalies may be due to heating effects or transient photochemical events in the sample. To avoid influences other than the diffusive contribution to grating decay, the values for $1/\tau$ and for $\mu_2/\bar{\Gamma}^2$ reported here are those obtained by fitting the data after the point where the signal had decayed by approximately 50% from its maximum.

3 Results

The decay shown in Figure 2a was obtained with a DABMI–BSA sample dialysed simply against buffer. Figure 2b shows a decay for a DABMI–BSA sample where the dialysis was carried out using activated charcoal, as described in the Materials and Methods Section. The best fits to equation (7) are shown, and the weighted residuals are also displayed. It appeared that DABMI–BSA as prepared initially, without the treatment using activated charcoal, gave rise to an FRS signal that corresponded to more than one diffusing species, since the FRS data does not give a good fit to equation (7) (Figure 2a). The simplest explanation of this is that there was still a significant amount of DABMI (or a hydrolysis product of DABMI, *e.g.* the corresponding maleamic acid) free in solution. Although the dialysis step would have ensured the removal of free dye, it might not have removed all dimethylaminophenylazophenyl derivatives non-covalently bound to BSA. (BSA is well known for its ability to bind hydrophobic molecules, including aminoazobenzene derivatives.[43,44]) We suggest, as a possible explanation of these results, that in the absence of the activated charcoal treatment there was a significant proportion of DAB derivatives bound to BSA. On photoexcitation, these dye molecules underwent *trans* to *cis* isomerization and were, at that point, released from the BSA. This supposes that the *cis*-isomer has a lower affinity for the BSA than the *trans*-isomer. Treatment with activated charcoal was tried to see whether this postulated non-covalently bound dye could be removed. The data in Figure 2b are fitted well by equation (7), corresponding to a single diffusing species (the covalently labelled BSA). All subsequent results reported here from

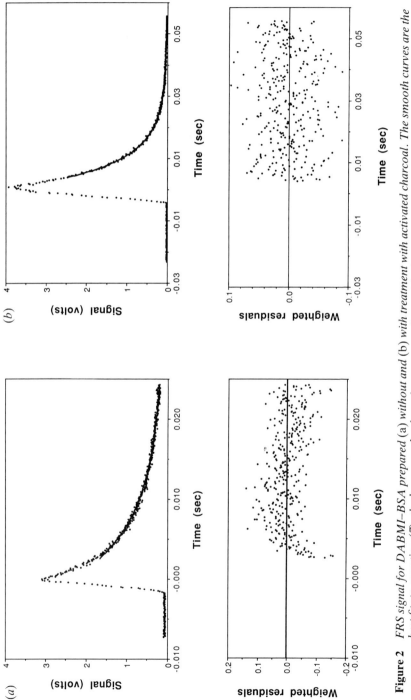

Figure 2   *FRS signal for DABMI–BSA prepared* (a) *without and* (b) *with treatment with activated charcoal. The smooth curves are the best fits to equation* (7); *the lower graphs show the corresponding residuals*

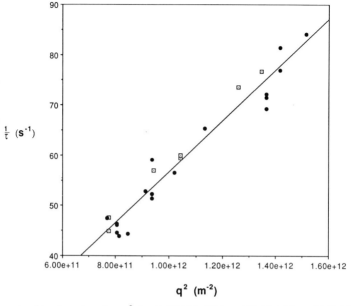

Figure 3 $1/\tau$ plotted against q^2 for DABMI–BSA at ($\square$) 18.0 °C and ($\bullet$) 18.6 °C.
The 18.0 °C data have been corrected as described in the text

DABMI–BSA are from preparations where the activated charcoal treatment was
included.

Figure 3 shows a plot of $1/\tau$ against q^2 for DABMI–BSA; data for two different
preparations of modified BSA are shown. Linear regression to the set of data
obtained at 18.0 °C gave a value for D_S of $(5.3 \pm 0.3) \times 10^{-11}$ m^2 s^{-1} (the slope);
$1/\tau_{\text{dye}}$ was 5 ± 3 s^{-1}. The values of D_S and $1/\tau_{\text{dye}}$ at 18.6 °C were
$(5.0 \pm 0.3) \times 10^{-11}$ m^2 s^{-1} and 6 ± 3 s^{-1} respectively. The 18.0 °C data were
corrected to 18.6 °C as follows. The value of $1/\tau_{\text{dye}}$ of 5 s^{-1} obtained at 18.0 °C was
subtracted from each value of $1/\tau$ to give a set of values of $D_S q^2$. The $\{D_S q^2\}$ were
corrected assuming that D_S over the range 18–20 °C varies with the viscosity of
water and the temperature according to the Stokes–Einstein equation. Thus,
$D_S = kT/6\pi\eta_T r_h$ where k is the Boltzmann constant, η_T is the viscosity at absolute
temperature T, and r_h is the hydrodynamic radius. Values of η_T for water were
taken from the literature.[45] We assume that r_h is independent of temperature in
the range 18–20 °C. The two sets of data were then amalgamated and fitted by
linear regression to give values of D_S and $1/\tau_{\text{dye}}$ of $(5.1 \pm 0.2) \times 10^{-11}$ m^{-2} s^{-1}
and 6 ± 3 s^{-1} respectively. The diffusion coefficient of native BSA (at an
approximate concentration of 17.2 mg ml^{-1}), measured at 20 °C by PCS was
found to be $(5.43 \pm 0.01) \times 10^{-11}$ m^2 s^{-1}. It was difficult to measure the diffusion
coefficient of DABMI–BSA by PCS because of the strong absorbance by the
sample. Values obtained by PCS for the two different DABMI–BSA samples
varied in the range $(4.6–5.7) \times 10^{-11}$ m^2 s^{-1}. The value of D_S obtained from FRS
for DABMI–BSA was corrected to 20 °C using the Stokes–Einstein equation. We

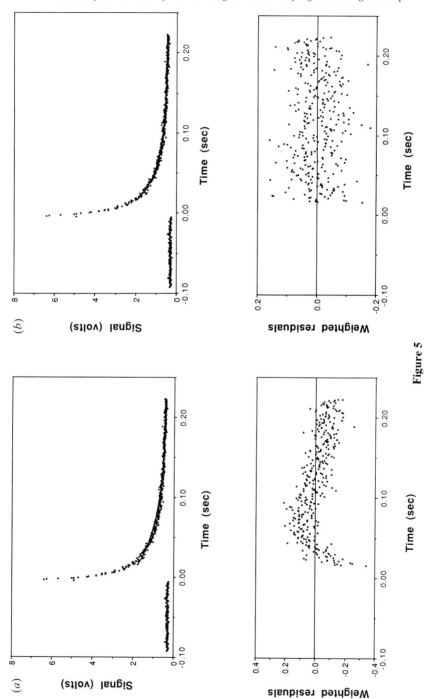

Figure 5

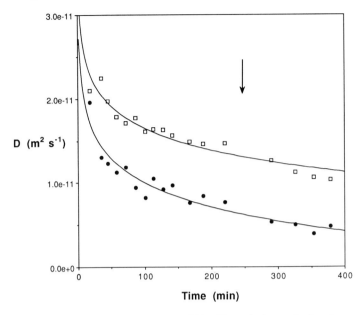

Figure 6 D_S *for DABMI–BSA in* 4.7 wt% *gelatin, calculated assuming* $1/\tau_{dye} = 6.0\,s^{-1}$, *plotted against time.* (●) *and* (□) *show* D_S *calculated from* $1/\tau$ *values obtained from the best fits to equations* (7) *and* (14), *respectively. The arrow shows where the sample was cooled to* 22.5 °C. *The lines are to guide the eye*

$|Q - 1| < 0.1$). For the most polydisperse data set (the last data set of the gelation experiment), it was only possible to obtain $|Q - 1| < 0.1$ with $0.35 < R < 0.6$. The polydispersity obtained was in the range 0.10–0.11. If we make the assumption that $1/\tau_{dye}$ takes a mean value of $6\,s^{-1}$ throughout the experiment, then values of $1/\tau$ may be converted to values of diffusion coefficient; these are plotted against time in Figure 6.

4 Discussion

It appeared from our experiments on DABMI–BSA that BSA would bind the DABMI azo dye non-covalently as well as covalently, and that the adsorbed dye could be ejected from its binding site on photoexcitation by light at 488 nm. Lovrien and colleagues[44,47] showed that BSA will catalyse the *cis → trans*

Figure 5   *FRS signal for DABMI–BSA in* 4.7 wt% *gelatin,* 354 *minutes after mixing. In* (a) *the smooth curve shows the best fit to equation* (7) *for a monodisperse decay. In* (b) *the smooth curve shows the best fit by the cumulants method, equation* (14). *For both* (a) *and* (b) *the lower graphs show the corresponding residuals*

11. 5th International Symposium on Migration, *Food Addit. Contam.*, 1988, **5** (Suppl. 1).
12. D. W. Pohl, S. E. Schwarz, and V. Irniger, *Phys. Rev. Lett.*, 1973, **31**, 32.
13. D. W. Pohl, *IBM J. Res. Develop.*, 1979, **23**, 604.
14. H. Hervet, W. Urbach, and F. Rondelez, *J. Chem. Phys.*, 1978, **68**, 2725.
15. H. Hervet, L. Léger, and F. Rondelez, *Phys. Rev. Lett.*, 1979, **42**, 1681.
16. L. Léger, H. Hervet, and F. Rondelez, *Macromolecules*, 1981, **14**, 1732.
17. M. Antonietti, J. Coutandin, R. Grütter, and H. Sillescu, *Macromolecules*, 1984, **17**, 798.
18. T. Chang and H. Yu, *Macromolecules*, 1984, **17**, 115.
19. H. Kim, T. Chang, J. M. Yohanan, L. Wang, and H. Yu, *Macromolecules*, 1986, **19**, 2737.
20. D. L. Ross and J. Blanc, in 'Photochromism', ed. G. H. Brown, 'Techniques of Chemistry', Wiley, New York, 1971, Vol. 3, Chap. 5, p. 471.
21. G. E. Lewis, *J. Org. Chem.*, 1960, **25**, 2193.
22. A. J. Bailey, in 'Advances in the Molecular Biology of Connective Tissue Fibrous Proteins' (Professor S. M. Partridge Festschrift Volume), Scottish Academic Press, 1987, p. 1.
23. J. Engel, H. P. Baechinger, P. Bruckner, and R. Timpl, in 'Protein Folding', ed. R. Jaenicke, Elsevier, Amsterdam, 1980, p. 345.
24. K. A. Piez, in 'The Protein Folding Problem', ed. D. B. Wetlaufer, *AAAS Symp. Ser.* **89**, 1984, Chap. 3, p. 47.
25. D. A. Ledward, in 'Functional Properties of Food Macromolecules', ed. J. R. Mitchell and D. A. Ledward, Elsevier Applied Science, London, 1986, Chap. 4, p. 171.
26. M. Djabourov, Thesis, Université Pierre et Marie Curie, Paris VI, 1986.
27. M. Djabourov, J. Leblond, and P. Papon, *J. Physique*, 1988, **49**, 319.
28. R. G. Reed, F. W. Putnam, and T. Peters, Jr., *Biochem. J.*, 1980, **191**, 867.
29. J. R. Brown, *Fed. Proc.*, 1976, **35**, 2141.
30. W. F. Harrington, P. Johnson, and R. H. Ottewill, *Biochem. J.*, 1956, **62**, 569.
31. T. Raj and W. H. Flygare, *Biochemistry*, 1974, **13**, 3336.
32. M. C. A. Griffin and W. G. Griffin, *Biochim. Biophys. Acta*, 1984, **789**, 87.
33. J.-Y. Chang, R. Knight, and D. G. Braun, *Biochem. J.*, 1983, **211**, 163.
34. T. Peters, Jr., in 'The Plasma Proteins', ed. F. W. Putnam, Academic Press, New York, 1975, Vol. 1, p. 183.
35. T. Tao, M. Lamkin, and S. S. Lehrer, *Biochemistry*, 1983, **22**, 3059.
36. G. A. Means and R. E. Feeney, 'Chemical Modification of Proteins', Holden-Day, San Francisco, 1971, p. 19.
37. M. C. A. Griffin, *J. Colloid Interface Sci.*, 1987, **115**, 499.
38. D. Koppel, *J. Chem. Phys.*, 1972, **57**, 4814.
39. J. C. Brown, P. N. Pusey, and R. Dietz, *J. Chem. Phys.*, 1975, **62**, 1136.
40. B. Chu, in 'The Application of Laser Light Scattering to the Study of Biological Motion', ed. J. C. Earnshaw and M. W. Steer, NATO A.S.I., A59, 1982, p. 53.
41. M.-F. Marmonier, Thesis, Université Pierre et Marie Curie, Paris VI, 1985.
42. Using NAG (Numerical Algorithms Group, Oxford, UK) routine E04FDF.
43. J. Steinhardt and J. A. Reynolds, 'Multiple Equilibria in Proteins', Academic Press, New York, 1969.
44. R. Lovrien, P. Pesheck, and W. Tisel, *J. Am. Chem. Soc.*, 1974, **96**, 244.
45. 'Handbook of Chemistry and Physics', ed. R. C. Weast, Chemical Rubber Co. Press, Cleveland, OH, 1975–1976.
46. M. Djabourov, J. Leblond, and P. Papon, *J. Physique*, 1988, **49**, 333.
47. R. Lovrien and T. Linn, *Biochemistry*, 1967, **6**, 2281.
48. A. M. Gurney and H. A. Lester, *Physiol. Rev.*, 1987, **67**, 583.
49. E. Dickinson, W. L.-K. Lam, and G. Stainsby, *Colloid Polym. Sci.*, 1984, **262**, 51.
50. P. Favard, J.-P. Lechaire, M. Maillard, N. Favard, M. Djabourov, and J. Leblond, *Biol. Cell*, 1989, **67**, 201.
51. U. A. Stewart, M. S. Bradley, C. S. Johnson, Jr., and D. A. Gabriel, *Biopolymers*, 1988, **27**, 173.

Interactions between Small Amphipathic Molecules and Proteins

By M. N. Jones and A. Brass

DEPARTMENT OF BIOCHEMISTRY AND MOLECULAR BIOLOGY, UNIVERSITY OF
MANCHESTER, MANCHESTER M13 9PT

1 Introduction

Food emulsions and foams are generally stabilized by the adsorption of surface-active materials at the aqueous–oil and aqueous–air interfaces, respectively. These materials are often proteins or low molecular weight amphipathic emulsifiers (surfactants) or a combination of both these species.[1,2] Proteins and emulsifiers not only compete for adsorption sites at the interfaces but interact in the bulk aqueous phase to form a range of protein–surfactant complexes which are themselves surface active. Thus, it is important for understanding the stabilization of food emulsions and foams that the interactions between the proteins and surfactants which lead to the formation of such complexes are characterized.

Protein–surfactant interactions have been extensively studied by a variety of experimental methods.[3–5] It is established that surfactants can be broadly divided into those which complex to proteins and initiate unfolding of the tertiary structure (denaturing surfactants) and those in which the tertiary structure is maintained (non-denaturing surfactants). The commonly used anionic surfactants, *e.g.* sodium n-dodecylsulphate (SDS) or n-dodecylsulphonate, fall into the former category; the nonionic surfactants, *e.g.* the Tritons or n-octyl-β-glucoside (OBG), fall into the latter category, and, when used to solubilize cellular systems, they disperse membrane lipids and membrane proteins without substantial loss of enzymic activity.[6,7] It should, however, be noted that there are significant exceptions to the above generalization. The anionic amphipathics, sodium cholate and deoxycholate, which are related to the 'biological surfactant' bile salts[8] are non-denaturing. There are also some proteins which are resistant to denaturation by even powerful denaturants such as SDS under certain conditions, *e.g.* papain, pepsin and bacterial catalase,[9,10] and there are cases of surfactant activation of enzymes, *e.g. Aspergillus niger* catalase is activated by SDS,[11] glucose-6-phosphatase by Triton X-100,[12] and phospholipase by deoxycholate.[13]

Apart from the above exceptions, the general pattern of protein–surfactant interactions can be broadly depicted as in Figure 1, in which the surfactant ligand

65

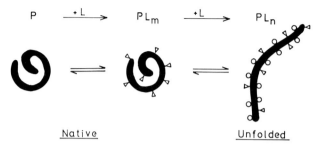

Figure 1   *A schematic representation of the binding of surfactant ligands* L *to the native state of a protein* P *and the subsequent unfolding process*

initially binds to sites on the surface of the native protein. For anionic surfactants, this initial interaction will involve the cationic amino-acid residues of lysine, histidine, and arginine, whereas for non-ionics the binding sites will be hydrophobic patches on the protein surface. In the case of non-ionics, binding ceases once such sites are occupied, but for ionic surfactants the protein unfolds exposing the hydrophobic interior and numerous potential binding sites. Saturation of all the binding sites generally occurs below the critical micelle concentration (CMC) of the surfactant, and on a weight basis this corresponds to approximately 1–2 grams of surfactant per gram of protein; the latter figure is found for reduced proteins (*i.e.* no disulphide bridges) at high ionic strength.[4] That initial binding of anionic surfactants to cationic residues occurs has been confirmed by chemical modification of the residues[14] and studies on polypeptides.[15] However, it should be noted that the ionic interaction by itself is insufficient to anchor the surfactant to the protein, and there must be an accompanying hydrophobic interaction between the alkyl chain of the surfactant and hydrophobic regions adjacent to the cationic sites on the protein surface, since the binding characteristics are dependent on the alkyl chain length.[14]

Theoretical Background—The pattern of protein–surfactant interaction is, from the theoretical viewpoint, one of multiple equilibria which can be written in terms of the protein (P), the surfactant (S), and the complexes (PS$_n$):

$$P + S \rightleftharpoons PS_1$$
$$PS_1 + S \rightleftharpoons PS_2$$
$$PS_2 + S \rightleftharpoons PS_3$$
$$PS_{n-1} + S \rightleftharpoons PS_n. \tag{1}$$

For such a series of equilibria, if the equilibrium constants K for each step are identical, then it follows that

$$K^n = \frac{[PS_n]}{[P][S]^n}, \tag{2}$$

and the average number of surfactant molecules bound per protein molecule $\bar{\nu}$ is given by

$$\bar{\nu} = \frac{n[\mathrm{PS}_n]}{[\mathrm{P}] + [\mathrm{PS}_n]} = \frac{n(K[\mathrm{S}])^n}{1 + (K[\mathrm{S}])^n}. \tag{3}$$

To take into account the fact that the equilibrium constants will in general not be identical, Hill[16] suggested the equation

$$\bar{\nu} = \frac{n(K[\mathrm{S}])^{n_H}}{1 + (K[\mathrm{S}])^{n_H}}, \tag{4}$$

where n_H is a co-operativity coefficient and K becomes an intrinsic binding constant. For $n_H < 1$, binding is negatively co-operative (*i.e.* the binding of a ligand weakens the binding of subsequent ligands); for $n_H > 1$, binding is positively co-operative (*i.e.* the binding of a ligand enhances the binding of subsequent ligands). For identical independent binding sites we have $n_H = 1$, and then equation (4) gives rise to the Scatchard equation

$$\bar{\nu}/[\mathrm{S}] = K(n - \bar{\nu}), \tag{5}$$

which has been extensively used by many workers despite its shortcomings as exposed and discussed by Klotz *et al.*[18-20]

Figure 2 shows model binding isotherms for a hypothetical molecule with 50 binding sites (intrinsic binding constant 10^4) for various degrees of co-operativity (n_H from 0.5 to 7.5). Apart from increasing steepness with increasing n_H, the

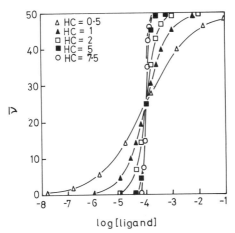

Figure 2 *Binding isotherms ($\bar{\nu}$ versus log [ligand]) calculated from the Hill equation for a protein with 50 binding sites (intrinsic binding constant 10^4) for a range of Hill coefficients from 0.5 to 7.5*

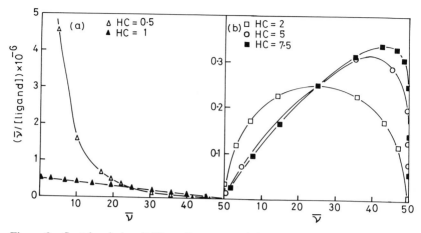

Figure 3 *Scatchard plots ($\bar{v}$/[ligand]$_{free}$ versus $\bar{v}$) for the isotherms of Figure 2 for a protein with 50 binding sites (intrinsic binding constant 10^4) for a range of Hill coefficients from 0.5 to 7.5*

curves are all qualitatively the same, *i.e.* sigmoidal. However, Scatchard plots derived from these isotherms are diagnostic of the type of co-operativity[21] (Figure 3), negative curvature and maxima being characteristic of negative and positive co-operativity, respectively. The description of multiple equilibria in terms of overall binding constants is inevitably an approximation since the binding of every surfactant ligand must change the binding constant for subsequent ligands. A procedure which enables binding constants to be determined as binding proceeds was proposed by Wyman[22] who introduced the binding potential concept.

The binding potential π $(p, T, \mu_1, \mu_2, \ldots)$ at pressure p and temperature T relates ligand binding v to chemical potential μ as follows

$$\bar{v} = \left(\frac{\partial \pi}{\partial \mu}\right)_{p,T} ; \tag{6}$$

and it can be calculated by integration under the binding isotherm assuming that the chemical potential of the ligand can be represented by the ideal solution expression:

$$\pi = 2.303RT \int_{\bar{v}\,=0}^{\bar{v}} \bar{v}\, d\log [S], \tag{7}$$

where R is the gas constant. Considering the formation of a specific complex (PS$_n$) by differentiating equation (3) with respect to ln [S] followed by substitution into equation (7) and integration, we have

$$\pi = RT \ln (1 + K[S]^n). \tag{8}$$

At a given [S] corresponding to a given $\bar{\nu}$, if $PS_{\bar{\nu}}$ is the predominant species, it follows that

$$\pi = RT \ln (1 + K_{app}[S]^{\bar{\nu}}). \qquad (9)$$

By calculating π from equation (7) and substituting into equation (9), the apparent binding constant K_{app} can be calculated for any given $\bar{\nu}$. Thus, the Gibbs energy 'per ligand bound' ($\Delta G_{\bar{\nu}}$) can be obtained from

$$\Delta G_{\bar{\nu}} = -\frac{RT}{\bar{\nu}} \ln K_{app}. \qquad (10)$$

The plot of $\Delta G_{\bar{\nu}}$ *versus* $\bar{\nu}$ shows how successive numbers of bound ligands affect the Gibbs energy of binding. Applying the treatment to the model isotherms in Figure 2 gives the Gibbs energy profiles shown in Figure 4a. The curves converge to the expected value of $\Delta G_{\bar{\nu}}$ of -22 kJ mol^{-1} (corresponding to $K_{app} = 10^4$) on saturation of the binding sites ($\bar{\nu} = 50$), but do not reflect the trends expected from the co-operativity coefficients. For $n_H = 1$, $\Delta G_{\bar{\nu}}$ should be independent of $\bar{\nu}$, whereas, for $n_H > 1$, $\Delta G_{\bar{\nu}}$ should become more negative with increasing $\bar{\nu}$ (*i.e.* positive co-operativity). These anomalies are due to the neglect of the statistical contributions to $\Delta G_{\bar{\nu}}$ which are very significant for large numbers of binding sites. For i ligands binding to n binding sites, the number of arrangements $\Omega_{n,i}$ is given by

$$\Omega_{n,i} = \frac{n!}{(n-i)!\, i!}, \qquad (11)$$

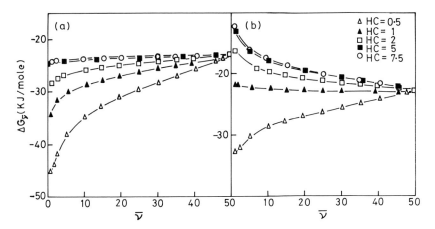

Figure 4   *Gibbs energies of binding per ligand bound ($\Delta G_{\bar{\nu}}$ versus $\bar{\nu}$) calculated by the Wyman binding potential method from the isotherms of Figure 2 for a protein with 50 binding sites (intrinsic binding constant 10^4) for a range of Hill coefficients from 0.5 to 7.5: (a) without statistical corrections, (b) with statistical corrections*

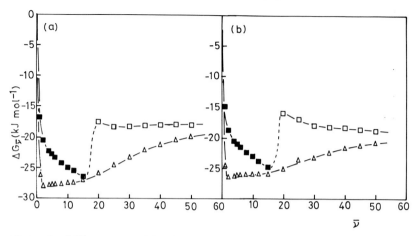

Figure 6 *Gibbs energy of binding per sodium n-dodecylsulphate ligand as a function of the number of SDS ligands bound to lysozyme in aqueous solution at 25 °C, pH 3.2:* △, *calculated from the binding potential without statistical corrections;* □,■, *calculated with statistical corrections for binding to the native* (■) *and unfolded* (□) *states;* (a) *ionic strength* 0.0119 M; (b) *ionic strength* 0.2119 M

close to the point of saturation of the specific binding sites. It seems reasonable to assume that surfactant remains bound to the specific sites after unfolding so that the statistical contributions to binding to the unfolded protein must be calculated from the difference between the number of specific binding sites and the number of binding sites at saturation. The binding isotherms suggest that the number of binding sites at saturation is *ca.* 60 (this figure corresponds to the binding of 1.2 g SDS per g of lysozyme, which is consistent with other saturation binding levels for native proteins[4]). Thus, to a first approximation, the statistical contributions to the Gibbs energies of binding to the unfolding protein should be calculated for *ca.* 45 binding sites.

Figure 6 shows that, when the statistical contributions are taken into account, the curves of $\Delta G_{\bar{\nu}}$ show a transition arising from protein unfolding. The change in $\Delta G_{\bar{\nu}}$ on unfolding can be related to the Gibbs energy of unfolding in the unliganded (ΔG_U) and liganded ($\Delta G_{U/SDS}$) states as follows:

$$N + \bar{\nu}SDS \rightleftharpoons N(SDS)_{\bar{\nu}} : \bar{\nu}\Delta G_{\bar{\nu}}^N, \qquad (12)$$

$$U + \bar{\nu}SDS \rightleftharpoons U(SDS)_{\bar{\nu}} : \bar{\nu}\Delta G_{\bar{\nu}}^U, \qquad (13)$$

where N and U are the native and unfolded protein respectively, and $\bar{\nu}$ corresponds to the number of ligands bound at the mid-point of the transition. Thus, we have

$$\bar{\nu}(\Delta G_{\bar{\nu}}^U - \Delta G_{\bar{\nu}}^N) = \bar{\nu}(\delta\Delta G_{\bar{\nu}}) = \Delta G_{U/SDS} - \Delta G_U. \qquad (14)$$

Table 2   *Thermodynamic parameters for the lysozyme–SDS interaction in aqueous solution (pH 3.2) at 25 °C*

Ionic strength (M)	$\bar{\nu}$ (transition pt)	$\delta\Delta G_{\bar{\nu}}$	$\bar{\nu}(\delta\Delta G_{\bar{\nu}})$	ΔG_U^* (kJ mol^{-1})	$\Delta G_{U/SDS}$
0.0119	17	10	170	46	216
0.0269	18	11.5	207	46	253
0.0554	17	10	170	46	216
0.1119	17	10.5	179	46	225
0.2119	18	10.5	189	46	235
Average					229 ± 16

* Ref. 26.

Table 2 shows the values of the parameters in equation (14) for the lysozyme + SDS system over a range of ionic strength. The Gibbs energy of unfolding of the liganded protein is considerably larger than that for the unliganded protein. Thus the initial binding of surfactant to the cationic sites stabilizes the native state complex, but as binding proceeds the decrease in Gibbs energy resulting from unfolding and exposure of a large number of hydrophobic binding sites more than compensates for the energy required to unfold the liganded native state. Figure 7 shows the total Gibbs energy of complex formation ($\bar{\nu}\Delta G_{\bar{\nu}}$) as a function of $\bar{\nu}$. The kink in the curves corresponds to unfolding.

Molecular Modelling—In order to gain a deeper understanding of the protein–surfactant binding process at the molecular level, computer simulations were performed of various lysozyme–SDS complexes. The dynamical behaviour of the

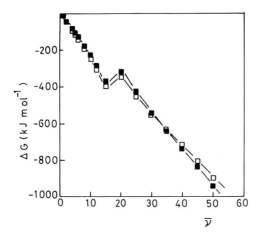

Figure 7   *Gibbs energy of formation of lysozyme–sodium-n-dodecylsulphate complexes as a function of the number of SDS ligands bound at 25 °C, pH 3.2: □, ionic strength 0.0119 M, ■, ionic strength 0.2119 M*

lysozyme–SDS complexes was modelled using the technique of molecular dynamics.[27] From a knowledge of the various potential functions describing the molecular interactions in a protein, the force on every atom in the protein at some time t can be calculated. Using Newton's equations of motion, it is therefore possible to calculate the acceleration on every atom and then to integrate iteratively the equations of motion to obtain the position of each atom at time $t + \delta t$, where δt is typically of the order 1 fs. By performing several tens of thousands of such iterations, the motion of a protein over a period of 10–1000 ps can be followed. Although this time period is very short, it is sufficient to calculate various thermodynamic properties of the lysozyme–SDS complexes.

The co-ordinates of lysozyme were taken from the Brookhaven database.[28] Because these co-ordinates are derived from X-ray studies, the positions of the hydrogens are not defined. Polar hydrogens were explicitly added to the structure and non-polar hydrogens were neglected. The CHARMM description of the protein potentials was used,[29] including modified potentials for carbon atoms with non-polar hydrogens. All the protein simulations were run using the POLYGEN[29] suite of programs on a Silicon Graphics 4D/240GTX graphics work station. Ideally, the lysozyme–SDS simulations would be performed in an aqueous environment by adding several thousand water molecules to the system. However, this would greatly increase the amount of time needed to perform the simulations. An aqueous environment was therefore approximated by using a radially dependent dielectric with a dielectric constant of 80 (to model the charge screening that would occur in a dielectric solvent).[30]

All simulation systems were gradually heated from 0 to 300 K in 10 ps. Each simulation was then run for a further 10 ps with the temperature maintained at 300 K in order to allow the system to equilibrate at this temperature. Each simulation was then run for a further 10 ps with no temperature rescaling to allow for further equilibration. The simulations were finally run for a further 40 ps during which time the average values of the various thermodynamic quantities were calculated.

Simulations were first made on a single molecule of lysozyme at pH 7. The structure of the lysozyme at the end of the simulation is shown in Figure 8. This structure was then protonated to match the charge state at pH 3 to correspond to the experimentally measured SDS binding data and the simulation was repeated using the final structure obtained from the pH 7 simulation as the starting configuration. Figure 9 shows the structure of lysozyme at pH 3 (averaged over the last 40 ps of the simulation). It is interesting to note that the cleft in the lysozyme closes at low pH. At no point in any of the subsequent simulations did the cleft re-open. A similar simulation was also made of an isolated SDS molecule.

A series of simulations was then made in which various numbers of SDS molecules were complexed to the lysozyme. The X-ray co-ordinates of lysozyme complexed with four SDS molecules are known[31] and could therefore be used to determine the initial configuration for the complexes with up to four SDS molecules. The position of other SDS binding sites could be determined by an electrostatic examination of the lysozyme molecule. It is assumed that the

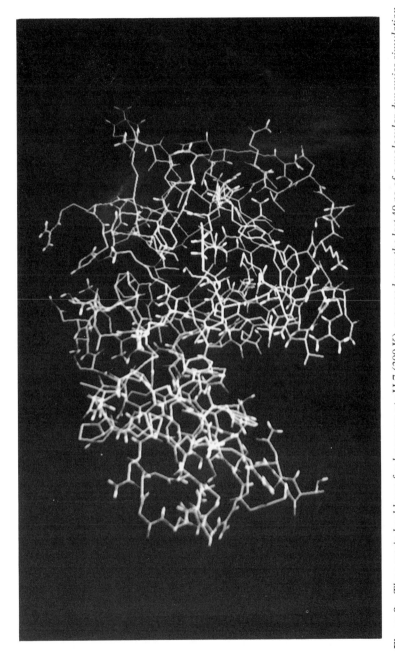

Figure 8   *The protein backbone for lysozyme at pH 7 (300 K) averaged over the last 40 ps of a molecular dynamics simulation*

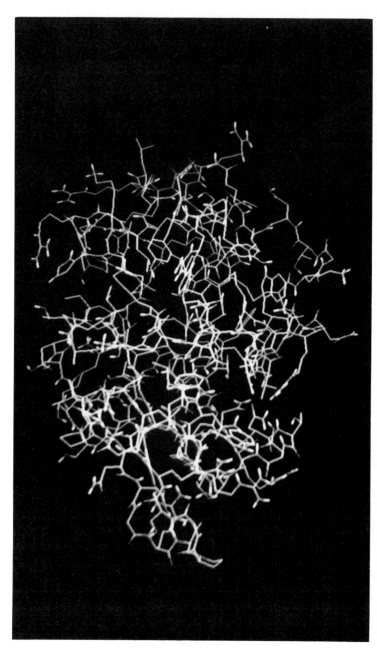

Figure 9   *The protein backbone for lysozyme at pH 3 (300 K) averaged over the last 40 ps of the simulation. The cleft seen in the structure at pH 7 (Figure 8, bottom left) has closed-up*

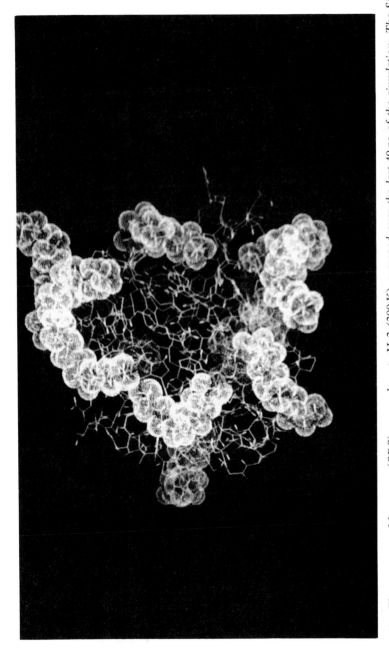

Figure 10 *The structure of lysozyme–(SDS)$_{10}$ complex at pH 3 (300 K) averaged over the last 40 ps of the simulation. The SDS molecules are depicted by van der Waals spheres*

Table 3   *The potential energy, kinetic energy, and total energy for a range of lysozyme–SDS complexes*

Complex	Total energy (kJ mol^{-1})	Potential energy (kJ mol^{-1})	Kinetic energy (kJ mol^{-1})
Lysozyme	6552	2029	4523
SDS	251	135	117
Lysozyme–(SDS)$_1$	6544	1948	4594
Lysozyme–(SDS)$_2$	6573	1933	4636
Lysozyme–(SDS)$_4$	7088	2151	4933
Lysozyme–(SDS)$_7$	7607	2330	5276
Lysozyme–(SDS)$_{10}$	8251	2556	5694

negatively charged head group of the SDS molecule interacts with positively charged residues, and the hydrophobic tail of SDS interacts with the hydrophobic region of the lysozyme surface. In order to determine potential SDS binding sites on the surface of lysozyme, a potential energy surface of lysozyme at pH 3 was generated using an electron as the probe charge. From this surface it was possible to predict the positions at which SDS can bind to the surface of lysozyme.

As expected, the hydrophilic head group of the SDS molecules bonded strongly to positively charged groups on the lysozyme surface, and the hydrophobic SDS tail oriented itself along hydrophobic channels on the protein surface (particularly favouring aromatic groups). The values of the potential energy, kinetic energy and total energy of the complexes with 1, 2, 4, 7, and 10 SDS molecules are shown in Table 3. Figure 10 shows a picture of the lysozyme–(SDS)$_{10}$ complex.

From the information given in Table 3, it is possible to calculate the binding energies of the SDS molecules to the lysozyme. The difference in energy between lysozyme and SDS separately and the complexes gives a measure of the binding energy of the complex [see equation (1)]. The binding energies per SDS molecule bound can then be calculated and are shown in Figure 11.

It is interesting that the shape of this simulated binding energy curve is similar to that obtained experimentally. The values of the binding energies are not, however, equivalent. This is not surprising, as the computer simulation measures the change in potential energy on binding and not the Gibbs free energy of binding which is measured experimentally. Also, the computer simulation does not explicitly include water, and so does not include the contribution to the binding energy from the making and breaking of hydrogen bonds as the surfactant molecules bind to the protein.

The distortion of the lysozyme structure by the addition of SDS molecules was measured by comparing the average protein backbone positions of the isolated lysozyme at pH 3, and the lysozyme with 10 SDS added. Figure 12 shows the protein backbone structure of lysozyme at pH 3 superimposed on the protein backbone of lysozyme with 10 SDS molecules attached. The RMS value for this displacement was found to be small, only 2.13 Å—the secondary structure of the lysozyme was maintained when 10 SDS molecules were bound to its surface. The molecular dynamics approach is now being extended to investigate the denaturation of lysozyme which is observed when larger numbers of SDS molecules are bound.

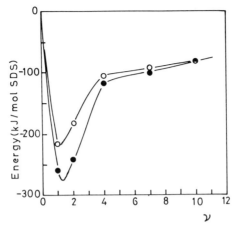

Figure 11 *Binding energy per SDS molecule calculated from the computer simulations of the energies of the lysozyme-SDS complexes as a function of the number of SDS molecules bound at* pH 3 (300 K). ●, *the binding energy measured from the differences in total energy;* ○, *the binding energies measured from the differences in potential energy*

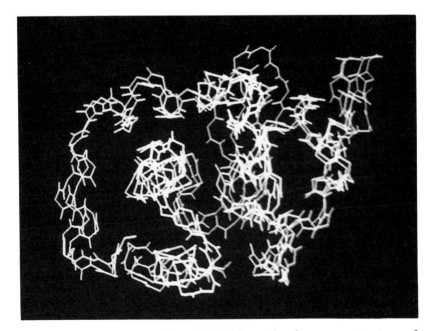

Figure 12 *The superimposed protein backbones for the average structures of lysozyme and the lysozyme–(SDS)$_{10}$ complex*

References

1. 'Food Emulsions and Foams', ed. E. Dickinson, Royal Society of Chemistry, London, 1987.
2. E. Dickinson and G. Stainsby, 'Colloids in Food', Applied Science, London, 1982.
3. J. Steinhardt and J. A. Reynolds, 'Multiple Equilibria in Proteins', Academic Press, New York, 1969.
4. M. N. Jones, 'Biological Interfaces', Elsevier, Amsterdam, 1975, p. 101.
5. M. N. Jones, in 'Biological Thermodynamics', ed. M. N. Jones, Elsevier, Amsterdam, 1988, p. 182.
6. A. Helenius and K. Simons, *Biochim. Biophs. Acta*, 1975, **415**, 29.
7. D. Lichtenberg, R. J. Robson, and E. A. Dennis, *Biochim. Biophys. Acta*, 1983, **737**, 285.
8. A. F. Hofmann and K. J. Mysels, *Colloids Surf.*, 1988, **30**, 145.
9. C. A. Nelson, *J. Biol. Chem.*, 1971, **246**, 3895.
10. M. N. Jones, P. Manley, P. J. W. Midgley, and A. E. Wilkinson, *Biopolymers*, 1982, **21**, 1435.
11. M. N. Jones, A. Finn, A. Mosavi-Movahedi, and B. J. Walker, *Biochim. Biophys. Acta*, 1987, **913**, 395.
12. F. E. Beyhl, *IRCS Med. Sci.*, 1986, **14**, 417.
13. M. Y. El-Sayert and M. F. Roberts, *Biochim. Biophys. Acta*, 1985, **831**, 133.
14. M. N. Jones and P. Manley, *J. Chem. Soc.*, *Faraday Trans 2*, 1980, **76**, 654.
15. M. I. Paz-Andrade, M. N. Jones, and H. A. Skinner, *J. Chem. Soc.*, *Faraday Trans. 1*, 1978, **74**, 2923.
16. A. V. Hill, *J. Physiol.*, 1910, **40**, 40P.
17. G. Scatchard, *Ann. N.Y. Acad. Sci.*, 1949, **51**, 660.
18. I. M. Klotz, *Science*, 1982, **217**, 1247.
19. I. M. Klotz and D. L. Hunston, *J. Biol. Chem.*, 1984, **259**, 10060.
20. H. A. Feldman, *J. Biol. Chem.*, 1983, **258**, 12865.
21. G. Schwarz, *Biophys. Struct. Mech.*, 1976, **2**, 1.
22. J. Wyman, *J. Mol. Biol.*, 1965, **11**, 631.
23. M. N. Jones and P. Manley, *J. Chem. Soc.*, *Faraday Trans. 1*, 1979, **75**, 1736.
24. J. Cordoba, M. D. Reboiras, and M. N. Jones, *Int. J. Biol. Macromol.*, 1988, **10**, 270.
25. M. N. Jones and P. Manley, in 'Surfactants in Solution', ed. K. L. Mittal and B. Lindman, Plenum Press, New York, 1984, Vol. 2, p. 1403.
26. W. Pfeil and P. L. Privalov, *Biophys. Chem.*, 1976, **4**, 41.
27. J. A. McCammon and S. C. Harvey, 'Dynamics of Proteins and Nucleic Acids', Cambridge University Press, Cambridge, 1987.
28. R. Diamond, D. C. Phillips, C. C. F. Blake, and A. C. T. North, *J. Mol. Biol.*, 1974, **82**, 371.
29. B. R. Brooks, R. E. Brucceroli, B. D. Olafson, D. J. States, S. Swaminathan, and M. Karplus, *J. Comp. Chem.*, 1985, **4**, 187.
30. J. A. McCammon, B. R. Gelin, and M. Karplus, *Nature (London)*, 1977, **267**, 585.
31. A. Yonath, A. Podjarny, B. Honig, A. Sielecki, and W. Traub, *Biochemistry*, 1977, **16**, 1418.

Surface Adsorption Studies of Amino-acids and Proteins at a Platinum Electrode

By Sharon G. Roscoe

DEPARTMENT OF FOOD SCIENCE, ACADIA UNIVERSITY, WOLFVILLE, NOVA
SCOTIA, CANADA B0P 1X0

1 Introduction

Surface adsorption of macromolecules has drawn considerable interest in recent
years, not only in the food processing industry confronted with the problems
associated with the fouling of metal surfaces, but also in the field of protein–metal
interaction such as occurs in biosensors and implant materials. The dairy proteins
β-lactoglobulin and κ-casein are particularly interesting because of their interactions in solution,[1,2] and their competitive adsorption behaviour at the metal
surface.[3] These well characterized milk proteins have been the subject of
numerous publications because of their important role in the fouling of metal
surfaces and in the formation and stabilization of dairy foams and emulsions.[1-6]

One of the factors contributing to the somewhat different behaviour of β-
lactoglobulin and κ-casein at interfaces is their structure. At pH 7, the whey
protein, β-lactoglobulin, is present as a dimer consisting of two globular units with
a molecular weight of 1.84×10^4 daltons each.[3,7] Because of its ordered globular
structure, β-lactoglobulin is able to form a tightly packed viscoelastic structure at
the air–water interface.[1] At the same pH, κ-casein exists as a mixture of polymers,
probably linked together by disulphide bonds with molecular weights ranging
from 6×10^4 to $>1.5 \times 10^5$ daltons.[3,7] The molecular weight of the κ-casein
monomer is 1.9×10^4 daltons. It has a more flexible structure, and, although
amphiphilic in character, it supports a hydrophilic negatively charged moiety
containing sugar groups.[3] Although complete dominance of one or more of the
proteins over the others does not occur, it appears generally that the caseins tend
to adsorb in preference to the whey proteins.[1]

Ellipsometry and infra-red reflection–absorption spectroscopy have been the
predominant techniques used in the investigations of the protein–metal
interface.[3-6] A variety of surfaces ranging from a hydrophilic chromium
surface,[3,4] to platinum,[5] titanium,[5] zirconium,[5] and gold[6] have been used with
these techniques. Arnebrant and Nylander,[3] using a combination of *in situ*
ellipsometry and radio-labelling, reported that, when κ-casein is adsorbed first on
to a metal surface at a high surface concentration (3–4 mg m^{-2}), the adsorbed

layer is unaffected by the subsequent addition of β-lactoglobulin. However, when β-lactoglobulin is adsorbed first at a surface concentration of about 1 mg m^{-2}, addition of κ-casein results in adsorption onto the monolayer of the β-lactoglobulin giving a final surface concentration of *ca.* 3 mg m^{-2} with almost no loss of the original adsorbed β-lactoglobulin.[3]

The present research has been undertaken to investigate the adsorption behaviour of β-lactoglobulin and κ-casein at the polarized platinum metal electrode. Experimental conditions similar to those used by Arnebrant and Nylander[3] were chosen in order to compare their results obtained by ellipsometry with those obtained by the electrochemical technique of cyclic voltammetry. Anodic reactions in aqueous solution at the platinum electrode have been well characterized and are controlled by processes involving the electrocatalytic surface oxide film which develops on the anodic metal surface.[8] Although the metal surface may be considered 'demetallized'[9] by its oxide film, a determination of the ability of the adsorbate to 'block' oxide film formation provides another measure of the efficiency of surface adsorption. The present research uses these electrochemical techniques.

2   Experimental

Methods—Cyclic voltammograms were obtained using the Hokuto Denko model HA-301 potentiostat and a Hokuto Denko model HB-111 function generator to produce a repeating triangular potential function. The sweep rate used throughout was 500 mV s^{-1}. The potential and the current were taken from the outputs of the potentiostat and measured by an Allen Datagraph model 720 M X–Y recorder/ plotter. The measurements were also made using a Nicolet model 310 oscilloscope with a digital output that enabled the data to be transferred directly to the computer for data analysis using waveform basic to obtain integrated areas. The rotating electrode system was made by Pine Instrument Company.

Elecrochemical Cells and Electrodes—Three-compartment, all-glass cells, provided with glass-sleeved stopcocks as two of the compartments, were used. Purified nitrogen was bubbled through the working and counter electrode compartments of the electrochemical cell to remove oxygen and to ensure well mixed conditions. Johnson Matthey and Mallory high-purity grade platinum wires were degreased by refluxing in acetone, sealed in soft glass, electrochemically cleaned by potential cycling in 1 M sulphuric acid, and stored in 98 wt% aqueous H_2SO_4. The reference electrodes were saturated calomel electrodes (E_{SCE}) made according to a standard procedure.[10] Their potentials were checked frequently against a standard hydrogen electrode and compared with the literature value.[11] The saturated calomel electrodes were found to be reproducible to within ± 1 mV. A platinum rotating electrode mounted in teflon, obtained from Pine Instrument Company (Grove City, PA), was used at a speed of 1600 r.p.m. The main purpose of using a rotating electrode was to eliminate or minimize the possiblity of mass transfer effects. For the experimental conditions used, very reproducible current–voltage relations could be established which were independent of any further increase of rotation frequency beyond the above rate. The estimation of the true surface areas of the platinum electrodes is based on the assumption that one atom of hydrogen is deposited on each metal atom of the surface, and that a monolayer of hydrogen is completed at the reversible potential of the hydrogen electrode.[12] The real area can be obtained from the charges under the hydrogen underpotential deposition (upd) peaks[13] determined by cyclic voltammetry in 0.5 M aqueous H_2SO_4 at 298 K, taking 210 μC cm^{-2} as the charge required for formation of a monolayer of H in the usual way.[14]

Solutions—The initial measurements were made with a 0.1 M phosphate buffer at pH 7.0 prepared from anhydrous potassium phosphate monobasic (KH_2PO_4) (cell culture tested) obtained from Sigma Chemical Company. Solutions were prepared from conductivity water (Nanopure water, resistivity = 18.3 MΩ cm^{-1}). Samples of κ-casein (C-0406, Lot 88F-9620) and β-lactoglobulin (L-0130, Lot 106F-8120, 3 × crystallized) were also obtained from Sigma Chemical Company.

3 Results and Discussion

Surface Adsorption Behaviour at a Platinum Rotating Electrode—In order to characterize the behaviour of the surface adsorption of β-lactoglobulin and κ-casein at the platinum rotating electrode, it was necessary first to determine the surface charge density of oxide deposition at the platinum rotating electrode in 0.1 M phosphate buffer at pH 7.0 under the same experimental conditions as those used with the proteins. Figure 1 shows the cyclic voltammogram for this phosphate buffer solution with incremental steps in anodic end potential. The cyclic voltammogram is similar to that observed for a 0.5 M aqueous H_2SO_4 solution at a platinum electrode.[8] A reversible deposition of oxide occurs at low anodic potentials which changes to an irreversible stage of oxide formation with increase in the anodic potential. This irreversibly deposited oxide is similar to that observed in the cyclic voltammograms of a sulphuric acid solution at platinum, which has been attributed to Pt–O or OH 'place-exchange' of the electrode surface.[8]

Cyclic voltammograms were recorded after each aliquot of β-lactoglobulin was added to the buffer solution in the electrochemical cell. An immediate change could be seen in the profile of the cyclic voltammogram, shown in Figure 2, with an addition of as little as 0.007 g l^{-1} β-lactoglobulin. The results of a subsequent aliquot of β-lactoglobulin giving a total amount of 0.014 g l^{-1} are shown in Figure 3. Further additions of the protein gave negligible change in the cyclic voltammograms.

In order to determine the surface charge density resulting from adsorption of the protein, integration of the current–potential response corresponding to anodic oxidation and to oxide reduction was carried out for each addition of β-lactoglobulin. The difference calculated between the surface charge density for anodic oxidation and that for oxide reduction is attributed to surface adsorption or oxidation of species other than O or OH present in these aqueous solutions. A

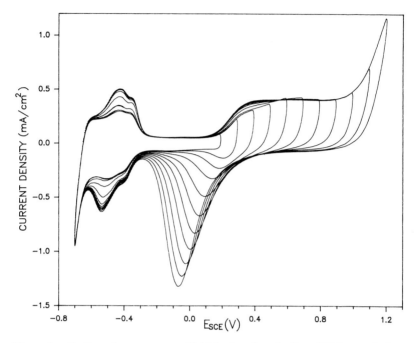

Figure 1   *Cyclic voltammogram of 0.1 M phosphate buffer pH 7.0 at a platinum rotating electrode (1600 r.p.m.) showing the current due to anodic oxidation and reduction as a function of anodic end potentials E_{SCE} at sweep rate 500 mV s^{-1}*

small difference in the integrated surface charge density from the cyclic voltam-mograms of the phosphate buffer was subtracted from those determined for the protein solutions. The remaining charge density was attributed to protein surface adsorption. This may, however, represent a lower limit, as protein adsorption may occur competitively with the phosphate buffer.

The charge associated with adsorption of β-lactoglobulin is shown in Figure 4 as a function of concentration of protein. A plateau in charge density of about $70\,\mu C\ cm^{-2}$ is reached with only $0.02\ g\ l^{-1}$ β-lactoglobulin. As the concentration

Figure 2   *Cyclic voltammogram of 0.1 M phosphate buffer pH 7.0 at a platinum rotating electrode (1600 r.p.m.) showing the effects of addition of $0.007\ g\ l^{-1}$ β-lactoglobulin to the buffer solution at sweep rate 500 mV s^{-1}*

Figure 3   *Cyclic voltammogram of 0.1 M phosphate buffer pH 7.0 at a platinum rotating electrode (1600 r.p.m.) showing the effects of addition of β-lactoglobulin to the buffer solution increasing the concentration from 0.007 to $0.014\ g\ l^{-1}$ β-lactoglobulin at sweep rate 500 mV s^{-1}*

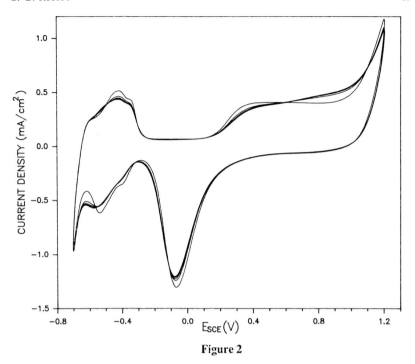

Figure 2

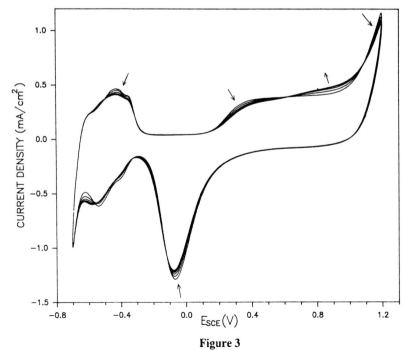

Figure 3

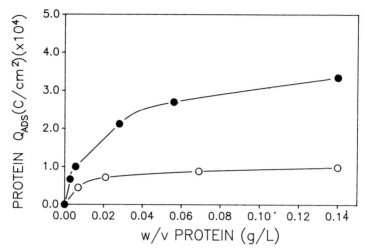

Figure 4 *Charge resulting from protein adsorption, Q_{ADS}, as a function of protein additions to the 0.1 M phosphate buffer pH 7.0 at a platinum rotating electrode (1600 r.p.m.): $\bigcirc$, β-lactoglobulin; $\bullet$, κ-casein*

of protein is raised in the bulk solution to a value of $0.14 \, \text{g} \, \text{l}^{-1}$, a corresponding gradual increase occurs in charge density reaching a value of $100 \, \mu\text{C} \, \text{cm}^{-2}$ for the anodic end potential of 1.2 V used in these measurements. This corresponds to a surface concentration of about $2 \, \text{mg} \, \text{m}^{-2}$ assuming a one-electron transfer process.

A similar set of experiments was carried out with additions of κ-casein to the electrochemical cell containing 0.1 M phosphate buffer at pH 7.0 in the absence of β-lactoglobulin. The charge density due to adsorption of κ-casein was enhanced compared with that obtained for adsorption of β-lactoglobulin. As little as $0.003 \, \text{g} \, \text{l}^{-1}$ κ-casein resulted in an adsorption charge density similar to that observed for the plateau value for adsorption of β-lactoglobulin (Figure 4). Again with increasing concentrations of κ-casein in the bulk solution, the onset of a plateau was reached at $0.06 \, \text{g} \, \text{l}^{-1}$ κ-casein. The charge density in the plateau region ranged from $300–400 \, \mu\text{C} \, \text{cm}^{-2}$ for a three-fold increase in κ-casein. The anodic end potential used for these measurements was 1.2 V. These values correspond to a surface concentration ranging from $6–8 \, \text{mg} \, \text{m}^{-2}$, significantly higher than that observed for the β-lactoglobulin.

The cyclic voltammogram with $0.14 \, \text{g} \, \text{l}^{-1}$ κ-casein in the 0.1 M phosphate buffer solution at pH 7.0 is shown in Figure 5. In the anodic oxidation profile, a diminished shoulder in the lower anodic potential regions indicates that there is some blocking of the initial oxide deposition up to a monolayer coverage. An enhanced adsorption charge density can be seen in the region of 0.6 V which was not present in the profiles for the phosphate buffer alone or for those containing β-lactoglobulin.

The surface adsorption charge density for a range of concentrations of each of the proteins in the phosphate buffer at the platinum rotating electrode is shown in Figure 6 as a function of anodic end potential. In the phosphate buffer the oxide

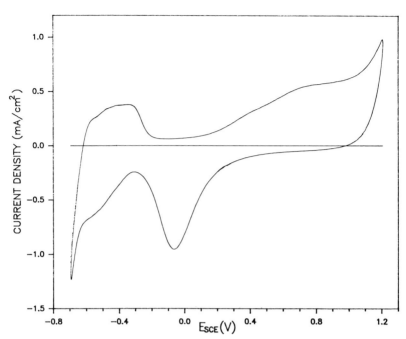

Figure 5 *Cyclic voltammogram of 0.1 M phosphate buffer pH 7.0 at a platinum rotating electrode (1600 r.p.m.) showing the effect of the addition of 0.14 g l^{-1} κ-casein to the phosphate buffer solution at sweep rate 500 mV s^{-1}*

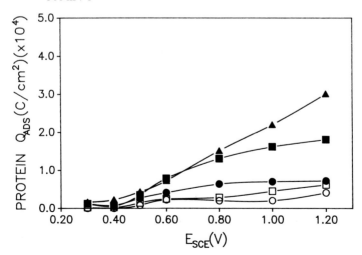

Figure 6 *Charge resulting from protein adsorption, Q_{ADS}, as a function of anodic end potential E_{SCE} for protein additions to the 0.1 M phosphate buffer pH 7.0 at a platinum rotating electrode (1600 r.p.m.): ○, 0.007 g l^{-1} β-lactoglobulin; □, 0.069 g l^{-1} β-lactoglobulin; ●, 0.006 g l^{-1} κ-casein; ■, 0.028 g l^{-1} κ-casein; ▲, 0.14 g l^{-1} κ-casein*

anodic profile at potentials reaching 0.55 V corresponds to a monolayer surface coverage of oxide. Over this same potential region, β-lactoglobulin adsorbs to give a surface charge density of about $20\,\mu C\,cm^{-2}$ for bulk solution concentrations up to $0.069\,g\,l^{-1}$. This surface charge density is maintained at a plateau level until potentials greater than 0.8 V are reached where additional adsorption occurs to give a value of $60\,\mu C\,cm^{-2}$ for the $0.069\,g\,l^{-1}$ concentration at 1.20 V. This corresponds to a surface concentration of about $1\,mg\,cm^{-2}$, which is similar to the results observed by Arnebrandt and Nylander.[3]

A similar behaviour can be seen for κ-casein at low concentrations $(0.006\,g\,l^{-1})$ in the phosphate buffer, but, with higher concentrations, surface adsorption as indicated by the surface charge density increases dramatically at anodic potentials more positive than 0.6 V and is dependent on the concentration of κ-casein in the bulk solution.

Surface Adsorption Behaviour at a Platinum Wire Electrode—A similar series of experiments was made using an electrochemical cell designed for use with a stationary platinum wire electrode. The cyclic voltammograms for a 0.1 M phosphate buffer at pH 7.0 are shown in Figure 7 with a stepwise increasing

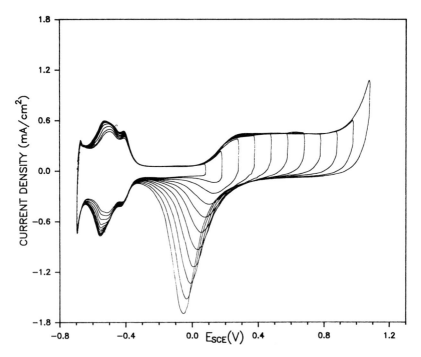

Figure 7   *Cyclic voltammogram of* 0.1 M *phosphate buffer pH* 7.0 *at a platinum wire electrode showing the current due to anodic oxidation and reduction as a function of anodic end potentials* E_{SCE}, *at sweep rate* $500\,mV\,s^{-1}$

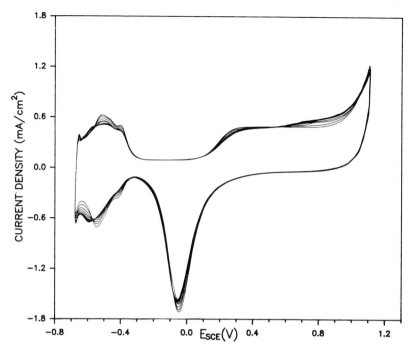

Figure 8   *Cyclic voltammogram of* 0.1 M *phosphate buffer pH* 7.0 *at a platinum wire electrode showing the effects of additions of β-lactoglobulin to the buffer solution*: 0.012 *and* 0.024 g l^{-1} *β-lactoglobulin in phosphate buffer at sweep rate* 500 mV s^{-1}

anodic end potential. The potentiodynamic profile at the platinum wire electrode is similar to that observed for the platinum rotating electrode. A small reversible oxide deposition occurs at the lower anodic end potentials, but is converted to the irreversible form as the potential is swept to more positive values. An immediate response was observed in the current–potential relationship in the cyclic voltam-mograms with the addition of 0.012 g l^{-1} β-lactoglobulin to the phosphate buffer solution (Figure 8). As the concentration of β-lactoglobulin is increased in solution, a plateau value in surface charge density of 70 μC cm^{-2} is reached at 0.024 g l^{-1} of protein, as is also observed at the platinum rotating electrode (Figure 9).

In order to investigate the competitive adsorption behaviour of κ-casein and β-lactoglobulin, aliquots of κ-casein were added to the solution of 0.1 M phosphate buffer at pH 7.0 containing 0.12 g l^{-1} β-lactoglobulin (Figure 9). Additions as small as 0.012 g l^{-1} κ-casein result in an enhancement in the surface charge density, which reaches a plateau level for concentrations of *ca.* 0.07 g l^{-1} κ-casein. Further increase in the bulk solution concentration of κ-casein in the presence of β-lactoglobulin results in surface charge densities of 300–400 μC cm^{-2}. These results compare favourably with those obtained with the platinum rotating

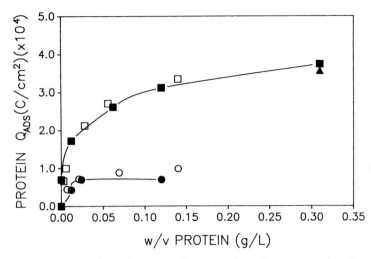

Figure 9   *Charge resulting from protein adsorption,* Q_{ADS}, *as a function of protein additions to the* 0.1 M *phosphate buffer pH* 7.0 *at a platinum wire electrode:* ●, *β-lactoglobulin;* ■, *κ-casein* + 0.12 g l^{-1} *β-lactoglobulin;* ▲, *κ-casein* + 0.36 g l^{-1} *β-lactoglobulin. Also shown are the data from Figure 4 for the platinum rotating electrode:* ○, *β-lactoglobulin;* □, *κ-casein*

electrode in the absence of β-lactoglobulin as shown in Figure 9 for both sets of data. Figure 10 shows the cyclic voltammogram for 0.062 g l^{-1} κ-casein in 0.12 g l^{-1} β-lactoglobulin in the phosphate buffer at the platinum-wire electrode.

An additional aliquot of β-lactoglobulin, bringing its concentration to 0.36 g l^{-1}, was added to the solution already containing 0.31 g l^{-1} κ-casein to study further the competitive adsorption behaviour of the two proteins. The surface charge density was found not to increase with this additional amount of β-lactoglobulin. Examination of the surface charge density suggests that, under the conditions of potentiodynamic sweeping of the platinum electrode, κ-casein adsorbs predominantly. Some co-adsorption with β-lactoglobulin may also occur. It is important to note that, with each potentiodynamic sweep, the electrode–metal surface is electrochemically cleaned by the oxidation and reduction processes as can be seen by the double-layer region. This allows competitive surface adsorption to occur during the positive going anodic sweep. In this respect, cyclic voltammetry differs from the experimental techniques used by Arnebrandt and Nylander,[3] who found that, when β-lactoglobulin was adsorbed first onto a metal surface with a surface concentration of about 1 mg m^{-2}, κ-casein adsorbed onto the monolayer of the β-lactoglobulin. But, when κ-casein was allowed to adsorb first with a surface concentration 3–4 mg m^{-2}, the surface adsorption was unaffected by addition of β-lactoglobulin. The results of this latter experiment are supported by the present work in which surface adsorption by κ-casein appears to be unaffected by the presence of β-lactoglobulin.

The surface charge resulting from protein adsorption as a function of anodic end potential (Figure 11) was found to be dependent on the concentration of

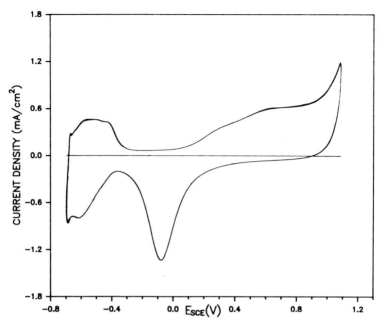

Figure 10 *Cyclic voltammogram of 0.1 M phosphate buffer pH 7.0 at a platinum wire electrode showing the effect of the addition of 0.062 g l^{-1} κ-casein to the phosphate buffer solution containing 0.12 g l^{-1} β-lactoglobulin at sweep rate 500 mV s^{-1}*

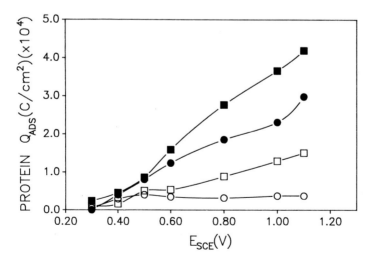

Figure 11 *Charge resulting from protein adsorption, Q_{ADS}, as a function of anodic end potential E_{SCE} for protein additions to the 0.1 M phosphate buffer pH 7.0 at a platinum electrode: $\bigcirc$, 0.012 g l^{-1} β-lactoglobulin; $\square$, 0.12 g l^{-1} β-lactoglobulin; $\bullet$, 0.12 g l^{-1} β-lactoglobulin + 0.062 g l^{-1} κ-casein; $\blacksquare$, 0.12 g l^{-1} β-lactoglobulin + 0.31 g l^{-1} κ-casein*

protein in the bulk solution, and showed a similar behaviour to the results obtained with the platinum rotating electrode (Figure 6). Significant increases in surface adsorption occur at potentials >0.55 V, which corresponds normally to a monolayer surface oxide coverage for the phosphate buffer solution in the absence of protein.

Oxide Growth Rates in the Presence of Adsorbed Proteins—Oxide growth rates were determined after a period of continuous cycling to establish the steady-state potential–current relationship in order to characterize both the anodic oxidation and reduction profiles. The potential was then held for a measured length of time, after which a negative going cathodic sweep allowed characterization of the oxide reduction profile, which became enhanced as a result of holding the anodic end potential. The potential was then allowed to sweep in a continuous manner to ensure return of the system to the potential–current steady-state before repeating the experiment for a different holding time. Figure 12 shows the oxide reduction charge for holding times of 10, 30, and 60 seconds at the anodic end potentials as shown for the solution containing 0.028 g l^{-1} β-lactoglobulin in 0.1 M phosphate buffer at the platinum rotating electrode. Similar sets of curves were obtained for the different solutions and all were found to be linear with respect to the logarithm of the holding time as described by the equation

$$Q = A \log (t + t_0), \tag{1}$$

where $t \gg t_0$ for the main region of the logarithmic growth, and t_0 corresponds to a

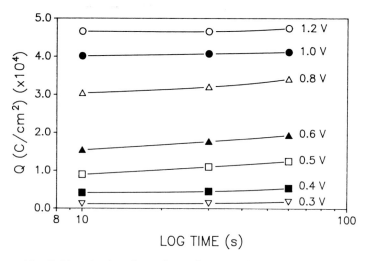

Figure 12   *Oxide reduction charge Q as a function of the logarithm of the holding time at the specified anodic end potential E_{SCE} for 0.028 g l^{-1} β-lactoglobulin in 0.1 M phosphate buffer pH 7.0 at a platinum rotating electrode (1600 r.p.m.)*

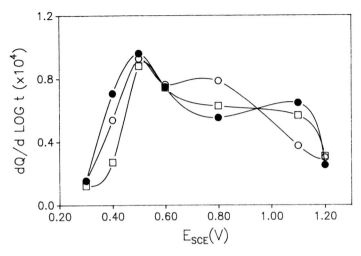

Figure 13   *Rate of oxide growth (dQ/d log* t*) as a function of the anodic holding potentials* E_{SCE} *for additions of protein to 0.1 M phosphate buffer pH 7.0 at a platinum rotating electrode (1600 r.p.m.):* ●, *0.1 M phosphate buffer pH 7.0;* ○, *0.007 g l^{-1} β-lactoglobulin;* □, *0.069 g l^{-1} β-lactoglobulin*

time of the order of a millisecond or less, the time required for adjustment of the potential.[15] The slopes of these lines, $dQ/d \log t$, which represent the rate of growth of the oxide film as a function of the anodic holding potentials at the platinum rotating electrode are shown in Figure 13 for additions of β-lactoglobulin to the phosphate buffer. At potentials between 0.2 V, which is the onset for oxide deposition, and 0.55 V, which represents deposition of a mono-layer of oxide on the surface of the electrode, a diminished oxide growth rate results from the addition of β-lactoglobulin. This is presumably due to a blocking effect by the macromolecule at the surface of the electrode. The effect, however, is not as strong as that seen with glycine at very low pH where the oxide is completely blocked in this area.[16] In the region between 0.6 V and 0.9 V where Pt–O or OH 'place-exchange' normally occurs,[8] the oxide growth rates appear to be enhanced somewhat in the presence of the β-lactoglobulin. This may be due to a rearrangement of the protein on the surface, which allows oxide growth to continue. Between 0.9 V and 1.20 V, there is again a diminished effect on the oxide growth rate, presumably resulting from the increased adsorption of β-lactoglobulin at these higher anodic potentials.

Figure 14 shows a similar set of oxide growth rates as a function of potential at the platinum rotating electrode for additions of κ-casein to the phosphate buffer. The results show a stronger blocking effect toward oxide formation by κ-casein than for β-lactoglobulin. In the potential region 0.6 to 0.7 V, the surface charge density increases (Figures 5 and 10), accompanied by a decrease in the oxide growth rates over this potential range. This differs from the potentiodynamic

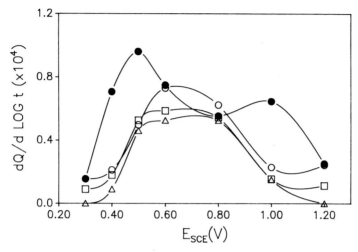

Figure 14   *Rate of oxide growth (dQ/d log t) as a function of the anodic holding potentials E_{SCE} for additions of protein to 0.1 M phosphate buffer pH 7.0 at a platinum rotating electrode (1600 r.p.m.): ●, 0.1 M phosphate buffer pH 7.0; ○, 0.006 g l^{-1} κ-casein; □, 0.028 g l^{-1} κ-casein; △, 0.14 g l^{-1} κ-casein*

profiles (Figures 2 and 3) and the enhanced growth rates obtained with β-lactoglobulin over this potential range. Again, there is an even more pronounced effect on the diminished oxide growths at the potentials between 0.8 V and 1.2 V indicative of the stronger adsorption of the κ-casein molecule compared with β-lactoglobulin.

4   Conclusions

The surface adsorption of β-lactoglobulin in 0.1 M phosphate buffer at pH 7 increases in a similar manner at both the platinum rotating electrode and the stationary platinum wire as the concentration of the protein in the bulk solution was increased. A plateau value in charge density of about $100\,\mu C\,cm^{-2}$, corresponding to a surface concentration of about $2\,mg\,m^{-2}$, was reached with 0.002–0.14 g l^{-1} of β-lactoglobulin at the anodic end potential of 1.2 V. Under similar experimental conditions, 0.06 g l^{-1} κ-casein reached a plateau level of about $300\,\mu C\,cm^{-2}$ which gradually increased to $400\,\mu C\,cm^{-2}$ as the concentration of κ-casein was increased. The corresponding surface concentration of 6–$8\,mg\,m^{-2}$ represents a three- to four-fold increase in surface adsorption by κ-casein in comparison with β-lactoglobulin.

Surface adsorption of both proteins is dependent on anodic end potential. The surface charge density resulting from adsorption of κ-casein increases markedly at potentials greater than 0.55 V, the potential corresponding to a monolayer surface coverage of oxide and a surface concentration for κ-casein of about

$1\,mg\,m^{-2}$. The surface charge density resulting from adsorption of κ-casein maintained a value of 3–4 times that obtained for β-lactoglobulin adsorption, measured separately in the phosphate buffer under similar conditions of anodic end potential and bulk solution concentration.

When both β-lactoglobulin and κ-casein are present in solution, the surface charge density and potentiodynamic profile suggest κ-casein adsorbs preferentially. Oxide growth rates substantiate this, as κ-casein was found to be more efficient at blocking oxide deposition at the anodic potentials studied.

References

1. E. Dickinson, A. Mauffret, S. E. Rolfe, and C. M. Woskett, *J. Soc. Diary Technol.*, 1989, **42**, 18.
2. A. R. Hill, *Can. Inst. Food Sci. Technol. J.*, 1989, **22**, 120.
3. T. Arnebrant and T. Nylander, *J. Colloid Interface Sci.*, 1986, **111**, 529.
4. T. Arnebrant, K. Barton, and T. Nylander, *J. Colloid Interface Sci.*, 1987, **119**, 383.
5. B. A. Ivarsson, P. O. Hegg, K. I. Lundstrom, and U. Jonsson, *Colloids Surf.*, 1985, **13**, 169.
6. B. Liedberg, B. Ivarsson, P. O. Hegg, and K. I. Lundstrom, *J. Colloid Interface Sci.*, 1986, **114**, 386.
7. P. Walstra and R. Jenness, 'Dairy Chemistry and Physics', Wiley, New York, 1984.
8. B. E. Conway and H. Angerstein-Kozlowska, *Acc. Chem. Res.*, 1981, **41**, 49.
9. A. K. Vijh, *J. Electrochem. Soc.*, 1972, **119**, 1498.
10. Polariter Instructions Manual, Radiometer Company.
11. 'CRC Handbook of Chemistry and Physics', 64th Edn., ed. R. C. Weast, Chemical Rubber Company, Cleveland, 1983.
12. A. N. Frumkin, in 'Advances in Electrochemistry and Electrochemical Engineering', ed. P. Delahay, Interscience, New York, 1963, Vol. 3, p. 278.
13. T. Biegler, D. A. J. Rand, and R. Words, *J. Electroanal. Chem. Interfacial Electrochem.*, 1971, **29**, 269.
14. H. Angerstein-Kozlowska, in 'Comprehensive Treatise of Electrochemistry', ed. E. Yeager, J. O'M. Brockris, B. E. Conway, and S. Sarangapani, Plenum Press, New York, 1984, Vol. 9, p. 15.
15. S. G. Roscoe and B. E. Conway, *J. Electroanal. Chem. Interfacial Electrochem.*, 1987, **224**, 163.
16. D. G. Marangoni, R. S. Smith, and S. G. Roscoe, *Can. J. Chem.*, 1989, **67**, 921.

Preliminary Studies of β-Lactoglobulin Adsorbed on Polystyrene Latex

By A. R. Mackie, J. Mingins, R. Dann

AFRC INSTITUTE OF FOOD RESEARCH, NORWICH LABORATORY, COLNEY LANE, NORWICH NR4 7UA

and A. N. North

PHYSICS LABORATORY, UNIVERSITY OF KENT, CANTERBURY, KENT CT2 7NR

1 Introduction

The hierarchy of structures seen with proteins in solution reflects the intra- and inter-molecular interactions amongst the protein molecules, and the interactions of the peptide chains with the solvent (usually water). Divergences from the native conditions in solution can engender large changes in structure, as seen, for example, by heating or cooling, or by changing the solvent composition through shifts in pH or ionic strength or the addition of ethanol or detergent. The introduction of an interface can likewise influence structure, and not always reversibly, as shown by the precipitation of proteins on shaking aqueous solutions with air, or the changes in circular dichroism (CD) spectra seen in proteins recovered from interfaces.[1-3] Interface-induced changes in protein conformation are difficult to quantify *in situ*, and a variety of methods have been applied with varying degrees of success. The disposition of the peptide groups within the adsorbed layer is not accessible. Probing has generally entailed seeing whether gross changes in structure can be deduced from spectroscopic measurements, *e.g.* FTIR, or from optical/hydrodynamic thicknesses often allied to measurements of adsorption density. With developments in powerful methods such as interfacial CD, scanning tunnelling microscopy, and the critical reflectance of neutrons or X-rays, however, a more detailed picture of adsorbed proteins should emerge. To this modern armoury of techniques, we wish to add that of small-angle X-ray scattering (SAXS) using synchrotron radiation. The benefits of synchrotron radiation have been seen in a host of applications, but the approach has been under-utilized so far in the colloid area.

   This paper is the first in a series devoted to the use of scattering techniques to characterize adsorbed layers and adsorbed-layer interactions, and as such is a vehicle for describing the SAXS technique and the method of analysis. The experiments are confined to bovine β-lactoglobulin adsorbed on polystyrene particles. Other proteins (globular and random coil), other surfaces (liquid and solid), and other scattering methods (light and ultra-SAXS) will be dealt with in subsequent papers and extension will be made to model particle interactions.

Polystyrene latex is now a standard model colloid system (see review article of Hearn et al.[4]). The particles are essentially smooth, hard, non-porous spheres and can be prepared with different types of charge at various charge densities. Size distributions can be very sharp for a range of diameters convenient for either SAXS or light scattering measurements. The primary charge density can be assessed by titration and the *net* charge density estimated from micro-electrophoresis. The high surface areas enable adsorption isotherms to be established by solution-depletion experiments. The β-lactoglobulin is a model food protein with known amino acid sequence and a recently published secondary structure for the crystalline state.[5] Its solution behaviour is well characterized and is summarized below to set the present experiments in proper context. The surface behaviour (also reviewed below) has received far less attention and no clear molecular picture has been established.

The Behaviour of β-Lactoglobulin in Aqueous Solution

Although the protein β-lactoglobulin is the major whey protein amounting to *ca.* 0.3 wt% of cow's milk, its function is not clear. At the natural pH of milk (6.7), it exists as a dimer* of relative molecular mass 36 700 daltons (see Swaisgood[6]). This quaternary structure, is, however, sensitive to both pH and temperature. In the pH range from 3.5 to just above the isoelectric point at 5.1 the dimer associates to give octamers[7-10] and as the pH is moved outside this range the dimer $\rightleftharpoons$ octamer equilibrium is increasingly attended by dimer $\rightleftharpoons$ monomer dissoci-ation/association.[11,12] At alkaline pH (>7.5), there is a time-dependent irrevers-ible denaturation of the protein. The dimer $\rightleftharpoons$ octamer reaction is rapid, the proportion of octamer increasing with decrease of temperature. The denaturation process is slow but accelerates in the cold or at high pH (typically >9.0).

The X-ray data of Green and Aschaffenburg[13] and Witz et al.[14] indicate that the monomer chains in the dimer are formed of spheres of 3.6 nm diameter that have merged at contact by 0.23 nm. The dimer contains two disulphide bonds and a free sulphydryl group per monomer which are masked,[15] being buried within the protein.[16] The octamer is composed of the four dimers associated symmetrically about a tetrad axis giving a closed ring and an overall decahedral shape based on small-angle X-ray scattering data.[10]

As the pH varies, not only do the association equilibria respond, but there are also conformational changes within the sub-units as evidenced by the exposure of titratable groups (in the early work of Tanford and his colleagues,[17-19] Timasheff

---

* It is necessary to point out that the dimer we refer to is the kinetic unit in the association over a wide pH range and as such is referred to as a monomer in the portfolio of papers from Timasheff and his colleagues for instance. Their tetramer is therefore the octamer in this paper.

et al.[20] and McKenzie[21]), and expansion of the molecule or increased reactivity of the sulphydryl groups (Zimmerman *et al.*[22]). Using optical rotatory dispersion (ORD) and proton binding measurements, Timasheff *et al.*[20] recognized two conformational changes as the pH is varied, one in the range pH 4–6 and the other in the range 6.5–9. Up to pH 7.5, the conformational changes are reversible. Several authors[13,23–26] have investigated the secondary structure of β-lacto-globulin, and their results indicate that at moderate pH a good proportion of the peptide chains have a β-structure in the native protein which unfolds to give a random structure as the solution is made alkaline. The CD and IR spectral data of Timasheff *et al.*[27] would seem to indicate that there is *ca.* 45% β-sheet and 10% α-helix at the lower pH range.

A detailed structure of the orthorhombic crystal form of β-lactoglobulin has recently been provided by Papiz *et al.*[5] where antiparallel β-sheets formed by nine β-strands account for *ca.* 50% of the peptide residues. Approximately 15% α-helix and reverse turns accounting for 20% leave about 15% for the 'random' structure. They also suggest that an exposed β-strand forms an intermolecular antiparallel β-sheet upon dimer formation. This is in keeping with the IR results of Casal *et al.*[28] who showed, using conformation-sensitive amide I bands, that, although the total amount of β-structure remains constant between pH 2 and 10, the proportions of the various β-components change. In particular, a loss of β-strand to antiparallel β-sheet occurs on the formation of dimers as the pH is changed from 2 to 3. Casal *et al.*[28] also show that the secondary structure of β-lactoglobulin does not change in the temperature range 20–50 °C at pH 7, but that there is a distinct change at 60 °C. Finally, on the question of dynamics of secondary structure formation the stopped-flow CD measurements of Kuwajima *et al.*[29] show that the formation of β-structure in β-lactoglobulin is very fast (<18 ms).

Understanding the solution behaviour of β-lactoglobulin is made more difficult by the finding that there are several genetic variants of this protein.[30,31] Most of the studies have dealt with β-lactoglobulin A and β-lactoglobulin B using samples prepared from milk from homozygous cows. Although the work of Tanford and Nozaki[17] using ORD and UV absorption leads them to conclude that the differences between the A and B forms were slight, and that the only significant difference was seen in titration curves, where the A form had one more titratable carboxyl group per chain than had B, it is clear that the association behaviour of the two forms is markedly different. Using ultracentrifuge, electrophoresis, and light scattering techniques, Timasheff and Townend[9] showed that β-lactoglobulin A forms octamers in the pH range 3.7–5.2, whereas β-lactoglobulin B cannot form aggregates greater than a tetramer. However, a mixture of A and B gives mixed octamers. Using the Green and Aschaffenburg[13] model of slightly impacted spheres for the dimer, Timasheff and Townend[9] argue persuasively that the reactive groups for further aggregation of the dimer must be symmetrically distributed on the double sphere, that there are two of them, and that they are located to give a compact structure to the cyclic octamer and steric restriction to the formation of *n*-mers with *n* > 8. Steric hindrance of bond formation at the second site in β-lactoglobulin B was suggested for the limit to tetramers with this variant. They also suggested that the carboxyl groups were likely to be implicated in the bond formation. Bonds involving sulphydryl oxidation and sulphydryl/disulphide exchange reactions were more recently proposed as partly responsible for the oligomer formation (*cf.* McKenzie *et al.*[16]).

3 The Behaviour of β-Lactoglobulin in Adsorbed Layers

Although there is now a substantial body of work on the adsorption of proteins on to polystyrene latex, there appears to be no study of β-lactoglobulin on such particles. Relevant adsorption data on other solids that we have been able to find are summarized in Table 1. Most of the experiments have been done on metal surfaces in mechanistic studies of protein fouling, particularly of the heat exchangers in milk processing plant. The majority of the results in Table 1 refer to experiments with a single bulk concentration of protein $\sim$1 mg ml^{-1}.

Table 1 *Published adsorption data for β-lactoglobulin on various surfaces*

Material	pH	T (°C)	Ionic strength (M)	Adsorption density (mg m^{-2})	Method	Ref.
Stainless steel	6.8	27	?	1.5–1.6	solution-depletion	*a*
Hydrophilic chromium	6.0	25	0.17	1.80	ellipsometry	*b,c**
	7.0	25	0.17	1.20	ellipsometry	*b*
	7.0	25	0.17	0.98	radiotracer	*d**
				0.75	ellipsometry	*d**
	7.0	25	0.21	1.20	ellipsometry	*e***
Hydrophobic chromium	7.0	25	0.17	1.20	radiotracer	*d**
				0.96	ellipsometry	*d**
	7.0	25	0.21	1.4–1.5	ellipsometry	*e***
Hydrophilic gold	4.5	?	~0.17	3.90	ellipsometry	*f****
	6.0	?	~0.17	2.10	ellipsometry	*f****
	10.0	?	~0.17	1.70	ellipsometry	*f****
Phospholipid monolayer at the air–water interface	4.4	?	?	low Π 2.3–2.8	UV absorption	*g*†
				high Π 2.4–3.0	UV absorption	*g*†
	7.0	?	?	low Π 0.3–0.4	UV absorption	*g*†
				high Π 0.3–0.4	UV absorption	*g*†

* Bulk protein concentration = 1 mg ml^{-1}.
** Bulk protein concentration = 1 mg ml^{-1}, but adsorption densities quoted after one sequential adsorption step. Adsorptions measured after rinsing with buffer.
*** Bulk protein concentration = 0.95 ± 0.02 mg ml^{-1}. The range in adsorption densities arises from the differing assumptions used to calculate the equivalent optical thickness from the ellipsometric readings Δ and Ψ. Adsorptions measured after rinsing with buffer.
† Adsorption densities obtained from UV absorption measurements on the film obtained by passing a quartz plate through the monolayer air–water interface. Bulk protein concentrations used are not clear.
a J. C. Kim and D. B. Lund, in 'Fouling and Cleansing in Food Processing', ed. H. G. Kessler and D. B. Lund, Druckerei Walch, Augsburg, 1989, p. 187.
b T. Nylander, Ph.D. Thesis, University of Lund, 1987.
c T. Arnebrant, K. Barton, and T. Nylander, *J. Colloid Interface Sci.*, 1987, **119**, 383.
d T. Arnebrant and T. Nylander, *J. Colloid Interface Sci.*, 1986, **111**, 529.
e T. Arnebrant, B. Ivarsson, K. Larsson, I. Lundstrom, and T. Nylander, *Prog. Colloid Polym. Sci.*, 1985, **70**, 62.
f B. Liedberg, B. Ivarsson, P. O. Hegg, and I. Lindstrom, *J. Colloid Interface Sci.*, 1986, **114**, 386.
g D. G. Cornell and D. L. Patterson, *J. Agric. Food Chem.*, 1989, **37**, 1455.

The results of Kim and Lund[32] on non-porous stainless steel microspheres (25.5 μm diameter) are the only ones in the form of an isotherm and indicate that a plateau in adsorption is reached in quite dilute solutions ($\sim$3 μg ml^{-1}). The other adsorptions from 1 mg ml^{-1} solutions, in that they are around the Kim and Lund value or lower at roughly the same pH, would imply that there is no further step in the isotherm up to 1 mg ml^{-1}, although in the absence of data at concentrations >1 mg ml^{-1} a bimodal adsorption isotherm cannot be ruled out. The agreement of the ellipsometric data with results from other methods on solutions in the same pH range is rather poor, but this is not unusual for proteins.

The data of Liedberg *et al.*[33] show an increase in adsorption as the pH is reduced from 10.0 to 4.5, and this finding receives support from the results of Nylander[34] with a far smaller pH drop. Again, with only the one bulk protein concentration examined, the nature of the adsorption is somewhat obscured. It is perhaps worth noting that the adsorptions measured by Liedberg *et al.*[33] apply to reasonably firmly bound protein, since the ellipsometric readings were taken on protein-covered surfaces that had been rinsed with pure buffer solutions subsequent to equilibration. Similarly, the measurements of Arnebrant and Nylander[35] and Arnebrant *et al.*[36,37] show that rinsing with buffer gives substantial losses of protein from hydrophilic chromium leaving behind firmly bound protein. Less protein desorption on rinsing is seen with hydrophobic chromium,[36,37] and Kim and Lund[32] judge that in dilute solutions the adsorption is irreversible at stainless steel surfaces. Although more protein would seem to be released from the hydrophilic chromium, indicating a possible difference in the mode of desorption, it would be difficult to say because of the spread of values in Table 1 whether there are significant differences in adsorption densities between hydrophilic and hydrophobic metals. Another feature that may indicate that the mode of adsorption is different stems from the time-course of the adsorption on the two types of surface. On hydrophilic chromium at pH 7.0, adsorption is completed within a few minutes,[35,36] whereas on hydrophobic chromium protein is still adsorbing after 60 minutes[35,37] although a large proportion goes on quickly in the first few minutes. The diffusion times are the same in the two instances. (The results on hydrophilic chromium in ref. 37 are out of line.) Plateau values are also quickly reached in dilute solutions on stainless steel,[32] but whether this surface can be considered hydrophilic is a matter of conjecture.

A few adsorption densities that we have been able to find for a liquid surface (Cornell and Patterson[38]) are also included for comparison. The increased adsorption with reduction in pH mirrors the Liedberg *et al.*[33] results but comparisons would benefit by duplication of one of the techniques on the two systems.

4   Materials and Methods

Materials and Sample Preparation—The polystyrene latex was obtained from Sigma Chemical Co. (sample LB-1, nominally 0.1 μm average diameter). Electron microscopy and photon correlation spectroscopy measurements have shown that the particles were highly monodisperse spheres with a mean diameter of 91 nm and an essentially smooth surface. The solids content was measured as

10 wt%, and samples were made up by weighing aliquots from this stock. The latex was used untreated, the charged sulphate groups on the surface maintaining stability throughout the whole pH and ionic strength range used. Buffers were obtained from BDH and were also used untreated. Water was a surface chemically pure sample. The β-lactoglobulin was a three-times crystallized and lyophilized sample from Sigma Chemical Co. (product No. L0130). The sample was stored at 4 °C, and, for the purposes of this work, was used without further purification.

Adsorption—Adsorption isotherms were obtained by the solution depletion method. This entails dissolving the β-lactoglobulin over a range of concentrations in the buffer appropriate to the pH of interest. Known quantities of latex were then added to the solutions which were allowed to adsorb overnight. The latex and attached protein were then removed by centrifugation, and the concentration of protein remaining in solution was calculated by assaying the supernatant by the Lowry method. This assay was used because the low concentration of protein left in the supernatant phase made concentration determination based on optical density too inaccurate. Adsorption densities were calculated using a surface area based on a radius of 45.5 nm. Adsorption measurements were done at a pH of 4.65 in citrate buffer, at pH 7.2 in phosphate buffer, and at pH 9.0 in a glycine buffer. The ionic strength of all three buffers was 4 mM.

Small-angle Scattering—Small-angle scattering of X-rays (SAXS) or neutrons (SANS) are techniques that are widely used to provide microstructural information over the size range of a few nanometres to a few hundred nanometres. A rigorous discussion of the theoretical and experimental aspects of small-angle scattering is not necessary, since this has previously been done by many authors.[39–41] However, it is useful to restate the main features.

The small-angle scattering profile from a particulate system may be written as

$$I(Q) = NP(Q)S(Q), \tag{1}$$

where N is the number of particles in the scattering volume, and Q is the scattering vector defined by the experimental conditions as

$$Q = \frac{4\pi}{\lambda} \sin \frac{\theta}{2}, \tag{2}$$

where λ is the wavelength of the incident beam, and θ is the angle of scatter. The particle form factor $P(Q)$ is a function of particle shape, size, and orientation, as well as of the scattering density contrast $\Delta\rho$ between the particles and the continuous phase. In systems for which there is some degree of ordering, interparticle interference terms contribute to the measured scattering profile, this contribution being given by the structure factor $S(Q)$. In the case of a dilute system of non-interacting particles, the structure factor is approximately unity for all Q values,[39] such that

$$I(Q) = NP(Q). \tag{3}$$

In the simplest case, for a homogeneous sphere of radius R, in a homogeneous continuous phase, $P(Q)$ is given by

$$P(Q) = 9V^2(\Delta\rho)^2 \left[\frac{\sin QR - QR \cos QR}{(QR)^3} \right]^2, \tag{4}$$

where V is the volume of the particle and $\Delta\rho$ is the scattering density contrast,

$$\Delta\rho = \rho_p - \rho_c. \tag{5}$$

Here, ρ_p and ρ_c are the scattering length densities of the particle and the continuous phase, respectively. The scattering length density ρ is the local average of the atomic form factors in a volume element. In the case of X-ray scattering this form factor is proportional to the number of electrons in a given atom. Therefore, in SAXS, the electron density is a convenient measure of the scattering potential.

In our simple case of an homogeneous particle, the scattering density contrast is a simple step function (Figure 1) and the form of the scattering profile but not the magnitude [equations (3) and (4)] is independent of the contrast. In more complicated systems, for instance, when large molecules are adsorbed on to the homogeneous particle, the scattering density will vary continuously with radius

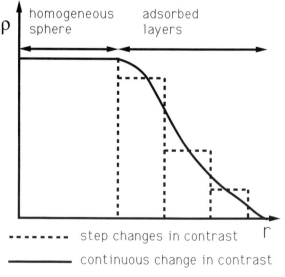

Figure 1 *A schematic plot of contrast ρ against radius r, showing the type of contrast profile expected in the samples (—) and that assumed in the model of equation (6) (---)*

(Figure 1). In such cases, it is possible to approximate this smooth change in scattering density with a finite number of spherical shells of varying scattering length density (Figure 1). The form factor for such a particle is

$$P(Q) = \left[\sum_{j=1}^{M} 4\pi\Delta\rho_j \left(\frac{\sin QR_j - QR_j \cos QR_j}{Q^3} \right) \right]^2, \qquad (6)$$

where M is the number of spherical shells, $\Delta\rho_j = \rho_j - \rho_{j+1}$ is the contrast between adjacent shells, and R_j is the outer radius of the jth shell. The scattering density of the continuous phase is ρ_{m+1}. The form and magnitude of the scattering profile now depends upon the contrast.

Figure 2 shows the theoretical scattering profile from a system of homogeneous particles in a continuous medium calculated using equation (4). The figure also shows a similar profile for a sphere of identical size, surrounded by a series of shells of different scattering length density similar to that in Figure 1. The major difference between the two profiles is the shift to lower Q values of the positions of the maxima and minima on going from the simple sphere to the multi-shell model. Such a shift is an indication of the larger size of the 'inhomogeneous' particle.

All SAXS measurements were made on station 8.2 (Figure 3) of the Synchrotron Radiation Source (SRS) at the SERC Daresbury Laboratory. Details of the operation of this instrument can be found in work by North et al.[39] The

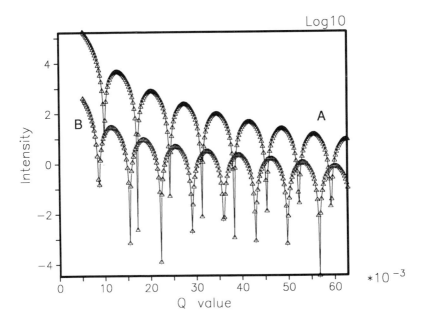

Figure 2   *Theoretical small-angle scattering curves for a single-step contrast* (A) *and for a multi-step contrast* (B)

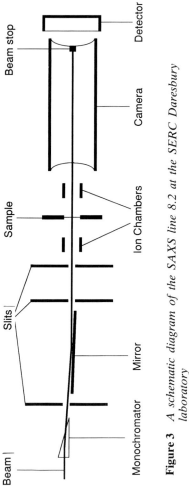

Figure 3 *A schematic diagram of the SAXS line 8.2 at the SERC Daresbury laboratory*

wavelength of the incident radiation was 0.15 nm and a sample-to-detector distance of 4 m was used giving a useful Q range of 7×10^{-2} to $1\,\mathrm{nm}^{-1}$. The incident and transmitted flux were measured using the ionization chambers before and after the sample. These were used to correct the data for changes in beam intensity during data collection and to facilitate accurate subtraction of the background scattering. There was no significant change in the absorption of X-rays by the samples during data collection, an indication that little radiation damage had occurred; nor were there any visible signs of such damage. Data collection times were sub-divided and the data from each sub-division compared. These showed no statistically valid change in the scattering profile, again suggesting little or no occurrence of radiation damage. Typical data collection times were 1–2 hours. The samples were contained in a 0.6 nm thick aluminium cell with nominally 30 μm thick mica windows. Experiments were carried out at room temperature, typically 21 °C.

Data were collected for 5% LB-1 with no protein, in order to characterize the substrate. A sample of 4 mM phosphate was also run to provide a background for the scattering on the bare and protein-covered particles. Measurements on other buffers and dilute protein solutions indicate that the form of the scattering is the same. An initial bulk concentration of β-lactoglobulin, corresponding to $1.7\,\mathrm{mg\,m}^{-2}$ if it all adsorbed, was chosen to give a steady-state adsorption at the start of the adsorption plateau for pH 7.2 and the scattering measured from protein-coated 5% LB-1. In the absence of a plateau at the other two pH values, the same initial protein concentration was also chosen for the scattering runs at those pH values again on 5% LB-1. As in the adsorption measurements, all the samples were allowed to equilibrate overnight.

5 Results

Adsorption—Adsorption isotherms for β-lactoglobulin on polystyrene latex are shown in Figure 4 for the three pH values used in the SAXS experiments, *i.e.* 4.65, 7.2, and 9.0. The isotherm at pH 7.2 is the only one to reach a plateau value in the adsorption. This amounts to about $1.4\,\mathrm{mg\,m}^{-2}$. The isotherm for pH 4.65 superimposes that for pH 7.2 in the initial region, but then diverges at higher concentrations to give consistently higher adsorptions. The isotherm at pH 9.0 is lower than the other two throughout virtually the whole concentration range studied. The adsorption densities at the initial bulk concentration corresponding to $1.7\,\mathrm{mg\,m}^{-2}$ selected for the SAXS runs can be estimated from Figure 4 as 1.6, 1.3, and $0.9\,\mathrm{mg\,m}^{-2}$ for the pH values 4.65, 7.2, and 9.0, respectively.

Small-angle Scattering Analysis—The scattering profile obtained from the bare latex is shown in Figure 5. The profile has been analysed using equation (4), and the particle radius is found to be 45.2 nm, which is in good agreement with the manufacturer's quoted value of 45 nm. The experimental profile does not have the deep minima (of the theoretical data Figure 2) owing to smearing of the profile which arises from the use of a finite sized incident beam and finite detector resolution. The least-squares fit to the profile from the bare latex (Figure 5) takes

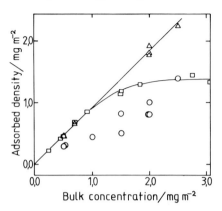

Figure 4   *Adsorption isotherms for β-lactoglobulin on polystyrene latex at pH 4.65 (○), pH 7.2 (△), and pH 9.0 (□) at room temperature and an ionic strength of 4 mM. Solution concentrations are plotted as equivalent surface coverages as if all the protein were adsorbed on the area available*

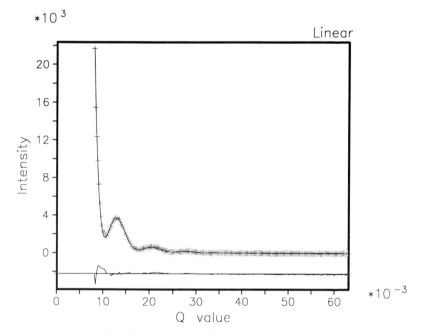

Figure 5   *The results of a least-squares fit to SAXS data from a bare latex sample using a particle radius of 45.2 nm and the particle form factor from equation (4)*

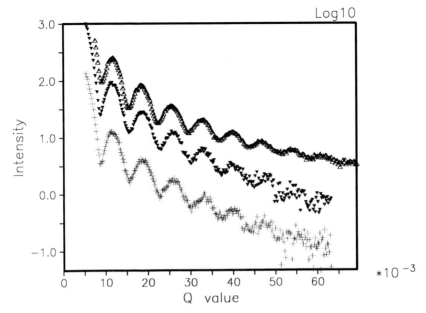

Figure 6 *Scattering curves for β-lactoglobulin adsorbed on to polystyrene latex at pH 4.65 (▼), pH 7.2 (△), and pH 9.0 (+). The curves have been artificially spaced apart for clarity*

into account these factors. The same smearing parameters were used in the analysis of the scattering profiles obtained from latex with adsorbed protein.

The measured scattering profiles from the latex with adsorbed protein at each different pH value are shown in Figure 6. Observation of the profiles clearly shows that the positions of the maxima and minima are different for all three pH values. In each case these positions are at lower Q values than for the bare latex, which indicates that as expected all the samples have adsorbed protein at the interface.

The SAXS data for the protein-coated latex were fitted to a model based on equation (6). The least-squares fitting routine produces electron density/scattering density profiles which are shown in Figure 7. These electron density profiles are not absolute and are relative to the buffer. It is clear from Figure 7 that all three samples show a significant protein layer of *ca.* 2.0 nm thickness. In addition, the sample at pH 4.65 also shows evidence that some of the protein extends to 3.8 nm.

6 Discussion

The isotherms in Figure 4 show that, at the higher bulk protein concentrations, the adsorption increases with decrease of pH in keeping with the findings on other interfaces as summarized in Table 1.

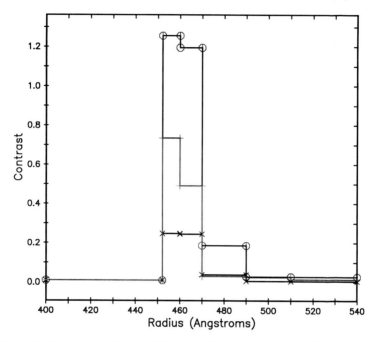

Figure 7   *Contrast profiles for β-lactoglobulin adsorbed on to polystyrene latex at pH 4.65 (○), pH 7.2 (+), and pH 9.0 (×)*

The adsorption density of *ca.* 1.4 mg m^{-2} seen as the plateau in our results at pH 7.2 compares very favourably with 1.5 mg m^{-2} seen for plateau adsorption of β-lactoglobulin adsorbed on stainless steel.[32] These are the only results available for adsorption on a solid at comparable ionic strength to ours (although not stated by Kim and Lund,[32] it must be low corresponding to the buffer concentration and any residual electrolyte in the Sigma sample of β-lactoglobulin, just as in our case). The results on chromium or gold derive from a single bulk protein concentration of *ca.* 1 mg ml^{-1} (where the low surface area would only marginally deplete the solution) and a high ionic strength. The measured adsorption densities at pH 7.2 are nevertheless compatible with our findings, but, as indicated earlier, the spread of results precludes quantitative comparison. The pH 7.2 is below the point for reversible transformation of β-lactoglobulin in solution (pH 7.5) and so the protein should be existing as dimers in solution. If we consider for the moment that the protein adsorbs as a monomer, and that the monomer adsorbs as a hard sphere of diameter 3.58 nm, then the adsorption density Γ_{max} for a hexagonal close-packed array of such spheres of molecular weight 18 350 daltons would be 2.8 mg m^{-2}. If the adsorbed spheres have no lateral movement in the surface, and there is no deformation or unfolding of the monomer, then adsorption of the protein reduces to the *random parking limit* in two dimensions which results in much reduced adsorption. No analytical solutions are available for this limit. Recent simple elegant experiments on the adsorption of polystyrene latex particles on to treated glass surfaces and their visualization using scanning electron microscopy[42] arrive at an area fraction of $55 \pm 1\%$ for the limit. This is in excellent agreement with the computer simulation value of 0.547 by Tanemura[43] and Feder.[44] Morrissey and Han[45] use a value of 0.56 taken as the square of the parking limit in one dimension. Applying the value of 0.547 to our β-lactoglobulin monomers would give a random adsorption value of 1.66 mg m^{-2}. Adsorption of the dimers in the end-on position would therefore give an adsorption of 3.32 mg m^{-2}. Sideways adsorption of the dimer would require a new limit. From examination of the radial distribution of single spheres from the computer simulation experiments, it can be seen that the particles tend to cluster, and the same can be seen in the electron micrographs of Onada and Liniger.[42] The anticipation is then that dimers adsorbing sideways on as a pair of slightly impacted spheres would only have a marginal influence on the parking limit.* The plateau adsorption of *ca.* 1.4 mg m^{-2} at pH 7.2 is therefore not an unreasonable figure. The bulk of the adsorption is over in a short time, and, if other processes such as unfolding at the interface are much slower, the parking limit may still be a valid parameter for the total adsorption provided further molecules do not adsorb in second layers. We may then have a set limit for the adsorption density, but still continue to have subsequent change in other features such as conformation, thickness, spreading exposure of groups, *etc.*

At pH 4.65, where octamer formation in solution is at its optimum, the adsorption is also seen to respond. Looking along the tetrad axis, the profile of the octamer is a good approximation to a circle, and, if the octamer adsorbs along this axis, the area occupied can be taken as that swept out by rotation of four contacting undeformed spheres about the tetrad axis. Again using a diameter of 3.58 nm for the monomer and a molecular weight now of $8 \times 18\,350$ daltons, Γ_{max} amounts to 1.6 mg m^{-2}. Applying the value of 0.547 for the area fraction from the random parking limit sets a ceiling of 0.9 mg m^{-2} on the adsorption. As can be seen from Figure 2, this adsorption is exceeded over a good part of the isotherm, including the protein concentration used for the SAXS measurements. The implication is that other smaller oligomers are adsorbing as well, or more protein is adsorbing on the first layer, or there is surface diffusion of the adsorbing species. If octamers are present and they collapse to smaller units on the surface, the parking of octamers is still a problem, since they have to find free space, and the adsorption would be reduced unless again they resorted to second layer adsorption. Preferential adsorption of dimers[46,47] or oligomers[48] over monomers on polystyrene latex has been seen with bovine serum albumin, but there is no evidence of multilayer formation.

In solutions at pH 9.0, β-lactoglobulin dimers dissociate rapidly and the monomers suffer an irreversible denaturation to a random coil. The much reduced adsorption at low bulk concentrations seen at this pH is readily appreciated with such an extended charged molecule.

Using a particle radius of 45.2 nm calculated for the bare latex and applying the form factor analysis for shells [equation (6)], the scattering data on protein-coated latex yields the contrast profiles shown in Figure 7. These profiles are the simplest

---

* We are grateful to Dr G. C. Barker for drawing this to our attention.

picture that gives a good fit to the complexities of the scattering data. Immediately, it can be seen that the bulk of the protein is confined to within 1.8 nm of the surface for all pH values, and that the protein content of the adsorbed layer increases with decrease of pH, in keeping with the adsorption densities at this bulk protein concentration. For pH 9.0 there is no substantial extended structure towards the solution. It would be impossible to see a few loops and tails, although they could possibly be picked up by measurements of the hydrodynamic radius. At pH 7.2, the thicknesses are substantially less than the diameter of the monomer (3.85 nm) indicating that either monomers or dimers on the surface must change conformation and substantially flatten. At pH 4.65, the octamers (if present) have clearly broken down and the contrast profiles would again indicate substantial flattening of the sideways-on dimer or the monomer. There is also protein in an additional shell of 2 nm, although of less contrast (therefore less density). The overall thickness is roughly that of the undeformed monomer or sideways-on dimer and the radial averaging of the scattering process would account for the lowered contrast from a few such particles on the surface.

Overall, then, we are seeing that the polystyrene latex surface is engendering dramatic changes in both association and conformation. Liedberg *et al.*[33] deduce a surface denaturation of β-lactoglobulin by gold surfaces at pH 6.0, but propose that octamers adsorb in the native form at pH 4.5. Their deduced values of thickness and adsorption density are consistent with this model. This adsorption density is much greater than the parking limit for octamers. Conformational changes have been seen by interfacial CD for β-lactoglobulin adsorbed at the quartz–water interface[49] with solutions at pH 7.0 using the technique described by Clark *et al.*,[50] but no such data are available at other pH values. Using photon correlation spectroscopy (PCS) on other globular proteins adsorbed on polystyrene latex, Fair and Jamieson[51] find no evidence of multilayer formation at higher coverages or 'conformational flattening' in the lower plateau régime. This would be similar to the plateau we see with pH 7.2 solutions. This latter finding of Fair and Jamieson runs counter to our *X*-ray evidence. It relies on the hydrodynamic radii calculated from the PCS data—such radii are frequently larger than those measured by SAXS for a variety of materials (see Pusey[52] and North *et al.*[53]). The hydrodynamic radius is calculated from the diffusion coefficient *D via* the Einstein equation. Values obtained for spherical systems differ from those obtained by SAXS by up to 1.5 nm. A small number of non-flattened particles on the polystyrene surface would have an effect on the diffusion coefficient measured by PCS, and hence on the hydrodynamic radius (especially since the coated particle is no longer strictly spherical), but probably not on the SAXS profile.

Acknowledgements. We should like to thank David Wilson for his help and advice with the adsorption experiments, and Dr Wim Bras for help on the SAXS station at the Daresbury Laboratory. Thanks are also due to Dr Richard Heenan for the use of his data analysis package 'FISH' for small-angle scattering data. Financial support from MAFF for part of the work is gratefully acknowledged. One of us (ANN) acknowledges the support of the SERC in terms of an Advanced Research Fellowship.

References

1. M. E. Soderquist and A. G. Walton, *J. Colloid Interface Sci.*, 1980, **75**, 386.
2. D. C. Clark, J. Mingins, L. J. Smith, F. E. Sloan, and D. R. Wilson, in 'Food Emulsions and Foams', ed. E. Dickinson, Royal Society of Chemistry, London, 1987, p. 110.
3. D. C. Clark, L. J. Smith, and D. R. Wilson, *J. Colloid Interface Sci.*, 1988, **121**, 136.
4. J. Hearn, M. G. Wilkinson, and A. R. Goodall, *Adv. Colloid Interface Sci.*, 1981, **250**, 173.
5. M. Z. Papiz, L. Sawyer, E. E. Elcopoulos, A. C. T. North, J. B. C. Findlay, R. Sivaprasadarao, T. A. Jones, M. E. Newcomer, and P. J. Kraulis, *Nature (London)*, 1986, **324**, 383.
6. H. E. Swaisgood, in 'Developments in Dairy Chemistry', ed. P. D. Fox, Applied Science, London, 1982, p. 1.
7. R. Townend, R. J. Winterbottom, and S. N. Timasheff, *J. Am. Chem. Soc.*, 1960, **82**, 3161.
8. R. Townend and S. N. Timasheff, *J. Am. Chem. Soc.*, 1960, **82**, 3168.
9. S. H. Timasheff and R. Townend, *J. Am. Chem. Soc.*, 1961, **83**, 464.
10. S. H. Timasheff and R. Townend, *Nature (London)*, 1964, **203**, 517.
11. R. Townend, L. Weinberger, and S. H. Timasheff, *J. Am. Chem. Soc.*, 1960, **82**, 3175.
12. C. Georges, S. Guinand, and J. Tonnelat, *Biochim. Biophys. Acta*, 1962, **59**, 737.
13. D. W. Green and R. Aschaffenburg, *J. Mol. Biol.*, 1959, **1**, 54.
14. J. Witz, S. N. Timasheff, and V. Luzzati, *J. Am. Chem. Soc.*, 1964, **86**, 168.
15. R. L. J. Lyster, *J. Dairy Sci.*, 1964, **31**, 41.
16. W. A. McKenzie, G. B. Ralston, and D. C. Shaw, *Biochemistry*, 1972, **11**, 4539.
17. C. Tanford and Y. Nozaki, *J. Biol. Chem.*, 1959, **234**, 2874.
18. C. Tanford, L. G. Bunville, and Y. Nozaki, *J. Am. Chem. Soc.*, 1959, **81**, 4032.
19. C. Tanford and V. G. Tuggart, *J. Am. Chem. Soc.*, 1961, **83**, 1634.
20. S. N. Timasheff, L. Mescanti, J. J. Bash, and R. Townend, *J. Biol. Chem.*, 1966, **241**, 249.
21. W. A. McKenzie, 'Milk Proteins, Chemistry and Molecular Biology', Academic Press, London, 1971, Vol. 2.
22. J. K. Zimmerman, I. M. Klotz, and G. Barlow, *Fed. Proc. Fed. Am. Soc. Exp. Biol.*, 1969, **28**, 914.
23. S. N. Timasheff, R. Townend, and L. Mescanti, *J. Biol. Chem.*, 1966, **241**, 1863.
24. H. Susi, S. N. Timasheff, and L. Stevens, *J. Biol. Chem.*, 1967, **242**, 5467.
25. W. J. Young, P. R. Griffiths, D. M. Byker, and H. Susi, *Appl. Spectrosc.*, 1985, **39**, 282.
26. H. Susi and D. M. Byler, *Biochim. Biophys. Res. Commun.*, 1983, **115**, 391.
27. S. N. Timasheff, H. Susi, R. Townend, L. Stevens, M. J. Gorbunoff, and T. F. Kumosinski, in 'Conformations of Biopolymers', ed. G. N. Ramachandran, Academic Press, London, 1967, Vol. 1, p. 173.
28. H. L. Casal, U. Kohler, and H. H. Mantsch, *Biochim. Biophys. Acta*, 1988, **957**, 11.
29. K. Kuwajima, H. Yamaya, S. Miwa, S. Sugai and T. Nagamura, *FEBS Lett.*, 1987, **221**, 115.
30. R. Aschaffenburg and J. Drewry, *Nature (London)*, 1955, **176**, 218.
31. R. Aschaffenburg and J. Drewry, *Nature (London)*, 1957, **180**, 376.
32. J. C. Kim and D. B. Lund, in 'Fouling and Cleaning in Food Processing', ed. H. G. Kessler and D. B. Lund, Druckerei Walch, Augsburg, 1989, p. 187.
33. B. Liedberg, B. Ivarsson, P.-O. Hegg, and I. Lundstrom, *J. Colloid Interface Sci.*, 1986, **114**, 386.
34. T. Nylander, Ph.D. Thesis, University of Lund, 1987.
35. T. Arnebrant and T. Nylander, *J. Colloid Interface Sci.*, 1986, **111**, 529.
36. T. Arnebrant, K. Barton, and T. Nylander, *J. Colloid Interface Sci.*, 1987, **119**, 383.
37. T. Arnebrant, B. Ivarson, K. Larrson, I. Lundstrom, and T. Nylander, *Prog. Colloid Polym. Sci.*, 1985, **70**, 62.

Preliminary Studies of β-Lactoglobulin Adsorbed on Polystyrene Latex

38. D. G. Cornell and D. L. Patterson, *J. Agric. Food Chem.*, 1989, **37**, 1455.
39. A. N. North, J. C. Dore, A. R. Mackie, A. M. Howe, B. H. Robinson, and C. Nave, *Nucl. Instrum. Meth.*, 1988, **B34**, 188.
40. 'Small-angle X-ray Scattering', ed. O. Glatter and O. Kratky, Academic Press, New York, 1984.
41. L. A. Feigin and D. I. Svergun, 'Structure Analysis by Small-angle X-ray and Neutron Scattering', Plenum, New York, 1987.
42. G. Y. Onada and E. G. Liniger, *Phys. Rev.*, 1986, **A33**, 715.
43. M. Tannemura, *Ann. Inst. Statist. Math.*, 1979, **31**, 351.
44. J. Feder, *J. Theor. Biol.*, 1980, **237**, 87.
45. B. W. Morrissey and C. C. Han, *J. Colloid Interface Sci.*, 1976, **65**, 423.
46. R. G. Greig and D. E. Brooker, *J. Colloid Interface Sci.*, 1981, **83**, 661.
47. G. G. W. Lensen, D. Bargeman, P. Bergveld, C. A. Smolders, and J. Feijen, *J. Colloid Interface Sci.*, 1984, **99**, 1.
48. R. L. J. Szom, *J. Colloid Interface Sci.*, 1986, **111**, 434.
49. D. C. Clark and L. J. Smith, to be published.
50. D. C. Clark, M. Coke, L. J. Smith, and D. R. Wilson, in 'Foams: Physics, Chemistry and Structure', ed. A. J. Wilson, Springer-Verlag, Berlin, 1989, p. 55.
51. B. D. Fair and A. M. Jamieson, *J. Colloid Interface Sci.*, 1980, **77**, 525.
52. P. N. Pusey, in 'Photon Correlation and Light Beating Spectroscopy', Plenum, New York, 1974.
53. A. N. North, J. C. Dore, A. Katsikides, J. A. McDonald, and B. H. Robinson, *Chem. Phys. Lett.*, 1986, **132**, 541.

Dimensions and Possible Structures of Milk Proteins at Oil–Water Interfaces

By Douglas G. Dalgleish and Jeffrey Leaver

HANNAH RESEARCH INSTITUTE, AYR, SCOTLAND KA6 5HL

1 Introduction

The determination of the dimension of proteins at interfaces may allow some indications of their conformations in the adsorbed state. Thus, albumin adsorbed on a polystyrene latex appears to occupy space little larger than its dimensions in solution,[1] but α_{s1}- and β-caseins appear to protrude from the interface to a considerable distance, forming a layer between 10 and 15 nm in thickness around the latex particle.[2] This is substantially larger than would be expected from a protein adsorbed in a globular form. The caseins, in contrast to albumin, possess apparently flexible structures, and appear to be able to displace one another from an oil–water interface,[3] which may suggest that they are not strongly adsorbed, although it is not easy for other proteins to displace them.[4] In this respect, it is clear that in all systems where mixtures of proteins are in contact with an interface, there appear to be unexplained factors, incorporating the changes in surface composition,[5] the behaviour of the surface viscosity[6] and possibly the formation of mixed multilayers.[7]

It is important to consider the applicability of results which have been obtained by using model systems to more realistic circumstances. For example, polystyrene latices have been used as models of emulsion droplets, and the adsorption of proteins to these materials has been studied.[1,8,9] The latices are excellent supports, inasmuch as they adsorb proteins readily; and they also have the advantage that, being monodisperse, they are ideal for the measurement of particle sizes, and the estimation of the thicknesses of the adsorbed layers of protein.[2,9] Also, it is possible to vary the surface concentration Γ, without greatly altering the stability of the particles. On the other hand, true oil-in-water emulsions are polydisperse, and generally contain a proportion of fairly large particles: this precludes their use in studies of the thickness of adsorbed layers using techniques such as light-scattering or photon correlation spectroscopy (PCS). It is also not possible to make a stable emulsion where the surface is less than completely covered by protein, in the absence of other adsorbing species. Therefore, when considering the thicknesses of adsorbed protein layers, it is not possible to observe the build-up of such layers in emulsions as it is in the latex model systems.

Nevertheless, it is possible to gain some information on the relative behaviour of emulsion and latex systems by studying their behaviour when the surface layer is being destroyed, *e.g.* by the attack of proteolytic enzymes on the adsorbed protein. It is possible to follow the proteolytic reaction chemically, so that in some cases the particular changes which occur in the layer of adsorbed protein may be defined in detail.[10,11] This paper deals with attempts to describe three aspects of the behaviour of adsorbed protein: (i) the effect of one protein on another when they are adsorbed at a latex surface, as defined by the thickness of the adsorbed layer; (ii) a comparison between a model system based on polystyrene latex and a β-casein stabilized oil-in-water emulsion in terms of the thickness of the adsorbed protein layer; and (iii) the relationship between the specific reactions of the protease trypsin and the possible conformation of β-casein molecules at an oil–water interface.[12]

2 Materials and Methods

Polystyrene (LB-1), soya oil, β-lactoglobulin, and TPCK-trypsin were obtained from Sigma Chemical Co. The β-casein was prepared in the laboratory by acid precipitation of whole casein from skim milk, dissolving the casein mixture in 6 M urea, 50 mM imidazole/HCl, pH 7.0, and carrying out chromatography of the protein mixture using Sepharose-Q Fast Flow ion-exchange material with a gradient of NaCl. The peak containing β-casein was collected, dialysed exhaustively against water, and freeze dried.

Emulsions of soya oil with β-casein were prepared by homogenizing a 20 wt% soya oil, 0.5 wt% protein solution in 20 mM imidazole buffer, pH 7.0, in a mini-homogenizer[7] at a pressure of 300 bar. To obtain emulsion particles which were approximately monodisperse, and which were small enough to have their diameters measured conveniently, the complete emulsion was centrifuged at $4200g$ in a 3×25 ml swing-out rotor for 15 min. Most of the emulsion droplets formed a layer at the top of the centrifuge tube, and the remaining subnatant layer was collected by piercing the centrifuge tube at the bottom and allowing the subnatant liquid to drip out. It was found that this consisted of a dilute emulsion containing particles of about 200 nm diameter. These particles were used for the studies of the changes in diameter when trypsin was added.

Complexes of β-casein or β-lactoglobulin with polystyrene latex were made simply by adding the appropriate volume of a solution of the protein in imidazole buffer to a suspension of the latex. The latex suspensions contained 8 μl of the original 10 wt% suspension in 3 ml of imidazole buffer.

Diameters of the emulsion droplets or latex complexes were measured by PCS, using a Log-Lin correlator (Malvern Instruments Ltd.). Measurements of the dynamics of the scattered light were made at a scattering angle of 90° only. Average diffusion coefficients were determined by the method of cumulants[13] and from these the particle diameters were calculated using the Stokes–Einstein relation for spheres. The detailed hydrodynamic behaviour of the casein-coated latex particles has already been described.[2]

To study the changes in diameter of latex or emulsion particles with adsorbed layers when treated with trypsin, a solution of trypsin (1.3 mg/10 ml) was made up, and diluted 50-fold with imidazole buffer. Of this diluted suspension, 50 μl was added to 3 ml of a latex suspension to which protein (0.1 mg ml^{-1}) had been added, or to 3 ml of a diluted suspension of the separated emulsion droplets. The diameters of the particles were then measured at ~1 min intervals for 1 hour.

The kinetics of the attack of trypsin on the intact emulsion were established by analysis of the peptides liberated from the complete emulsion when treated with trypsin.[12]

3 Results and Discussion

Thickness of Adsorbed Layers—The increase in latex particle diameter (originally 90 nm) as increasing amounts of β-casein bind to the particle is shown in Figure 1. From the results, it is clear that an appreciable increase in the diameter of the particle occurs as the protein is adsorbed, consistent with the formation of a layer of β-casein around the latex particle. The thickness of this layer was found not to

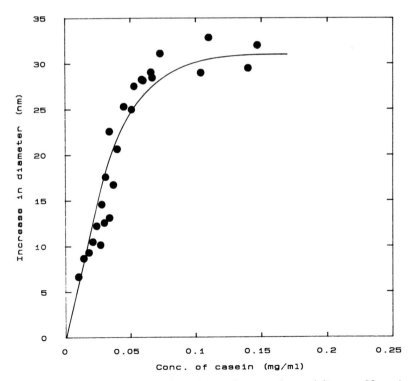

Figure 1 *Increase in particle radius when polystyrene latex of diameter* 90 nm *is treated with increasing concentrations of β-casein, in buffer of* 20 mM *imidazole/HCl* pH 7.0. *See ref. 2 for further details*

be affected by the presence of high concentrations of urea, nor by the presence of moderate amounts of NaCl (50 mM),[2] as long as there was sufficient protein on the latex surface to give almost complete coverage.[14] Much the same behaviour has been observed for the other caseins, α_s-casein[2] and κ-casein, with some minor variations, and it has been found that all of the caseins behave very differently from globular proteins, *e.g.* β-lactoglobulin. This latter protein was found to give a layer whose thickness was almost within the experimental error for the measurement, *i.e.* about 2 nm, in contrast to the caseins, where the layer thickness appears to be 10–15 nm.

A simple interpretation of these results is that, whereas some sections of the adsorbed caseins are sufficiently free in loops or tails[15] so as to protrude into solution, none of this is possible for β-lactoglobulin. Not even a denatured form of the β-lactoglobulin, prepared by carboxymethylation of the cysteinyl residues of the protein,[6] showed behaviour similar to the caseins, perhaps because it may (*a*) retain some amounts of the β-structure present in the original protein,[16] or (*b*) bind, as has been supposed, flat to the latex–water interface, with no protruding loops or tails. (It is conceivable that a third option, where a very freely draining layer of protein is formed at the interface, exists; but this is considered unlikely.)

Nevertheless, it is possible for the presence of some β-lactoglobulin to interfere with the binding of casein. When latex is treated with β-casein, the diameter increases, as has been shown, and addition of β-lactoglobulin to this complex makes little difference to the thickness of the layer of β-casein. However, if the latex is first exposed to β-lactoglobulin, and then treated with β-casein, then, although the diameter of the complex increases, the size of the increase is significantly less than that observed when only β-casein is used (Table 1). Thus, the established interfacial layer of β-lactoglobulin interferes with the adsorption of a layer of β-casein, as has been suggested to be the case for the two proteins at the oil–water interface.[6] As before, little difference between the native and modified forms of β-lactoglobulin was observed in respect of their abilities to interfere with the formation of the layer of adsorbed β-casein. The similarity of these two forms of β-lactoglobulin in terms of interfacial viscosity has already been noted,[6] which is consistent with the observation that they affect the binding of β-casein in a similar manner.

Table 1 *Increase in diameter, Δd, of polystyrene latex (nominal diameter 90 nm) when proteins are added, in the order shown*

System	Δd (nm)
Latex (no protein)	0.0
Latex + β-casein	19.3
Latex + β-lactoglobulin + β-casein	14.5
Latex + modified β-lactoglobulin + β-casein	15.2
Latex + β-casein + β-lactoglobulin	20.0
Latex + β-casein + modified β-lactoglobulin	20.4

When an emulsion formed from n-tetradecane and β-lactoglobulin is treated with β-casein, some of the β-lactoglobulin is displaced and then gradually readsorbs to the interface, at the same time as some β-casein is adsorbed.[4] The amounts involved suggest that β-casein will not completely cover the emulsion surface but will alter the binding of the β-lactoglobulin, and that the presence of the β-lactoglobulin will modify the binding of β-casein. On the other hand, although β-lactoglobulin will bind to the surface of β-casein-coated droplets, it will not displace any of the casein.[4] These results are in accord with the study of the diameters of the protein-coated latex particles described above.

Effect of Trypsin on Diameters of Latices and Emulsion Droplets—The effect of trypsin on β-casein in solution is to break the protein into a number of peptides. Trypsin attacks the polypeptide at lysine and arginine residues, of which there are a total of 15 in β-casein (Figure 2). It is therefore to be expected that the adsorbed layers of β-casein on latices and emulsion droplets will be broken down when the particles are treated with trypsin. Such a reaction appears to offer an opportunity to study the thicknesses of the adsorbed layers of β-casein in emulsions, since the nature of emulsions precludes the observation of the protein being gradually adsorbed at the interface. The breakdown of the surface layer, however, will only allow the estimation of a minimum thickness of the adsorbed layer provided that

Figure 2 *Amino acid sequence of bovine β-casein A^2 (ref. 18). Arrows show potential sites for attack by the proteolytic enzyme trypsin*

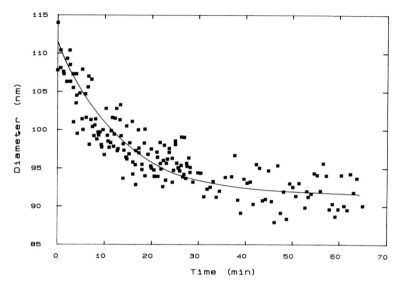

Figure 3   *Change in the diameter of a β-casein–latex complex as a function of time when treated with trypsin in the ratio β-casein:trypsin of* 1250:1

aggregation of the particles does not immediately occur as the adsorbed layer is broken down. Although emulsion particles do coagulate when treated with trypsin, it is possible to control the reaction simply by reducing the amount of protease used.

Results of trypsin treatment on the diameter of a β-casein–latex complex are shown in Figure 3. There was found to be a rapid decrease of the diameter from its initial value of 112 nm down to a value of 92.5 nm (Table 2). This final value was not that measured for the original latex (85.9 nm); so it appeared that an adsorbed layer still remained on the latex, albeit one which had been considerably altered. The difference in layer thickness induced by trypsin treatment of the β-casein adsorbed on the latex was about 9.5 nm. When the experiment was repeated using the separated small emulsion droplets, the magnitude of the decrease in the thickness of the β-casein layer (9 nm) was very close to that found with the latex. Although the emulsion droplets were appreciably larger than the latex, this had no effect upon the result (Table 2). These similarities suggest that the adsorbed layers of β-casein in the emulsion and on the latex are similar in their response to

Table 2   *Effect of trypsin treatment on diameter* d *of latices and emulsion droplets*

System	d (nm)	System	d (nm)
Latex	85.9	Emulsion	201.5
Latex + β-casein	112.0	After trypsin treatment	183.4
After trypsin treatment	92.5	Change	18.1
Change	19.5		

proteolytic attack, and therefore it is reasonable to postulate that the conformations of β-casein on latex and oil–water interfaces are similar, and in that
respect the latex–β-casein complex is a valid model for the oil-in-water emulsion.
This may argue against models of protein adsorption in which portions of the
polypeptide may actually penetrate the oil–water interface, rather than simply
resting on the interface.

Proteolysis of the Casein and Removal of Peptides—The action of trypsin on
β-casein in solution is relatively random. That is, none of the possible sites for
proteolysis appears to be especially favoured, so that the proteolysis results in a
complex mixture of peptides. On the other hand, in emulsions the β-casein is
broken down in a highly specific manner.[12] Analysis of the peptides produced
have shown that the adsorbed β-casein is initially attacked by trypsin at one of two
positions, both of which are close together in the amino-acid sequence (*i.e.*
positions 25 and 28).

   This effectively splits the β-casein into two peptides, the N-terminal phospho-
peptide and the remaining 'macropeptide'. None of the other sites for proteolysis
is attacked significantly in the early stages of the proteolysis of adsorbed β-casein.
Thus, the fact that the β-casein is adsorbed at an oil–water interface has a most
significant effect on the positions at which the trypsin attack can occur. Once the
initial attack has occurred, there is a period during which little or no proteolysis
appears to occur, after which proteolysis restarts with attack on sites in the middle
of the remaining macropeptide, at residues in the region of position 100.

   We have interpreted these results as showing that the conformation of the
adsorbed β-casein is such that the rather hydrophilic region of the molecule (*i.e.*
residues 1–48 of the protein) forms either a loop or a tail on the oil–water
interface, rendering the sites of attack by trypsin on that peptide (*i.e.* residues 25
and 28) highly accessible to the protease. Conversely, none of the other potential
sites are attacked, either because the relevant part of the β-casein molecule is in
an unsuitable conformation (flat on the interface) or the trypsin is prevented from
approaching by the protruding N-terminal peptides. Once the phosphopeptides
have been removed, the remaining macropeptides can change their conformation
so as to allow tryptic attack in the moderately hydrophilic portion of the molecule
in the region of residue 100.

   The initial loss of the peptide from the N-terminal portion of the β-casein is the
reason for the rapid decay of the radius of either the latex particle or the emulsion
droplet when treated with trypsin. The size of the change (*ca.* 9 nm in the effective
radius of the particle) reflects the extent to which the peptide protrudes from the
surface. There seem to be two possibilities: (i) that the peptide in the region of
residues 1–48 forms a loop, which is split in the middle by the trypsin, one half of
the loop then breaking away and the other half collapsing nearer to the oil surface,
or (ii) that the peptide sequence 1–30 forms a tail which is detached by the trypsin
action. Some estimate of the probable conformation may be assessed by plotting
the hydrophobicity of the protein side-chains as functions of the distance along
the chain, and then averaging over a number of residues (seven in this case), to
provide a map of local hydrophobicity. This is shown in Figure 4, and the
suggestion can be made that, since β-casein binds to the oil–water interface via its

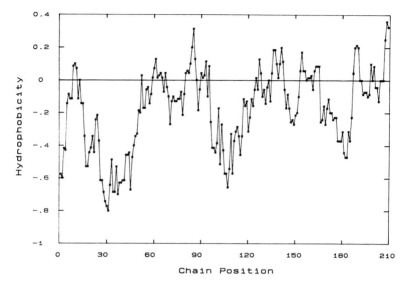

Figure 4   *Plot of hydrophobicity of the polypeptide chain in β-casein as function of position. The calculated value of residue hydrophobicity (free energy difference between buried and exposed residues[19]) is averaged over the seven residues centred at each position*

hydrophobic portions and is regarded as being a fairly flexible molecule, the representation in Figure 4 may not be unrealistic, as shown in Figure 5. For example, the loop formed between residues 1 and 48 will provide for the observed results, since it suggests that the residues 25 and 28 will be presented to the enzyme with minimal hindrance. It also explains the observation that the *N*-terminal arginine residue is not attacked by trypsin when the β-casein is adsorbed. The difficulty in the interpretation is that it is not clear why the breaking of bond 25 or 28 should result in the large decrease in the radius of the surface layer. If, as is shown, the loop is anchored at both ends, it might be suggested that breakage in the middle of the loop would not affect the structure of the surface layer to a great extent. It may be that the two halves of the loop together form a rather rigid structure, so that the splitting of the apex of the loop causes both sides to collapse.

It is certain that the breakdown of the *N*-terminal peptide and the decrease in particle radius are closely connected, since both occur within a very short time of adding trypsin to the latex or to the emulsion. Since any other positions for proteolysis are not affected in the short time available, it is probable that the decrease seen represents the length of the *N*-terminal peptide. In previous work, we estimated that the related protein κ-casein formed a layer about 12 nm thick when in its native position on the surface of the casein micelle:[17] this layer is probably formed from the 63-amino acid peptide forming the *C*-terminal of κ-casein. In the case of β-casein, the 1–25 peptide must apparently have a length of about 9 nm, *i.e. ca.* 0.36 nm per peptide segment. While this is possible, it is clear that the peptide must exist in a rather extended conformation. The layer of β-

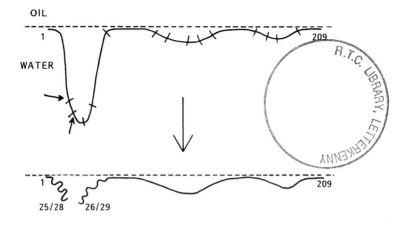

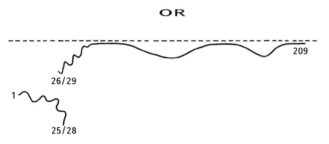

Figure 5 *Representation of the conformation of β-casein at an oil–water interface from the information provided by proteolysis and change in radius. Two possible situations following splitting of the chain at position 25 or 28 are indicated*

casein does not change its dimensions when treated with 6 M urea,[2] and so there is no evidence that there is any regular structure involved in the *N*-terminal loop.

Despite the problems of the geometry of the loop, the model shown in Figure 5 appears to be preferable to a model where the entire region 1–48 forms a tail extending away from the surface. While this would offer more segments, and therefore more readily explain the change in thickness of the adsorbed layer, it creates problems when the tryptic activity is considered. The susceptible bonds would be rather inaccessible to the trypsin in the proposed conformation, so that the rapid breakdown which is observed could hardly be expected to occur. In fact, the β-casein peptides would form a 'hairy' layer preventing close approach of the trypsin. The surface coverage by β-casein of latex and emulsion appears to be about 3 mg m^{-2} and this leads to an area per protein molecule of about 12.8 nm^2. On this basis, the 'hairs' will be separated by about 3.6 nm, which would make it very difficult for the trypsin to attack the labile site on the *N*-terminal peptide.

Moreover, in none of the studies did we detect proteolysis of the β-casein at position 1, which would be rather likely if the N-terminal peptide were in an extended conformation.

4 Conclusions

We have established two effects of milk proteins at an oil–water interface. The first is that the serum proteins, represented by β-lactoglobulin, bind relatively close to the interface and do not protrude into solution to a marked extent. The second is that the caseins bind to the interface in such a way as to leave considerable portions of their structures protruding. In the case of β-casein, it is possible to map the molecular structure by considering the hydrophobicity of the amino acid side-chains, and drawing the structure on a graph of hydrophobicity against position in the chain. While this procedure may be permissible for β-casein, which is considered to be a rather flexible protein, and may possibly be valid for α_s- and κ-caseins also, it is probably unlikely that it will be applicable to all adsorbed proteins, which have more secondary structure originally, and which may retain at least some of it when adsorbed to the oil–water interface. The β-lactoglobulin is a case in point where the original β-sheet structure of the protein may give it a rather specific conformation when adsorbed on the interface.

References

1. E. E. Uzgiris and H. P. M. Fromageot, *Biopolymers*, 1976, **15**, 257.
2. D. G. Dalgleish, *Colloids Surf.*, 1990, **46**, 141.
3. E. Dickinson, S. E. Rolfe, and D. G. Dalgleish, *Food Hydrocolloids*, 1988, **2**, 397.
4. D. G. Dalgleish, S. E. Euston, J. A. Hunt, and E. Dickinson, this volume, p. 485.
5. E. Dickinson, S. E. Rolfe, and D. G. Dalgleish, *Food Hydrocolloids*, 1989, **3**, 193.
6. E. Dickinson, S. E. Rolfe, and D. G. Dalgleish, *Int. J. Biol. Macromol.*, 1990, **12**, 189.
7. E. Dickinson, A. Murray, B. S. Murray, and G. Stainsby, in 'Food Emulsions and Foams,' ed. E. Dickinson, Royal Society of Chemistry, London, 1987, p. 86.
8. P. van Dulm and W. Norde, *J. Colloid Interface Sci.*, 1983, **91**, 248.
9. B. W. Morrissey and C. H. Han, *J. Colloid Interface Sci.*, 1978, **65**, 423.
10. M. Shimizu, A. Ametani, S. Kaminogawa, and K. Yamauchi, *Biochim. Biophys. Acta*, 1986, **869**, 259.
11. S. Kaminogawa, M. Shimizu, A. Ametani, S. W. Lee, and K. Yamauchi, *J. Am. Oil Chem. Soc.*, 1987, **64**, 1688.
12. J. Leaver and D. G. Dalgleish, submitted for publication.
13. D. E. Koppel, *J. Chem. Phys.*, 1972, **15**, 4814.
14. E. Dickinson, E. W. Robson, and G. Stainsby, *J. Chem. Soc. Faraday Trans. 1*, 1983, **79**, 2937.
15. B. Vincent and S. Whittington, in 'Surface and Colloid Science', ed. E. Matjevic, John Wiley, New York, 1982, Vol. 12, p. 1.
16. M. Z. Papiz, L. Sawyer, E. E. Eliopoulos, A. C. T. North, J. B. C. Finlay, R. Silvaprasadarao, T. A. Jones, M. E. Newcomer, and P. J. Kraulis, *Nature (London)*, 1986, **324**, 383.
17. C. Holt and D. G. Dalgleish, *J. Colloid Interface Sci.*, 1986, **114**, 513.
18. F. Grosclaude, M.-F. Mahé, J.-C. Mercier, and B. Ribadeau Dumas, *Eur. J. Biochem.*, 1922, **26**, 328.
19. J. Janin, *Nature (London)*, 1979, **277**, 491.

Phase Diagrams of Mixed Soybean Phospholipids

By Björn Bergenståhl

INSTITUTE FOR SURFACE CHEMISTRY, BOX 5607, 114 86 STOCKHOLM, SWEDEN

1 Interparticle Forces and Adsorbed Layers

Emulsion stability is to a large extent determined by the interparticle force balance. These forces are created by adsorbed layers—the surface materials on the emulsion droplets.[1] Similar forces clearly also act between molecular aggregates in solution, and will thereby also determine the phase behaviour of the corresponding bulk system. Hence, it is clear that there is a connection between the phase diagram of an emulsifier and how it functions as a stabilizing agent. This relation was first demonstrated by Friberg and co-workers.[2-5] They stated, after series of experiments in several different environments, that the existence of a lamellar liquid crystalline phase gives an indication for a suitable balance of an emulsifier. (Of course, a similar relationship also exists for macromolecular stabilizers,[6] where the phase diagrams will show if the stabilizer has a suitable solubility.) However, the type of experiment that has been used to justify this statement does not clearly separate between properties creating good stability (low probability for coalescence) and properties that facilitate particle disruption during the emulsification process. The latter are surface tension effects due to the presence of the emulsifier and Gibbs–Marangoni effects due to the surface elasticity of the emulsifier layer at the interface when the new oil–water interphase is created.[7]

2 Molecular Interactions and Phase Diagrams

A clear case of how repulsive forces are reflected in the phase diagram is illustrated by sodium dioctylsulphosuccinate. Sodium dioctylsulphosuccinate (Aerosol OT) is an ionic double-chained surfactant that forms bilayers[8] (Figure 1). The phase diagram is dominated by a large lamellar phase which has a maximum swelling in pure water well above 10 nm. In excess water, the swelling of the lamellar phase continues until the repulsive hydration force is balanced by the attractive van der Waals force. The extended swelling is a result of long-range repulsive double-layer forces. The swelling of the lamellar phase is strongly reduced[9,10] when the counter-ions are changed from Na^+ to divalent ions like Mg^{2+} or Ca^{2+}. The reduced swelling is clearly a result of the diminished electrostatic repulsion with the divalent counter-ion.

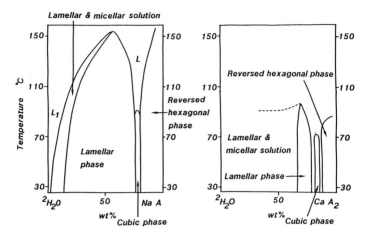

Figure 1   *The phase diagram of sodium dioctylsulphosuccinate (Aerosol OT) and calcium dioctylsulphosuccinate [redrawn after refs 8, 9, and 10]*

The analysis of inter-aggregate forces and phase equilibria is more complicated in systems where several liquid crystalline phases are formed. However, the effective geometrical shape of the surfactant molecules in the different types of aggregates[11] can be used to describe the hydrophilic lipophilic character (Figure 2). When water is added to a surfactant system, the effective volume of the polar group increases as the surfactant becomes more hydrated. This proceeds until the upper limit of the net repulsive interaction between neighbours has been reached. The type of aggregates formed in diluted surfactant dispersions will reflect how able the molecules are to create surface forces compared with the size of their hydrophobic tails.

In the traditional view of emulsifiers, the properties of surfactants are described as a balance between the hydrophilic and the hydrophobic parts of the molecule, described as HLB numbers.[12] Hence, it is clear that there should be a connection between the HLB number and the phase formation. In Figure 2, we present the characteristic shapes for different liquid crystalline aggregates compared with the corresponding HLB numbers. Micellar aggregates correspond to high HLB numbers above 12, and reversed aggregates correspond to low numbers below 6. Lamellar phases correspond to HLB numbers around 7.

3   The Phase Diagrams of Different Phospholipids

These can be applied to get a better understanding of the hydrophilic–lipophilic properties of the pure and mixed soybean phosphatides, and how they reflect the abilities of the different lipids to generate repulsive forces.

The binary phase diagram of pure soybean phosphatidylcholine[13] is shown in Figure 3. The phase diagram is characterized by a large swelling lamellar phase.

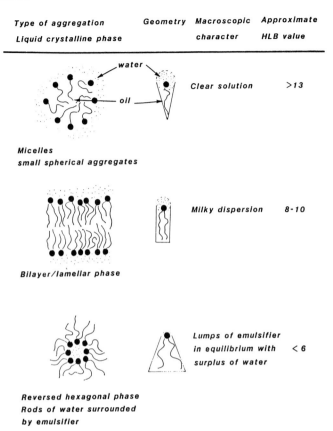

Type of aggregation Liquid crystalline phase	Geometry	Macroscopic character	Approximate HLB value
Micelles small spherical aggregates		Clear solution	>13
Bilayer/lamellar phase		Milky dispersion	8-10
Reversed hexagonal phase Rods of water surrounded by emulsifier		Lumps of emulsifier in equilibrium with surplus of water	< 6

Figure 2   *Structures formed with amphiphilic lipids of different hydrophilic–lipophilic balance. [The Figure is constructed after ref. 1]*

The upper limit of the swelling of the lamellar phase is *ca.* 40% water*
corresponding to a water layer of about 22 Å. Above 40% of water the lamellar
phase exists in equilibrium with almost pure water (the solubility of lecithin is
estimated[15] to be about 10^{-14} M). If the lamellar phase is dispersed in water,
liposomes are formed with a long-term stability. With low concentrations of water
(below 8 wt%), a cubic phase is formed. Below 3 wt% the crystals remain stable
at room temperature. When samples with low water content (3–10 wt%) are
heated, a reversed hexagonal phase is formed at about 120 °C. This phase melts at
about 250 °C. A similar melting point is also valid for the lamellar phase when
heated in sealed samples. Similar phase diagrams have been noted for egg lecithin
by Small,[16] and by Luzzati *et al*.[17]

---

* This value is difficult to determine accurately. Different authors differ significantly.
Compare, for instance, estimations of the swelling of DOPC (dioleoylphosphatidylcholine)
which range between 42–55% water as estimated by different experimentalists and
methods.[14]

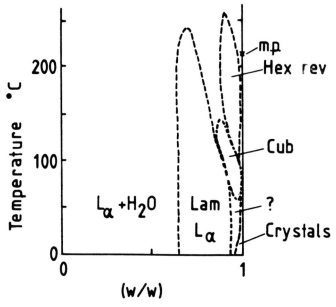

Figure 3   *The phase diagram of soybean lecithin (pure phosphatidylcholine, Epikuron 200 from Lucas Meyer).* [*From ref. 13*]

In the study reported here, we make a preliminary investigation of the phases formed by three different soybean phospholipids when mixed with water. The results are expressed as three-component diagrams.

4   Materials and Methods

The phase determination in this study has mainly been performed through X-ray diffraction experiments. The camera used was a slit-collimated Luzzati–Baro camera equipped with a position-sensitive electronic detector (PSD 100, Tennelec, Oak Ridge, TN, USA). The radiation was Cu K_α with $\lambda = 1.54$ Å. The distance between detector and sample was kept at 43 cm. The detector had a slit width of 6 cm. The exposure times were chosen, depending on the sample, to be between two and ten minutes. The spectra were divided into 6000 separated channels, whose values were calibrated repeatedly with crystalline sodium octanoate ($d = 23$ Å). The spacings have been calculated according to Fontell *et al.*[18]

The samples were made by first dissolving the lipids in diethylether and mixing them in weighted test tubes. After evaporation of the ether at room temperature under mild vacuum, the samples were carefully dried under high vacuum (10^{-2} mm Hg) for about 10–20 h. After drying the weight was checked and the aqueous phase was added. The samples were flame sealed. The samples were mixed by several centrifugations (4000 r.p.m.) in both directions and stored for *ca*. 24–48 h.

The soybean phospholipids were chromatographically purified (CPL lipids) by Lipidteknik, Stockholm. The water was double distilled, and the ether was of AnalaR quality.

5 Results for Mixed Soybean Systems

Figure 4 is the phase diagram of the mixture of soybean phosphatidylcholine (PC) + soybean phosphatidylinositol (PI) + water at 20 °C. As shown by Söderberg and Larsson,[19] the PI shows a large swelling lamellar phase. This is comparable to several other charged surfactant systems (Aerosol OT,[8] dicetyltrimethylammonium bromide with hexanol, monoglycerides with sodium soaps[20,21]) with large hydrophobic moieties. An interesting observation is that, above *d* values of 140 Å, the first reflection is cancelled, and only the second, third, and fourth reflections are observed. A similar observation has been published for Aerosol OT by Fontell.[22]

The mixture PI + PC, in different ratios, also gives a large swelling lamellar phase dominated by the character of ionic PI. The upper limit of the swelling has not been determined as it is above the limit of the technique (in this case *ca*. 200 Å). The results, in terms of repeat distances, are shown for pure PI and PI + PC mixtures in Figure 5. The curves clearly show the extended swelling that remains down to at least 16 wt% of PI in the PI + PC mixture. The repeat distances for the pure PC are also included in Figure 5. The swelling is limited to about 64 Å. This result is somewhat different to that found with pure soybean PC from Lucas Meyer, Epikuron 200,[13] where we see an upper limit of the swelling at 56 Å.

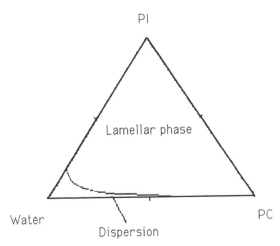

Figure 4   *The three-component phase diagram of water + soybean phosphatidylcholine + soybean phosphatidylinositol at* 20 °C

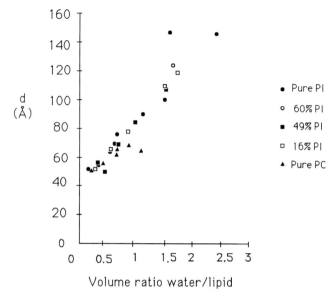

Figure 5   *The repeat distance from X-ray diffraction experiments with different ratio between PC and PI. All samples except pure PC show an extensive swelling*

Figure 6 shows the phase diagram of soybean phosphatidylcholine (PC) + soybean phosphatidylethanolamine (PE). Pure PE forms a reversed hexagonal phase, as has been shown previously.[23] The swelling extends to about 30 wt% of water with a repeat distance of 70 Å. The X-ray spacings of the different PC + PE mixtures are shown in Figure 7. The reversed hexagonal phase is virtually

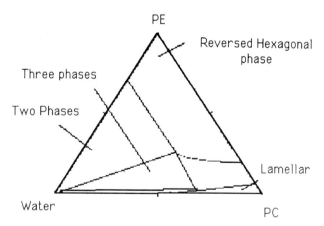

Figure 6   *The phase diagram of soybean phosphatidylcholine + soybean phosphatidylethanolamine + water at 20 °C*

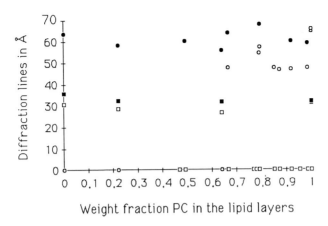

Figure 7 *The X-ray spacings of the different PC + PE mixtures with water contents of* 50–60 wt% (*above the upper swelling limit*)

unchanged until about 60 wt% of PC. Above 60 wt% the system separates into a three-phase area with a reversed hexagonal phase in equilibrium with a lamellar phase. Traces of the hexagonal phase remain until almost no PE remains in the sample.

Figure 8 shows the phase diagram of soybean phosphatidylethanolamine + soybean phosphatidylinositol. The X-ray spacings of the mixtures are shown in Figure 9. The results show that the lamellar phase remains undisturbed down to PI concentrations of about 30 wt%. Below 30 wt% of PI, the lamellar phase co-exists with a hexagonal phase.

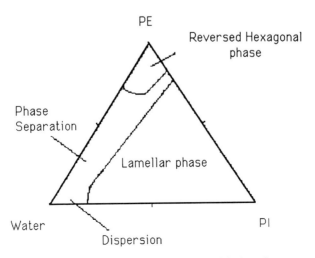

Figure 8 *The phase diagram of soybean phosphatidylethanolamine + soybean phosphatidylinositol + water at* 20 °C

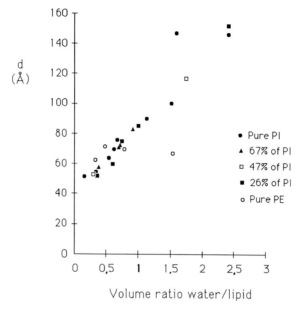

Figure 9 *The repeat distance from X-ray diffraction experiments of different mixtures of soybean phosphatidylethanolamine + soybean phosphatidylinositol*

When the three different soybean phospholipids are compared, PC can be described as the balanced emulsifier with intermediate properties relative to the other two. The PE has a more hydrophobic character, forming a reversed hexagonal phase. It is interesting to note that the behaviour of this phase is not changed when PC is included into it until PC reaches about 60% of the mixture. The PI is, through its electrostatic charge, the most hydrophilic component. The potential is clearly shown when mixed with PE. The PI continues to keep the mixture in a lamellar phase until only 30 wt% of PI remains in the lipid mixture.

Technical soybean lecithin contains a mixture of different phospholipids.[4,5] In most cases, the rather hydrophobic phosphatidylethanolamine dominates and makes this type of lecithin suitable for inverse emulsions such as margarines. More hydrophilic soybean lecithins suitable for oil-in-water emulsions are obtained by partial hydrolysis to form lysolecithins.[24]

References

1. B. A. Bergenståhl and P. M. Claesson, in 'Food Emulsions', ed K. Larsson and S. E. Finberg, Marcel Dekker, New York, 1990, p. 41.
2. S. Friberg and L. Rydhag, *Kolloid Z. Polym.*, 1971, **244**, 233.
3. S. Friberg and I. Wilton, *Am. Perfum. Cosmet.*, 1970, **85**, 27.
4. L. Rydhag, *Fette, Seifen. Anstrichm.*, 1979, **81**, 168.
5. L. Rydhag and I. Wilton, *J. Am. Oil Chem. Soc.*, 1981, **58**, 830.

6.   E. Dickinson, *Food Hydrocolloids*, 1986, **1**, 3.
7.   P. Walstra, in 'Encyclopedia of Emulsion Technology', ed. P. Becher, Marcel Dekker, New York, Vol. 1, p. 57.
8.   J. Rogers and P. A. Winsor, *Nature (London)*, 1967, **216**, 477.
9.   A. Khan, K. Fontell, and B. Lindman, *J. Colloid Interface Sci.*, 1984, **101**, 193.
10. A. Khan, K. Fontell, G. Lindblom, and B. Lindman, *J. Phys. Chem.*, 1982, **86**, 4266.
11. J. N. Israelachvili, D. J. Mitchell, and B. W. Ninham, *J. Chem. Soc.*, *Faraday Trans. 2*, 1976, **72**, 1525.
12. W. C. Griffin, in 'Kirk-Othmer Encyclopedia of Chemical Technology', 1979, Vol. 8.
13. B. A. Bergenståhl and K. Fontell, *Prog. Colloid Polym. Sci.*, 1983, **68**, 48.
14. B. A. Bergenståhl and P. Stenius, *J. Phys. Chem.*, 1987, **91**, 5944.
15. C. Tanford, 'The Hydrophobic Effect', Wiley, New York, 1973.
16. D. M. Small, *J. Lipid Res.*, 1967, **8**, 551.
17. V. Luzzati, T. Gulik-Krzgwicki, and A. Tardieu, *Nature (London)*, 1968, **218**, 1031.
18. K. Fontell, L. Mandell, H. Lehtinen, and P. Ekwall, 'Chemistry Series, Acta Polytechnica Scandinavica', 1968, Chap. 2, **74**, III, 2.
19. I. Söderberg and K. Larsson, to be published.
20. P. Ekwall, L. Mandell, and K. Fontell, *J. Colloid Interface Sci.*, 1969, **29**, 639.
21. K. Larsson and N. Krog, *Chem. Phys. Lipids*, 1973, **10**, 177.
22. K. Fontell, *J. Colloid Interface Sci.*, 1973, **44**, 318.
23. P. R. Cullis and B. de Kruijff, *Biochim. Biophys. Acta*, 1978, **513**, 31.
24. 'Emulfluid,' Lucas Meyer, Elbdeich 62, Hamburg, Germany.

Stability of Food Emulsions Containing both Protein and Polysaccharide

By Eric Dickinson and Stephen R. Euston

PROCTER DEPARTMENT OF FOOD SCIENCE, UNIVERSITY OF LEEDS,
LEEDS LS2 9JT

1 Introduction

An important functional property of food polymers is in the stabilization of food colloids.[1] The two main types of food polymers found in oil-in-water emulsions are proteins and polysaccharides, and some food emulsion products contain both types of macromolecule. This article is concerned with the way in which the nature of the protein–polysaccharide interaction affects the behaviour of such colloidal systems.

It is useful to distinguish between the terms 'emulsifier' (emulsifying agent) and 'stabilizer' (stabilizing agent), though in practice many food polymers are able to execute both functional roles. We can define an 'emulsifier' as a single chemical component, or mixture of components, having the capacity for promoting emulsion formation and stabilization by interfacial action, and a 'stabilizer' as a chemical component, or mixture of components, which can confer long-term stability on an emulsion, possibly by a mechanism involving adsorption, but not necessarily so.[2] To be a good emulsifier, then, a macromolecular species should have the capacity to adsorb rapidly at the nascent oil–water interface created during emulsification, and so protect the newly formed fine droplets against immediate recoalescence. On the other hand, the role of a good stabilizer is to keep droplets apart in the emulsion once it has been formed, thereby retarding or inhibiting creaming, flocculation and coalescence during long-term storage. The traditional view is that proteins make good emulsifying agents because of their substantial hydrophobicity and molecular flexibility which allows rapid adsorption and rearrangement at the interface to give a coherent macromolecular protective layer.[3] Polysaccharides make good stabilizing agents because of their hydrophilicity, high molecular weight, and gelation behaviour which leads to the formation of a macromolecular barrier in the aqueous medium between dispersed droplets.[4] One might imagine that the combination of protein + polysaccharide would give optimum results by bringing together the emulsifying role of the protein with the stabilizing role of the polysaccharide.

There are many ways in which complex biopolymers can interact at the molecular level—covalent linking, ionic bonding, hydrogen bonding, hydrophobic interaction, van der Waals interaction, and physical entanglement. Some of these interactions are specific to the particular biopolymers involved, whereas others are more general. In attempting to embrace the whole range of different protein + polysaccharide combinations, it is convenient for us to distinguish between three situations.

(i)   *Strong association.* Here the protein and polysaccharide are assumed to be irreversibly bound together over the experimental timescale. This includes protein (polypeptide) covalently bound to polysaccharide (oligosaccharide) in glycoproteins and hydrocolloid gums, as well as strong electrostatic complexes (*e.g.* anionic polysaccharide + protein at pH < pI).

(ii) *Weak association.* Here the protein–polysaccharide attraction is relatively weak and hence the association is potentially reversible. This includes non-ionic and weak electrostatic complexes (*e.g.* anionic polysaccharide + protein at pH > pI).

(iii) *No association.* Here the protein–polysaccharide interaction is either repulsive or negligible. This includes solutions containing mixtures of protein + non-ionic or charged polysaccharides exhibiting thermodynamic incompatibility.

We shall see that these three categories are useful in understanding the different types of physico-chemical phenomena that are exhibited by colloids containing protein + polysaccharide, though it must be recognized that there are no sharp divisions between categories, and that borderline cases exist between weak and strong association and between association and no association.

Droplets in a stable food emulsion are prevented from coming close together by a combination of electrostatic stabilization and steric stabilization.[1] Surface-active polyelectrolytes such as proteins can provide both kinds of stabilization, with electrostatic stabilization being favoured by low ionic strength and a pH well away from pI. However, the optimum characteristic of a good steric stabilizer is that it should not only be strongly attached to the surface, but should also protrude significantly into the aqueous medium so as to produce a macromolecular layer of appreciable thickness.[4] It would seem that the conflicting requirements of substantial hydrophobicity for strong adsorption and substantial hydrophilicity for thick steric layers can best be reconciled by covalent linkage of protein and polysaccharide moieties to form a single hybrid amphiphilic macromolecule. The main advantage of a covalent linkage over other types of non-covalent bonding is the retention of solubility and molecular integrity over a wide range of solution conditions (temperature, pH, ionic strength). An example of a natural protein–polysaccharide hybrid is gum arabic. The presence of just 2% protein in this polysaccharide gum is responsible for conferring upon gum arabic its special emulsifying properties, together with its solubility in water at concentrations up to 50–55 wt%.[5] Synthetic protein–polysaccharide hybrids with good emulsification properties can be prepared by chemical reaction of proteins and polysaccharides in the laboratory.[6,7] Optimum functionality is generally found with protein contents in the range 10–30%.

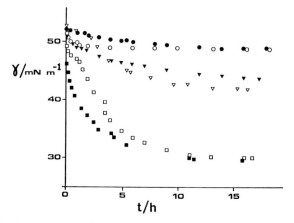

Figure 1   *Surface activity of various Acacia gum samples measured by the Wilhelmy plate method[23] at the n-hexadecane–water interface (10^{-3} wt% gum, pH 7, 25 °C). The interfacial tension γ is plotted against the adsorption time t: $\bigcirc$, sample A; $\bullet$, sample B; $\triangledown$, sample C; $\blacktriangledown$, sample D; $\square$, sample E; $\blacksquare$, sample F*

protein contents in the range 0.6–47%). Figure 1 shows time-dependent interfacial tensions $\gamma(t)$ for 10^{-3} wt% aqueous solutions (pH 7, ionic strength 0.05 M) adsorbing at the hydrocarbon oil–water interface.[23] The six samples are (A) *A. kirkii* (0.09% N), (B) *A. senegal* (0.33% N), (C) *A. resinomarginea* (0.83% N), (D) *A. irrorata* (1.57% N), (E) *A. gerrardii* (1.86% N), and (F) *A. difficilis* (7.5% N). After 15–20 hours, the $\gamma(t)$ plots have more or less levelled off. From the limiting values of γ, we can distinguish three categories of behaviour: low surface activity (A, B), intermediate surface activity (C, D), and high surface activity (E, F). This correlates with a classification of the samples into low, intermediate, and high nitrogen contents.

Studies of droplet-size distributions and creaming behaviour of hydrocarbon oil-in-water emulsions have shown[23] that gum samples which rapidly lower the interfacial tension give emulsions with small droplets and good stability. *Acacia* gum samples B and C, which show the slowest lowering of γ in the early stages of adsorption (see Figure 1), are the poorest emulsifiers of the six gum samples. The very poor performance of sample C may be due to its especially low average molecular weight as compared with the other samples investigated. Similar general trends of emulsifying behaviour for these *Acacia* gum samples at neutral pH are also found when the hydrocarbon oil (n-hexadecane) is replaced by a pure citrus oil (D-limonene), although the mechanism of droplet growth is different for the two cases—mainly Ostwald ripening of the D-limonene droplets, as compared with coalescence of the hydrocarbon droplets.[24]

The above results indicate an apparent correlation between *Acacia* gum nitrogen content and limiting long-time surface activity, and also between emulsifying capacity and short-time surface activity. On the other hand, there seems to be no direct relationship between nitrogen content and emulsion stability. Recent experiments[25] with three different samples of gum arabic (*A. senegal*) of known analytical composition, each having *ca.* 0.3% N (*ca.* 2 wt% protein), have shown substantial differences between samples, in terms of emulsifying capacity and emulsion stability, which cannot be attributed to protein content alone. Quantitative differences have also been found[25] between emulsions made with hydrocarbon oil and those made with D-limonene or orange oil. The gum arabic sample showing the most rapid lowering of the tension at the hydrocarbon oil–water interface was found to give the most stable hydrocarbon oil-in-water emulsion, as well as the smallest droplets with the other two oils. But the same gum gave the worst stability for the D-limonene-in-water and orange oil-in-water emulsions. On this evidence, it would seem that D-limonene is a better model system for simulating commercial flavour oil emulsions than is n-hexadecane.

Results reported elsewhere in this volume[26] show that a 10% fraction of a gum arabic sample corresponding to the highest molecular weight (0.38% N), separated by gel permeation chromatography, gives a more stable emulsion than the residual 90% fraction (0.35% N). In separate experiments,[26] samples of a different natural gum arabic (0.35% N) subjected to increasing degrees of controlled degradation by irradiation gave decreasing emulsion stability as the weight-average molecular weight was gradually reduced from 3.1×10^5 daltons to 2.2×10^5 daltons. The results from these two sets of experiments are consistent with the findings of Nakamura[27] showing a correlation between emulsion stability and molecular weight based on a study of 100 separate samples of gum arabic with molecular weights in the range $2–3 \times 10^5$ daltons. Higher molecular weight protein–polysaccharide hybrids may give better steric stabilization due to the formation of a thicker adsorbed layer. A mixture of small and large arabinogalactan–protein complexes may be a desirable combination, the former providing the greater emulsifying capacity (faster diffusion to the interface) and the latter conferring long-term emulsion stability by better steric stabilizing power.

3 Interaction of Proteins with Propylene Glycol Alginate

Propylene glycol alginate (PGA) is used industrially in the stabilization of foams (beer froth), emulsions (salad dressings), and particulate suspensions (fruit drinks). The replacement of some of the charged carboxyl groups in sodium alginate by the bulkier uncharged propylene glycol groups hinders aggregation of the polysaccharide chains and increases tolerance to calcium ions and low pH. In addition, the ester groups are slightly hydrophobic, which gives the polymer some surface activity at oil–water and air–water interfaces, though it cannot be used alone to make stable emulsions and foams. Commercial PGA samples are graded according to the degree of esterification and the viscosity/shear-rate behaviour. The degree of esterification is the percentage of carboxylic acid groups that have reacted with propylene oxide; it typically lies in the range 50–85%.

Covalent cross-linking of proteins with PGA may be achieved[6] without the use of undesirable chemicals by simply taking the mixture of protein + PGA to alkaline conditions (pH 9–11) for a short time, and subsequently reducing the pH again to the neutral or acidic conditions found in food. The cross-linking reaction

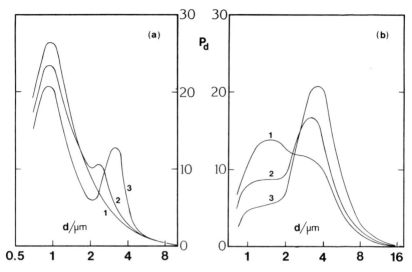

Figure 2   *Time-dependent droplet-size distribution of n-tetradecane-in-water emulsions* (10 wt% *oil,* 0.72 wt% *emulsifier,* pH 3) *made with* (a) *whey protein–PGA covalent hybrid* (1:10) *and* (b) *whey protein + PGA mixture* (1:10). *The volume-weighted percentage droplet probability* P_d *as given by the Coulter counter is plotted against the droplet diameter* d: (1) *fresh emulsion;* (2) *after* 48 h; (3) *after* 128 h

is considered to occur between the mannuronic acid ester group on the PGA and the ε-amino group of lysine to give an amide-type linkage,[28] though it is possible also that transesterification could occur via the hydroxyl group of serine. The positive effect of protein–PGA cross-linking on emulsification behaviour is illustrated in Figure 2 for the case of n-tetradecane-in-water emulsions made with a 1:10 mixture of whey protein isolate and a 'low viscosity' PGA (degree of esterification 55%). Emulsions were prepared using a high-pressure mini-homogenizer[29] operating at 300 bar. Compositions were as follows: 10 wt% oil, 0.65 wt% PGA, 0.065 wt% protein, and 89.3 wt% aqueous citrate buffer (0.05 M, pH 3) containing benzoic acid as preservative. Figure 2a shows time-dependent Coulter counter droplet-size distributions for an emulsion made with the protein–PGA covalent complex ('reacted' sample), and Figure 2b shows the same data for an emulsion made with a simple mixture of protein + PGA ('unreacted' sample) of the same composition. We observe that the droplets produced by the reacted sample are smaller and more stable with respect to coalescence than those produced by the unreacted sample. The former emulsion exhibits no visible signs of creaming or serum separation when stored at 25 °C for 5 days, whereas the latter emulsion shows clearly discernible cream and serum layers after less than 24 hours.

The results in Figure 2 show that, in a system of low protein content, the formation of a covalent protein–polysaccharide complex leads to a substantial improvement in emulsifying behaviour. In systems of higher protein content, however, simple (unreacted) mixtures of PGA with milk proteins (casein, β-

lactoglobulin or α-lactalbumin) are very satisfactory emulsifiers. At pH 3 or pH 7, emulsions made with protein + PGA using hydrocarbon oil or orange oil are at least as stable as equivalent gum arabic emulsions made with five or ten times as much emulsifying agent.[30] Good stability is obtained both with the emulsion concentrate and after extensive dilution. An important advantage of using the milk protein + PGA combination as emulsifier instead of just protein alone is that the emulsions produced are much more resistant to flocculation by simple electrolytes or ionic colouring agents.[30] This behaviour is consistent with the presence of a protein–polysaccharide complex at the surface of the emulsion droplets. We believe that the primary emulsifying agent in these systems is the milk protein, and that the complexed polysaccharide acts as a secondary steric-stabilizing component, protecting the adsorbed protein against electrostatically induced (self-)aggregation with protein on other droplets.

A simple illustration of film formation by non-covalent protein–PGA complexes at pH 3 is provided by observations[30] of wrinkling of the surface of a macroscopic oil droplet upon reducing its volume after exposure to, first, the protein in solution, and then, the polysaccharide. Caseinate + PGA or whey protein + PGA give visible 'stable' wrinkling behaviour down to a total macro-molecular concentration of 1×10^{-3} wt%. (A stable wrinkle is one that reappears on reducing the droplet volume again after the shrunken droplet has been returned to its full initial volume.) With whey protein alone or gum arabic, it was found[30] that stable wrinkling occurs only down to a concentration of 1×10^{-1} wt%. PGA or caseinate alone do not give wrinkling at any concentration.

A more quantitative indication of non-covalent complex formation between protein and PGA at the oil–water interface is provided by surface rheological measurement.[31] Figure 3 shows the effect of PGA addition on the surface

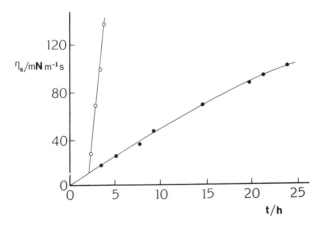

Figure 3 *Surface viscosity of caseinate + PGA at the n-hexadecane–water interface* (pH 3, 25 °C). *The apparent surface shear viscosity* η_s *is plotted against the adsorption time* t: ●, 10^{-3} wt% *caseinate from* t = 0 h; ○, 10^{-3} wt% *caseinate from* t = 0 h + 8 × 10^{-3} wt% *PGA introduced at* t = 2 h

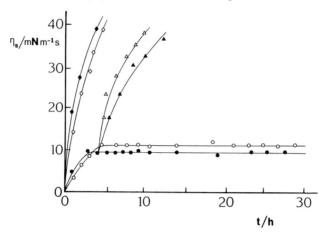

Figure 4   *Surface viscosity of PGA + caseinate at the n-hexadecane–water inter-face* (pH 3, 25 °C). *The apparent surface shear viscosity η_s is plotted against the adsorption time* t: $\bigcirc$, $\bullet$, 10^{-3} wt% *PGA from* t = 0 h; $\diamondsuit$, $\blacklozenge$, 10^{-3} wt% *PGA from* t = 0 h + 10^{-3} wt% *caseinate from* t = 0 h; $\triangle$, $\blacktriangle$, 10^{-3} wt% *PGA from* t = 0 h + 10^{-3} wt% *caseinate introduced at* t = 4 h. *The open and filled symbols denote results from duplicate sets of independent experiments*

viscosity of a film adsorbed at the oil–water interface from a 10^{-3} wt% solution of sodium caseinate at pH 3. The time-dependent surface shear viscosity was measured under standard conditions as described previously.[32] Following intro-duction of PGA after 2 hours into the aqueous sub-phase at a concentration of 8×10^{-3} wt%, the rate of increase of the surface viscosity increases sharply as compared with caseinate alone. The formation of a more highly structured film with much greater resistance to flow is attributed to strong complexation of protein and PGA at the hydrocarbon oil–water interface. Experiments in which PGA is introduced to the interface before caseinate, or both are introduced together, also provide evidence for strong casein–PGA complex formation. Figure 4 shows duplicate sets of data at pH 3 for (i) PGA alone (10^{-3} wt%), (ii) PGA (10^{-3} wt%) with caseinate (10^{-3} wt%) added after $t = 4$ h, and (iii) PGA (10^{-3} wt%) + caseinate (10^{-3} wt%) present together from $t = 0$. While PGA gives an adsorbed film of considerable viscosity in its own right, the presence of protein as well in the system leads to a much greater interfacial viscosity. These results for PGA + caseinate, together with similar behaviour for PGA + α-lactalbumin and PGA + whey protein isolate,[31] can be attributed to a strong electrostatic interaction between the anionic polysaccharide and the protein molecule having a net positive charge at pH < pI.

At neutral pH also, however, there is clear evidence for complex formation between protein and PGA. Figure 5 shows the effect on the surface viscosity of introducing PGA (10^{-3} wt%) into the aqueous sub-phase below a four hour film adsorbed from a solution of sodium caseinate (10^{-3} wt%). The protein + poly-

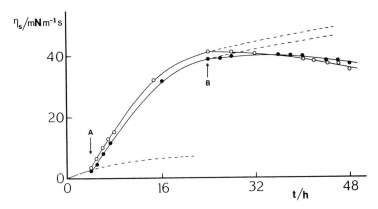

Figure 5   *Surface viscosity of caseinate + PGA at the n-hexadecane–water inter-*
face (pH 7, 25 °C). The apparent surface shear viscosity η_s is plotted
against the adsorption time t. The interface is exposed to 10^{-3} wt%
caseinate at t = 0 h. Arrow A denotes point of addition of 10^{-3} wt%
PGA at t = 4 h. Arrow B denotes point of dilution (1:5) of aqueous sub-
phase at t = 24 h. The open and filled symbols denote results from two
independent experiments. Dashed lines denote expected behaviour in
the absence of any addition or dilution

saccharide system definitely gives a higher surface viscosity than the protein (or
polysaccharide) alone; the viscosity is, however, much lower than under acidic
conditions, but similar to that obtained with a 10^{-3} wt% solution of gum arabic at
pH 7.[4] Even though both macromolecules carry the same net charge, electrostatic
interaction may still occur at pH 7 between the negatively charged polysaccharide
chain and positively charged patches on the protein. Electrostatic interaction may
be reinforced by hydrophobic association between PGA ester groups and non-
polar protein side-chains, as well as by hydrogen-bonding interactions between
—OH and —NH_2 groups on the two macromolecules. The complex formed at pH
7 must be reasonably strong, since it is not readily dissociated by extensive
dilution of the aqueous bulk phase, as indicated in Figure 5. After 24 hours, when
the caseinate–PGA complex surface viscosity was *ca.* 40 mN m^{-1} s and still rising
slowly, the sub-phase was subjected to a 1:5 dilution with respect to polymer
concentration using a procedure described previously.[17] The effect of the bulk
phase dilution was to arrest the increase in surface viscosity after a few hours, and
then possibly to reduce the viscosity (by *ca.* 10%) over a further 24-hour period.
This means that, if PGA desorption from the interface does occur, it happens only
slowly and to a limited extent.

   Additional evidence for interfacial casein–PGA association at neutral pH is
provided by particle electrophoresis measurements, as shown in Figure 6.
Emulsions containing 20 wt% n-tetradecane and 0.5 wt% caseinate in aqueous
phosphate buffer (0.5 M, pH 7) were prepared using the mini-homogenizer,[29] and
then centrifuged and washed (twice) to remove unadsorbed proteinaceous
emulsifier. After equilibration of emulsion droplets at high dilution for 24 h in

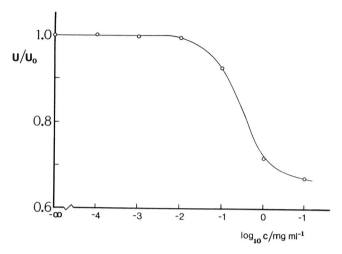

Figure 6   *Effect of PGA on the electrophoretic mobility of casein-coated n-tetradecane emulsion droplets (ionic strength 0.05 M, pH 7, 20 °C). The relative mobility u/u_0 is plotted as a function of the logarithm of the added polymer concentration c. The mobility of the droplets in the absence of polymer (c = 0) is $u_0 = 2.13 \times 10^{-8} \, V^{-1} \, m^2 \, s^{-1}$*

PGA solutions of known concentration at pH 7, the mobilities were determined by laser Doppler electrophoresis as described in previous papers.[33-35] The data in Figure 6 show that the introduction of casein-coated droplets into a 0.1 wt% solution of PGA leads to a reduction in electrophoretic mobility to a value *ca.* 70% of that measured in the absence of polysaccharide. This change in mobility is attributed to adsorption of PGA molecules from solution on to the casein-coated droplets to form a secondary stabilizing layer—such behaviour was previously observed[35] to an even stronger extent on addition of gelatin to milk protein-coated emulsion droplets.

4   Destabilization of Emulsions by Polysaccharides

Most non-ionic polysaccharides and many charged polysaccharides do not show any obvious signs of complex formation with food proteins over a wide range of solvent conditions. What we have in these protein + polysaccharide systems is either very weak association or net repulsive protein–polysaccharide interactions compared with protein–protein and polysaccharide–polysaccharide interactions. In protein-stabilized emulsions, this means either weak adsorption of polysaccharide at the droplet surface or depletion from the vicinity of the surface (*negative* adsorption). Either of these situations may lead to flocculation—in the former case by a bridging mechanism and in the latter case by a depletion mechanism.

High molecular weight water-soluble polymers (hydrocolloids) are added to suspensions and emulsions in order to act as gelling or thickening agents. That is, the processes leading to instability are inhibited or retarded by the change in rheology of the aqueous dispersion medium caused by the presence of the network-forming polymer. However, while the advantageous stabilizing effects of these polysaccharides are undoubtedly predominant at high polymer concentrations, there is usually an unfavourable destabilization at low concentrations. With casein-stabilized oil-in-water emulsions at neutral pH, this is manifest in a significantly enhanced creaming rate on addition of the non-ionic polysaccharide, dextran,[36] or the anionic polysaccharides, xanthan, carboxymethylcellulose, or succinoglycan.[37] Figure 7 shows the effect of xanthan on the time-dependent concentration–height profile of a casein-stabilized emulsion (15 wt% n-tetradecane, 0.75 wt% protein, pH 7).[38] Creaming profiles were determined at 20 °C over a period of several days by measuring the velocity of ultrasound through the emulsion at various heights within the sample using equipment similar to that developed by Robins and co-workers.[39,40] Figure 7a gives results for the original protein-stabilized emulsion, and Figure 7b gives results for the same emulsion containing 0.05 wt% xanthan. At time zero, the two samples have a uniform volume fraction $\phi = 0.2$ at all heights. We see from Figure 7a that, in the absence of any added polysaccharide, the volume fraction in the middle three-quarters of the sample still remains at a value very close to $\phi = 0.2$ after nine days of storage. At the bottom of the sample there develops a clear serum layer ($\phi \approx 0$) separated by a diffuse boundary from the main body of the emulsion; a concentrated cream layer ($\phi \approx 0.6$) develops at the top of the sample. We see from Figure 7b that, in the presence of 0.05 wt% xanthan, creaming is much faster, the moving boundary between the serum layer ($\phi \approx 0$) and the opaque emulsion is sharper, and the creaming is close to completion after nine days of storage. On increasing the polysaccharide concentration to 0.15 wt%, the creaming rate was found[38] to be similar to that with no added polysaccharide (*i.e.* like Figure 7a). On increasing the xanthan content to 0.25 wt%, no discernible creaming was detectable over the storage period (*i.e.* ϕ remains at 0.2 throughout the whole sample). Similar behaviour was observed at higher xanthan concentrations.

In contrast to caseinate + PGA discussed earlier, the system caseinate + xanthan shows no evidence for protein–polysaccharide complex formation, either at the planar oil–water interface (based on surface rheology) or at the emulsion droplet surface (based on particle electrophoresis). This means that the likely explanation for the enhanced creaming rate is depletion flocculation of the emulsion by non-adsorbing polymer.[1,4] We have also found enhanced creaming rates with other microbial polysaccharides, dextran[36] and succinoglycan,[37,38] neither of which shows any evidence for direct association with protein, or for adsorption at the surface of protein-stabilized droplets. In fact, the phenomenon is not just limited to emulsions containing protein and polysaccharide. It has been observed on addition of xanthan to droplets emulsified with surfactant (Brij 35)[39] and dextran to droplets emulsified with gum arabic.[17]

It is perhaps worth pointing out here that, though the actual term 'depletion flocculation' is relatively new, the phenomenon which it describes—enhancement of creaming or sedimentation by non-adsorbing polymer—has been recognized in the technological literature for well over 50 years, most especially in connection with the concentration of natural rubber latex (the milky sap from *Hevea*

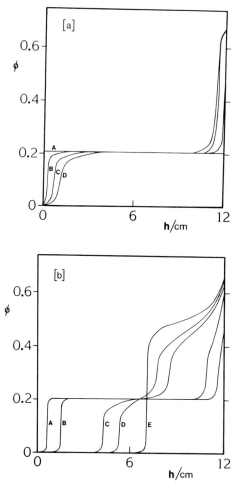

Figure 7   *Effect of xanthan on the rate of creaming of a n-tetradecane-in-water emulsion* (15 wt% *oil,* 0.75 wt% *caseinate,* pH 7) *as determined by ultrasonic velocity measurement at* 20 °C.[38] *The oil volume fraction ϕ is plotted as a function of height* h: (a) *emulsion with no added polymer stored for* (A) 0 h, (B) 72 h, (C) 168 h, *and* (D) 216 h; (b) *emulsion containing* 0.05 wt% *polymer stored for* (A) 18 h, (B) 43 h, (C) 127 h, (D) 154 h, *and* (E) 223 h

brasiliensis).[41] The enhanced rate of creaming was attributed[42] by Baker in 1937 to reversible flocculation of latex particles by hydrophilic colloids, and the following year in *Faraday Transactions*[43] Bondy postulated that flocculation of rubber latex particles or protein-stabilized emulsion droplets by sodium alginate is due to a non-adsorbing mechanism.

In the above discussion, it is assumed that there is no protein–polysaccharide association. Were there to be any tendency towards weak association, this would imply weak polysaccharide adsorption at the surface of protein-coated droplets, and hence enhanced creaming caused not by depletion flocculation but by bridging flocculation.[1,4] A recent computer simulation has shown[44] that, when the effective pair interaction between dispersed particle and polymer is relatively weak, either attractive or repulsive, it may be difficult to distinguish between flocs caused by polymer bridging and flocs arising from polymer depletion. Small changes in polymer structure or solvent conditions could easily tip the balance one way or the other. Either way the result is aggregation, which implies enhanced creaming or sedimentation at particle or polymer concentrations significantly below the gelation threshold. On this basis, we conclude that it is unlikely for small polysaccharide additions to an emulsion system *not* to induce flocculation of one sort or another. Whether the flocculation is due to depletion or bridging will depend on the delicate balance of protein–polysaccharide interactions in the particular system under investigation.

References

1. E. Dickinson and G. Stainsby, 'Colloids in Food', Applied Science, London, 1982.
2. E. Dickinson, in 'Food Structure—Its Creation and Evaluation', ed. J. M. V. Blanshard and J. R. Mitchell, Butterworth, London, 1988, p. 41.
3. E. Dickinson, B. S. Murray, and G. Stainsby, in 'Advances in Food Emulsions and Foams', ed. E. Dickinson and G. Stainsby, Elsevier Applied Science, London, 1988, p. 123.
4. E. Dickinson, in 'Gums and Stabilisers for the Food Industry', ed. G. O. Phillips, D. J. Wedlock, and P. A. Williams, IRL Press, Oxford, Vol. 4, p. 249.
5. L. G. Enriquez, J. W. Huang, G. P. Hong, N. A. Bati, and G. J. Flick, in 'Food Emulsifiers', ed. G. Charalambous and G. Doxastakis, Elsevier, Amsterdam, 1989, p. 335.
6. G. Stainsby, *Food Chem.*, 1980, **6**, 3.
7. A. Kato, T. Sato, and K. Kobayashi, *Agric. Biol. Chem.*, 1989, **53**, 2147.
8. H. G. Bungenberg de Jong, in 'Colloid Science', ed. H. R. Kruyt, Elsevier, Amsterdam, 1949, Vol. 1, p. 232.
9. F. A. Bettelheim, T. C. Laurent, and H. Pertoft, *Carbohydr. Res.*, 1966, **2**, 391.
10. T. H. M. Snoeren, T. A. J. Payens, J. Jeunink, and P. Both, *Milchwissenschaft*, 1975, **30**, 393.
11. A. P. Imeson, D. A. Ledward, and J. R. Mitchell, *J. Sci. Food Agric.*, 1977, **28**, 661.
12. C. J. van Oss, *J. Dispersion Sci. Technol.*, 1989, **9**, 561.
13. V. B. Tolstoguzov, in 'Functional Properties of Food Macromolecules', ed. J. R. Mitchell and D. A. Ledward, Elsevier Applied Science, London, 1986, p. 385.
14. E. Dickinson and G. Stainsby, in 'Advances in Food Emulsions and Foams', ed. E. Dickinson and G. Stainsby, Elsevier Applied Science, London, 1988, p. 1.
15. V. B. Tolstoguzov, V. Ya. Grinberg, and A. N. Gurov, *J. Agric. Food Chem.*, 1985, **33**, 151.
16. E. Shotton and K. Wibberley, *Boll. Chim. Farm. (Milan)*, 1961, **100**, 802.
17. E. Dickinson, D. J. Elverson, and B. S. Murray, *Food Hydrocolloids*, 1989, **3**, 101.
18. C. A. Street and D. M. W. Anderson, *Talanta*, 1983, **30**. 887.
19. D. M. W. Anderson, J. F. Howlett, and C. G. A. McNab, *Food Addit. Contam.*, 1985, **2**, 159.
20. G. B. Fincher, B. A. Stone, and A. E. Clarke, *Ann. Rev. Plant Physiol.*, 1983, **34**, 47.
21. M.-C. Vandevelde and J.-C. Fenyo, *Carbohydr. Polymers*, 1985, **5**, 251.
22. R. C. Randall, G. O. Phillips, and P. A. Williams, *Food Hydrocolloids*, 1988, **2**, 131.

23. E. Dickinson, B. S. Murray, G. Stainsby, and D. M. W. Anderson, *Food Hydro-colloids*, 1988, **2**, 477.
24. E. Dickinson, V. B. Galazka, and D. M. W. Anderson, in 'Gums and Stabilisers for the Food Industry', ed. G. O. Phillips, D. J. Wedlock and P. A. Williams. IRL Press, Oxford, 1990, Vol. 5, p. 41.
25. E. Dickinson, V. B. Galazka, and D. M. W. Anderson, *Carbohydr. Polymers*, in press.
26. E. Dickinson, V. B. Galazka, and D. M. W. Anderson, this volume, p. 490.
27. M. Nakamura, *Yukagaku*, 1986, **35**, 554.
28. E. L. Wilson, 'Studies of polyuronide–protein interactions', Ph.D. Thesis, University of Leeds, 1978.
29. E. Dickinson, A. Murray, B. S. Murray, and G. Stainsby, in 'Food Emulsions and Foams', ed. E. Dickinson, Royal Society of Chemistry, London, 1987, p. 86.
30. E. Dickinson and H. Randall, unpublished results.
31. E. Dickinson, A. Harrison, and J. Purewal, unpublished results.
32. E. Dickinson, B. S. Murray, and G. Stainsby, *J. Colloid Interface Sci.*, 1985, **106**, 259.
33. D. G. Dalgleish, E. Dickinson, and R. H. Whyman, *J. Colloid Interface Sci.*, 1985, **108**, 174.
34. E. Dickinson, R. H. Whyman, and D. G. Dalgleish, in 'Food Emulsions and Foams', ed. E. Dickinson, Royal Society of Chemistry, London, 1987, p. 40.
35. E. Dickinson, S. E. Rolfe, and D. G. Dalgleish, *Food Hydrocolloids*, 1989, **3**, 193.
36. S. Bullin, E. Dickinson, S. J. Impey, S. K. Narhan, and G. Stainsby, in 'Gums and Stabilisers for the Food Industry', ed. G. O. Phillips, D. J. Wedlock, and P. A. Williams, IRL Press, Oxford, 1988, Vol. 4, p. 337.
37. Y. Cao, E. Dickinson, and D. J. Wedlock, *Food Hydrocolloids*, 1990, **4**, 185.
38. Y. Cao, E. Dickinson, and D. J. Wedlock, unpublished results.
39. D. J. Hibberd, A. M. Howe, A. R. Mackie, P. W. Purdy, and M. M. Robins, in 'Food Emulsions and Foams', ed. E. Dickinson, Royal Society of Chemistry, London, 1987, p. 219.
40. P. A. Gunning, D. J. Hibberd, A. M. Howe, and M. M. Robins, *J. Soc. Dairy Technol.*, 1989, **42**(3), 70.
41. I. Traube, *Gummi Ztg*, 1925, **39**, 434.
42. H. C. Baker, *Trans. Inst. Rubber Ind.*, 1937, **13**, 70.
43. C. Bondy, *Trans. Faraday Soc.*, 1939, **35**, 1093.
44. E. Dickinson, *J. Colloid Interface Sci.*, 1989, **132**, 274.

Function of α-Tending Emulsifiers and Proteins in Whippable Emulsions

By J. M. M. Westerbeek

PRODUCT RESEARCH AND ANALYSIS LABORATORY, DE MELKINDUSTRIE
VEGHEL, P.O. BOX 13, 5460 BA VEGHEL, THE NETHERLANDS

and A. Prins

DEPARTMENT OF FOOD SCIENCE, WAGENINGEN AGRICULTURAL UNIVERSITY,
P.O. BOX 8129, 6700 EV WAGENINGEN, THE NETHERLANDS

1 Introduction

Whipped emulsions should always have solid-like properties, because the aerated product has to be stable against flow for a long period of time. In the case of whipped products, a yield stress, caused by the presence of a particle network, is very often the way to prevent the product from flowing. The formation of such a network is frequently induced by the addition of small-molecule emulsifiers. Lipophilic, so-called α-tending emulsifiers, such as propylene glycol monostearate (PGMS), acetylated monoglycerides (ACTM), or lactylated monoglycerides (GLP), are especially effective in promoting aggregation of fat globules.[1,2] These α-tending emulsifiers may be characterized by the fact that they are non-polymorphic, they can only exist in the α-crystalline form below the melting point of the hydrocarbon chains, and they are practically insoluble in water.[2,3]

It is not understood what mechanism is responsible for this extensive fat particle aggregation phenomenon. Recently, Buchheim and Krog[4] suggested that clumping may well be the mechanism which causes the occurrence of fat particle aggregation in whippable emulsions. It is indeed true that crystallization phenomena play an important role in the formation of a particle network in these emulsions. As in whipped cream, the network probably consists of partly crystallized oil droplets. However, since whippable emulsions which contain relatively large amounts of α-tending emulsifier do not churn during whipping,[5] it seems likely that another mechanism is responsible for the instability of these emulsions.

It is well known that mixtures of water and emulsifiers like phospholipids[6,7] or monoglycerides[8,9] may form mesomorphic phases.[10] Above the crystallization

147

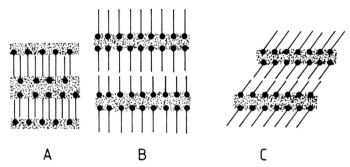

A   B   C

Figure 1   *The schematic structure of three possible lamellar gel phases formed by surfactants in contact with water below the crystallization temperature of their hydrocarbon chains*

temperature of the hydrocarbon chains of the emulsifier, a lamellar phase, a cubic phase, or an hexagonal phase may be formed. When the systems are cooled down, a lamellar gel phase may be formed. This gel phase is characterized by a lamellar structure of alternating layers of emulsifier molecules and water molecules.[10,11] The lipid molecules are crystallized in the α-polymorphic form. In Figure 1 three types of α-gel phase structures are represented.[10]

In the literature, it has often been stated that α-tending emulsifiers do not show lyotropic mesomorphism.[12–14] However, the physical properties of GLP + water mixtures and glycerol monostearate + water mixtures are very similar below the crystallization temperature of the hydrocarbon chains of the lipid molecules.[1,15] Depending on the amphiphile concentration, both these binary systems gel under these conditions, and they both show excellent whipping properties.

Our hypothesis is that α-tending emulsifiers are able to form an α-gel phase with water at the interface of fat particles below the crystallization temperature of the hydrocarbon chains of the emulsifier. The aggregated fat particles are linked to each other by means of this gel phase. Proteins like sodium caseinate are essential in order to be able to prepare first relatively stable emulsions at a temperarure between 60 and 80 °C. However, the proteins should at least partly desorb from the oil–water interface below the crystallization temperature of the emulsifier.

In this paper, we present a study on the physical behaviour of glycerol lactopalmitate (GLP), as an example of an α-tending emulsifier. Our aim is to propose a satisfactory mechanism for structure formation in emulsions containing an α-tending emulsifier. The most appropriate techniques to study lipid polymorphism and mesomorphic behaviour are a combination of small and wide angle X-ray diffraction (SAXD and WAXD), and differential scanning calorimetry (DSC). Furthermore, neutron diffraction has proved to be a useful technique here for the structural analysis of mesomorphic phases, as was the case previously for phospholipids.[15–17]

2 Experimental

Materials—Glycerol lactopalmitate (GLP) was purchased from Grindsted, Denmark. The sample is a very complex mixture of mono-, di-, and triglycerides of lactic and palmitic acids. According to the supplier, the product contains *ca.* 15% lactic acid. Sodium caseinate was obtained from De Melkindustrie Veghel, The Netherlands. This sample is a spray-dried milk protein in powder form containing 94.5 wt% protein (N × 6.38) on a moisture-free basis, 5.2 wt% moisture, 4.1 wt% ash and 0.8 wt% fat. Glucose syrup was obtained from Cerestar, The Netherlands. It is a combined acid/enzyme-hydrolysed corn starch with a mean DE = 35.

Preparation of Spray-dried GLP Powders—Prior to spray-drying, a concentrated emulsion of GLP particles in water was prepared containing 30 wt% GLP, 5 wt% sodium caseinate, 15 wt% glucose syrup, and 50 wt% demineralized water. The dry protein powder was dispersed directly in the melted emulsifer at 70 °C, and the protein/emulsifier mixture was added to the water at 70 °C. Subsequently, this dispersion was homogenized at a constant pressure of 100 bar at 70 °C in a high-pressure homogenizer (Rannie, 100 l h^{-1}). The emulsion was spray-dried using an A/S NIRO Atomizer as described previously,[18] and the powders were stored at room temperarure for DSC and diffraction studies.

Preparation of Freeze-dried GLP Powders—Melted GLP was mixed with demineralized water at a temperature of about 60 °C using a Sorvall mixing apparatus. During the mixing, the sample was gradually cooled to a temperature below the crystallization point of the emulsifier mixture. The resulting GLP gel was freeze-dried and stored at room temperature.

Differential Scanning Calorimetry—DSC was performed with a Mettler TA-3000 system. In the DSC-cup, the sample amount varied between 10–20 mg, depending on the concentration of potentially crystallizing matter present in the sample. The heating curves were not corrected for differences in the amount of sample in the DSC-cup. Therefore, the peak areas below the curves cannot be interpreted as being representative of the heat values per unit amount of sample.

X-Ray Diffraction—SAXD measurements were performed with the spray-dried GLP powder samples. Part of the samples was dispersed in demineralized water the day before carrying out the X-ray diffraction experiments. The SAXD measurements were conducted using a Kratky camera, manufactured by A. Paar. The camera was equipped with a Braun one-dimensional position-sensitive detector connected to a Braun multi-channel analyser. The radiation source was a PW-1729 X-ray generator, producing Ni-filtered Cu K$_\alpha$-rays (wavelength 1.54 Å). Each channel of the multi-channel analyser corresponded to a certain diffraction angle. The samples were put into small glass capillaries (diameter 1.0 mm, wall thickness 0.01 mm). Calibration of the apparatus was carried out with lead stearate. A set of measurements was usually completed within 30 minutes. Corrections were made for background noise, and the curves were subsequently desmeared according to a program based on Lake's theory.[19]

Neutron Diffraction—Neutron diffraction experiments were performed at ISIS (Rutherford Appleton Laboratory, Didcot, England). The small-angle diffraction measurements were conducted with the LOQ spectrometer. The spallation neutron source, ISIS, produces intense neutron bursts 50 times per second by means of collisions of a highly energetic, pulsed proton beam with a uranium target. The fast neutrons from the target station are slowed down in a hydrogen moderator at a working temperature of 25 K in order to obtain a cold neutron enhanced thermal neutron spectrum. A rotating disk chopper at 25 Hz removes alternate pulses from ISIS to avoid frame overlap from adjacent pulses. Neutrons of wavelengths varying from 2 to 10 Å are recorded by a two-dimensional position sensitive neutron detector. The area detector is a multiwire (128×128), $^{10}BF_3$ filled, proportional detector. It has 64×64 channels, its resolution being 1 cm in both directions, each one of them containing about 80 time-of-flight channels. They are all handled by an in-house data acquisition system and a Microvax computer. Reduction of the raw LOQ time-of-flight data to a composite cross section $I(Q)$ is done by accurate transmission corrections over a wide range of wavelengths.

Dry and hydrated GLP powder samples, containing variable amounts of D_2O were introduced into circular shaped quartz cuvettes. The weight of the samples inside the cells was determined in order to be able to calculate the average cross section of each sample. The cell thickness could be either one or two millimetres, depending on the amount of D_2O, and the amount of air present in each gel sample. The measurement lasted between one and three hours for each sample depending upon whether 1×10^6 or 3×10^6 counts were required for acceptable data statistics.

3   Results

Diffraction Studies—The existence of an α-gel phase can readily be established by diffraction techniques. The swelling of the crystal lattice caused by the uptake of water should clearly be apparent from an increase of the long spacing which is easily measured by SAXD. The swelling of crystallized GLP was studied by means of hydration experiments with spray-dried GLP samples. To this end, the powder, containing crystallized GLP particles with an average size smaller than 1 μm, was brought into contact with demineralized water at room temperature. After about 24 hours, we performed SAXD experiments on both the dry and the wet samples. In Figure 2, two examples of SAXD spectra are represented. Comparing the two diffraction spectra, it is obvious from the increase of the long spacing that some hydration has indeed occurred at room temperature.

Comparable information may also be obtained with neutron diffraction. In principle, water molecules can be located within the structure of a multilayer if hydration is performed with D_2O.[15,17] In order to obtain improved knowledge of the hydration properties of GLP, we performed neutron diffraction experiments

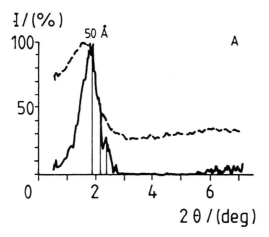

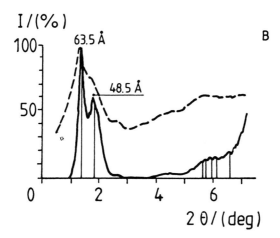

Figure 2 *SAXD curves obtained at* 20 °C *for different GLP samples. Both smeared* (– – –) *and desmeared* (——) *curves are represented. Intensity* I *is plotted against twice the scattering angle* θ: A, *dry, spray-dried GLP powder sample*; B, *hydrated, GLP powder sample containing 67 wt% water*

on three different types of GLP samples which were first hydrated with variable amounts of D_2O. The first series was prepared from the spray-dried GLP powder. These samples were hydrated at ambient temperature. The second series of experiments was performed with a freeze-dried GLP powder sample, free of any other additional component. This powder sample was also hydrated at ambient temperature. The third series of samples consisted of GLP gels, which were obtained by heating variable amounts of D_2O and GLP to a temperature of about 60 °C in sealed bottles, followed by cooling to ambient temperature under

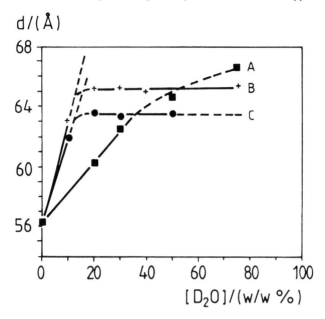

Figure 3   *The long spacing* d *of three different GLP samples as function of the weight percentage deuterium oxide as determined by neutron diffraction at ambient temperature*: A, *spray-dried GLP powder containing* 60 wt% *GLP*; B, *freeze-dried GLP powder containing* 100 wt% *GLP*; C, *melted GLP*

vigorous mixing. The differences between the three samples were thus related to composition, average particle size, and the temperature at which the samples were hydrated with D_2O.

The results of the neutron diffraction measurements are shown in Figure 3. The measured long spacing of the gel phase of GLP is plotted as a function of the D_2O concentration. It is obvious that the values for the long spacings of the fully hydrated gel phases depend on the preparation method. The differences are not due to experimental error, because this figure clearly shows that the values of the long spacing for samples B and C are quite constant at high D_2O concentrations. Our explanation for this effect is that fractionation of specific GLP components at the interface of the particles probably causes differences in the composition of the gel phase, which, in turn, may cause differences in the value of the long spacing at the point of maximum hydration.

Temperature Dependence—A proper technique to detect phase transitions in a system like a mixture of an α-tending emulsifier and water is differential scanning calorimetry. The influence of water on the phase behaviour of α-tending emulsifiers has been studied by means of both cooling and heating experiments.[5] In Figure 4, the influence of the presence of water on the melting properties of spray-dried GLP is shown. It is obvious that the complete melting curve is shifted a few

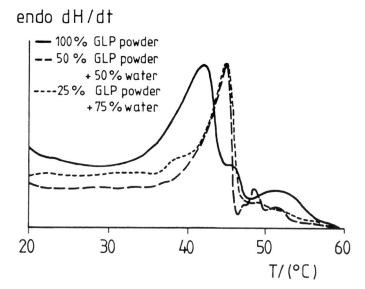

Figure 4 *DSC heating curves of completely crystallized dry and hydrated spray-dried emulsifier powders (heating rate = 1 °C min⁻¹)*

degrees towards a higher temperature when the GLP particles are brought into contact with water at ambient temperature. This experiment confirms our hypothesis of hydration of α-tending emulsifiers below the crystallization temperature of the hydrocarbon chains of the amphiphilic components.

The heating curves suggest that a large part of the emulsifier present participates in the hydration process, because the shift observed almost accounts for the complete melting curve. This effect was not expected to occur so explicitly. If the gel phase were to consist of only part of the total emulsifier sample, it would have been likely that the melting peak of the hydrated emulsifier, in comparison with the dry sample, would have been broader, or would even have been split into two peaks. Instead, the endothermic peaks of the hydrated samples are significantly sharper than those of the dry samples, which may be indicative of a better fitting of molecules with bulky head groups into the crystal lattice of the α-gel phase.

Finally, we present data, based on temperature-dependent SAXD measurements, from which it can be proved that the long spacings are indeed related to a gel phase structure. In Figure 4, we see that the main GLP fraction of the hydrated sample melts at a temperature of about 46 °C, whereas the dry sample melts at a temperature of about 43 °C. Figure 5 shows SAXD curves of the hydrated GLP powder at different temperatures. (These curves are not corrected in such a way that the peak heights can be compared with each other.) The long spacing of about 62 Å disappears with increasing temperature, indicating that the α-gel phase melts over a wide temperature range. At 47 °C the long spacing is no longer present in the SAXD spectrum. Furthermore, this experiment gives evidence for the previous assertion that GLP does not show mesomorphism above the melting

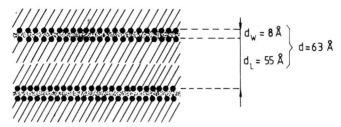

Figure 6   *A schematic molecular model for the α-gel phase of glycerol lactopalmi-tate hydrated above the crystallization temperature of the emulsifier, showing thicknesses of water (d_W) and lipid (d_L) layers ($d = d_W + d_L$)*

other hand, the melting temperature of monoglycerides[22] or phospholipids[25] decreases in the presence of water. It is obvious that the melting temperature of crystallized hydrocarbon chains of an amphiphile depends not only on the water concentration and the hydrocarbon chain length, but also on the nature of the polar head-group. In our opinion, hydration of such a group may cause either an increase or a decrease in melting temperature of the chains, dependent on whether hydration leads to an increase or a decrease in the lattice energy of the crystalline amphiphile. In the case of GLP, a decrease in steric head-group repulsion between molecules in adjacent layers, and also the formation of hydrogen bonds, may play an important role in the raising of the melting point.

What does this phase behaviour of GLP mean for the stability of emulsions which contain both protein, like sodium caseinate, and an α-tending emulsifier? In the case of GLP-containing emulsions, we suggest a temperature-dependent model for the structure of the oil–water interface of the emulsion droplets as represented in Figure 7. At a temperature exceeding the melting of the amphi-phile mixture, the emulsion is stabilized mainly by the proteins, and perhaps also by the relatively more hydrophilic components of the added emulsifier. Depen-dent on its concentration, below the melting point of the emulsifier an α-gel phase is formed at the oil–water interface. It is likely that the stabilizing proteins will spontaneously become partly or completely desorbed from the oil–water interface,[26] and that flocculation will then occur.

The particles do not coalesce because of the stabilizing hydration force exerted between droplets by the hydrated emulsifier molecules. The last ten years has seen greater insight into hydration forces.[27] It appears that hydration forces are only active over a very short range (at most *ca.* 30 Å). When electrostatic repulsion is of no importance, as with the case of neutral amphiphiles or ionic ones at high salt levels, surfactants may form mesomorphic structures as a result of van der Waals forces. It is believed that the van der Waals forces are counterbalanced by the short-range hydration forces. In this respect, it is significant to mention that it has been shown that hydration forces may form a strong barrier to both bilayer aggregation and fusion of phospholipid membranes.[28,29] Presumably the α-gel phase of GLP is also stabilized by this repulsive hydration force. Hence, it is very likely that the flocculated particles in a whipped emulsion, containing a relatively

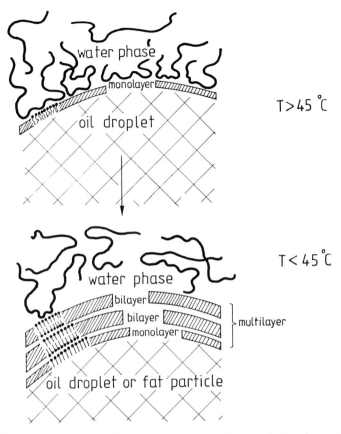

Figure 7 *Schematic model of the temperature-dependent α-gel phase formation at the oil–water interface of fat particles in an emulsion which contains both sodium caseinate and GLP*

high level of GLP in relation to fat, are prevented from coalescing by these repulsive hydration forces.

Acknowledgements. The authors gratefully acknowedge the help provided by P. Aarts (Unilever, Vlaardingen) in carrying out the SAXD experiments, Dr J. van Tricht and Dr T. Rekveldt (IRI, Delft) for their advice and help with the neutron diffraction, and J. van Rijswijck (DMV-Veghel) in performing the DSC measurements.

References

1. N. Krog, *J. Am. Oil Chem. Soc.*, 1977, **54**, 124.
2. J. Andreasen, *Deutsche Molkereizeitung*, 1981, **36**, 1161.

3. N. Krog, in 'Water Relations of Foods', ed. R. B. Duckworth, Academic Press, London, 1975, p. 587.
4. W. Buchheim, N. M. Barfod, and N. Krog, *Food Microstructure*, 1985, **4**, 221.
5. J. M. M. Westerbeek, 'Contribution of the α-gel phase to the stability of whippable emulsions', Ph.D. Thesis, University of Wageningen, 1989.
6. V. Luzatti, in 'Biological membranes', ed. D. Chapman, Academic Press, London, 1968, p. 71.
7. R. M. Williams and D. Chapman, in 'Progress in the Chemistry of Fats and Other Lipids', ed. R. T. Holman, Pergamon Press, Oxford, 1970, Vol. 11, p. 3.
8. E. S. Lutton, *J. Am. Oil Chem. Soc.*, 1965, **42**, 1068.
9. N. Krog and K. Larsson, *Chem. Phys. Lipids*, 1968, **2**, 129.
10. G. J. T. Tiddy, *Phys. Rep.*, 1980, **57**, 1.
11. J. M. Vincent and A. Skoulios, *Acta Crystallogr.*, 1966, **20**, 432.
12. J. Andreasen, 'The efficiency of emulsifiers in whipped topping', Lecture at the international symposium on emulsions and foams in Food Technology, Ebeltoft, Denmark, 1973.
13. J. V. Boyd, N. Krog, and P. Sherman, in 'Theory and Practice of Emulsion Technology', ed. A. L. Smith, Academic Press, London, 1976. p. 99.
14. G. Schuster, 'Emulgatoren für Lebensmittel', Springer-Verlag, Berlin, 1985.
15. G. Büldt, H. U. Gally, J. Seelig, and G. Zaccai, *J. Mol. Biol.* 1979, **134**, 673.
16. D. L. Worchester, in 'Biological Membranes', ed. D. Chapman and D. F. H. Wallach, Academic Press, London, 1976, Vol. 3, p. 1.
17. D. L. Worchester and N. P. Franks, *J. Mol. Biol.*, 1976, **100**, 359.
18. N. M. Barfod and N. Krog, *J. Am. Oil Chem. Soc.*, 1987, **64**, 112.
19. J. A. Lake, *Acta Crystallogr.*, 1967, **23**, 191.
20. N. Krog and A. P. Borup, *J. Sci. Food Agric.*, 1973, **24**, 691.
21. N. H. Kurt and R. A. Broxholm, US Pat. 3 388 999, 1968, Eastman Kodak Co., Rochester.
22. N. Krog and J. B. Lauridsen, in 'Food Emulsions', ed. S. Friberg, Marcel Dekker, New York, 1976, p. 67.
23. A. S. C. Lawrence, M. A. Al-Mamum, and M. P. McDonald, *Trans. Faraday Soc.*, 1967, **63**, 2789.
24. T. de Vringer, 'Physicochemical aspects of lamellar gel structures in nonionic O/W creams', Ph.D. Thesis, University of Leiden, 1987.
25. H. Hauser, in 'Reversed Micelles', ed. P. L. Luisi and B. E. Straub, New York, 1984, p. 37.
26. N. Krog, N. M. Barfod, and W. Buchheim, in 'Food Emulsions and Foams', ed. E. Dickinson, Royal Society of Chemistry, London, 1987, p. 144.
27. J. N. Israelachvili, *Chem. Scr.*, 1985, **25**, 7.
28. V. A. Parsegian, N. Fuller, and R. P. Rand, *Proc. Natl. Acad. Sci. USA*, 1979, **76**, 2750.
29. L. J. Lis, M. McAlister, N. Fuller, R. P. Rand, and V. A. Parsegian, *Biophys. J.*, 1982, **37**, 657

Importance of Peptides for Food Emulsion Stability

By Zahur U. Haque

DEPARTMENT OF DAIRY SCIENCE AND SOUTHEAST DAIRY FOODS RESEARCH
CENTER, MISSISSIPPI AGRICULTURAL AND FORESTRY EXPERIMENT STATION,
MISSISSIPPI STATE UNIVERSITY, MISSISSIPPI 39762, USA

1 Introduction

The modern food industry is being constantly influenced by the informed public, and is becoming more and more dependent on food functional ingredients for food formulation. Because of rigorous requirements, the functionality of these ingredients has to be consistent.

Milk is a good source of food functional ingredients. It is rich in peptides that constitute about 2–6 wt% of total protein in bovine milk.[1] Normally these are produced by endogenous proteolytic activity. However, refrigerated storage ($<7\,^{\circ}$C) is optimum for the selective growth of psychotropic bacteria which release the extracellular proteases that produce peptides.[2] Even though much has been done to study the organoleptic properties of milk peptides,[3,4] there is a need for more information about the physico-chemical properties of these peptides and how they influence the functionality of the total milk protein. Recently, we have observed that crude peptide mixtures from milk improve the emulsifying activity of β-lactoglobulin and that delipidated total milk protein was negatively affected.[5]

The present study is an effort to understand more fully the influence of milk peptides, produced by controlled enzyme hydrolysis, on the functionality of milk proteins.

2 Experimental

Materials—Trypsin (EC 3.4.21.4) and chymotrypsin (EC 3.4.21.1) were purchased from Calbio-chem Company (La Jolla, CA). Rhozyme-41 (RH-41) (a mixture of various proteolytic enzymes used for accelerated cheese maturation) was obtained from Genencor, Inc. (San Francisco, CA). Aminopropyl glass beads (pore size 500 Å) and glutaraldehyde (25 wt% solution) were purchased from Sigma Chemical Company (St. Louis, MO). VHR–PAGE gradient gel (10–20%) (cal #GG50200) of polyacrylamide containing SDS was purchased from Geltech (Salem, OH). Dye for protein dye binding assay was from Bio-Rad Chem. Div. (Richmond, CA). All other reagents were analytical grade.

Whey protein concentrate (WPCX3) was obtained from fresh sweet whey (Edam) by ultra-filtration using a Romicon commercial scale unit (10 K cut) at the MSU Dairy Plant. The protein content as estimated by the Kjeldahl method (N to protein conversion factor = 6.38) was 18.7 wt%. Casein was made from fresh skim milk by isoelectric precipitation. The protein content determined by the same method was 86.2 wt%.

Preparation of Immobilized Enzymes—Trypsin and chymotrypsin in 0.05 M phosphate buffer, pH 6.0, were immobilized by adsorption of the enzymes on to aminopropyl glass beads for 16 h (4 °C) and then cross linked with glutaraldehyde (final concentration, 3 vol%) for 2 h at 4 °C.[6] The immobilized-trypsin (IM-TRY) and immobilized-chymotrypsin (IM-CHY) thus produced were kept wet at 4 °C until needed.

Determination of Immobilized Enzyme Activity—A measured amount of the wet immobilized enzyme (IME) was incubated with 0.5 wt% casein that had been heated in a boiling water bath for 15 minutes to inactivate the endogenous proteolytic enzymes and for better dispersibility of casein in 0.05 M Tris–HCl buffer, pH 8.0 (assay buffer) at 37 °C with gentle stirring. At various times, 2 ml of the reaction mixture was removed, deproteinized by adding 3 ml of 5 wt% trichloroacetic acid (TCA), kept one hour at room temperature and filtered. Absorbance of the filtrate was read at 280 nm (Computerized Spectronic 1201 dual beam spectrophotometer, Milton Roy Company, USA) against water. The IM-Enzyme unit was defined as the increase in absorbance at 280 nm of TCA soluble products, using casein as substrate, per minute per mg of immobilized enzyme at pH 8.0 and 37 °C.

Hydrolysis of Casein—Casein (10 ml of 0.5 wt%) was dispersed in the assay buffer and placed along with the IME in a batch reactor at 37 °C. The IM-enzyme activity (as defined above) was kept constant at 0.051 units. Aliquots of the hydrolysate (0.5 ml) were removed at given time intervals (from 5 to 120 min).

Hydrolysis of Whey Protein Concentrate—The WPCX3 was hydrolysed using a batch reactor as mentioned above for the casein hydrolysis. The total volume was 10 ml containing 1 g WPCX3, *i.e.* 187 mg protein (see materials section). The activity of the IM-enzymes was kept constant at 0.0004 units (as defined above). The method of monitoring was as given below.

Continuous Reaction with Immobilized Enzymes for the Production of Functional Peptides—The continuous hydrolysis of casein with immobilized enzymes[7] was accomplished using a Diaflo cell (Amicon) of 10 ml volume, with Diaflo VF YM-100 membrane, as a bio-reactor. Stirring in the cell was accomplished with a magnetic stirrer suspended from the top of the cell. Casein dispersed in the assay buffer was continuously fed at 4 °C through the inlet of the reactor and the effluent was collected from the bottom. Monitoring was effected by studying the absorbance at 280 nm after TCA precipitation of the effluent at different time intervals as mentioned above, by VHR-PAGE and HPLC as explained below. Continuous production of various functional peptides was carried out by changing the retention times of the substrate in the reactor. The enzyme activity was kept constant at 0.02 units.

Monitoring of Hydrolysis—Very high-resolution polyacrylamide gel electrophoresis (VHR-PAGE) was carried out in the presence of SDS using a 10–20% gradient gel.[7]. High performance liquid chromatography (HPLC) of the hydrolysates was performed with a Bondaclone 10 C_{18} Column (60 cm × 0.7 cm) (Phenomenex, CA) as described elsewhere.[8] Samples were eluted with a linear gradient where buffer A was 0.11 wt% trifluoroacetic acid in double distilled water, and B was 90 wt% acetonitrile in water containing 0.1 wt% trifluoroacetic acid. An ISCO ChemResearch System containing a model 2360 gradient programmer (ISCO, Inc. USA) coupled with pressure regulators (Model 2350), V^4 absorbance detector (ISCO), and refrigerated ISIS autoinjector, was used for the separation and for monitoring the hydrolysis.

Determination of Emulsifying Activity Index (EAI)—Emulsifying activity of the casein and its hydrolysates was determined as described earlier.[9] The protein dispersion (0.8 ml) was mixed with peanut oil (0.2 ml) and sonicated for 20 s over cold water to maintain temperature at around 25 °C (Sonics and Materials, model C1A; 600 W, r.m.s.; titanium tip diameter 1.2 cm; attenuated to 50% power output). Turbidity of the emulsion was measured at 600 nm after dilution (×1000) with 10 mM imidazole buffer, pH 7.0, containing 0.1 wt% SDS. The result was expressed as surface area per unit mass of proteins used in preparing the emulsion.

Emulsion Stability (ES)—The comparative stability of the different emulsions was determined following centrifugation of the freshly prepared emulsion. The emulsion (1.0 ml) was centrifuged in 1.5 ml tapered micro-tubes using a Millipore Personal Centrifuge at 2000g (6400 r.p.m.) for 30 min (25 °C). The apparent emulsion stability was calculated by subtracting the volume of the 'drainage' from the initial volume of the emulsion.[8] Protein concentration was estimated according to Bio-Rad Protein Assay method (dye binding).

Calculated Surface Load—The surface load was obtained as described elsewhere.[10] Surface area, as determined from the spectrophotometric data, was calculated for unit mass of the hydrolysate in the interface. The amount of hydrolysate in the interface was quantitated by difference following the determination of the protein content in the serum layer after centrifugation at 2000g for 30 min at 25 °C.

Surface Energy (SE)—The surface energy was calculated according to the method of van Oss *et al.*[11] which employs an extended version of the Young equation. The contact angle for the determinations was obtained using a Contact-Angle Viewer as described elesewhere.[8] The liquids (of known properties) that were used for calibration were water and alpha-bromonaphthalene.[11]

3 Results and Discussion

The hydrolysis of the whey protein that had been three-fold concentrated (WPCX3) was slow and incomplete under the conditions employed. The HPLC data indicate that, within eight hours, the extent of hydrolysis of WPCX3 by IM-CHY is greater than by IM-TRY (Figure 1). In terms of the emulsifying properties (at 1 wt% solids), IM-TRY depressed the emulsifying activity (EAI) and initially (<2 h) decreased the emulsion stability (ES) as well (Figure 2). After eight hours of incubation in the batch reactor, the ES value was found to be markedly increased to 41% from 23% after 30 minutes. The activity at this stage was at a minimum. The calculated load appeared inversely related to the EAI.

For the case of IM-CHY, the functionality improved to an optimum level following four hours of incubation (Figure 3). Even though there was an improvement in the emulsifying activity, the most noticeable improvement was in stability. During the initial stages of the hydrolysis (<2 h), the ES value almost quadrupled after one hour to 63% from 16% in the control. The 'optimum' degree of hydrolysis, *i.e.* the point at which both the emulsifying activity and emulsion stability were at their highest values, was at *ca.* four hours of incubation. Again, the estimated load was low when the activity was high, and *vice versa*. In analysing these data, it should be noted that the protein content in the WPCX3 is

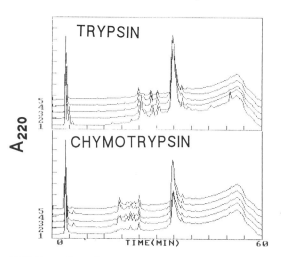

Figure 1   *HPLC profiles of WPCX3 at different stages of hydrolysis in a batch reactor at 4 °C. The abscissa is the retention time in the HPLC column and the ordinate is the relative intensity A of detection at 220 nm. The smaller peaks starting around 25 min represent the peptides. The HPLC was carried out at 25 °C using a linear gradient buffer system. Numbers 1–5 represent HPLC runs of the samples hydrolysed for 1/2, 1, 2, 4, and 8 hours, respectively*

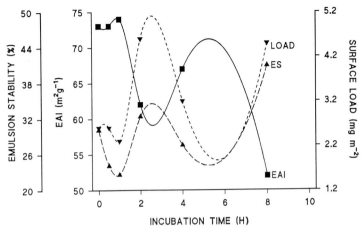

Figure 2 *The functional properties of the whey protein concentrate following batch hydrolysis using immobilized trypsin. The abscissa represents the incubation time with a constant enzyme activity of 0.0004 units. Plotted along the ordinate are the emulsifying activity index (—), the calculated surface load (– – –), and the emulsion stability (— — —). The smoothing regression curve connecting the data points is cubic*

low $(1.8\ mg\ ml^{-1})$ (see materials section), and that the formation of a thick interface further limits the amount of protein that can spread and stabilize the interface.

The surface energy was calculated in order to study its relationship to emulsifying activity and emulsion stability (Figure 4a). IM-TRY was used with a short incubation period to test the effect of small amounts of peptides. There was a decrease in the surface energy with incubation time, at all the WPCX3 concen-

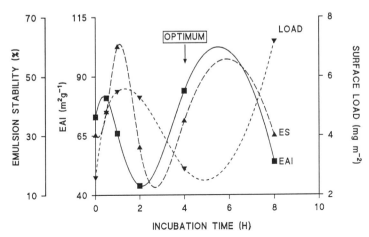

Figure 3 *As Figure 2, except with immobilized chymotrypsin*

(*a*)

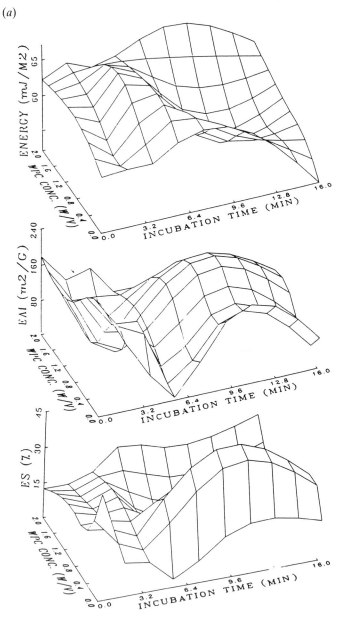

Figure 4   *The surface property of whey protein concentrate (WPCX3) following
short-time controlled hydrolysis using* (a) *immobilized trypsin and* (b)
immobilized rhozyme 41. *In each of the three-dimensional plots, two of
the axes represent the incubation time in a batch reactor and the
concentration of whey protein concentrate. The vertical axis varies in the*

(b)

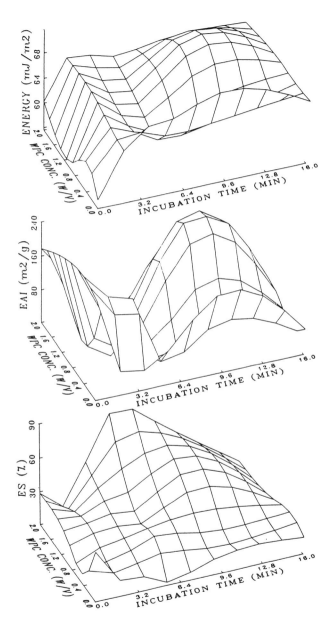

different plots as follows: *surface energy* (*top*), *emulsifying activity index* (*middle*), *and emulsion stability* (*bottom*). *The plots are fitted using spline functions. The proteolytic activity in each case is constant at* 0.0004 *units for* 10 ml of *WPCX3* (187 mg *protein*)

is unnecessary. Timely stoppage of the reaction is necessary because, as observed, even limited hydrolysis causes marked changes in functionality. In general, IM-TRY produces peptides that enhance emulsifying activity, whereas IM-CHY causes hydrolysis which is more beneficial for the stability of the emulsions. Extensive hydrolysis in all cases causes an increase in the surface load. However, the interface is very 'loose', leading to decreased stability.

Walstra and Oortwijn[13] have reported that the Tween 20 depresses the emulsifying property of proteins by displacing them from the interface. This effect, however, was found to vary between emulsifiers. The present observations may partly be attributed to a similar phenomenon. However, in a closely packed interface, that is stabilized by strong protein–protein interactions, the movement of the peptides to the interface (and consequent displacement of the proteins) is conceivably more difficult than when the interface is loosely packed (*i.e.* in the presence of hydrophobic interaction disrupters). Based on personal observation and published data,[14,15] the author is of the opinion that protein–protein hydrophobic interactions, which play a major role in protein-stabilized emulsions, are hampered by small soap-like peptides. Displacement of the macromolecular layer may follow the initial loosening of the layer.

Acknowledgements. The work was supported by a grant from the Southeastern Dairy Foods Research Center. This article is approved as Journal Series #BC-7553 of the Mississippi Agricultural and Forestry Experiment Station. The author appreciates the hard work and expert technical support of Mr L. G. Bohoua and Dr Zahid Mozaffar.

References

1. C. V. Morr, in 'Functionality and Protein Structure', ed. A. Pour-El, ACS Symposium Series, American Chemical Society, Washington, D. C., 1979, Chap. 4.
2. B. A. Law, *J. Dairy Sci.*, 1979, **46**, 573.
3. J. Edwards and F. V. Kosikowski, *Milchwissenschaft*, 1987, **42**, 139.
4. S. Visser, G. Hup, F. A. Exterkete, and J. Stadhouders, *Neth. Milk Dairy J.*, 1983, **37**, 169.
5. Z. U. Haque and J. E. Kinsella, *Milchwissenschaft*, 1988, **43**, 226.
6. Z. Mozaffar and Z. U. Haque, *J. Appl. Microbiol. Biotechnol.*, 1990, submitted.
7. Z. U. Haque and Z. Mozaffar, *Z. Food Sci.*, 1990, submitted.
8. Z. U. Haque and J. E. Kinsella, *J. Food Sci.*, 1989, **54**, 1341.
9. L. G. Bohoua and Z. U. Haque, Annual Meeting of the IFT, Anaheim, CA, 1990, presentation number 45.
10. Z. U. Haque and J. E. Kinsella, *J. Food Sci.*, 1988, **53**, 416.
11. C. J. van Oss, R. J. Good, and M. K. Chaudhury, *J. Colloid Interface Sci.*, 1986, **111**, 378.
12. Z. U. Haque and M. Kito, *Agric. Food Chem.*, 1983, **31**, 1231.
13. P. Walstra and H. Oortwijn, *Neth. Milk Dairy J.*, 1975, **29**, 263.
14. Z. U. Haque and M. Kito *Agric. Food Chem.*, 1984, **32**, 1392.
15. Z. U. Haque and J. E. Kinsella, *Agric. Biol. Chem.*, 1988, **52**, 1141.

Monitoring Crystallization in Simple and Mixed Oil-in-Water Emulsions using Ultrasonic Velocity Measurement

By Eric Dickinson, Michael I. Goller, D. Julian McClements, and Malcolm J. W. Povey

PROCTER DEPARTMENT OF FOOD SCIENCE, UNIVERSITY OF LEEDS, LEEDS LS2 9JT

1 Introduction

Emulsion stability is an important consideration in the food industry, having a bearing on the shelf life and the microbiological quality of many food products. The degree of crystallinity of fat droplets in food oil-in-water emulsions is an important factor affecting the stability of the colloidal system and the distribution of solute molecules between dispersed and continuous phases. Solid fat content is relevant to the texture of food colloids such as ice-cream, whipped cream, margarine, and to the stability of food emulsions in shear flow.[1-3] Changes in the phase state of emulsion droplets can also be an effective way of regulating interfacial reactions in emulsions.[4] Oil and fat crystallization in food emulsions is still poorly understood, although it is well known that it is affected by particle size, the presence of surfactants, the temperature history of the emulsion, and the nature of the oil phase.[5] Furthermore, the chemical nature of the low molecular weight surfactants and macromolecular emulsifiers used in emulsions is also important.[6]

The ultrasonic pulse-echo technique[7] provides a rapid and accurate[8] way of monitoring crystallization processes. We have recently shown[9] that the ultrasonic velocity method gives very good agreement with density measurements for an n-hexadecane-in-water emulsion by comparing ultrasonic velocity data with densities over the temperature range 0–30 °C. The ultrasonic technique is a simple and reliable method for studying food emulsions and dispersions,[10] and especially for monitoring crystallization and melting of the dispersed phase in emulsion droplets.[9] It has an advantage over pulsed NMR in that the contribution of water to the signal can be accurately measured and accounted for. The sensitivity of the technique arises from the fact that there is a large change in ultrasonic velocity as the oil passes from liquid to solid, or *vice versa*, and this change is opposite in sign to the temperature coefficient of velocity in both liquid and solid phases.[8]

171

The normal alkanes, hexadecane and octadecane, are chosen as the dispersed phase in our oil-in-water emulsions because they do not, unlike triglycerides, exhibit polymorphism, they are relatively easy to obtain in a pure state, and they have melting points conveniently located between room temperature and 0 °C. In addition, they have been studied in emulsions by other workers using dilatometric[11] and electron spin resonance[4] techniques.

2 Materials and Methods

Samples of n-hexadecane and n-octadecane (>99% pure) were obtained from the Sigma Chemical Company (St. Louis, USA). Commercial-grade polyoxyethylene (20) sorbitan monolaurate (Tween 20), provided by Unilever Research (Bedford, UK), or sodium caseinate were used as emulsifiers. Oil-in-water emulsions were prepared using a high-pressure laboratory-scale homogenizer or a jet homogenizer[12] to a formula of 20 wt% hydrocarbon in distilled water containing 2 wt% emulsifier for the Tween 20 emulsion. The aqueous phase of the sodium caseinate emulsions contained 0.5 wt% sodium caseinate, sodium azide as preservative, and a phosphate buffer (pH 7). Average droplet diameters (d_{32}) were determined using a Malvern Mastersizer (Model S2.01).

The velocity of ultrasound was measured at a frequency of 1.25 ± 0.15 MHz using a pulse-echo technique.[7] Samples were contained in small glass cuvettes with a path length of *ca.* 10 mm. Electrical pulses generated using a Sonatest Ultrasonic Flaw Detector (UFD1) stimulated ultrasound pulses in a transducer, and these were then transmitted into the sample, and the subsequent echos detected by the same transducer. A Tekronix 468 digital storage oscilloscope was used to display the received echos and to make time measurements. The cuvettes containing the samples were placed in a thermostatically controlled water bath (± 0.1 °C), together with the transducer, and measurements were made between 0 and 40 °C. Ultrasonic velocity was measured to within 1.5 m s^{-1}.

Instantaneous pulse power levels were at least an order of magnitude below the 10 kW m^{-2} at which cavitation begins in water,[13] and were at least three orders of magnitude less than power levels found in low-power ultrasonic baths. Temperature excursions associated with the ultrasonic wave were at most a few millikelvins, and pressure excursions at most a few tens of millibars. Average power levels were, moreover, two hundred times less than these instantaneous figures. Samples were exposed to the ultrasound field only during measurements which lasted a few minutes. Consequently, we would not expect the ultrasound measurements themselves to affect the crystallization and melting process. Agreement between our ultrasonic velocity results and those of other workers obtained using dilatometric techniques[11] (see later), as well as with our own density measurements,[9] confirms to us that ultrasound energy at these power levels leaves the emulsion samples unaffected.

Samples were first cooled from above the melting point of the oil down to 0 °C, and then heated back to the starting temperature. To check that the sample properties had not changed as a result of the temperature cycle, the velocity of sound was measured carefully both before and after crystallization. Approximately half an hour intervals elapsed between each measurement as the samples were cooled or heated. This was the minimum time necessary for the sample to reach thermal equilibrium after a temperature change. However, once the oil had frozen, the emulsions were held overnight, prior to commencing the heating phase. Particle size was checked at the beginning and again at the end of each experiment.

Temperatures were measured using type-T thermocouples and Comark digital thermometers standardized to an accuracy of $\pm 0.2\,^{\circ}C$ against an NPL traceable, calibrated, Hewlett Packard Quartz thermometer.

3 Results and Discussion

Single Oil Emulsions—Figure 1 presents results for an n-hexadecane-in-water emulsion prepared with Tween 20, having a droplet diameter d_{32} of 0.36 μm. Oil droplets appear to supercool down to 3.6 °C and then rapidly and completely crystallize. The behaviour of the liquid emulsion is accurately predicted by ultrasonic scattering theory[14] over the entire temperature range of interest. The

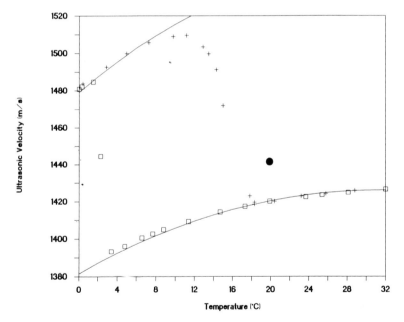

Figure 1 *Ultrasonic velocity plotted against temperature for an n-hexadecane-in-water emulsion prepared with Tween 20. The droplet size is 0.36 μm. The emulsion was first cooled then heated. The unbroken curves are the velocities in the wholly liquid and wholly solidified oil droplets in water as predicted from scattering theory. The theory uses the measured particle-size distribution rather than the Sauter mean diameter d_{32}. Key: $\square$, cooling; +, heating; ●, 0.8 μm diameter emulsion droplets at 20 °C, plotted to demonstrate the effect of particle size on the velocity of ultrasound*

same theory also predicts exactly the velocity of sound in the emulsion containing wholly solid droplets. This confirms that the technique provides an extremely precise method for estimating the onset of freezing and thawing. One point is plotted on the graph for an n-hexadecane-oil-in-water emulsion produced using sodium caseinate with a larger mean droplet diameter of $d_{32} = 0.76\,\mu$m. The difference demonstrates the substantial effect of particle size on ultrasonic velocity in the submicron particle size range.

The wide melting range (24.8–28.5 °C) of emulsified hydrocarbon was first observed by Turnbull and Cormia.[11] Our results for the n-octadecane emulsions (see Figures 2 and 3) are similar to theirs, apart from the somewhat wider dispersion in the melting temperatures obtained here, which is not surprising since our droplet diameters are approximately ten times smaller. Turnbull and Cormia[11] demonstrated that the freezing behaviour in their emulsions was consistent with that expected if homogeneous nucleation preceded the freezing of each droplet. Homogeneous nucleation depends solely on the probability that a crystal, whose size exceeds a critical amount, will form by chance in the liquid. Heterogeneous nucleation, on the other hand, is promoted by a foreign object, either adsorbed at the surface of the droplet, or contained as adventitious impurity within the droplet. Turnbull and Cormia proposed that the wide dispersion in melting points in their emulsions was due to the inclusion of surfactant at or near the surface of droplets during the freezing process. This would suppress the dispersed phase melting point near the surface, but leave the bulk of the droplet to melt at the ordinary thermodynamic melting point of the bulk material.

Our ultrasonic measurements are certainly consistent with homogeneous nucleation, since it is difficult to explain such sharp, spontaneous freezing behaviour from a highly supercooled liquid in any other way. Nevertheless, during freezing, some surfactant may be incorporated into some of the crystals, suppressing their melting point, and producing the overall dispersion in droplet melting points observed experimentally.

Another point to note is that mean particle size did not change during the experiment. There was no difference in the Mastersizer results recorded at the beginning and at the end of our experiments. Anyway, we would have expected changes in particle size to have affected the ultrasonic velocity, yet the velocities for the melted emulsion were found to coincide exactly with those for emulsion at the start of the experiment, at the same temperature. This was the case for all of the experiments described here.

The significant difference between the two curves in Figure 1 reflects the great sensitivity to the freezing and melting transition exhibited by the ultrasonic data. This is because ultrasonic velocity v is related to both adiabatic compressibility β and density ρ through the relation $v^2 = 1/(\beta\rho)$. Freezing simply increases further the magnitude of the temperature coefficient of density, whilst the temperature coefficient of the adiabatic compressibility reverses its sign during the freezing process. The velocity of sound in the wholly liquid and wholly solid emulsions can be predicted very precisely using ultrasonic scattering theory.[14] Consequently, the onset of freezing or melting can be detected very precisely.

Mixed Oil Emulsions—When an emulsion of n-hexadecane oil-in-water droplets was mixed with a separately prepared emulsion of n-octadecane oil-in-water droplets, we obtained the results shown in Figures 2 and 3. These emulsions were produced with sodium caseinate as the emulsifier; average droplet diameters in the two emulsions were identical ($d_{32} = 0.76\,\mu$m). Figure 2 shows that, in the mixed emulsion, firstly all of the n-octadecane droplets freeze, and then all of the n-hexadecane droplets freeze. Curves for the single pure n-hexadecane and pure n-octadecane emulsions are presented for comparison. In Figure 3 the melting behaviour of these same three emulsion systems are presented. Again, the melting behaviour of the n-hexadecane droplets is quite distinct from that of the n-octadecane droplets in the mixture, but similar to the behaviour of each emulsion containing only one sort of oil. There is clearly no synergistic interaction between the droplets that significantly effects the crystallization behaviour of individual droplets in these caseinate-stabilized systems.

In Figures 4 and 5, data are presented for the same experiment performed exactly as above with the one exception that Tween 20 was used as the emulsifier

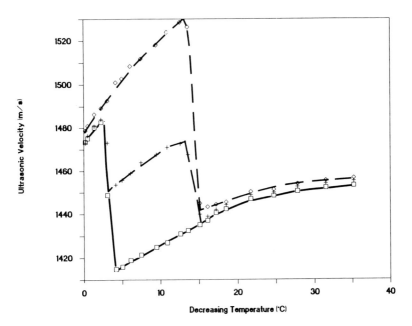

Figure 2 *Ultrasonic velocity plotted against decreasing temperature for 20 wt% hydrocarbon-in-water emulsions composed of: (a) —□—, n-hexadecane oil-in-water emulsion produced with sodium caseinate, particle size 0.76 μm; (b) —◇—, n-octadecane oil-in-water emulsion produced with sodium caseinate, particle size 0.76 μm; and (c) —+—, a 1:1 mixture of emulsion (a) with emulsion (b) to produce an emulsion containing both discrete n-hexadecane droplets and discrete n-octadecane droplets*

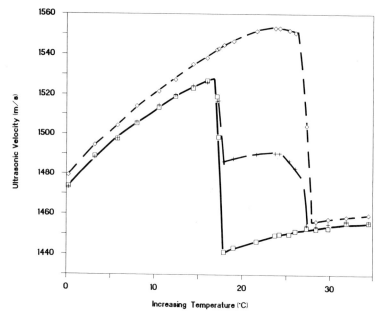

Figure 3   *Ultrasonic velocity plotted against increasing temperature for the three emulsions described in Figure 2; (a) —□—, n-hexadecane oil-in-water; (b) —◇—, n-octadecane oil-in-water; and (c) —+—, a 1:1 mixture of initially discrete n-hexadecane droplets and n-octadecane droplets*

instead of sodium caseinate. Figure 4 demonstrates that, in the mixture of the n-hexadecane droplets with the n-octadecane droplets, the behaviour differs from that in the single oil emulsions. The freezing point of the n-octadecane droplets has been depressed, and the freezing process takes place over a range of several degrees. The n-octadecane droplets do not wholly freeze, however, as can be seen by comparing Figure 4 with Figure 2. The velocity at 5 °C in Figure 4 is approximately 8 m s^{-1} below that at the same temperature in Figure 2, whose value is equivalent to the complete freezing of the n-octadecane droplets. Secondly, the freezing of the n-hexadecane droplets in the mixture is more diffuse and total freezing is not achieved. These results can only be explained by assuming that relatively rapid mixing is occurring between the contents of n-hexadecane droplets and the n-octadecane droplets.

Mixing the oils, occurring gradually as freezing proceeds, would begin to depress the freezing points of both the n-octadecane droplets and the n-hexadecane droplets, progressively with time, *i.e.* as n-hexadecane is transferred into the n-octadecane droplets and n-octadecane is transferred into the n-hexadecane droplets, the freezing point of the respective sorts of droplets would be increasingly depressed.

Intermixing of the hydrocarbon oils between the two kinds of droplets during the time period (a few hours) between making the emulsions and measuring the

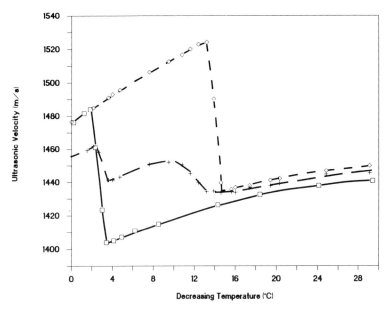

Figure 4   *Ultrasonic velocity plotted against decreasing temperature for* 20 wt%
hydrocarbon oil-in-water emulsions. The emulsions and their anno-
tation are the same as in Figure 2, but here the system is emulsified with
Tween 20 (particle diameter $d_{32} = 0.76\,\mu m$)

ultrasonic velocity would depress the freezing points of both kinds of droplets
progressively with time. As n-hexadecane is transferred into the n-octadecane
droplets, and *vice versa*, the freezing point of the resulting mixed oil droplets
becomes gradually reduced in comparison to pure n-octadecane or pure n-
hexadecane droplets. There are two things to explain. Firstly, why does this effect
occur with the low molecular weight emulsifier, Tween 20, but not with the
proteinaceous emulsifier, sodium caseinate? And, secondly, why does oil transport between droplets occur at a rapid rate in the mixed oil emulsion system,
whereas there is no apparent manifestation of oil transport betwen droplets (*i.e.*
no discernible Ostwald ripening) in the surfactant-stabilized pure hydrocarbon
oil-in-water emulsions?

We can explain the apparently anomalous results in Figures 4 and 5 as follows.
Excess non-ionic surfactant, Tween 20, exists as a micellar solution in the aqueous
dispersion medium. These surfactant aggregates are able to solubilize hydrocarbon oil into the continuous phase, and the swollen micelles act as a mode of
transport of n-hexadecane and n-octadecane molecules between emulsion droplets. In addition, some direct transport of oil molecules may occur on droplet
collision driven by Brownian motion. This mechanism of oil transport between
droplets does not occur with sodium caseinate as emulsifier because the protein is
not able to solubilize individual hydrocarbon oil molecules in solution to any

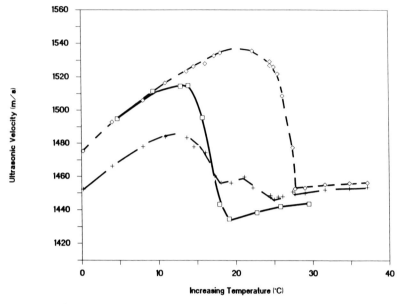

Figure 5   *Ultrasonic velocity plotted against increasing temperature for* 20 wt%
*hydrocarbon oil-in-water emulsion. Emulsions and annotation as in
Figure* 4

significant extent, and because the adsorbed protein forms a thicker sterically
stabilizing layer around the emulsion droplets, thereby inhibiting direct oil
transport on droplet collision by not allowing pairs of diffusing particles to get
very close together. In terms of their liquid solution thermodynamics, the two n-
alkanes considered here are chemically very similar, and so we would expect them
to be solubilized to a relatively similar extent by the aqueous surfactant solution.
This implies that we would expect similar transport coefficients for n-hexadecane
and n-octadecane in the aqueous phase. The associated similar mass transport
rates for the two hydrocarbon components between droplets would lead to
substantial symmetric changes in droplet compositions without any significant
change in droplet size.

Having established that there is a viable molecular mechanism for intermixing
oils between the two kinds of droplets, it is necesary only to confirm that there is a
strong thermodynamic driving force for such mass transport to occur, since the
statistical aspects of molecular diffusion of n-alkanes between emulsion droplets
must occur at essentially the same rate with pure n-hexadecane or pure n-
octadecane as with the mixture. The driving force is, of course, the entropy of
mixing, which, for the binary system n-hexadecane + n-octadecane, would be
expected to be very close to the statistical entropy of ideal mixing. Elementary
solution thermodynamics shows that the difference in chemical potential between
the two kinds of droplets (even after slight intermixing) is three or four orders of
magnitude larger than that between droplets of different size in a pure hydro-

carbon oil-in-water emulsion. This means that the thermodynamic driving force for intermixing in our mixed oil emulsion is 10^3 or 10^4 times larger than the driving force for Ostwald ripening in a pure oil emulsion. Moreover, Ostwald ripening may be retarded by the surface dilatational properties of the surfactant adsorbed layer, whereas the intermixing phenomenon would be unaffected by such considerations since it does not involve any change in droplet surface area.

4 Conclusions

Ultrasonic velocity measurement is a rapid, simple, and reliable way of following crystallization and melting in emulsion droplets. In mixed oil systems, the sensitivity of the technique allows the identification of crystallization in different kinds of oil droplets, whose composition may be changing with time due to rapid mass transport between droplets, notably when the emulsion is stabilized by water-soluble low molecular weight surfactant. The techniques described here are, in principle, applicable to the study of crystallization in triglyceride emulsions, although the situation is complicated in such systems due to the effects of polymorphism, impurities, and the wide range of lipid components present in even the purest of food oils.

References

1. E. Dickinson and G. Stainsby, 'Colloids in Food', Applied Science, London, 1982.
2. P. Walstra and R. Jenness, 'Dairy Chemistry and Physics', Wiley, New York, 1984.
3. 'Food Emulsions and Foams', ed. E. Dickinson, Royal Society of Chemistry, London, 1987.
4. O. I. Mikhalev, I. N. Karpov, E. B. Kazarova, and M. V. Alfimov, *Chem. Phys. Lett.*, 1989, **164**, 96.
5. P. Walstra, in 'Food Structure and Behaviour', ed. J. M. V. Blanshard and P. Lillford, Academic Press, London, 1987, p. 67.
6. W. Skoda and M. Van den Tempel, *J. Colloid Sci.*, 1963, **18**, 568,
7. M. J. W. Povey, in 'Advances in Food Emulsions and Foams', ed. E. Dickinson and G. Stainsby, Elsevier Applied Science, London, 1988, p. 285.
8. D. J. McClements and M. J. W. Povey, *Int. J. Food Sci. Technol.*, 1988, **23**, 159.
9. E. Dickinson, M. I. Goller, D. J. McClements, S. Peasgood, and M. J. W. Povey, *J. Chem. Soc., Faraday Trans.*, 1990, **86**, 1147.
10. D. J. McClements, M. J. W. Povey, M. Jury, and E. Betsanis, *Ultrasonics*, 1990, **28**, 266.
11. D. Turnbull and R. L. Cormia, *J. Chem. Phys.*, 1961, **34**, 820.
12. I. Burgaud, E. Dickinson, and P. V. Nelson, *Int. J. Food Sci. Technol.*, 1990, **25**, 39.
13. M. J. W. Povey and D. J. McClements, *J. Food Eng.*, 1988, **8**, 217.
14. D. J. McClements and M. J. W. Povey, *J. Phys. D*, 1988, **22**, 38.

Reactivity of Food Preservatives in Dispersed Systems

By B. L. Wedzicha, A. Zeb, and S. Ahmed

PROCTER DEPARTMENT OF FOOD SCIENCE, UNIVERSITY OF LEEDS,
LEEDS LS2 9JT

1 Introduction

Food preservatives are substances added to foods to inhibit the growth of micro-organisms. This action usually requires that the preservative be absorbed by the organism in question, and thus the chemical structure must be such as to allow passage through the microbial cell wall. It is seen that the effective antimicrobial agent is usually a protonated acid (*e.g.* benzoic, propionic, sorbic) or an ester (*e.g.* hydroxybenzoate).[1] It is important that the preservative has some solubility in both water and non-aqueous phases (*i.e.* oil), so as to make it available in the aqueous phase where micro-organisms exist, but yet be sufficiently non-polar to traverse the relatively non-polar regions of microbial cell membranes by passive transport (diffusion).

Foods are multiphase systems where one of the phases is often oil. Numerous surfactants are also likely to be present. For example, phosphoglycerides are normal components of biological systems and will be found in intact and denatured cell membranes. A wide variety of surfactants may also be added to foods; these are generally non-ionic. The known tendency for solutes which are sparingly soluble in water to become associated with surfactant micelles or aggregates[2] leads one to expect that food preservatives may also be found associated with micellar structures in foods; this has undoubted consequences for the activity (and reactivity) of these solutes.

This paper describes a stage in our development of a model for the distribution of food preservatives in multiphase foods; we consider here the quaternary system water + surfactant + oil + preservative to gain understanding of the affinity of surfactants for benzoic and sorbic acids. The implications of preservative–surfactant interactions are considered for the specific case of the reaction between sorbic acid and thiols. The latter are potentially the most reactive species towards sorbic acid in foods.

2 Description of the System

The distribution of a carboxylic acid HA in the system water + surfactant + oil is described by at least the following three simultaneous equilibria,

$$\{HA\}_{aq} \rightleftharpoons \{HA\}_{oil} \tag{1}$$

$$\{HA\}_{aq} \rightleftharpoons \{HA\}_{mic} \tag{2}$$

$$2\{HA\}_{oil} \rightleftharpoons \{(HA)_2\}_{oil} \tag{3}$$

where the subscripts aq, oil, and mic refer to the species dissolved in the aqueous phase, in the oil phase, and present within micelles, respectively. Equations (1) and (2) refer to solute partitioning behaviour, whilst equation (3) accounts for the known dimerization of carboxylic acids in non-aqueous solvents.[3] Solute partitioning is conventionally measured in terms of the partition coefficient which is the ratio of equilibrium concentrations (or mole fractions) of the solute in the two phases (for oil + water), or the micellar environment and the surrounding water in the case of solute–surfactant interactions.[4] Strictly, the ratio is defined in terms of identical species, but in practice one measures the total concentration of the solute in each phase or environment, and conventionally the *apparent* ratio of concentrations or mole fractions of acid would be reported. The task is then to resolve from such apparent partition coefficient data any solute–solute or solute–solvent interactions which may be of interest.

3 Equilibrium Measurements

The effect of benzoic acid concentration on its distribution in the system sunflower oil + water is illustrated in Figure 1. The equation of the straight line is[3]

$$c_o/c_a = P + 2P^2K_dc_a, \tag{4}$$

where c_o and c_a are the total analytical concentrations of benzoic acid species (*i.e.* as benzoic acid and dimers) in the oil and aqueous phases, respectively, P is the concentration partition coefficient in terms of identical species (*e.g.* isolated benzoic acid molecules), and K_d is the dimerization constant of the acid. Equation (4), which is obtained by simultaneous solution of law of mass action expressions for equations (1) and (3), does not allow for the fact that the solute is solvated to different extents in the two phases, or the effect of solute on the solubility of water in the oil phase. Partitioning data such as those illustrated here are rarely of sufficient accuracy to permit solute–solvent interactions to be seen as deviations from equation (4).

The effect of adding a surfactant (Tween 80) to the aqueous phase is predictably to lower the apparent partition coefficient, c_o/c_a, and, therefore, to increase the bias of the solute for the aqueous phase (Figure 1). Simultaneous solution of the equilibria shown in equations (1)–(3) allows a measure of the concentrations of total (free + micelle-bound) and free benzoic acid to be calculated. The essential assumption is that only free (but solvated) benzoic acid is in equilibrium with

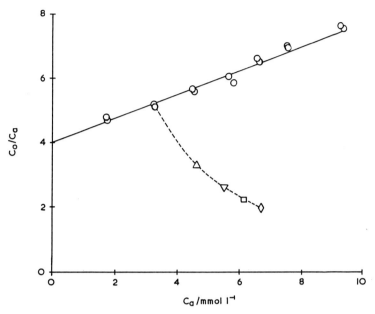

Figure 1 *Distribution of benzoic acid in the system sunflower oil + Tween 80 + water, where c_a and c_o are the total concentrations of sorbic acid species in the aqueous and oil phases, respectively. Tween 80 concentrations are as follows: $\bigcirc$, 0 wt%; $\triangle$, 2 wt%; ∇, 4 wt%; $\square$, 6 wt%; $\diamondsuit$, 8 wt%. The pH of the aqueous phase is 2.2, and all measurements are at $25.0 \pm 0.5\,°C$*

monomeric benzoic acid in the oil phase. We see from Table 1 that the addition of Tween 80 results in only a small change in the concentration of the free solute in comparison with the amount of solute which becomes associated with the micelles. This behaviour is, of course, due to the solute in the oil phase acting as a reservoir of preservative to buffer the concentration of free preservative as it is bound to surfactant, and its success depends on the preservative molecule having

Table 1 *Total benzoic acid concentration in aqueous phase, c_a, and concentration of free benzoic acid, $[BzOH]_{\text{free}}$, when a 20 mM solution of the acid in sunflower oil is allowed to equilibrate with solutions of Tween 80 at 25 °C*

Tween 80/wt%	c_a/mmol l^{-1}	$[BzOH]_{\text{free}}$/mmol l^{-1}
0	3.22	3.22
2	4.65	2.98
4	5.50	2.84
6	6.15	2.74
8	6.70	2.65

a relatively high bias for the oil phase in oil + water systems. A similar situation exists when oil-in-water emulsions containing a predominantly oil-soluble flavour are diluted with water; the intensity of the flavour is relatively unaffected.[5] It is well known that surfactants reduce the antimicrobial activity of food preservatives in water + surfactant systems, presumably by reducing the concentration of the free preservative.[6,7] Indeed, it is solute activity which is most closely related to antimicrobial action.[8,9] We speculate here that perhaps the effect of surfactant is not as great in the presence of an oil phase.

Micelle partition coefficients are usually obtained from critical micelle concentration measurements,[10] and from diffusion coefficient measurements using either NMR[11] or one-dimensional diffusion techniques.[12] There is, however, a lack of data on specific solute–surfactant combinations and one has to resort to semi-empirical correlations between two-phase partition coefficients (*e.g.* in the system octanol + water) and micelle partition coefficients[13,14] if direct measurement is not feasible. Table 2 shows a comparison of mole-fraction micelle partition coefficients for benzoic and sorbic acids in two non-ionic surfactants, calculated from the correlations reported by Treiner's group[13] and obtained in the present work from two-phase distribution studies. The agreement between the calculated and measured values is most encouraging, particularly when one considers that the values of constants used in the semi-empirical correlations were originally obtained for dodecyltrimethylammonium bromide (DoTAB). Also, the correlations used take no account of possible effects of solute concentration, and it is seen from Table 1 that these could well be of significance.

A more detailed investigation of the effect of concentration is summarized in Figures 2 and 3 for non-ionic, and anionic and cationic surfactants, respectively. In order to avoid being constrained to the mole-fraction micelle partition

Table 2 *Comparison of calculated and measured mole-fraction micelle partition coefficients, P(x)$_{mic}$, for benzoic and sorbic acids in Brij 99 and Tween 80 micelles. Calculated values were obtained from B. L. Wedzicha and A. Zeb, in 'Food Colloids', ed. R. D. Bee, P. Richmond, and J. Mingins, Royal Society of Chemistry, Cambridge, 1989, p. 193, and are based on semi-empirical predictions for ionic surfactants. Measured values were calculated from the distribution of the acids in the system sunflower oil + surfactant + water at 25 °C as described by B. L. Wedzicha, S. Ahmed, and A. Zeb, Food Additives and Contaminants, 1990, in press. The concentration of free benzoic acid in the aqueous phase is given in brackets*

		$P(x)_{mic}$	
		Measured	
	Calculated	*Tween 80*	*Brij 99*
Benzoic acid	1580	1040 (1.5 mM)	1770 (1.4 mM)
		1947 (11 mM)	2480 (18 mM)
Sorbic acid	1260	1470 (3.6 mM)	
		1750 (4.4 mM)	

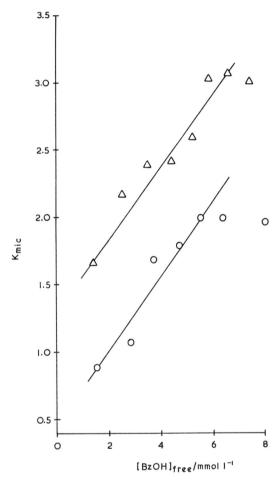

Figure 2   *The effect of free benzoic acid concentration, [BzOH]$_{free}$, on its distri-*
bution between Tween 80 (○) and Brij 99 (△) micelles and the
surrounding water. K$_{mic}$ is the number of moles of benzoic acid
associated with the surfactant divided by the number of moles of benzoic
acid in the surrounding water. All measurements were made at pH 2.2
and 25.0 ± 0.5 °C

coefficient approach, the data are shown in an elementary form as the number of
moles of solute associated with the surfactant divided by the number of moles of
solute in the surrounding water. Thus, it is clear that there is an increased bias of
the solute for the surfactant with increase in solute concentration, except for
DoTAB where no trend could be seen at the concentrations of solute tried.
Phenomenologically, this behaviour is similar to the distribution of the solute in
question in oil + water systems and is indicative of solute–solute interactions.

Table 4   Concentrations of reactants required to saturate dodecyltrimethylammonium bromide (DoTAB) micelles, determined kinetically for reactions of sorbic acid with mercaptoacetic acid, cysteine, and glutathione at
the concentrations of DoTAB shown. For all experiments in which the
concentration of thiol was varied (1.25–25 mM), we have [sorbic
acid] = 20 mM. The runs in which sorbic acid concentration was varied
(1.25–25 mM) contained either cysteine or glutathione (25 mM). Other
reaction conditions: pH 5.0; 0.2 M acetate buffer; 80 °C. (Reproduced
from B. L. Wedzicha and A. Zeb, Int. J. Food Sci. Technol., 1990, **25**,
167, with permission.)

	Concentration of reactant required to saturated micelles/mmol l^{-1}					
	[DoTAB]/mol l^{-1}					
Reactant varied	0.05	0.1	0.2	0.3	0.4	0.5
Mercaptoacetic acid	5.0	5.7	4.5	4.0	4.0	
Cysteine		7.5		7.3	5.0	
Glutathione	3.0	3.5	2.7	3.5	3.0	
Sorbic acid		4.2^a		2.5^a	2.0^a	2.0^a
	13.5^b			7.5^b		6.0^b

^{a} Reaction with cysteine.
^{b} Reaction with glutathione.

similarly small. It is interesting to see that, for this solute, different stoichiometries are found according to whether the thiol is cysteine or glutathione. The
reason for this is probably that the stoichiometry refers to the thiol or sorbic acid
molecules which are in a kinetically significant location which does not necessarily
represent the total amount of that reactant which exists within the micelles. Thus,
different amounts of sorbic acid could be present at the different locations of
cysteine and glutathione within the micelles.

The possible existence of different sites for the different thiols investigated
prompted a consideration of the behaviour when two thiols are allowed to react
simultaneously. Figure 6 shows the effect on the initial rate of reaction with sorbic
acid of the concentration of cysteine and mercaptoacetic acid. When present
individually, both thiols show the expected saturation behaviour and a rate,
independent of concentration, is obtained at [thiol] > 30 mM. On the other hand,
if an increasing concentration of cysteine is added to a mixture containing 50 mM
mercaptoacetic acid, the total rate of loss of thiol exceeds that which would have
been obtained with either thiol alone; the rate of loss of mercaptoacetic acid is
reduced to a small extent, and we see that both reactions are proceeding
simultaneously, with the cysteine and mercaptoacetic acid reactions proceeding
at 65 and 77% of their maximum possible rates, respectively. The effect of these
two thiols together is, therefore, partly additive and partly competitive. A feature
of this system is that surfactant already saturated with mercaptoacetic acid
becomes saturated with cysteine at a much lower concentration than if only
cysteine were present. Similar behaviour is found in cysteine + mercaptoethanol
mixtures, and when DoTAB is the surfactant. The rates of reaction of the other

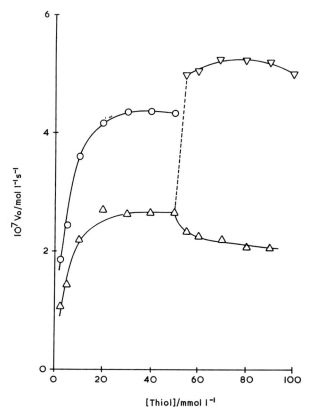

Figure 6   *Effect of concentration on the initial rate, v_o, of reaction of cysteine ($\bigcirc$) and mercaptoacetic acid (MAA) ($\triangle$) with sorbic acid. Reaction conditions: [sorbic acid] = 20 mM; [cysteine] = 0–50 mM; [MAA] = 0.50 mM; [Tween 80] = 4 wt%; 0.2 M acetate buffer, pH 5.0; 80 °C. Mixtures with [thiol] > 50 mM contained 50 mM MAA + 0–50 mM cysteine and ($\triangledown$) represents the total rate loss of thiol ($-d$[cysteine]/dt + $-d$[MAA]/dt)*

thiols taken in pairs have not yet been measured individually, but it is seen from Table 5 that all combinations of the four thiols investigated here, in pairs, give overall rates of reaction which are greater than the maximum rates of the reactions of the thiols on their own. Such partly additive behaviour suggests that the different thiols might occupy different locations.

5   Discussion

Whereas ionic species tend not to partition into non-aqueous solvents,[19] it is well known that surfactant micelles can bind a variety of ionic species; *e.g.* cationic surfactants bind inorganic[20] and organic[21] anions, whilst non-ionic surfactants are

Table 5   *Maximum initial rates of reaction, v_1^{max} and v_2^{max}, of individual thiols 1 and 2 (cysteine, cys; gluathione, glut; mercaptoethanol, ME; mercapto-acetic acid, MAA), and of a mixture of the two thiols, v_{1+2}^{max}, with sorbic acid. R is v_{1+2}^{max} expressed as a percentage of $v_1^{max} + v_2^{max}$. Reaction conditions: [sorbic acid] = 20 mM; 0.2 M acetate buffer, pH 5.0; [Tween 80] = 4 wt%; 80 °C*

Thiol 1	Thiol 2	$10^7 v_1^{max}/$ $M^{-1}s^{-1}$	$10^7 v_2^{max}/$ $M^{-1}s^{-1}$	$10^7 v_{1+2}^{max}/$ $M^{-1}s^{-1}$	R
cys	glut	4.3	3.9	5.7	69
cys	ME	4.3	3.4	5.6	72
cys	MAA	4.3	2.6	5.0	72
glut	ME	3.9	3.4	5.2	71
glut	MAA	3.9	2.6	4.7	72

known to bind organic anions including sorbate ion.[22] In the former example, electrostatic interactions are likely to play an important role, whilst interactions involving non-ionic surfactants presumably depend on the presence of a hydro-phobic moiety attached to the ionized group. The work reported here suggests that non-ionic surfactants are also capable of binding complex ionic species such as cysteine and glutathione which have very limited hydrophobic regions (*e.g.* alkyl carbon chains).

Unlike the experiments on the partitioning of benzoic and sorbic acids in two-phase systems, where the medium was acidified to avoid complications from ionization of the acid, the kinetic experiments were carried out at pH 5.0 in order for them to be more applicable to foods. At pH 5.0, sorbic acid is ionized to the extent that the ratio [sorbate]/[sorbic acid] is 3.1, and the concentration of the kinetically important thiolate anion is of the order $10^{-5}c$, where c is the total concentration of thiol.[18] Thus, if one were to express the data in Table 4 in terms of the kinetically significant species, all concentrations of thiols would be some five orders of magnitude smaller, and that for sorbic acid one quarter of the value shown. The extent to which the undissociated thiol associates with the micelles cannot be obtained from these data. The true kinetically derived thiol:surfactant stoichiometry of between 1×10^4 and 1×10^5 is much smaller than that for the binding of undissociated sorbic acid.

The data used to plot the graphs in Figure 5 were obtained with the concen-tration of thiol (25 mM) well above that required to saturate the surfactant (*ca.* 3 mM from Table 4), and the concentration of sorbic acid was increased to well above that for saturation. At the highest concentration, one would expect there to be present reactants which are not associated with surfactant, and so it is surprising that rate *versus* concentration curves reach asymptotes. If the results shown in Table 4 are used to estimate the concentration of reactant outside the micelles, the rate of the uncatalysed reaction is expected to be at least 2×10^7 mol l^{-1} s^{-1} for total sorbic acid and glutathione concentrations of 25 mM, and the rate would increase linearly with concentration. The fact that no uncatalysed reaction seems to be proceeding suggests that the reactants in question are, in fact, unavailable, and perhaps larger amounts than suggested may be accommodated within the micelles, but not at kinetically significant locations.

Micelles are visualized as dynamic entities with surfactant molecules in a state of constant motion, and with hydrophobic parts being continually exposed to the surrounding aqueous environment.[23] Molecules of solute associated with the micelles exchange rapidly with the water,[24] and one needs to imagine a time-averaged situation with respect to the distribution of polar and non-polar groups within the micelle and the concentration of solute at various locations. Thus, the simplest model we can envisage is that the thiols, on average, penetrate the micelles to differing extents, dependent on their ionic character. Glutathione, perhaps, spends the least time at a given penetration depth whilst mercaptoacetic acid or mercaptoethanol spends the greatest. If the micelles have a limited capacity for these solutes, the idea of different locations based on a time-averaged penetration depth reconciles the partly additive behaviour which we have observed when two thiols are present together. For reaction, these solutes would interact with sorbic acid, which likewise would exist distributed in a particular way within the micelles.

The two-phase distribution experiments illustrate a simple method of obtaining unambiguous surfactant–solute binding data in which the total binding capacity is being measured. We are currently using this technique to compare the kinetically derived binding stoichiometries with the actual capacity of the surfactant for the reactants under the conditions of the kinetic measurements.

In surfactant-stabilized emulsions, the presence of an oil phase will tend to reduce the concentration of reactant in the aqueous phase, and one expects a reduction in the rate of reaction between the components in the aqueous phase. Catalysis of the reaction is possible if there is an excess of surfactant present, but there may also be a need to consider the behaviour of reactant associated with surfactant adsorbed at the oil–water interface. This possibility is currently under investigation.

Acknowledgements. The authors are indebted to the Agricultural and Food Research Council for a Research Fellowship to one of us (S.A.), to the Government of Pakistan for a grant to A.Z., and to the Ministry of Agriculture, Fisheries and Food for generous support of research into the chemical reactivity of sorbic acid with food components. We are also grateful to Dr D. J. McWeeny for his continuing interest.

References

1. 'Mechanism of Action of Food Preservation Procedures', ed. G. W. Gould, Elsevier Applied Science, London, 1989.
2. A. M. Blokhus, M. Hoiland, and S. Backlund, *J. Colloid Interface Sci.*, 1986, **114**, 9.
3. K. Ezumi and T. Kubota, *Chem. Pharm. Bull.*, 1980, **28**, 85.
4. B. L. Wedzicha, in 'Advances in Food Emulsions and Foams', ed. E. Dickinson and G. Stainsby, Elsevier Applied Science, London, 1988, p. 329.
5. P. B. McNulty and M. Karel, *J. Food Technol.*, 1973, **8**, 309.
6. N. K. Patel and H. B. Kostenbauder, *J. Am. Pharm. Assoc.*, 1958, **47**, 289.
7. F. D. Pisano and H. B. Kostenbauder, *J. Am. Pharm. Assoc.*, 1959, **48**, 310.
8. N. A. Allawala and S. Riegelman, *J. Am. Pharm. Assoc.*, 1953, **42**, 267.

B. L. Wedzicha, A. Zeb, and S. Ahmed 193

9. N. A. Allawala and S. Riegelman, *J. Am. Pharm. Assoc.*, 1953, **42**, 396.
10. C. Treiner, *J. Colloid Interface Sci.*, 1982, **90**, 444.
11. P. Stilbs, *Prog. Nucl. Magn. Reson. Spectrosc.*, 1987, **19**, 1.
12. L. Castle and J. Perkins, *J. Am. Chem. Soc.*, 1986, **108**, 6381.
13. C. Treiner and M.-H. Mannebach, *J. Colloid Interface Sci.*, 1987, **118**, 243.
14. A. Leo, C. Hansch, and D. Elkins, *Chem. Rev.*, 1971, **71**, 525.
15. 'Reaction Kinetics in Micelles', ed. E. Cordes, Plenum Press, New York, 1973.
16. B. L. Wedzicha and A. Zeb, *Int. J. Food Sci. Technol.*, 1990, **25**, 167.
17. G. D. Khandelwal and B. L. Wedzicha, *Food Chem.*, 1990, **37**, 159.
18. B. L. Wedzicha and A. Zeb, *Int. J. Food Sci. Technol.*, 1990, **25**, 230.
19. R. F. Rekker, 'The Hydrophobic Fragmental Constant', Elsevier, Amsterdam, 1977, p. 15.
20. H. A. Al-Lohedan, *Tetrahedron*, 1989, **45**, 1747.
21. C. A. Bunton and C. P. Cowell, *J. Colloid Interface Sci.*, 1988, **122**, 154.
22. D. L. Reger and M. M. Habib, *J. Phys. Chem.*, 1980, **84**, 177.
23. M. C. Woods, J. M. Haile, and J. P. O'Connel, *J. Phys. Chem.*, 1986, **90**, 1875.
24. D. J. Jobe, V. C. Reinsborough, and P. J. White, *Can. J. Chem.*, 1982, **60**, 279.

Biochemical and Physical Analysis of Beers—Roles for Macromolecular Species in Foam Stabilization at Dispense

By Tom J. Leeson, Maria Velissariou, and Andrew Lyddiatt

BIOCHEMICAL RECOVERY GROUP, SCHOOL OF CHEMICAL ENGINEERING, UNIVERSITY OF BIRMINGHAM, EDGBASTON, BIRMINGHAM B15 2TT

1 Introduction

Current studies in our group, which are directed at the foaming characteristics of biological fluids, arise out of practical interests in the exploitation of simple unit operations of foam fractionation for the recovery and concentration of target products from complex feedstocks.[1]

Foam formation is generally regarded as a negative feature of bioprocessing requiring mechanical or chemical suppression.[2] However, beer manufacture, perhaps uniquely, must accommodate foam suppression during fermentation to maximize process efficiency, yet provide for subsequent enhancement at product dispense to satisfy consumer expectations.[3] Such specifications recommend the study of those molecular characteristics important to the stability and quality of dispensed beer foam as an example of the practical design requirements of a positive foaming process. Brewing research has therefore inevitably focused on the influence of manufacturing conditions upon beer foam quality. Many attempts have been made to identify the presence or absence of key components associated with foam formation and stability in beer.[4] However, the published data are largely inconclusive. This may result from a combination of the low concentrations of key components present in beers, and the adoption of methods of foam generation suited to bulk production rather than to mimicry of beer dispense.[5]

In the present study, these problems are addressed through the design and operation of a continuous foam tower capable of generating foams representative of beer dispense, but in small quantities (ca. 10 g) suited to a wide range of biochemical and physical analyses. A preliminary report is made here of the molecular composition of such foams, and comparisons are drawn with the foaming behaviour of biochemical mixtures having defined molecular compositions.

2 Materials and Methods

Foam Production—A single-stage, co-current, gas–liquid contactor design (see Figure 1) was employed for bulk foam production at 8 °C. Unless stated otherwise, the experimental material was Heldenbrau lager, supplied by Whitbread and Company of Luton, UK. Beer was degassed by standing at atmospheric pressure for 16 h at 8 °C. Foam tower dimensions and operating flow rates (0.05 to 0.2 l min^{-1}) were chosen to yield an average of 0.5 l of collapsed foam per 45 l barrel, which was judged to be representative of commercial dispense in terms of volume and bubble quality. Average tower residence times for beer were 3–5 minutes, whilst sparging rates for carbon dioxide gas (0.02 ml min^{-1}) through a standard Rudin sinter[6] offered an average drainage time of 4 minutes per bubble. Foam was collected by drainage or deionized water washings under a blanket of nitrogen gas and collapsed by centrifugation. Materials were dialysed exhaustively against running tap water at 8 °C with final equilibration against deionized water. Dialysis sacs were characterized by 10^4 dalton molecular rejection. Dialysed samples were lyophilized to dryness, and the solids content compared with original beers by gravimetric analysis. The foam quality of original beers, collected foams, and beer residues (dialysed, non-dialysed, and/or lyophilized) was estimated from the head retention values (HRV) determined by Rudin analyses.[6]

Biochemical Analyses—The protein contents of the original beer, the foam, and the residual beer were determined by the Pierce modification of the Bradford assay[7] using bovine serum albumin (BSA) as standard. Devor's modification of the Molisch procedure[8] was used for the determination of carbohydrate content. Concentrations of isomerized hop acids were determined by chromatography using a C_{18} Novapack column (10×0.8 cm; courtesy Whitbread and Company) calibrated with standard iso-α-acids. The mobile phase was 60 wt% methyl cyanide in water containing 1 wt% phosphoric acid.

  Samples were subjected to isoelectric focusing, or electrophoresis under native or denaturing conditions (SDS with or without 2-mercaptoethanol), using a Pharmacia Phast System with protocols described by the manufacturer.[9] Dialysed samples were also fractionated by ion-exchange or gel-permeation chromatography. The former used columns of DEAE-cellulose (Whatman; 10×1.6 cm) equilibrated in 20 mM sodium phosphate pH 8.0, and eluted in a linear gradient of NaCl (0 to 0.5 M in 500 ml 20 mM sodium phosphate pH 8.0). Gel permeation exploited a 50×1 cm column of Sephadex G50 equilibrated and eluted in phosphate buffered saline (PBS; 10 mM sodium phosphate containing 0.1 M NaCl at pH 7.4) at 20 ml h^{-1}.

Immunochemical Analysis—Various preparations of beer antigens (A1–5) were used in the preparation of sheep polyclonal antibodies. Antigen 1 comprised whole foam, whilst antigen 2 comprised a discrete fraction selected from the ion-exchange chromatography of whole foam. Antigen 3 was the 4×10^4 dalton protein manufactured by preparative SDS–polyacrylamide gel electrophoresis

(SDS–PAGE) of lyophilized foam. The method of Laemmli[10] was adapted for slab gels ($10 \times 13 \times 0.15$ cm). Protein was located by rapid staining of vertical guide strips with Coomassie blue, and selected fractions were recovered by electrophoretic elution of relevant horizontal zones of SDS–PAGE gels.[11] Antigen 4 comprised a clarified extract of Heldenbrau yeast disrupted by wet-milling in PBS, whilst antigen 5 was prepared from clarified, hot water extracts of malt grist specific to the manufacture of that lager.

All antigens (A1–5) were exhaustively dialysed against PBS, centrifuged, and used with standard adjuvants in a sheep immunization programme supervised by Binding Site Limited, University of Birmingham. Plasmophoresis yielded approximately 1 litre volumes of each anti-serum (S1–5) which were character-ized by immunoaffinity chromatography. Selected antigens (A1, 4, and 5) were immobilized upon cyanogen bromide activated Sepharose CL-4B (Pharmacia) at protein concentrations of 1–5 mg protein ml^{-1} agarose. Covalent coupling was proven by the spectrophotometric analysis of acid and alkaline washes of freshly coupled adsorbents,[12] and of blank elution cycles. Columns (5 ml) were chal-lenged with antisera and washed in PBS. Bound protein was desorbed by successive treatments with 15 ml volumes of 1 M NaCl in PBS at pH 7.4, 10 mM glycine–HCl at pH 3.0, and 3 M KSCN in PBS at pH 7.4. Eluted fractions were monitored by continuous UV spectrophotometry and analysed retrospectively by SDS–PAGE under reduced and unreduced conditions.

Anti-foam IgG, specifically purified from S1 using immobilized antigen A1, was immobilized upon cyanogen bromide activated Sepharose, and applied to the molecular dissection of beer, foam, and residual beer samples characterized by predetermined HRV.

Foaming of Defined Biochemical Mixtures—Single and multi-component sol-utions containing BSA, lysozyme, dextran (7×10^4 daltons), and isomerized hop acids in citrate buffer (50–100 mM; pH 3.2 to 6.4) were characterized in respect of HRV by Rudin analyses.[6]

3   Results and Discussion

Foam Production—The operation of the foam tower (Figure 1), under controlled conditions of beer in-flow, gassing rate, and height of the liquid–foam interface, permitted the reproducible bulk production of foams having uniform physical and chemical characteristics. Experiments with Heldenbrau lager were designed to generate 1–2 vol% of the total feed as collapsed foam (*i.e.* representative of commercial dispense). Similar conditions generated foam volumes of 1–8% of the original volume for a wider range of beers and lagers (see ref. 5)

Collapsed Heldenbrau foam, after exhaustive dialysis against deionized water (10^4 dalton rejection membrane) and subsequent lyophilization, yielded 7–9 g dry matter l^{-1}, equivalent to a 50 wt% concentration over the original beer. For other beers, wetter foams (>2% of original volume) routinely yielded lower degrees of concentration of the solid matter.

Dialysed and lyophilized preparations of whole beers, collapsed foams, and beer residues were analysed for protein, carbohydrate, and hop acid content (see Table 1). Kjeldahl and Lowry Folin–Phenol analyses proved unreliable for the

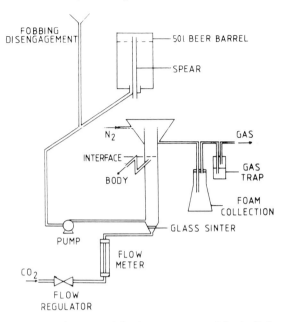

Figure 1 *Schematic diagram of the foam tower used for bulk foam production. Beer (45–50 l) is degassed at 8 °C for 16 h at atmospheric pressure, foamed under defined conditions of liquid and gas (carbon dioxide) flow, and collected under a blanket of nitrogen by processes of natural collapse, washing with deionized water, and centrifugation*

determination of total high molecular weight protein (HMW $> 10^4$ daltons) in these complex materials.[5] Coomassie blue determinations[7] calibrated against BSA standards, yielded consistent results with all samples and indicated average concentrations of $0.38 \, \text{g} \, \text{l}^{-1}$ for Heldenbrau foam. This represented a two-fold concentration over the original beer, equivalent to a 5 wt% content in lyophilized foam. Bulk carbohydrate concentrations only marginally increased in collapsed foam to represent 66 % of the dry weight. The iso-α (isomerized hop) acid content of lyophilized beer foam averaged $48 \, \text{mg} \, \text{l}^{-1}$, equivalent to a four-fold concentration over the original beer.

Table 1 *Biochemical analysis of beers and beer foams. Analytes were estimated by procedures described in the text, and values represent averages of three determinations. Concentrations in the dialysed foam relate to the volume of collapsed material collected from the foam tower*

	Dialysed beer	Dialysed foam
Protein content ($\text{g} \, \text{l}^{-1}$)	0.17	0.38
Carbohydrate content ($\text{g} \, \text{l}^{-1}$)	3.10	3.50
Iso-α-acid content ($\text{g} \, \text{l}^{-1}$)	0.011	0.048

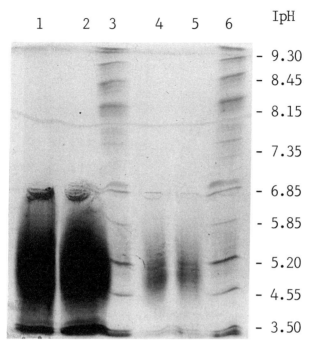

IpH

- 9.30
- 8.45
- 8.15
- 7.35
- 6.85
- 5.85
- 5.20
- 4.55
- 3.50

Figure 3   *Isoelectric focusing of beer and beer foam. The samples of dialysed, lyophilized beer foam (tracks 1, 4) and original beer (tracks 2, 5) were analysed on Phast isoelectric focusing gels (pH 3–10).[9] Protein concentrations of samples are 40 mg ml^{-1} (tracks 1, 2), and 5 mg ml^{-1} (tracks 4, 5). The standard proteins (tracks 3, 6), in order of decreasing isoelectric point, are trypsinogen, lentil lectin (3 forms), myoglobin (two forms), carbon anhydrase (human and bovine forms), β-lactoglobulin, soybean trypsin inhibitor, and amyloglucosidase*

Chromatographic Analyses—Gel permeation of samples of lyophilized beer foam on Sephadex G50 (3×10^4 dalton exclusion for globular proteins)[16] was undertaken to search for low molecular weight, native peptides unlikely to be identified in denaturing conditions on SDS–PAGE. However, experiments with this matrix and crude lyophilized foam materials proved inconclusive[5] (data not shown). The behaviour of relatively weak molecular associations in gel permeation would be expected to be complex. However, analyses in PBS in the presence or absence of 0.2 wt% SDS yielded significant protein peaks in the void volume in close agreement with SDS–PAGE (see Figure 2) and native gel electrophoresis[5] (data not shown).

In similar fashion, ion-exchange fractionation of lyophilized foam samples on DEAE–cellulose yielded complex profiles in respect of the Coomassie blue positive and UV absorbing materials contained in fractions eluted in linear salt gradients (see Figure 4). The observed behaviour was typical of that expected of a

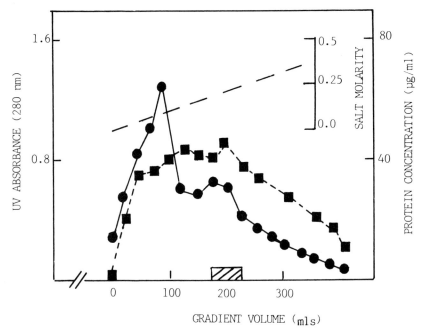

Figure 4 *Gradient elution of foam bound to DEAE cellulose. A sample of dialysed, lyophilized foam in 20 mM sodium phosphate pH 8.0, was bound to DEAE cellulose (20 ml) and eluted in a linear gradient of NaCl (0–0.5 M in 400 ml of loading buffer). Fractions were monitored for UV absorbance (●) and protein content (■). The fraction marked by hatching was used as antigen A2 in the production of polyclonal antibodies in sheep*

molecular complex loosely associated through ion-exchange and hydrophobic interactions. The isoelectric points seen for components in Figure 3 predicted the adsorption of lyophilized foam samples to an anion exchanger at pH 8.0. Elution at constant pH in a linear gradient of salt (0–0.5 M NaCl) generated a number of protein peaks, one of which was used for the generation of polyclonal antibodies (Figure 4; see later). However, the influence of increasing ionic strength, which would be expected to dissociate electrostatic interactions incrementally whilst enhancing hydrophobic associations between foam components, was not clearly defined by this methodology.

Immunochemical Analyses—Anti-sera raised in sheep against lyophilized foam antigens (A1), ion-exchange fractions (A2; see Figure 4), SDS–PAGE 4×10^4 dalton protein (A3; see Figure 2), clarified, wet-milled Heldenbrau yeast (A4), and hot water extracts of Heldenbrau malt grist (A5) were recovered by plasmophoresis. The immunochemical activity of anti-sera was tested by contacting S1 to S3 with lyophilized whole foam immobilized upon cyanogen bromide activated Sepharose. Assessments of S4 and S5, respectively, exploited immobi-

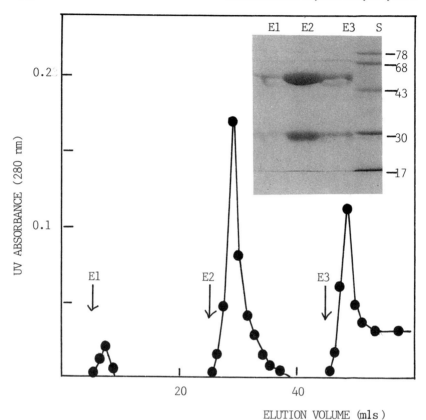

Figure 5   *Characterization of sheep anti-foam polyclonal antibodies binding to dialysed, lyophilized foam covalently immobilized on sepharose. Cyanogen bromide activated Sepharose C1-4B was reacted with dialysed, lyophilized foam to yield an adsorbent (5 ml; 3 mg protein per ml) which was contacted with 20 ml sheep anti-serum taken by plasmophoresis from an animal previously immunized with whole foam antigen. Bound protein was eluted in successive volumes of 1 M NaCl (E1), 10 mM glycine–HCl (E2), and 3 M KSCN (E3). Protein-rich fractions, as determined by UV spectrophotometry, were analysed on SDS–PAGE (see insert) along side the standard molecular weight markers (S) listed in Figure 2*

lized yeast (A4) and malt antigens (A5). After washing in PBS, adsorbed material was desorbed in successive volumes of 1 M NaCl in PBS at pH 7.4, 10 mM glycine–HCl at pH 3.0, and 3 M KSCN in PBS at pH 7.4. This protocol was initially selected to safeguard the adsorbent, but served to illustrate the relative strengths of interactions for bound materials. Figure 5 illustrates the desorption of specifically bound protein from sheep anti-foam anti-sera to immobilized foam.

SDS–PAGE analyses in reducing conditions of protein-rich fractions yielded the characteristic heavy (50 k) and light (25 k) chain pattern of sheep immunoglobulin G (IgG) accompanied by only trace amounts of sheep serum protein.[5]

Anti-sera S1 to S3 all reacted specifically (but with varying degrees of strength) with immobilized foam. S4 (anti-yeast) and S5 (anti-malt) both reacted specifically with immobolized foam (A1), and with their respective immobilized antigens (A4 and A5). The presence of antigens in foam (A1), having both yeast and malt origins, was also indicated by specific cross-reaction with S4 and S5. Ion-exchange antibodies (S3) cross-reacted with immobilized foam (A1), but the anti-SDS–PAGE 40 k anti-sera (S3) reacted only weakly with that antigen. This latter observation may reflect differences in conformation between the administered antigen A3 (a denatured preparation recovered from SDS–PAGE) and that in 'less denatured' whole foam (A1). Evaluation of the interaction of S3 against A3 was not possible because of a lack of sufficient material.

Western blot analyses[17] of SDS–PAGE of lyophilized foam were undertaken using nylon membranes, anti-foam antibodies (S1), and porcine anti-sheep IgG conjugated with horseradish peroxidase.[5] Preliminary studies indicate that the 40 k antigen is not strongly antigenic to anti-foam sera (S1), in contrast to other peptides not visualized in silver staining but exhibiting electrophoretic mobilities equivalent to molecular masses of 20 and 80×10^3 daltons. Similar outline results have been reported for a different beer.[18] The phenomena observed here may reflect poor recognition by anti-foam antibodies of denatured 40 k peptides.

Foaming of Defined Mixtures—Superficial analyses of the physical and biochemical composition of representative beer foam invites the conclusion that quality and stability is inherently dependent *inter alia* upon occupancy of the bubble lamellae by a complex of HMW proteins ($>10^4$ daltons), carbohydrates, and hop acids associated through a combination of electrostatic, hydrophobic, and other molecular interactions.

In order to mimic this situation, the study was made of the foaming qualities of mixtures of BSA (68×10^3 daltons; pI = 4.8), lysozyme (14×10^3 daltons; pI = 10.5), dextran (70×10^3 daltons), and isomerized hop acids in 0.1 M citrate buffer at pH 6.0 containing 4 wt% ethanol. Such an experimental system was designed to have oppositely charged, moderately sized proteins (well characterized and supplied commercially) in the presence of defined carbohydrate and organic acid components. This experimental system has been the subject of intensive study in the Biochemical Recovery Group and is discussed elsewhere.[20] However, a few key points are summarized in Figure 6. The foaming of 0.2 wt% BSA, which was shown to yield a maximum HRV in 4 wt% ethanol, was found to be enhanced to varying degrees by the addition of 0.5 wt% dextran, 40 p.p.m. isomerized hop acids, or 0.1 wt% lysozyme. Such concentrations were selected on the basis of relevance to beer composition and the absence of significant individual foaming properties. Mixtures of all four components, or proteins + dextran, yielded the highest foam stability in terms of HRV. The influence of pH, ionic strength, protein concentration, and molecular properties upon foam stability is the subject of current study.[20–22]

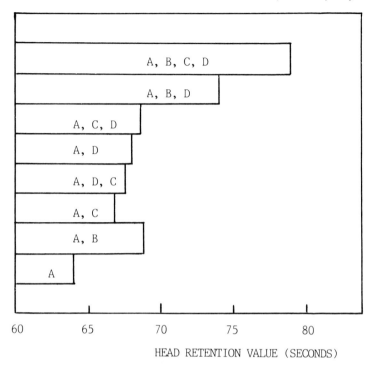

HEAD RETENTION VALUE (SECONDS)

Figure 6 *Head retention values (HRV) of defined biochemical mixtures by Rudin analysis: A, 0.2 wt% BSA; B, 0.5 wt% dextran; C, 40 p.p.m. hop acids; D, 0.1 wt% lysozyme. Solutions were prepared in 0.1 M citrate buffer at pH 6.0 containing 4 wt% ethanol*

4 Conclusions

We conclude that the bulk production of beer foams representative of commercial dispense is essential to ensure the success of subsequent studies of the role of component molecules whose interactions are fundamental to the quality and stability perceived by the consumer. Foams generated by shaking, stirring, or batch sparging were found to be too wet and too critically contaminated with residual beer for useful study.[5]

Gravimetric analysis has indicated that protein and hop acids are preferentially partitioned into representative foams. The importance of complex interactions between molecules is underpinned by the stability of foaming qualities exhibited by 'reconstituted beers' assembled from dialysed and lyophilized foam in 4 wt% ethanol. The varied characteristics of foam proteins in respect of molecular weight and isoelectric point reflect the extreme physical and biochemical environments associated with mashing and fermentation in beer manufacture, and highlight the complex response of beer foam quality to varied process conditions.[15]

The complex nature, and low protein content, of beer foam is not suited to the simple fractionations offered by ion-exchange or gel-permeation chromatography. This has invited immunochemical dissection with polyclonal antibodies. These experiments have demonstrated the strong antigenicity of certain foam components, and indicated routes to the development of diagnostic tests (ELISA, affinity HPLC, *etc.*)[5,21] for the bioquantitation of foam-positive components in raw materials, process streams, and finished products of beer manufacture. Cross-reaction of yeast and malt anti-sera with whole foam antigens confirms the presence of yeast and malt antigens in representative beer foams, whilst preliminary Western blotting analyses confirms the complexity of their interactions. Present work is concerned with the detailed immunochemical analysis of molecular components implicated in positive foaming reponses.[22]

Simple interacting systems involving oppositely charged proteins, neutral polysaccharides, and organic acids have been demonstrated to exhibit many of the foaming properties of dispensed beers. It is proposed that further study of the response of such defined systems in respect of molecular characteristic and solvent conditions, plus concomitant immunochemical dissection of the structure of representative foams, will contribute to a deeper understanding of the formation and stabilization of beer foam as well as of other biological foams.

Acknowledgments. We acknowledge the helpful comments of Dr Sudesh Mohan of the Biochemical Recovery Group. TJL acknowledges the generous support of Whitbread and Company through an SERC CASE studentship. The work of MV was funded by the SERC Rolling Programme in Biochemical Engineering at the University of Birmingham.

References

1. F. Uraizee and G. Narsimhan, *Enzyme Microb. Technol.*, 1989, **12**, 232.
2. A. Prins and K. van t'Riet, *Trends Biotechnol.*, 1987, **5**, 296.
3. L. R. Bishop, A. L. Whitear, and W. R. Inman, *J. Inst. Brew.*, 1974, **80**, 68.
4. C. W. Bamforth, in 'Food Colloids', ed. R. D. Bee, P. Richmond, and J. Mingins, Royal Society of Chemistry, Cambridge, 1989, p. 48.
5. T. J. Lesson, Ph.D. Thesis, University of Birmingham, 1989.
6. A. D. Rudin, *J. Inst. Brew.*, 1957, **63**, 1255.
7. M. M. Bradford, *Anal. Biochem.*, 1976, **72**, 249.
8. D. T. Plummer, 'Introduction to Practical Biochemistry', 2nd edn, McGraw-Hill, London, 1978.
9. Phast System Electrophoresis Manual, Pharmacia Fine Chemicals, Uppsala, Sweden, 1988.
10. U. K. Laemmli, *Nature (London)*, 1970, **227**, 680.
11. F. Kalousek, M. D. Dartigo, and L. E. Rosenberg, *J. Biol. Chem.*, 1980, **50**, 1526.
12. P. G. Dean, W. S. Johnson, and F. A. Middle, 'Affinity Chromatography—A Practical Approach', IRL Press, Oxford, 1985, Chap. 2.
13. S. Johnson and B. Skoog, 'Rapid staining of polyacrylamide gels', cited in Phast System Owners Manual, Development Technique File 210, Technical Note 1, Pharmacia LKB, Biotechnology AB, S-775182, Sweden, 1987.
14. J. Hejgaard and P. Kaersgaard, *J. Inst. Brew.*, 1983, **89**, 402.
15. J. S. Hough, 'The Biotechnology of Malting and Brewing', Cambridge Univ. Press, Cambridge, 1985.
16. A. Z. Preneta, in 'Protein Purification Methods—A Practical Approach', IRL Press, Oxford, 1990, Chap. 6.

17. J. Renart, *Proc. Natl. Acad. Sci. USA*, 1979, **76**, 3116.
18. M. Hollemans and A. R. J. M. Tonies, in 'Proceedings of European Brewing Convention—Twenty Second Congress', ed. M. van Wijngaarden, IRL Press, Oxford, 1989, p. 561.
19. M. Velissariou, M.Sc. Project Thesis, University of Birmingham, 1988.
20. M. Velissariou and A. Lyddiatt, this volume, p. 477.
21. R. Patel, M.Sc. Project Thesis, University of Birmingham, 1989.
22. S. B. Mohan and A. Lyddiatt, unpublished results.

On Static Drainage of Protein-Stabilized Foams

By Ganesan Narsimhan

DEPARTMENT OF AGRICULTURAL ENGINEERING, PURDUE UNIVERSITY,
WEST LAFAYETTE, INDIANA 47907, USA

1 Introduction

A foam is a high volume fraction dispersion of gas in a liquid. Because of the high volume fraction of the gas phase, the gas bubbles in a foam are distorted in the form of polyhedra separated by thin liquid films. Foam is colloidal in the sense that the liquid films separating the gas bubbles are of colloidal dimensions. Foam deformation necessitates the adsorption of a surface-active component at the gas–liquid interface so as to retard the rate of drainage of thin liquid films. The liquid from thin films drains into the neighbouring plateau borders under the action of plateau border suction and the disjoining pressure due to van der Waals attraction, double-layer, and steric repulsions. This is followed by the drainage of the continuous phase liquid through the network of plateau borders under the action of gravity. In addition, larger bubbles grow at the expense of smaller ones because of the diffusion of the inert gas from the smaller to the larger as a result of the differences in their capillary pressures. Collapse of the foam occurs due to the rupture of thin liquid films as a result of the instability caused by thermal and mechanical perturbations. Foam stability is intricately related to the kinetics of liquid drainage. Fundamental understanding of foam drainage is, therefore, important in order to characterize its stability.

Foamability and foam stability are important functional properties of proteins essential in many food formulations. Foamability of protein solutions has been inferred through the experimental measurement of over-run. Bubbling, whipping, and shaking are the three different methods employed for foam formation. It is customary to infer the stability of foam from the transients of drainage and collapse. Several experimental techniques such as static drainage,[1,2] surface decay,[3-5] half-life of foam,[6] and half-life of drainage[7] have been developed for this purpose. Extensive experimental investigations[8] on the effect of different variables on the drainage and stability of protein-stabilized foams have been undertaken. Despite the considerable volume of work elucidating the qualitative effects of different variables, both on the formation and stability of foams, studies on quantitative prediction of foam drainage are limited. Considerable attempts have been made to model the hydrodynamics of a foam bed. The effects of (a) gravity drainage from the plateau border,[1,9,10] (b) drainage of liquid from the plateau border as well as from thin films due to plateau border suction,[11] (c) surface viscosity,[12–14] and (d) inter-bubble gas diffusion[15,16] have been accounted for in the calculation of foam density. Extensive reviews on the stability of thin films are available.[17,18] Thin film drainage and stability have been coupled to the hydrodynamics of the foam bed in order to predict the collapse of the foam.[19] General equations for the prediction of aeration ratio,[20] hydraulic conductivity,[21] and syneresis[22] of foams and concentrated emulsions have also been presented.

2   Model for Drainage of a Standing Foam

Consider a standing foam. The dispersed bubbles of the foam, on the average, can be considered to be regular pentagonal dodecahedrons separated by thin films. Three adjacent thin films intersect in a plateau border and the continuous phase is interconnected through a network of plateau borders.[23] As time progresses, the liquid in thin films drains into the neighbouring plateau borders under the action of plateau border suction. This is followed by the drainage of the continuous phase liquid through the network of plateau borders due to gravity, eventually resulting in the accumulation of the drained liquid at the bottom of the standing foam. Consequently, a profile of liquid holdup and film thickness is set up, both increasing from the top to the bottom of the foam bed. The subsequent drainage of the foam is influenced by the evolution of this liquid holdup profile.

   A schematic diagram illustrating the evolution of liquid holdup profile in a static draining foam bed is shown in Figure 1. The foam bed is assumed to consist of dodecahedral bubbles of the same size. Since the drainage occurs in the vertical direction (direction of gravity), the foam bed can be assumed to be uniform across the bed cross section. The z co-ordinate is taken to be along the direction of gravity with $z = 0$ referring to the top of the foam and $z = L_0$ referring to the foam–liquid interface. As pointed out earlier, the liquid is distributed between the films and plateau borders. The liquid in the films drains into the neighbouring plateau borders under the action of plateau border suction, whereas the liquid in plateau borders drains under the action of gravity. For a dodecahedral arrangement of bubbles, the co-ordination number is 12. Therefore, the number of films n_f per bubble is 6. Since a plateau border is formed by the interaction of three adjacent thin films, the number of plateau borders per bubble n_p is equal to 10, and the number of plateau borders per bubble on a horizontal plane n_p' is 2 ($= n_p/5$). Denoting the film thickness by x_f, the area of the film by A_f, and the number of bubbles per unit volume by N, we have,

$$N = \frac{(1 - \varepsilon)}{v}, \tag{1}$$

where ε is the liquid holdup, and v ($= 4\pi R^3/3$) is the volume of a bubble of radius R. An unsteady-state material balance for the liquid in a film yields

$$\frac{\partial}{\partial t} x_f = -V_f, \tag{2}$$

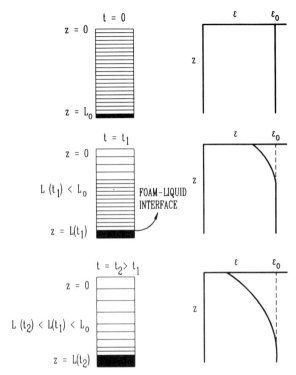

Figure 1 *Schematic diagram describing the evolution of liquid holdup profile in a static foam bed. For illustrative purposes, the initial liquid holdup profile is shown to be uniform*

where V_f is the velocity of film drainage. The total number of plateau borders intercepted by a unit cross-sectional area of the foam bed is $2NRn_p'$.[19] If the plateau borders are randomly oriented, the average volumetric flow rate per plateau border is $2a_p u/3$,[19] where a_p is the area of cross-section of a plateau border and u is the velocity of gravity drainage. An unsteady-state material balance for the liquid in the plateau borders over a volume element between z and $z + dz$ yields,

$$\frac{\partial}{\partial t}(Nn_p a_p l) = -\frac{4}{15}\frac{\partial}{\partial z}(Nn_p a_p u R) + Nn_f A_f V_f, \tag{3}$$

where l is the length of a plateau border. For a dodecahedral arrangement, we have $l = \delta R$, where δ is a constant (0.816).[14] In the above equation, the first term refers to the accumulation, whereas the second and third terms refer to the change in the volume of liquid in plateau borders due to gravity drainage and drainage from films, respectively. In order to obtain the evolution of the liquid holdup profile, the above two coupled partial differential equations have to be solved with the appropriate initial and boundary conditions, and the liquid holdup ε is related to the film thickness x_f and the plateau border area a_p by

$$\varepsilon = Nn_f A_f x_f + Nn_p a_p l. \tag{4}$$

In order to solve equations (2) and (3), however, one should have knowledge of u and V_f. The evaluation of these quantities is discussed in the subsequent sections.

Velocity of Gravity Drainage through Plateau Border—The velocity of drainage u through a plateau border can be obtained by solving the Navier–Stokes equations for the plateau border geometry with the appropriate boundary conditions which account for the surface mobility. In order to simplify the analysis, the actual shape of plateau border cross section has been idealized by various geometries.[1,12,13,24] In what follows, the solution given by Desai and Kumar,[24] who have approximated the cross-section by an equilateral triangle, is modified. Their expression for the velocity of drainage u is given by

$$u = \frac{c_v a_p}{20\sqrt{3}\mu} \rho g, \tag{5}$$

where ρ is the density of the liquid, g is the acceleration due to gravity, μ is the viscosity of the liquid, and c_v, the velocity coefficient, is defined as the ratio of the average velocity through the plateau border to that for infinite viscosity. In their analysis, Desai and Kumar[24] have expressed the general solution in terms of a pressure gradient, which, in their subsequent calculation for the liquid holdup profile,[14] is assumed to be that due to gravity. In other words, they have assumed that the force per unit volume responsible for flow of liquid in the plateau border is ρg. However, there will, in general, be an additional force other than gravity because of the gradient of pressure within the liquid in the plateau border along the z direction (vertical direction) resulting from the variation of plateau border suction. The force per unit volume responsible for the flow of liquid in the plateau border should therefore be $[\rho g - (dp/dz)]$.[22] The pressure within the plateau border p can be related to the pressure within the gas bubbles p_d through

$$p = p_d + p_c, \tag{6}$$

where the plateau border suction p_c is given by

$$p_c = -\sigma/r, \tag{7}$$

σ and r being the surface tension and the radius of curvature of the plateau border, respectively. The surface tension σ of the solution is related to the surface tension of the pure liquid σ_0 by

$$\sigma = \sigma_0 - \Pi_s, \tag{8}$$

where Π_s is the surface pressure due to the adsorption of protein employed to stabilize the foam. Since the dispersed phase (the gas bubbles) can be assumed to be in hydrostatic equilibrium, the variation of the dispersed phase pressure is given by

$$\frac{dp_d}{dz} = \rho_d g, \tag{9}$$

ρ_d being the density of the dispersed phase. Combining equations (6), (7), and (9), one obtains:

$$\frac{dp}{dz} = \rho_d g - \sigma \frac{d}{dz}\left(\frac{1}{r}\right). \tag{10}$$

The expression for the velocity of drainage, u, through the plateau border in terms of the modified driving force is therefore given by:

$$u = \frac{c_v a_p}{20\sqrt{3}\mu}\left[(\rho - \rho_d)g + \sigma \frac{d}{dz}\left(\frac{1}{r}\right)\right]. \tag{11}$$

Since $\rho_d \ll \rho$, the above equation can be approximated by

$$u = \frac{c_v a_p}{20\sqrt{3}\mu}\left[\rho g + \sigma \frac{d}{dz}\left(\frac{1}{r}\right)\right]. \tag{12}$$

From geometric considerations,[12] the radius of curvature of the plateau border r can be related to the area of the plateau border a_p and the film thickness x_f by

$$r = \frac{-1.732 x_f + [(1.732 x_f)^2 - 0.644(0.433 x_f^2 - a_p)]^{1/2}}{0.322}. \tag{13}$$

The velocity coefficient c_v is a function of the inverse of the dimensionless surface viscosity γ defined as[24]

$$\gamma = \frac{0.4387 \mu a_p^{1/2}}{\mu_s}, \tag{14}$$

where μ_s is the surface viscosity. The functional dependence of c_v on γ is expressed through a spline fit as[14]

$$c_v = b_{i0} + b_{i1}(\gamma - \gamma_i) + b_{i2}(\gamma - \gamma_i)^2 + b_{i3}(\gamma - \gamma_i)^3 \tag{15}$$

where $\gamma_i \leqslant \gamma \leqslant \gamma_{i+1}$, $i = 1, 2, \ldots, 5$. The values of the coefficients $\{b_{ij}, j = 0-3\}$ and the constants γ_i can be found elsewhere.[14]

Velocity of Drainage of a Foam Film—The velocity of drainage of a circular plane parallel film of radius R_f and thickness x_f is given by the Reynolds equation,

$$V_f = \frac{2\Delta p x_f^3}{3\mu R_f^2}, \tag{16}$$

assuming that the film surface is immobile, *i.e.* the surface viscosity is very large. For a dodecahedral arrangement, the radius of the film R_f can be related to the radius of the bubble R through geometric considerations by [14]

$$R_f = 0.808R. \tag{17}$$

The pressure drop Δp responsible for the thinning of the film is the sum of the contributions from plateau border suction and the disjoining pressure Π, and is given by

$$\Delta p = \sigma/r - \Pi. \tag{18}$$

It is worth noting that the disjoining pressure Π due to van der Waals, double-layer and steric interactions is important only when the film thickness is of the order of a thousand angstroms or smaller. When the film surface is mobile, *i.e.* the surface viscosity is not very large, the rate of film thinning is larger than that predicted by the Reynolds equation, and can be obtained by solving the Navier–Stokes equation with appropriate boundary conditions accounting for the mobility of the film surface as well as the effects of surface and bulk diffusion of the surfactant.[25]

Comparison of the Timescales of Film and Plateau Border Drainage—A comparison of typical timescales of film and plateau border drainage is shown in Figure 2. Drainage time is plotted against the fraction of liquid in films, ϕ_f, for different values of liquid holdup. As can be seen from the graph, the timescale of film drainage, τ_f, is a strong function of ϕ_f (and therefore of x_f), increasing by several orders of magnitude as the film thickness decreases. On the other hand, the timescale of plateau border drainage, τ_p, is fairly insensitive to ϕ_f. Consequently, τ_f is much smaller than τ_p for thick films, and much larger than τ_p for thin films, the cross-over film thickness being dependent on the liquid holdup. Therefore, thick films tend to drain extremely rapidly into the neighbouring plateau borders, thus resulting in a rapid decrease in the film thickness. As the film thickness, because of rapid drainage, decreases below a critical cross-over thickness, the rate of film drainage becomes extremely small compared to that of plateau border drainage. Consequently, the velocity of film drainage can be neglected in the unsteady-state equations for foam drainage. The relaxation time for this initial rapid drainage of liquid from thick films into the plateau borders would be expected to be extremely small. Moreover, the fraction of liquid in the films, ϕ_f, would decrease rapidly during this initial relaxation time. It is, therefore, reasonable to neglect the liquid in the films, and also the liquid drainage from the films, in the analysis of unsteady-state foam drainage.

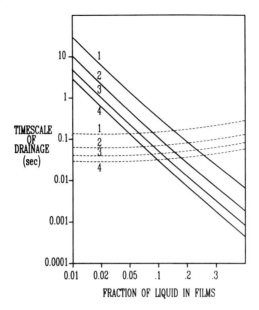

Figure 2 *Plot of timescales of drainage versus fraction of liquid in films for different liquid holdups for* R = 0.075 cm, $\sigma = 50$ mN m^{-1}, *and* $\mu = 0.01$ P. *Solid curves refer to the timescale of film drainage and dashed curves refer to the timescale of plateau border drainage. Liquid holdups corresponding to different curves are*: 1, 0.05; 2, 0.10; 3, 0.15; 4, 0.20

Simplified Equations for Drainage of a Standing Foam—In the following analysis, it is assumed that the fraction of liquid in the thin films is negligible. This assumption, of course, implies that the velocity of film drainage is negligibly small (because of the small film thickness) and therefore can be neglected. Consequently, the unsteady state drainage of a standing foam can be described by a single equation corresponding to the gravity drainage of liquid in the plateau border given by

$$\frac{\partial}{\partial t}(Nn_p a_p l) = -\frac{4}{15}\frac{\partial}{\partial z}(Nn_p a_p u R),\qquad(19)$$

where the velocity of gravity drainage u is given by equation (12). The simplified expressions for the liquid holdup ε and the radius of curvature of the plateau border r are given by

$$\varepsilon = Nn_p a_p l,\qquad(20)$$

and

$$r = \alpha a_p^{1/2},\qquad(21)$$

where α is a constant ($= \sqrt{0.644}/0.322$). From equations (1), (20), and (21), the area of the plateau border a_p and its radius of curvature r can be related to the liquid holdup ε as follows:

$$a_p = \left(\frac{v}{n_p l}\right)\left(\frac{\varepsilon}{1-\varepsilon}\right),\tag{22}$$

and

$$r = \alpha\left(\frac{v}{n_p l}\right)^{1/2}\left(\frac{\varepsilon}{1-\varepsilon}\right)^{1/2}.\tag{23}$$

Using equations (12) and (20), the unsteady-state drainage equation (19) can be recast in terms of the liquid holdup ε.

Boundary Conditions—If the dispersed phase bubbles are sufficiently small, then the capillary pressure inside a (spherical) bubble would far exceed the variation of hydrostatic pressure over a bubble diameter. This would be valid so long as the bubble radius R satisfies the following inequality:

$$R \ll \left[\frac{\sigma}{(\rho - \rho_d)g}\right]^{1/2}.\tag{24}$$

In such a case, the bubbles at the bottom of the foam bed, *i.e.* at the foam–liquid interface, would essentially be spherical[26] ('Kugelschaum'), so that the continuous phase liquid holdup at the interface ε_0 can be taken to be that corresponding to closed pack spheres. Therefore, we have

$$\varepsilon = 0.26, \quad \text{at} \quad z = L(t),\tag{25}$$

where $L(t)$ is the length of the foam bed at time t. It is to be noted that the height of the foam bed decreases with time because of the continuous drainage and accumulation of liquid at the bottom. In other words, the foam–liquid interface boundary is moving with respect to time. The actual location of the boundary at different times is to be determined through the material balance of liquid. Since there is no input of liquid at the top of the foam bed ($z = 0$), the velocity of gravity drainage of the plateau border should be zero at that location. This condition would be satisfied if the net driving force for plateau border drainage at the top of the foam bed is zero. The second boundary condition is, therefore, given by

$$u = 0 \quad (\text{at } z = 0)\tag{26}$$

which implies that

$$\sigma\frac{\partial}{\partial z}\left(\frac{1}{r}\right) = -\rho g \quad (\text{at } z = 0).\tag{27}$$

A material balance for the liquid in the foam yields

$$\int_0^{L(t)} \varepsilon(z, t) \, dz + L_0 - L(t) = \int_0^{L_0} \varepsilon_0(z) \, dz, \tag{28}$$

where $\varepsilon_0(z)$ is the initial liquid holdup profile, and L_0 and L refer to the foam heights at times zero and t, respectively. The first term in equation (28) refers to the total amount of liquid contained in the draining foam at time t, and the right hand side of the equation refers to the initial amount of liquid contained in the foam. Differentiating the above equation, and combining with the unsteady-state drainage equation (19) and the boundary conditions (25) and (26), one obtains the following expression for the rate of movement of the foam–liquid interface:

$$\frac{dL}{dt} = -\frac{1}{0.76}\frac{4}{15} Nn_p a_p uR \bigg|_{z=L(t)}. \tag{29}$$

In order to obtain the evolution of the liquid holdup profile and the foam–liquid interface for a standing foam, the unsteady state balance equation (19) has to be solved along with the boundary conditions (24) and (25) as well as equation (29) describing the movement of the interface. Of course, the initial distribution of the liquid holdup profile should also be known.

The Initial Condition—The initial liquid holdup profile in a standing foam depends on the manner in which the foam is generated. In the present analysis, it is assumed that the foam is generated by bubbling an inert gas through a liquid pool. As the inert gas is sparged into the liquid pool, the inert gas bubbles rise up to form a foam at the top of the liquid pool. The foam moves up the column entraining some liquid from the liquid pool. The entrained liquid in the plateau borders drains under the action of gravity through the moving foam, thus resulting in a profile of the liquid holdup (which increases from the top to the bottom of the moving foam). If the residence time of the rising gas bubbles in the liquid pool is sufficiently large, the surface concentration of protein can be assumed to be the equilibrium surface concentration. It will also be assumed here that the foam consists of bubbles all of the same size. If the superficial gas velocity is G, and the number of bubbles that flow per unit area of the foam bed per unit time is η, then

$$\eta = G/v. \tag{30}$$

If the superficial gas velocity of the moving foam is sufficiently small, the rates of entrainment and drainage of liquid can be assumed to be in quasi-steady state. A material balance for the liquid in the plateau border over a volume element between z and $z + dz$ would, therefore, yield:

$$\frac{d}{dz}(\eta a_p n_p l) - \frac{4}{15}\frac{d}{dz}(Nn_p a_p uR) = 0. \tag{31}$$

Using equations (20) and (30), the above equation can be recast in terms of liquid holdup ε as

$$G\frac{d}{dz}\left(\frac{\varepsilon}{1-\varepsilon}\right) - \frac{4}{15}\frac{d}{dz}\left(\frac{\varepsilon u}{\delta}\right) = 0. \tag{32}$$

As the foam moves up, the height of the foam bed increases. At the foam–liquid pool interface, the liquid holdup is taken to be that corresponding to closed packed spheres. If $z = 0$ refers to the top of the moving foam, we have

$$\varepsilon = 0.26, \quad \text{at} \quad z = L, \tag{33}$$

where L is the instantaneous height of the moving foam. The velocity of the top of the moving foam is $G/(1 - \varepsilon)$. A material balance for the liquid with respect to an observer moving with the top of the foam yields

$$u = 0, \quad \text{at} \quad z = 0. \tag{34}$$

In order to generate a standing foam of initial height L_0, the inert gas is considered to be shut off at the instant the height of the moving foam reaches L_0, so that the initial liquid holdup profile can be obtained by solving equation (32) with the boundary conditions (33) and (34).

3 Materials and Methods

The experimental arrangement consisted of a glass frit attached to the bottom of a graduated cylindrical glass column. A protein solution of known concentration was charged into the column. Foam was generated by bubbling nitrogen through the frit at a constant flow-rate. Nitrogen entrained some protein solution, thereby forming a foam which moved up the column. Gas supply was shut off at the instant the foam surface reached the required height to form a static foam. As the liquid drained and accumulated at the bottom of the column, the foam–liquid interface movement was monitored as a function of time. Average bubble size was varied by using glass frits of different sizes, namely, a fine frit (4–$5.5\,\mu$m), a medium frit (10–$15\,\mu$m), and a coarse frit (40–$60\,\mu$m). The bubble size distribution within the foam was determined by high magnification macro-photography and image analysis. Photographs of the foam were taken with Canon FE-1 35 mm camera with autobellows. The photographs were analysed using Lemont Scientific image analyser in order to obtain the bubble size distribution. All experiments were performed with bovine serum albumin (BSA) supplied by Sigma (96–99 wt% albumin, with the remainder mostly globulins). The buffer was made with $Na_2HPO_4 \cdot 7H_2O$ and $H_3C_6H_5O_7 \cdot H_2O$ at pH of 4.8 and an ionic strength of 0.085 M. Xanthan gum (supplied by TIC Gums Inc.) was dissolved in the buffer at different concentrations in order to vary the viscosity of the solution. A Cannon–Fenske capillary viscometer was used to measure the viscosity of the solution. The surface tension of the solution was measured with a DüNouy Tensiometer supplied by Central Scientific Co.

4 Transients of Foam–Liquid Interface and Liquid Holdup Profile for a Standing Foam

Transients of liquid holdup profile were calculated by solving equation (19) with the boundary conditions (25) and (27). The location of the foam–liquid interface was updated using equation (29). The initial liquid holdup profile for a standing foam of specified height was obtained by solving equation (32) with the boundary conditions (33) and (34). Calculations were performed to investigate the effects of surface tension σ, bubble radius R, superficial gas velocity G employed for the formation of the standing foam, viscosity μ, surface viscosity μ_s, and the foam height L_0. The complete results of the calculations can be found elsewhere.[27]

Typical evolution of the liquid holdup profile is shown in Figure 3 as a plot of foam height *versus* liquid holdup for different times. As time progresses, the liquid holdup profile becomes steeper with the liquid holdup at the top of the foam bed ($z = 0$) decreasing with time. However, the liquid holdup at the bottom of the foam bed remains the same (0.26). Moreover, the height of the foam bed continuously decreases with time because of drainage. The change in the liquid holdup profile is rapid at short times. Because of the increase in the steepness of liquid holdup profile (and hence the increase in the gradient of plateau border suction), and because of the decrease in the liquid holdup, subsequent rates of drainage at larger times become smaller, which results in a slower change in the liquid holdup profiles. Eventually, the gradient of plateau border suction exactly

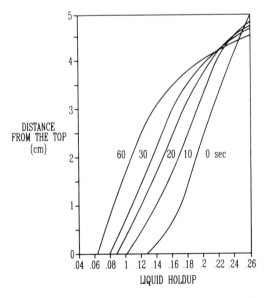

Figure 3 *Evolution of liquid holdup profile in a draining foam for* $L_0 = 5$ cm, $R = 0.02$ cm, $G = 0.1$ cm s^{-1}, $\sigma = 50$ mN m^{-1}, $\mu = 0.01$ P, *and an immobile gas–liquid interface*

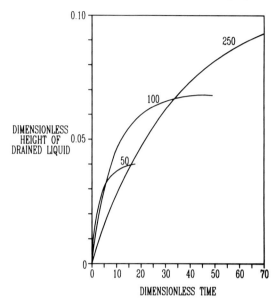

Figure 7   *Plot of dimensionless height of drained liquid* versus *dimensionless time for different initial dimensionless foam heights for* R = 0.02 cm, G = 0.1 cm s^{-1}, σ = 50 mN m^{-1}, μ = 0.01 P, *and an immobile gas–liquid interface*

The rate, as well as the extent of foam drainage, is found to be larger when larger superficial gas velocities are employed for the formation of the standing foam. At smaller superficial gas velocities, the initial liquid holdup profiles are steeper because of the smaller relative rates of entrainment compared to drainage. Consequently, the gradient of plateau border suction is steeper for smaller values of G, and this results in smaller rates of drainage.

The effect of initial foam height on foam drainage is shown in Figure 7 as a plot of the height of drained liquid *versus* time for different initial foam heights. The rate and the extent of drainage are found to be larger for larger initial foam heights. The total amount of liquid entrained in the foam is more for larger foam heights. Even though the liquid holdup at the top of the foam is smaller for a taller foam bed, the gradient of the liquid holdup is less steep, thereby providing less gradient of the plateau border suction and hence faster rates of drainage. Moreover, it takes longer for the liquid holdup profile to propagate through a taller foam bed, and therefore the foam bed drains longer before reaching the equilibrium liquid holdup profile.

In order to investigate the effect of viscosity on foam drainage, two sets of calculations were performed. In the first, the superficial gas velocity was varied during foam formation for solutions of different viscosity in order to produce foams of the same initial liquid holdup profile. It is to be noted that the rate of drainage of plateau border decreases as the viscosity of the solution increases.

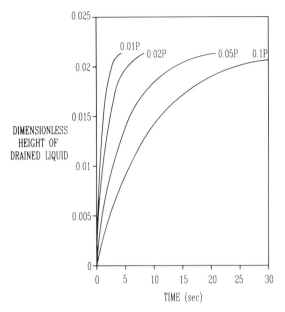

Figure 8 *Plot of dimensionless height of drained liquid versus time for different viscosities for the same initial liquid holdup profile for* R = 0.05 cm, $L_0^* = L_0/R = 20$, $\sigma = 50$ mN m^{-1}, *and an immobile gas–liquid interface*

Consequently, the rate of entrainment (and therefore the superficial gas velocity) has to be decreased in proportion to the increase in the viscosity in order to maintain the same ratio of the rates of entrainment and drainage so as to produce foams of the same initial liquid holdup profiles. For this case, the effect of viscosity on the rate and the extent of drainage are shown in Figure 8 as a plot of dimensionless height of drained liquid *versus* time. As expected, an increase in the viscosity decreases the rate of drainage. However, the extent of drainage is found to be independent of viscosity. This, of course, is a consequence of maintaining the same initial liquid holdup profile during foam formation. In the second set of calculations, the effect of viscosity on the rate and the extent of drainage is studied when the same superficial gas velocity is maintained during foam formation. The rate of entrainment of liquid compared to the rate of drainage during foam formation was found to be higher for higher viscosity solutions, thereby resulting in a more unifom initial liquid holdup profile. The predictions for the drainage of foams with different viscosities are shown in Figure 9. As can be seen from the graph, the rate of drainage is not very sensitive to the viscosity. But, the extent of drainage is found to be greater for more viscous solutions. Such behaviour can be explained in terms of the difference in the initial liquid holdup profiles. For the more viscous solution, the contribution to the driving force from the gradient of the plateau border suction (which opposes gravity) is smaller because of a more uniform initial liquid holdup profile. As a

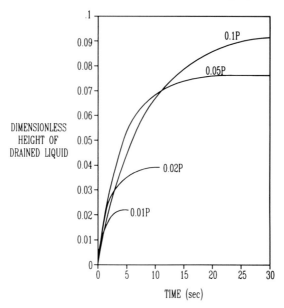

Figure 9 *Plot of dimensionless height of drained liquid* versus *time for different viscosities for* R = 0.05 cm, G = 0.1 cm s^{-1}, L$_0^*$ = L$_0$/R = 20, σ = 50 mN m^{-1}, *and an immobile gas–liquid interface*

result, the net driving force for plateau border drainage (gravity minus the gradient of plateau border suction) is larger. However, the rate of drainage for a specified driving force would be smaller because of higher viscosity of the solution. Because of these opposing effects, the rate of drainage is found to be fairly insensitive to the viscosity as can be seen from Figure 9. The drainage is predicted to continue until the net driving force for plateau border drainage becomes zero, *i.e.* gravity is counterbalanced by the gradient of plateau border suction, after which time the foam bed reaches equilibrium. The initial liquid holdup profile for the more viscous solution, being less steep than that for a less viscous solution, is further from the equilibrium liquid holdup profile, since the equilibrium liquid holdup profile for solutions of different viscosities is the same. Consequently, we see that (i) more drainage has to occur before the foam bed reaches equilibrium, and (ii) drainage takes longer (because of the longer timescale of foam drainage) for liquids of higher viscosity.

The effect of surface tension σ on foam drainage is shown in Figure 10 as a plot of dimensionless height of drained liquid *versus* dimensionless time. Drainage is found to be much faster, and the extent of drainage larger, for smaller values of surface tension. At smaller surface tensions, the driving force due to the gradient of plateau border suction (which opposes gravity) is smaller, and consequently the rate of foam drainage is larger. The effect is predicted to be quite pronounced for a small bubble size of 0.02 cm (Figure 10). It is to be noted, however, that the effect of surface tension σ on foam drainage is less pronounced for larger bubble

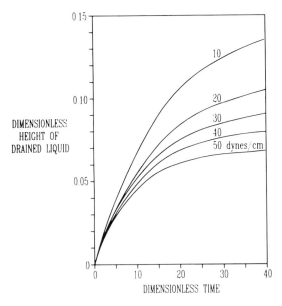

Figure 10 *Plot of dimensionless height of drained liquid versus dimensionless time for different surface tensions for* R = 0.02 cm, G = 0.1 cm s^{-1}, $L_0^* = L_0/R = 100$, $\mu = 0.01$ P, *and an immobile gas–liquid interface*

sizes because of the small contribution of the driving force due to the gradient of plateau border suction to plateau border drainage compared to gravity.

In view of the simplifying assumption that the dispersed phase bubbles are of the same size, the above analysis does not account for (i) the effect of bubble size distribution on the structure of the foam and (ii) Oswald ripening due to inter-bubble gas diffusion. Oswald ripening may not be important, except for very large times, as the timescale of inter-bubble gas diffusion, for typical bubble sizes and values of Henry's constant, has been shown to be very large.[15] Moreover, the present analysis does not account for bubble coalescence, and the resulting collapse of the foam due to the rupture of thin films. It is difficult to predict the foam collapse due to thin film rupture in a real system since the nature of the external disturbance a film would be subjected to is not known *a priori*. Experimental investigations of foam drainage indicate that it is extremely difficult to prevent collapse of bubbles at the top of the draining foam at long times. It is, therefore, important to note that the predictions of the present model are applicable only until the onset of foam collapse. Experimental measurements of the evolution of the foam–liquid interface for static foam stabilized by BSA at pH of 4.8 and an ionic strength of 0.1 M have been made for different bubble sizes, superficial gas velocities, foam heights, and viscosities. The experimental data are not reported in detail here. However, the dependence of foam drainage on bubble size, superficial gas velocity, foam height, and viscosity are consistent with the model predictions as discussed above. A typical comparison of experimental

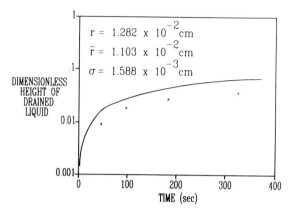

Figure 11   *Comparison of the model predictions with the experimental data. Data points are for foam drainage for a BSA concentration of 0.1 wt% and a xanthan gum concentration of 0.01 wt% at 25 °C (L_0 = 15.63 cm, G = 0.0693 cm s^{-1}, σ = 54.3 mN m^{-1}, μ = 1.3 cp, $\bar{R}$ = 1.103 × 10^{-2} cm, and standard deviation = 1.588 × 10^{-3} cm). The solid line refers to the model prediction for R = 1.282 × 10^{-2} cm*

data with the model predictions is shown in Figure 11. The model predictions seem to agree fairly well with the experimental data at short times, but the deviation is more pronounced at longer times. However, the characteristic bubble sizes for which the model predictions agree with the experimental data are found to be much higher than the average. This is believed to be due to a broad inlet bubble size distribution as can be seen from the typical histogram of foam bubble sizes in Figure 12. As pointed out earlier, for larger bubbles, the rate of plateau border drainage increases because of the larger area of cross section of plateau border, whereas steeper initial liquid holdup profile tends to slow down the rate of drainage. The latter effect seems to predominate over the former; this results in a slower rate of drainage for the broad experimental inlet bubble size distribution. Consequently, the characteristic bubble size is found to be larger than the mean. The deviation between the theory and the experimental data at very long times may be due to an increase in the bubble size as a result of coalescence processes not accounted for in the model.

5 Conclusions

A model for the unsteady-state drainage of a standing foam has been proposed. The model assumes that the foam bed consists of dodecahedral bubbles of the same size, and that the surface concentration of protein is the equilibrium concentration, implying thereby that the surface tension is constant. Based on timescale arguments, the assumption that there is a negligible fraction of liquid in the thin films in a draining foam is justified, and a simplified equation for the evolution of liquid holdup profile accounting only for liquid drainage from plateau borders is developed. The standing foam is assumed to be formed by

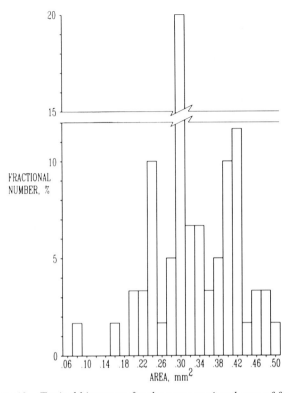

Figure 12   *Typical histogram for the cross-sectional area of foam bubbles*

bubbling an inert gas through a liquid pool in the form of bubbles of the same size at a constant superficial gas velocity. The unsteady-state drainage equation is then solved with appropriate boundary conditions accounting for the movement of the foam–liquid interface due to drainage in order to predict the evolution of the liquid holdup profile as well as the foam–liquid interface. Drainage is found to be faster and the extent of drainage larger for smaller surface tensions, and larger initial foam heights, as well as for larger superficial gas velocities. The effect of surface tension on foam drainage is found to be less pronounced for larger bubble sizes. The rate and the extent of foam drainage are found to be greatest at an optimum bubble size when a foam of different bubble sizes is generated using the same superficial gas velocity. As expected, an increase in the viscosity of the liquid decreases the rate of foam drainage. When the initial liquid holdup profile for foams of different viscosity liquids are kept the same, the extent of drainage is also found to be the same, with the rate of drainage being inversely proportional to the viscosity. When the superficial gas velocity is maintained the same for foam formation, however, the rates of foam drainage are found to be fairly insensitive to the viscosity of the liquid with the extent of foam drainage being larger for more viscous liquids. Model predictions compare favourably with experimental measurements of foam drainage for a static foam stabilized by BSA. Because of the broad bubble size distribution, the experimental foam drainage is predicted by the model for a characteristic bubble size much larger than the mean. On the whole, foam drainage is found to be sensitive to the initial liquid holdup profile, and therefore to the manner in which the foam is generated. Consequently, it is important to standardize the method of foam formation, if foam drainage is to be employed as a diagnostic test to infer foam stability.

Acknowledgements. The author acknowledges the technical assistance of Robert J. Germick in carrying out foam drainage experiments and financial assistance from the Agricultural Experimental Station, Purdue University.

References

1. G. D. Miles, L. Shedlovsky, and J. Ross, *J. Phys. Chem.*, 1945, **49**, 93.
2. A. P. Brady and S. Ross, *J. Am. Chem. Soc.*, 1944, **66**, 1348.
3. S. Ross and M. J. Cutillas, *J. Phys. Chem.*, 1955, **59**, 863.
4. G. Nishioka and S. Ross, *J. Colloid Interface Sci.*, 1981, **81**, 1.
5. A. Monsalve and R. S. Schechter, *J. Colloid Interface Sci.*, 1984, **97**, 327.
6. M. K. Sharma, D. O. Shah, and W. E. Brigham, *A.I.Ch.E.J.*, 1985, **31**, 222.
7. P. B. Rand and A. M. Kraynik, *Soc. Pet. Eng. J.*, 1983, **23**, 152.
8. P. C. Halling, *Crit. Rev. Food Sci. Nutr.*, 1981, **15**, 155.
9. W. H. Jacobi, K. E. Woodcock, and C. S. Grove, *Ind. Eng. Chem.*, 1956, **48**, 9046.
10. L. Steiner, R. Hunkeler, and S. Hartland, *Trans. Inst. Chem. Eng.*, 1977, **55**, 153.
11. S. Hartland and A. D. Barbaer, *Trans. Inst. Chem. Eng.*, 1974, **52**, 43.
12. R. A. Leonard and R. Lemlich, *AIChE.J.*, 1965, **11**, 18.
13. P. A. Haas and H. F. Johnson, *Ind. Eng. Chem. Fund.*, 1967, **6**, 225.
14. D. Desai and R. Kumar, *Chem. Eng. Sci.*, 1983, **38**, 1525.
15. G. Narsimhan and E. Ruckenstein, *Langmuir*, 1986, **2**, 494.
16. R. Lemlich, *Ind. Eng. Chem. Fund.*, 1978, **17**, 89.
17. R. K. Jain, I. B. Ivanov, C. Malderalli, and E. Ruckenstein, *Lect. Notes Phys.*, 1979, **105**, 140.
18. B. P. Rodev, A. D. Scheludko, and E. D. Manev, *J. Colloid Interface Sci.*, 1983, **95**, 254.
19. G. Narsimhan and E. Ruckenstein, *Langmuir*, 1986, **2**, 230.
20. V. V. Krotov, *Colloid J. USSR*, 1980, **42**, 903.
21. V. V. Krotov, *Colloid J. USSR*, 1980, **42**, 912.
22. V. V. Krotov, *Colloid J. USSR*, 1981, **43**, 33.
23. J. J. Bikerman, 'Foams', Springer-Verlag, New York, 1973.
24. D. Desai and R. Kumar, *Chem. Eng. Sci.*, 1982, **37**, 1361.
25. I. B. Ivanov and D. S. Dimitrov, *Colloid Polym. Sci.*, 1974, **252**, 982.
26. H. M. Princen, *Langmuir*, 1986, **2**, 519.
27. G. Narsimhan, *J. Food Eng.*, submitted.

Factors affecting the Formation and Stabilization of High-Fat Foams

By E. C. Needs and B. E. Brooker

AFRC INSTITUTE OF FOOD RESEARCH, READING LABORATORY, SHINFIELD, READING, BERKSHIRE RG2 9AT

1 Introduction

The interfacial changes which occur during the formation of whipped cream have been described previously.[1] The surfaces of the air bubbles are stabilized by adsorbed fat globules and partially by protein adsorbed at areas of air–water interface. The air bubbles are held in place by a rigid matrix of coalesced fat globules. Physical properties of these high fat foams may be affected by cream processing, especially separation and homogenization conditions, and composition, especially of the available surface active material.

2 Materials and Methods

Dairy creams were prepared by separating milk obtained from either the Institute farm bulk tank or from a commercial dairy.

Model creams were prepared which contained milk fat and non-ionic emulsifier dispersed in simulated milk ultra-filtrate.[2] The fat for these models was collected from washed cream as milk-fat globules. Cream was diluted in simulated milk ultra-filtrate and re-separated. The process was repeated three times until the milk protein had been washed out. The fat content was adjusted to 35 wt%, and 0.5 wt% sucrose ester (Mitsubishi Kasei Food Corporation, Japan) was added. The cream was homogenized at 3.5 MPa using a compressed air-driven laboratory-scale valve homogenizer (Microfluidics H5000).

Both the dairy and the model creams were stored at 4 °C for 24 h before whipping. Stiffness was measured either directly from the load on the whipper motor[3] or by a penetration test using a Lloyd Materials Testing machine. Seepage was measured as the volume of liquid which drained from a fixed weight of whipped cream during 24 h at 4 °C.

Freeze fracturing and replication were performed using a Balzer BAF 400D unit. Replicas were examined in an Hitachi H-600 transmission electron microscope.[4]

3 Results and Discussion

Cream Separation and Whipping Properties—The properties of a high fat foam are influenced by a number of factors. We have shown previously[5] that when the same milk is put through different separators the stability of the resulting whipped creams can be very different. Cream produced by one separator was found to have acceptable whipping properties, but with cream from the other the foam collapsed, and the fat globules showed extensive clustering, clumping, and coalescence.

In the present work, we examine cream which appears to have been severely damaged during separation. The effects of this damage on the whipped cream properties are shown in Table 1. The over-run was very low and in some cases no stable foam structure was formed.

The surface of the air bubble was examined by transmission electron microscopy (TEM) using the freeze-fracture technique. The creams were found to contain large aggregates of crystalline coalesced fat. Adsorbed at the air interface were large numbers of fat crystals. Figure 1 shows the air interface of damaged whipped cream, and the size of the fat crystals can be compared with the fat globules. It is assumed that fat aggregates were disrupted as a result of shear forces occurring during whipping and that the released crystals were rapidly adsorbed at the air surface.

It is well established[6] that the formation of whipped cream structure requires that the fat within the globules must contain a high proportion of solid fat. Darling[7] has described the mechanism of fat globule adsorption and coalescence. He suggests that the crystals act as points of weakness in the surface layer of the globules, and that this increases the probability of permanent interaction resulting from collision between fat globules or between fat globules and air bubbles.

It is not obvious how the observed fat crystals adsorbed to the air surface are related to the poor whipping performance of the present creams. Buchheim *et al.*[8] have shown fat crystals at the surface of air bubbles in whipped topping. This type of emulsion was rather different and contained high levels of selected low molecular weight emulsifiers. In shape, the crystals were small platelet-like structures, unlike the large needle-like crystals seen in Figure 1. These large crystals were, perhaps, adsorbed more rapidly than fat globules. This would result in a lack of fat network formation and therefore a lack of structure. It may be that these large crystals produce a rather brittle interface which cannot withstand shear forces during whipping, causing the bubbles to collapse.

Table 1 *Effect of fat damage during separation on cream properties. Quoted values are means of six whippings*

	Normal cream	Damaged cream
Whipping time (s)	101	196
Over-run %	101	22
Stiffness (N)	0.82	0.54

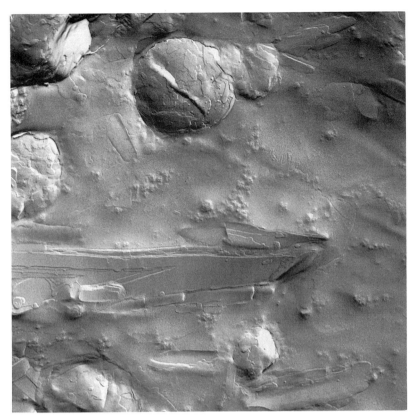

Figure 1 *An* en face *view of the air–water interface of a bubble from a defective whipped cream. In addition to adsorbed fat globules, large crystals of free fat are visible. The smooth background is the protein stabilized air–water interface. The small indentations are the impressions of casein micelles caused by the expansion of the aqueous phase during freezing*

Effect of the Composition of the Surface-active Material on Foam Properties—The compositions of the air–water and the oil–water interfaces are of considerable importance in the formation of a high-fat foam. A model cream was therefore produced by homogenizing milk-fat globules dispersed in simulated milk ultra-filtrate containing 0.5 wt% sucrose esters of stearic acids. Sucrose esters provide a series of chemically related emulsifiers covering a very wide range of HLB values.[9] Figure 2 shows the effect of increasing the HLB value of the emulsifier on the over-run and the stiffness of whipped cream. At very low HLB, over-run is less than 100% and stiffness is high. As HLB is increased, stiffness falls sharply and over-run increases to between 200 and 350%. The appearance also alters, starting as a dry, lumpy, stiff whip, but becoming smooth, light and glossy at high

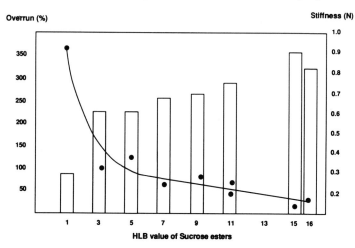

Figure 2   *Effect of the HLB values of sucrose esters on over-run (□) and stiffness*
(●)

HLB. Seepage is low (12%) at the lowest HLB, and averaged 36% (±7.8%) in the remaining samples.

The greatest difference in foaming properties is between creams containing sucrose esters with an HLB value of 1 and 3. The HLB1 sucrose esters contain no monoester—only di, tri-, and poly-esters. As HLB increases from 3 to 16, the monoester content increases from 20 to 75%. Values obtained for over-run suggest that the monoester is mainly responsible for stabilizing the air–water interface. Esters with a higher number of fatty acid residues per molecule are more likely to be associated with the fat phase at the oil–water interface.

A previous study[10] of the effect of HLB on whipped topping properties involved mixtures of Tweens covering the HLB range 10.5–15. Both over-run and stiffness were found to rise with increased HLB. The emulsifiers used were mixtures of oleate and stearate esters of polyoxyethylene sorbitan.

The effect of HLB was confounded by the effect of the different fatty acid esters. Polyoxyethylene sorbitan mono-oleate resulted in a lower over-run than the equivalent monostearate. When comparing stearate and oleate esters of sucrose, we found that the volume of foam stabilized by the sucrose oleate was significantly less than that for the equivalent sucrose stearate. This effect may be related to large differences in melting point between the two fatty acids.

4 Conclusions

We have shown that whipping cream which contains large fat aggregates, resulting from disruption to the fat phase during separation, has poor whipping properties. Adsorbed at the air bubble surface is a large number of large needle-like fat crystals. What has still to be confirmed is whether these crystals are directly responsible for the reduced foam formation in these creams.

In a model cream system, differences in foaming performance are attributable to the degree of substitution and to the type of fatty acid residue predominating in the sucrose ester emulsifier.

References

1. B. E. Brooker, M. Anderson, and A. T. Andrews, *Food Microstructure*, 1986, **5**, 277.
2. R. Jenness and J. Koops, *Ned. Melk Zuiveltijdschr.*, 1962, **16**, 3.
3. P. G. Scurlock, 'Whipping cream: effect of varying the fat and protein contents on functional properties', M. Phil. Thesis, University of Reading, 1983.
4. B. E. Brooker, *Food Structure*, 1990, **9**, 9.
5. M. Anderson, B. E. Brooker, and E. C. Needs, in 'Food Emulsions and Foams', ed. E. Dickinson, Royal Society of Chemistry, London, 1987, p. 100.
6. H. Mulder and P. Walstra, 'The Milk Fat Globule', Commonwealth Agricultural Bureau, Farnham Royal, 1974, p. 219.
7. D. F. Darling, *J. Dairy Res.*, 1982, **49**, 695.
8. W. Buchheim, N. M. Barfod, and N. Krog, *Food Microstructure*, 1985, **4**, 221.
9. Mitsubishi-Kasei Food Corporation 'Ryoto Sugar Ester' Technical Information booklet.
10. D. S. Min and E. L. Thomas, *J. Food Sci.*, 1977, **42**, 221.

Behaviour of Low-Calorie Spreads Based on Protein-Stabilized Oil-in-Water Systems

By G. Muschiolik

CENTRAL INSTITUTE OF NUTRITION POTSDAM–REHBRÜCKE, ACADEMY OF SCIENCES OF THE GDR, A.-SCHEUNERT-ALLEE 114, BERGHOLZ-REHBRÜCKE, DDR-1505, EAST GERMANY

1 Introduction

There is a trend towards new types of spreads with lower than 40 wt% of fat. Brands of such low-fat spreads, especially low-fat dairy spreads, based on either water-in-oil (w/o) or oil-in-water (o/w) emulsions, are already known. There is, however, a lack of low-calorie sweetened spreads, e.g. chocolate hazelnut spread. There is an opportunity, therefore, for improving such food products nutritionally by reducing the fat and carbohydrate contents, making them more suitable for diabetics, and exchanging cariogenic sugars.

One way to prepare spread systems with low fat and carbohydrate contents is by exploiting the emulsifying and structuring properties of proteins in o/w systems. Using acetylated field bean protein isolate (AFBPI) for producing o/w emulsions, a spread-like texture can be enhanced by adding calcium salts and by heating.[1] The favourable functionality of AFBPI in relation to emulsification and consistency formation has been described previously.[2–5]

Many patents have been produced for using milk or vegetable proteins as emulsifying agents in o/w spread formulations. This study was undertaken to test commercial whey protein and soy protein products as emulsifying, thickening, and texture forming agents, and to compare them with AFBPI in the formulation of a fluid cream model system and a spread model system.

2 Materials and Methods

Materials—Neutralized and spray-dried acetylated field bean protein isolate MOVICIA® (MOV, Central Institute of Nutrition, DDR) with a level of acetylation of 97% was prepared as described previously.[6] The following protein samples were donated by foreign manufacturers: soy protein isolate PP 500 E (PP, Protein Technologies International, Belgium); soy protein concentrates, PROMINE-HV (PRO, Central Soya Overseas BV, The Netherlands), and

DANPRO-S (DAN, Aarhus Oliefabrik A/S, Denmark); whey protein products, LACPRODAN 60 and 80 (LAC 60, LAC 80) and NUTRILAC CO 7601 (NU 01, Denmark Protein A/S, Denmark). Properties of the protein samples are recorded in Table 1. Sucrose and soybean oil were obtained from a local supermarket. The oil was plant fat NUGAMIN 204 (Walter Rau Lebensmittelwerke, FRG). Potato starch syrup powder (DE 30) was purchased from VEB Stärkefabrik Kyritz, DDR, and cocoa powder (10 wt% fat) from VEB Leipziger Süsswarenbetrieb, DDR. Sorbitol was purchased from VEB Deutsches Hydrierwerk Rodleben, DDR, and fructose from VEB Berlin-Chemie, DDR. Xanthan was donated by Jungbunzlauer Xanthan, Austria. Sodium chloride was analytical grade reagent.

Emulsion Preparation—Cream model emulsions (100 g) were prepared with composition as indicated in Table 2. Protein, salt, and sucrose were dissolved in distilled water by stirring for 1 hour with a magnetic stirrer. Emulsions were made using a Universal Laboratory Aid type-309 blender (Mechanika Precyzyjna, Poland) at a power setting of 100. Soybean oil was added while blending for 1 minute. Emulsions with whey protein were prepared at 50 °C. After suspending the potato starch syrup powder, the emulsion was blended for a further 5 minutes.

Spread model emulsions (100 g) were prepared with composition as indicated in Table 2. Protein samples and potassium sorbate were dissolved in distilled water by stirring for 1 hour with a magnetic stirrer. The aqueous protein phase was heated to 50 °C, mixed with liquid fat, and blended for 1 minute with the type-309 blender at a power setting of 100. The mixture was then blended with sucrose (or sucrose substitute) and cocoa powder, and stirred for a further 5 minutes.

One part of the emulsion sample cream model or spread model, was heated at 90 °C for 10 minutes, cooled to 20 °C, and stored at 5 °C for 20 hours. The other part was not heated, but stored at 5 °C for 20 hours.

Protein Solubility—A sample (1 g) of protein product was dispersed in 50 ml of distilled water for 1 hour using a small-scale laboratory stirrer LKR 1 (VEB MLW Labortechnik, DDR). After centrifuging at 7200g for 15 minutes, the protein concentration in the supernatant was analysed by the Kjeldahl technique.

Emulsifying Activity and Emulsion Stability—Emulsifying properties were determined as described by Yasumatsu et al.[7] A 2.5 g protein sample was dispersed in 50 ml of distilled water for 1 hour with a magnetic stirrer. The protein solution was mixed with 50 ml of rapeseed oil, and blended for 1 minute at ca. 10^3 r.p.m. using a Universal Laboratory Aid type-302 blender (beaker 200, blade 200). One part of the mixture was centrifuged at 1000g for 10 minutes. The Emulsifying Activity (EA) was expressed as the volume percentage of the emulsified phase after centrifugation. The other part of the mixture was allowed to stand at 80 °C for 15 minutes followed by centrifugation at 1000g for 10 minutes using 10 ml graduated centrifuge tubes. The Emulsion Stability (ES) was expressed as the volume percentage of the stable emulsion layer.

Table 1   *Characteristics of protein samples (S = Solubility, EA = Emulsifying Activity, ES = Emulsion Stability, GP = Gel Penetrometer Parameter)*

Sample	Protein (%)	pH	S (%)	EA (%)	ES (%)	GP (PU[d])	Gel property
MOVICIA (MOV)	78[a]	6.8	98	87	86	17	strong, springy
PP 500 E (PP)	83[b]	7.0	64	91	84	60	weak
PROMINE HV (PRO)	67[b]	6.9	22	68	76	35	weak
DANPRO-S (DAN)	64[c]	6.8	27	83	80	58	weak
LACPRODAN 60 (LAC 60)	64[c]	6.4	92	64	72	>200	no gel, like curd cheese, watery
LACPRODAN 80 (LAC 80)	79[c]	6.4	100	63	76	47	like curd cheese
NUTRILAC 7601 (NU 01)	78[c]	6.9	100	70	69	26	strong, springy

[a] Defined as Kjeldahl-N × 5.9.
[b] N × 6.25.
[c] N × 6.38.
[d] Defined as 0.1 mm = 1 PU.

Table 2   *Composition of o/w systems*

	Formulation/wt%		
		Spread model	
Ingredient	Cream model	1	2
Protein sample	4.0	2.0 and 4.0	3.0–6.0
Soybean oil	36.0	—	—
Plant fat (NUGAMIN 204)	—	18.0	18.0
Sucrose	6.0	—	40.0^a
Sorbitol	—	25.0	—
Fructose	—	15.0	—
Potato starch syrup powder	6.0	—	—
Cocoa powder	—	10.0	10.0
NaCl	1.5	—	—
Potassium sorbate	—	0.3	0.3
Distilled water	46.5	(completed to 100 wt%)	

^{a} Replaced by sorbitol or fructose.

Gelation Behaviour—A 14 g protein sample was dispersed in 86 ml of distilled water for 1 hour with a magnetic stirrer. The evacuated solution was poured into a glass cylinder (diameter 19 mm, height 20 cm) sealed with a rubber plug. The sample was heated to 90 °C for 15 minutes, and then cooled with tap water. Gelled samples (20 mm high) were analysed by a penetration test using a Penetrometer AP 4/2 (VEB Feinmess Dresden, DDR) equipped with a 'half-ball' device (diameter 20 mm, weight 19 g, measuring time 1 s).

Rheology of Cream Model Emulsions—Flow behaviour was investigated at 20 °C using a Type RV 2 Rheotest rotating viscometer (VEB MLW Prüfgerätewerk Medingen, DDR) equipped with cylinder system S1 or S2 (depending on the viscosity). Values of a flow index n and a consistency index k were fitted to a power-law model of Ostwalde–de Waele.[8] The emulsions were pre-sheared at 437 s^{-1} (cylinder S2) or 1312 s^{-1} (cylinder S1) for 20 minutes before rheological examination.

Consistency of Spread Model Emulsions—Extrusion force (EF) was determined using an extrusion cell (cylinder diameter 11.5 mm) with a hole of 1.5 mm as described by Prentice.[9] Measuring was done with a Firmness Testing Machine model BPG 50 (R. Kögel, Leipzig, DDR) with a downward cross-head speed of 74.4 mm min^{-1}. A Penetration Test (PT) was carried out using the Penetrometer AP 4/2 (see above). All measurements were made at 21.0 ± 1.5 °C. Recorded values are averages of five replicates.

Stickiness was determined using a stainless steel circular plate of 30 mm diameter (see Figure 1). The metal plate was positioned at the surface of the spread sample filled in a circular metal dish (height 8 mm). The plate was pulled up at a constant cross-head speed of 1 mm s^{-1} using the Firmness Testing

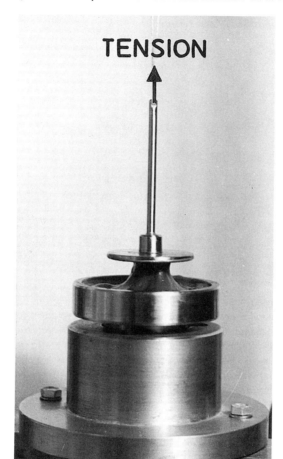

Figure 1 *Photograph of device used for determining adhesiveness and cohesiveness of spread model systems*

Machine. Adhesiveness (ADH) was expressed as the maximum measured tension force (see Figure 2). The maximum extensions reached by the stretched sample under tension between dish and plate before separation was taken to represent the Cohesiveness (COH).

3 Results and Discussion

Table 1 gives the protein content, the solubility, and the emulsifying and gelation properties of each of the tested protein products. The MOV and the whey protein products have good solubility, but there is no correlation with emulsification and gelation behaviour. The highest EA and ES values are given by PP, MOV, and DAN. Very strong and springy gels are formed with MOV and NU 01. Gels made

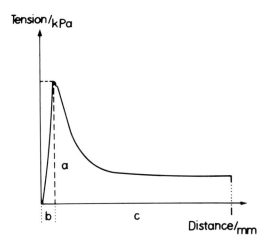

Figure 2 *Curve illustrating the typical consistency behaviour of a spread type o/w system under tension. Tension force is plotted against distance: a, Adhesiveness; b + c, Cohesiveness*

with the soy proteins (PRO, DAN, and PP) are weak. There is no gelation with LAC 60 (*i.e.* dispersion with 4 wt% product).

Table 3 gives data relating to the influence of the different protein products on the behaviour of the o/w systems, including both the cream and the spread models. Data for the cream model do not show a good correlation between emulsifying properties and stability. Non-heated cream emulsions (EAC) have a high stability with MOV, whereas heated cream emulsions (ESC) show better stability with NU 01, MOV, or LAC 80.

Emulsions made with whey proteins tend closely towards Newtonian flow behaviour ($n > 0.9$) and have a low consistency index k, as compared with those made with MOV or soy protein products (PP, DAN, PRO) which are more viscous. PP gives the lowest flow index n; values of k lie in the order PRO < PP < DAN < MOV. The firmness of the spready-type o/w systems decreases in the order MOV > PP > DAN > PRO. Non-heated emulsions with 4 wt% whey protein products have a consistency comparable to very thick syrup. Emulsions made with NU 01 are turned into a spreadable paste by heating. With plant proteins, the spread-type o/w systems show differences in firmness, adhesiveness, and cohesiveness. High values of firmness and low values of cohesiveness are found with PP and MOV, as compared with lower firmness and high cohesiveness with PRO. The results show the favourable functionality of MOV and PP with regard to consistency.

Figure 3 illustrates the small difference in values of consistency for 3 wt% MOV emulsions containing fructose, sorbitol, and sucrose. The consistency of emulsions made with 4 wt% PP is, however, more sensitive to sugar type. Of the results presented in Figure 3, the greatest sensitivity to sugar type is shown by emulsions made with 6 wt% NU 01 (non-heated emulsions are high viscosity liquids, and not pastes). The presence of sucrose leads to enhanced cohesiveness and adhesive-

Table 3 *Emulsion stability and rheological behaviour of cream and spread model systems made with 4 wt% protein*

Protein	Emulsion pH	Cream model				PT (PU)	Spread model 1			
		Stability[a]		Rheological behaviour[b]			Consistency			
		EAC/%	ESC/%	Flow index n	Consistency index k/Pa s^n		EF (N)	ADH (kPa)	COH (mm) b	c
MOVICIA	6.5	90	90	0.86	0.190	4	4.8	1.9	1.6	7.4
PP 500 E	6.2	58	70	0.69	0.130	4	4.0	1.6	1.4	5.5
PROMINE HV	6.5	48	46	0.83	0.060	26	1.8	0.7	1.6	17.3
DANPRO-S	6.3	44	47	0.75	0.150	6	1.9	1.6	1.9	14.4
LACPRODAN 60	6.2	56	65	0.96	0.028	400	fluid	fluid	fluid	
LACPRODAN 80	6.0	61	87	0.92	0.037	86	fluid	fluid	fluid	
NUTRILAC CO 7601	6.1	50	96	0.93	0.016	83	fluid, highly viscous	0.2	2.1	13.9

[a] No oil separated.
[b] $p < 0.01$.

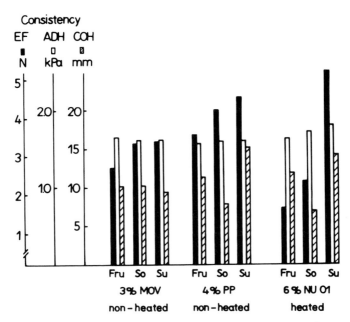

Figure 3 *Influence of different sugars on the consistency of spread type o/w systems with 3 wt% MOVICIA (MOV), 4 wt% PP 500 E (PP), and 6 wt% NUTRILAC CO 7601 (NU 01). Emulsions with MOV and PP were not heated; the emulsion with NU 01 was heated (10 min, 90 °C) (Fru = Fructose; So = Sorbitol; Su = Sucrose; Spread model 2; EF = Extrusion force; ADH = Adhesiveness; COH = Cohesiveness)*

ness, as well as enhanced firmness (as measured by the EF) due to the increased stickiness.

Figure 4 illustrates the influence of xanthan on the consistency of o/w paste systems with 3 wt% MOV. Cohesiveness increases and firmness decreases with increasing xanthan concentration. Emulsions with 0.2 wt% xanthan show a high degree of cohesiveness; they are more ductile, and not so spreadable. Such a consistency is suitable for fillings (*e.g.* biscuits, chocolate).

Another way of changing the consistency of these o/w systems is indicated by the results in Figure 5. For emulsions made with 2 wt% MOV, increasing the degree of acetylation of the protein increases the firmness and the adhesiveness, and decreases the cohesiveness. Emulsions made with highly acetylated MOV are less ductile. There is a big difference in consistency between non-acetylated MOV emulsions (Figure 5) and highly acetylated MOV emulsions with added xanthan (Figure 4). The latter are not so spreadable if the cohesiveness parameter exceeds 18 mm.

The results presented here clearly show the importance of testing the functionality of protein preparations in different model systems. The results also show the importance of lipid–protein–carbohydrate interactions on the functionality. The

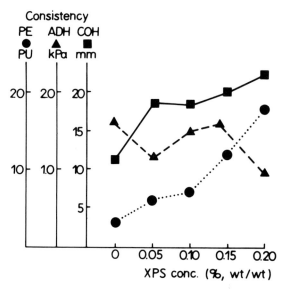

Figure 4   *Consistency of spread type o/w systems with* 3 wt% MOVICIA *depending on the xanthan* (XPS) *concentration* [PU = Penetrometer Unit (1 PU = 0.1 mm). *Spread model 2 with* 40% *sucrose*; PE = Penetration Test; ADH = Adhesiveness; COH = Cohesiveness]

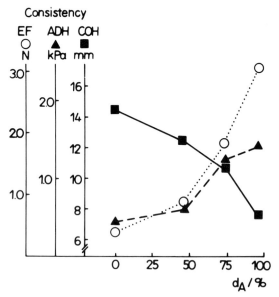

Figure 5   *Consistency of spread type o/w systems with* 2 wt% MOVICIA *depending on the degree of AFBPI acetylation. The consistency parameters* EF, ADH, *and* COH *are plotted against* d_A, *the degree of acetylation.* (*Spread model 1*, EF = Extrusion Force).

Table 4 Correlation coefficients between characteristics of protein samples (Table 1) and consistency characteristics of cream and spread model systems (Table 3)

	Protein 1	S 2	EA 3	ES 4	GP 5	Cream model EAC 6	ESC 7	n 8	k 9	PT 10	Spread model EF 11	ADH 12	COH 13
1. Protein	1.000												
2. S	0.196	1.000											
3. EA	0.286	0.000	1.000										
4. ES	0.312	-0.259	0.687	1.000									
5. GP	-0.179	-0.357	-0.143	-0.116	1.000								
6. EAC	0.643	0.500	0.071	0.402	-0.214	1.000							
7. ESC	0.562	0.848	0.205	0.000	-0.527	0.562	1.000						
8. n	-0.250	0.679	-0.679	-0.705	-0.071	0.179	0.312	1.000					
9. k	0.054	-0.357	0.643	0.955	-0.179	0.179	-0.134	-0.714	1.000				
10. PT	-0.304	0.205	-0.902	-0.821	0.312	-0.134	-0.089	0.812	-0.812	1.000			
11. EF	0.232	-0.179	0.857	0.937	-0.179	0.250	0.027	-0.679	0.929	-0.866	1.000		
12. ADH	0.143	-0.125	0.893	0.839	-0.339	0.089	0.098	-0.679	0.875	-0.911	0.964	1.000	
13. COH	-0.429	-0.152	-0.937	-0.545	0.312	-0.170	-0.384	0.580	-0.473	0.875	-0.670	-0.759	1.000

coefficients in Table 4 show the positive correlation between EA and ES and the firmness of the spread model emulsions. ES shows a good correlation with the consistency factor k of the cream model, and k shows a good correlation with the firmness (EF) of the spreads. There is, however, no clear relationship between protein characteristics and the consistency parameters of protein-stabilized emulsions.

It has been shown[4] that increasing the degree of acetylation of MOV leads to an increase in the surface viscosity at the oil–water interface. An enhanced surface viscosity, enhanced gelation, and enhanced formation of an inter-linking network structure of the more unfolded and more hydrophobic acetylated protein could explain the positive influence of MOV on the texture of such o/w systems. On the other hand, the favourable texture with MOV may be associated with calcium ion interactions.[3] Lack of information on the calcium content of the emulsion consistuents and on the sensitivity of all used protein samples to calcium prevents a deeper discussion of this matter.

The various proteins studied here differ in their influence on water-binding and rheology in the emulsion continuous phase. Only cream and serum phases were observed after centrifuging the emulsions—there was no oil separation. Effects of droplet size on the differences in rheological behaviour can be excluded since the microscopically determined average droplet sizes for all the emulsions were about the same (*ca.* 10 μm). Differences in hydrophobic interactions, and their effects on protein folding, association, and rheology at interfaces, are known only for pure proteins[10–13] and not for commercial soy and whey protein products used here.

Functionality can be influenced and improved for soy and whey proteins by chemical modification.[12] In the present study, the question arises as to what is the optimum degree of modification. Predictions are made more difficult in real food emulsion formulations by the presence of many extra components capable of interacting with the proteins. According to Dickinson and Stainsby,[13] 'a certain caution should be exercised . . . in extrapolating from single-interface experiments to the properties of food emulsions'. Interface experiments should be combined with experiments on model systems where the influence of carbohydrate (sugars, polysaccharides) and lipid is taken into consideration.

The experiments with xanthan demonstrate the opportunity to tailor the consistency of emulsions by combining the functionalities of a surface-active protein (like MOV) and a non-self-gelling polysaccharide. Xanthan restricts the formation of new inter-linkages between droplets in emulsions made with MOV.[2] This is consistent in the present study with the reduction in firmness and the increase in cohesiveness in paste o/w emulsions on addition of xanthan.

Heating improves the functionality of whey proteins in paste systems (Figure 3). Samples NU 01 and LAC 80 contain *ca.* 49% β-lactoglobulin, but they differ in relation to thermal denaturation and gelation in bulk solution (Table 1). Figure 3 shows that, in comparison with MOV, double the concentration of NU 01 (as well as heating) is required to get the same texture of spread-type o/w system. LAC 60 and LAC 80 are both poor in creating the required texture.

These results demonstrate the usefulness of acetylated field bean protein isolate MOVICIA® in imparting desirable consistency to spread-type o/w emulsion systems, and the importance of realistic food models for studying the functionality of protein preparations. The present work should be complemented by a characterization of surface behaviour in protein/lipid/carbohydrate systems, taking into account different sugars, and both non-gelling and self-gelling polysaccharides.

The results of this study on different commercial samples do not in any way reflect or suggest any preference by the author of one product over another in commerce or marketability.

Acknowledgement. The author is indebted to Mrs K. Lengfeld for skillful technical assistance.

References

1. G. Muschiolik, K. Ackermann, and Ch. Schneider, GDR Patent DD 232 191 A1, 1984.
2. G. Muschiolik, *Food Hydrocolloids*, 1989, **3**, 225.
3. G. Muschiolik, G. Schmidt, O. Andersson, Ch. Schneider, and H. Schmandke, in 'Gums and Stabilisers for the Food Industry', ed. G. O. Phillips, D. J. Wedlock, and P. A. Williams, Elsevier Applied Science, London, 1986, Vol. 3, p. 419.
4. G. Muschiolik, E. Dickinson, B. S. Murray, and G. Stainsby, *Food Hydrocolloids*, 1987, **1**, 191.
5. G. Muschiolik, A. Dahme, G. Schmidt, and H. Schmandke, in 'Gums and Stabilisers for the Food Industry', ed. G. O. Phillips, D. J. Wedlock, and P. A. Williams, IRL Press, London, 1988, Vol. 4, p. 347.
6. Ch. Schneider, M. Schultz, and H. Schmandke, *Nahrung*, 1985, **29**, 785.
7. K. Yasumatsu, K. Sawada, S. Moritaka, M. Masaru, J. Toda, T. Wada, and K. Yshii, *Agric. Biol. Chem.*, 1972, **36**, 719.
8. G. Schmidt, H. Schmandke, and R. Schöttel, *Acta Alimentaria*, 1986, **15**, 175.
9. J. M. Prentice, *Laboratory Practice*, 1954, 186.
10. M. C. Phillips, *Food Technol.*, 1981, **35** (1), 50.
11. J. Leman and J. E. Kinsella, *CRC Crit. Rev. Food Sci. Nutr.*, 1989, **28**, 115.
12. J. E. Kinsella and D. M. Whitehead, in 'Advances in Food Emulsions and Foams', ed. E. Dickinson and G. Stainsby, Elsevier Applied Science, London, 1988, p. 163.
13. E. Dickinson, B. S. Murray, and G. Stainsby, in 'Advances in Food Emulsions and Foams', ed. E. Dickinson and G. Stainsby, Elsevier Applied Science, London, 1988, p. 123.

Creaming in Flocculated Oil-in-Water Emulsions

By Sarah J. Gouldby, Paul A. Gunning, David J. Hibberd, and
Margaret M. Robins

AFRC INSTITUTE OF FOOD RESEARCH, NORWICH LABORATORY, COLNEY LANE,
NORWICH NR4 7UA

1 Introduction

Many foods are emulsions during or after manufacture.[1] In oil-in-water emulsions, the dispersed oil droplets generally possess a lower density than the continuous aqueous phase. Unless the droplets are very small or the emulsion is very concentrated, the density difference leads to the gradual accumulation of the droplets at the top of the container (creaming) with consequent loss of perceived quality.

Polysaccharide stabilizers are frequently used to reduce the rate of creaming, as well as to impart the required mouth-feel properties to a food product. We are interested in the mechanisms by which these polymers influence the separation process. In particular, previous work[2,3] has shown that if insufficient polymer is added, the droplets become flocculated and cream faster than in the absence of the stabilizer. In this paper, we present creaming results for n-alkane-in-water emulsions containing the polysaccharide hydroxyethylcellulose. Our analysis of the data draws heavily on the work of Michaels and Bolger,[4] who investigated the sedimentation behaviour of flocculated kaolin suspensions. Their system differed in two main respects from ours: their particles were rigid, and they were very strongly flocculated. In contrast, emulsion droplets are deformable, and, in our systems, the flocs formed are weak and easily disrupted upon dilution. However, the treatment of Michaels and Bolger includes general descriptions of flocculated systems, and we have found it to be a useful starting point for the preliminary analysis of our results.

The simplest theoretical treatment of creaming (or sedimentation) considers the balance of buoyancy and drag forces on a single particle moving in a viscous liquid. The terminal velocity v_s of the particle is given by Stokes' Law

$$v_s = \frac{\Delta \rho d^2 g}{18 \eta_c},\tag{1}$$

where $\Delta \rho = \rho_c - \rho_d$ is the density difference between the continuous and disperse phases, d is the droplet diameter, g is the acceleration due to gravity, and η_c is the viscosity of the continuous phase. In a non-dilute emulsion, the creaming rate of each droplet will be affected by the presence of the other droplets, and an empirical term[5] may be included in equation (1) to give the initial meniscus velocity, v, of a system of monodisperse particles at an initial concentration ϕ_0:

$$v = v_s(1 - \phi_0)^{4.65}. \tag{2}$$

Michaels and Bolger[4] studied kaolin suspensions, which were strongly floccu-lated and formed very open, rigid aggregates. Sedimentation rates were calcu-lated by measuring the rate of fall of the upper meniscus. They described three types of sedimentation behaviour (we will call them types I, II, and III) depending on particle concentration and degree of flocculation. Type I behaviour occurs at low concentrations where sedimentation proceeds at a constant rate until the upper meniscus reaches the sediment, whereupon it stops. Type II behaviour occurs at higher concentrations. The sedimentation rate increases to a constant velocity, then decelerates before the meniscus reaches the sediment. The initial accelerating régime is interpreted as the formation of channels in the flocculated network, through which the continuous phase flows as the sedimentation pro-ceeds. The régime of decreasing rate is attributed to the ability of the particles near the sediment to withstand a significant amount of compressive stress, and so offer some resistance to compaction. Type III behaviour occurs for high effective concentrations where the system merely compresses under gravity.

In type I sedimentation, Michaels and Bolger observed that the flocs moved individually. Their velocity may be obtained from equations (1) and (2) by substituting the floc density, concentration, and diameter for those of the individual particles, i.e.

$$v = \frac{\phi_m \Delta \rho d_f^2 g}{18\eta_c}(1 - \phi_f)^{4.65}, \tag{3}$$

where ϕ_m is the particle volume fraction in the floc, taken here to be the maximum packing fraction, d_f is the effective floc diameter, and ϕ_f is the effective floc concentration, actually ϕ_0/ϕ_m. The size of the flocs in the kaolin[4] was deduced to be several hundred micrometres.

In type II sedimentation, the system is modelled as a porous bed through which the background liquid flows. The value of the maximum sedimentation rate, v, enabled Michaels and Bolger to estimate the effective hydrodynamic diameter of the flow channels (pores), assumed to be smooth straight cylinders, i.e.

$$v = \frac{\phi_0 \Delta \rho d_p^2 g}{32\eta_c}(1 - \phi_f), \tag{4}$$

where d_p is the effective hydrodynamic pore diameter. This diameter was observed to increase with strength of flocculation.

We have used the above analysis in an attempt to characterize the structures in our flocculated emulsions. Since the flocs break up when diluted, it is not possible to examine them after removal from the emulsion. Workers such as Russel[6] have formulated more extensive models to predict the shape of concentration profiles during the sedimentation process, but the parameters required to fit the more sophisticated theories to our data are not readily obtainable. At present, therefore, we confine ourselves to general trends, with the emphasis on the mechanism of flocculation rather than the kinetics of the creaming behaviour.

2 Materials and Methods

Emulsion Composition and Preparation—All the emulsions contained n-alkane droplets, stabilized against coalescence by the non-ionic surfactant Brij 35, in an aqueous continuous phase containing 0.2 wt% sodium metabisulphite as preservative. The polysaccharide hydroxyethylcellulose (Natrosol 250HR, Hercules Ltd.) was added to the continuous phase at a range of concentrations. Two methods of preparation were used. In Method 1, the oil was n-heptane, n-decane, n-dodecane, or n-hexadecane (all >99 wt%, Sigma Ltd.) and the concentration of surfactant was 1.4 wt% in the continuous phase. An aqueous solution of the surfactant, polymer, and preservative was prepared by stirring at 80 °C until homogeneous. The solution was then added to the required volume of n-alkane (to a total of volume of 100 ml), and subjected to a set programme of shear cycles in a Waring commercial blender. Since the emulsion formation process is dependent on the physical properties and concentrations of the two phases, this method resulted in variable droplet-size distributions in the final emulsions, as shown in Table 1. Additionally, the emulsions of the lighter oils (n-heptane, n-decane, and n-dodecane) showed a tendency to instability via Ostwald ripening, resulting in significant size changes during the experiments. Figure 1 shows the size changes observed in a typical emulsion of 20 vol% n-decane without polymer. In order to obtain a reproducible size distribution, Method 2 was used. This involved the preparation of a concentrated emulsion pre-mix (containing 60 vol% oil and all the surfactant) which was emulsified as before. It was then diluted with a solution of polymer and preservative to obtain the required oil concentration. In these latter emulsions, the oil used was a mixture of 90 vol% n-heptane and

Table 1 *Typical average droplet-size parameters measured one hour after formation of emulsions of volume fraction $\phi_0 = 0.2$ by two different methods (see text)*

Oil	Weight mean diameter (μm)	10% dectile (μm)	90% dectile (μm)
Method 1			
n-heptane	2.42	0.61	5.65
n-decane	1.32	0.33	2.88
n-dodecane	1.25	0.33	2.69
n-hexadecane	1.56	0.46	3.03
Method 2			
n-heptane + n-hexadecane (90:10)	1.69	0.43	3.73

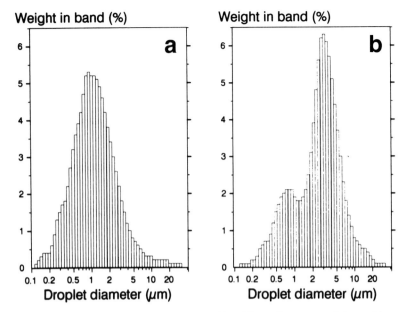

Figure 1 *Droplet-size distribution for a 20 vol% n-decane emulsion without polymer showing the effect of storage time at 20 °C: (a) 1 hour, (b) 14 days*

10 vol% n-hexadecane, and the concentration of surfactant was 0.35 wt% in the continuous phase. All the emulsions prepared by Method 2 possessed essentially the same droplet-size distribution, and the mixture of oils proved to be stable to Ostwald ripening while preserving a reasonably large density difference between the dispersed and continuous phases. The emulsions were all stored at 20 °C, either in 100 ml measuring cylinders or in perspex cells for use in the ultrasonic creaming monitor.

Characterization of Emulsions and Component Phases—The density of the oils and the continuous phases were measured using a Paar 602 density meter at 20 °C. Droplet-size distributions were determined using a Malvern Mastersizer. The viscosity (at low applied shear stress) of the continuous phases containing polymer was measured using a Bohlin Rheologi CS rheometer in a double-gap measuring system. Over a period of six weeks, we observed a decrease of *ca.* 30% in the viscosity of the polymer continuous phases. However, the changes were not significant during the initial creaming period.

After creaming, the emulsions formed two distinct layers: the concentrated cream above the clear serum sub-cream. The concentration of polymer in the sub-cream layers was determined using a spectrophotometric method.[7] The sub-cream layers were diluted to a nominal concentration of 0.006 vol% and 0.5 ml was pipetted into a stoppered test-tube, to which was added 1.5 ml of 3 wt%

phenol solution and 5 ml of concentrated sulphuric acid. The mixture was shaken and left to stand for 20 minutes, before samples were transferred to cuvettes for measurement of absorbance at 488 nm. Standard solutions of 0.005 and 0.007 wt% polymer were also measured for calibration purposes.

Measurement of Creaming Rate—Initially, creaming was measured by observing the rate of rise of a visible meniscus when the emulsions were stored in 100 ml measuring cylinders. Measurements of cream volume in millilitres were made once a day and converted to meniscus heights in millimetres. This method was used to determine creaming rates in the n-decane and n-dodecane emulsions. The other emulsions were investigated using the ultrasonic creaming monitor, and complete oil volume fraction profiles were obtained. The meniscus position was defined as the height of the contour of concentration $\phi_0/2$, where ϕ_0 is the initial oil concentration.

Determination of Oil Concentration Profiles—An instrument[8] based on the determination of the velocity of ultrasound through an emulsion was also used to monitor creaming at 20 °C. The instrument produces profiles of oil concentration as a function of height above the base of a sample cell of typical dimensions 24 mm wide × 32 mm deep × 160 mm high. Resolution is better than 2 mm in height, and 0.2 vol% in concentration. In polydisperse emulsions, if the droplets are unflocculated, creaming often occurs with a very diffuse meniscus that is not visible by eye. By collecting the concentration profiles, we detect the creaming at an early stage, and can estimate the droplet-size distribution from changes in the shape of the profiles during creaming.[9] In a flocculated system, the meniscus is usually visible, but the profiles give information on the detailed mechanisms of separation.[3]

3 Results and Discussion

Meniscus Creaming Rates—Figure 2 shows the height of the meniscus during creaming of emulsions containing 1 wt% polymer and initial concentrations (ϕ_0) of 5 vol% or 20 vol% n-hexadecane. At low oil concentrations, the meniscus rises steadily until it reaches the cream layer. This is similar to type I sedimentation described by Michaels and Bolger.[4] However, since the boundary is very sharp, it seems unlikely that the flocs formed are creaming individually, because their natural polydispersity would cause a more diffuse meniscus to develop, as is observed for individual droplets when no polymer is present.[9] When the oil concentration is increased to 20 vol%, the creaming rate shows several régimes, as in type II sedimentation.[4] There is a delay of over two days before creaming starts, and the cream layer undergoes significant compaction.

The effect of oil type on the meniscus creaming behaviour is shown for n-heptane, n-decane, and n-dodecane in Figure 3. It is clear that the lighter oils cream faster, and with less delay before creaming starts.

For type II creaming or sedimentation, the flocculated system is modelled in terms of a porous assembly of particles containing vertical channels through which the continuous phase flows. The channels take some time to develop

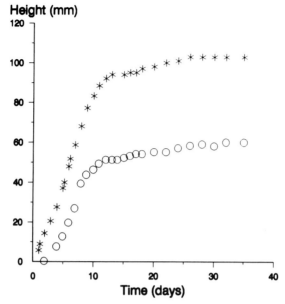

Figure 2 *Meniscus height* versus *storage time for* n-*hexadecane emulsions with* 1 wt% *polymer:* ∗, $\phi_0 = 5\%$; ○, $\phi_0 = 20\%$

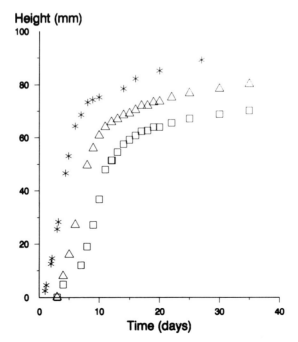

Figure 3 *Meniscus height* versus *storage time for emulsions of* n-*heptane* (∗), n-*decane* (△) *and* n-*dodecane* (□) *with* 1 wt% *polymer and* $\phi_0 = 20\%$

(hence the initial régime of increasing creaming rate) up to a maximum size, corresponding to the fastest creaming rate. The channels may be modelled to a first approximation by smooth straight cylinders of diameter d_p. The maximum creaming/sedimentation rate is then given by equation (23) in ref. 4, rewritten using our nomenclature as equation (4) above.

Making emulsions with the different oils enables us to vary ρ_d independently of the polymer concentration. However, the droplet-size distributions vary with oil type, and for n-heptane, n-decane, and n-dodecane also change with time. If the size of the vertical channels is related directly to the polymer concentration, and is not dependent on the size of the individual emulsion droplets, the creaming rates of the various oils in the same continuous phase should be directly related to the density difference $\Delta\rho = \rho_c - \rho_d$. Figure 4 shows the maximum creaming rates as a function of oil density for four concentrations of polymer. The variation with density is approximately linear, indicating that for a given polymer concentration the creaming rate is affected more by the density than by the other factors (*e.g.* particle size) which vary for the different oils. This implies that, once flocculated, the sizes of the primary droplets are of secondary importance. The emulsions containing n-heptane and the mixture of n-heptane + n-hexadecane contain particles of different size, but they show similar creaming rates. However, the intercept on the density axis is not as expected. According to equation (4), the

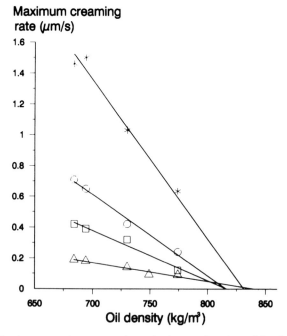

Figure 4   *Maximum meniscus creaming velocity as a function of the density of the oil phase for emulsions with $\phi_0 = 20\%$ and various polymer concentrations:* *, 0.35 wt%; ○, 0.5 wt%; □, 0.7 wt%; △, 1.0 wt%

creaming rates should be zero when the density of the oil equals that of the continuous phase, which in these systems ranges from 1001 to 1005 kg m^{-3}. All four polymer concentrations in Figure 4 show an intercept in the range 820–830 kg m^{-3}. The implication here is that the effective density of the droplets is considerably higher than that of the pure oil. This is not unreasonable, since surfactant layers may have a significant effect on the overall density of small droplets. Allowing for this effective density, the gradient of the dependence of the creaming rate on density for each concentration of polymer enables us to infer the size of the channels in the porous model, as shown below.

Effect of Polymer Concentration—The emulsions containing the mixture of n-heptane + n-hexadecane were found all to possess the same particle-size distribution, enabling us to study the effect of polymer concentration independently. However, the polymer had a large effect on the viscosity of the continuous phase, η_c. Visually, emulsions containing less than 0.04 wt% polymer appeared to be stable for several days, and the eventual creaming rate was slow, with a very diffuse meniscus. Figure 5 shows the rate of creaming versus polymer concentration for the higher concentration, where the meniscus was sharp and moved very fast. For comparison, the change in the continuous phase viscosity η_c is also

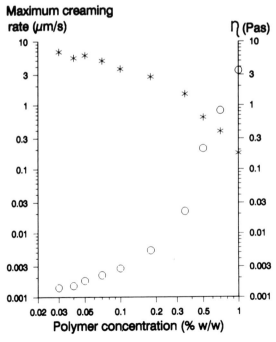

Figure 5   *Maximum creaming velocity (*) as a function of the polymer concentration for mixed-n-alkane emulsions, as compared with the variation in continuous phase viscosity η (O)*

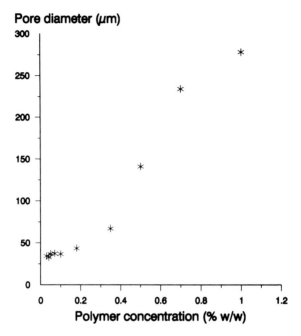

Figure 6 *Estimated diameter of pores in flocculated emulsions as a function of the polymer concentration*

shown. If the decrease in creaming rate with higher polymer concentrations were simply a result of increased viscosity, the two sets of data in Figure 5 would show the same variation with polymer concentration. Up to about 0.3 wt%, the changes with concentration are indeed similar (though, of course, in opposite directions). At higher polymer concentrations, the viscosity increases dramatically, but the creaming rate shows less change. Applying the pore/channel model, we can estimate the diameter of the pores, d_p, in equation (4) from the creaming rate v, the viscosity η_c, and the oil concentrations ϕ_0 and ϕ_m. We assume that the droplets within the flocs are close-packed, with $\phi_m = 0.7$ the final concentration in the cream layers without polymer. Figure 6 shows the estimated pore diameter as a function of polymer concentration. At low concentrations, the effective diameter of the pores is constant, at about 37 μm. At higher concentrations, there is a large increase with concentration up to about 300 μm in 1 wt% polymer. This is consistent with stronger flocculation at higher polymer concentrations.[4]

Oil Concentration Profiles—Figure 7 shows a set of oil concentration profiles for an emulsion made from a mixture of n-heptane + n-hexadecane, with no polymer in the continuous phase. The droplet-size distribution is the same as for the other mixed-oil emulsions. Figure 7a shows that initially the oil is uniformly dispersed in the container at a concentration $\phi_0 = 20$ vol%. After a few days (Figure 7b), the droplets have started to rise up the cell; so the concentration at the base has fallen,

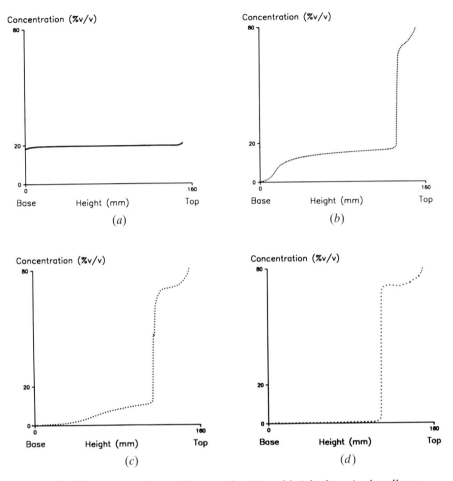

Figure 7 *Oil concentration profiles as a function of height for mixed-*n-*alkane emulsions without polymer:* (a) 0.02 *days,* (b) 3.23 *days,* (c) 12.94 *days,* (d) 48.25 *days*

and there is a concentrated cream layer at the top. With time, all the droplets arrive at the top (Figure 7d), having travelled individually at a speed consistent with their diameter and Stokes's Law [equation (1)]. It is possible to use the observation that they move at different speeds (resulting in the hazy, diffuse meniscus rising up the cell) to obtain the effective hydrodynamic size distribution.[9] (If the emulsion were exactly monodisperse, the meniscus would be sharp because they would all cream at the same speed.)

Figure 8a shows a set of superimposed concentration profiles for an emulsion containing 0.02 wt% polymer. The creaming was not visible to the eye until it was nearly complete, and the profiles are very similar to those without polymer. In fact, the creaming rates were reduced slightly due to the viscosity of the polymer solution, but no flocculation was observed. At a polymer concentration of 0.03 wt%, two distinct types of creaming behaviour are seen, as illustrated in Figure 8b. The majority of the oil droplets move up the cell rapidly, with a sharp meniscus, indicating that they are flocculated. However, about 5 vol% oil remains to cream slowly with a diffuse meniscus, as in the emulsions with zero or 0.02 wt% polymer.

When the polymer concentration is increased further (Figures 8c to 8h) several effects are apparent. As already discussed, the creaming rate of the meniscus does not entirely reflect the increasing viscosity of the continuous phase, due to changes in pore size. An increased strength of the flocs at higher concentrations of polymer is inferred from the estimate of pore diameter, and this is also evident in the reluctance of the cream layer to become closely packed. The oil concentration at the top of the samples in the early stages of creaming is indicative of the resistance of the flocs to compaction under gravity. With less than 0.03 wt% polymer, the cream builds to 70 vol% concentration right from the start, but in the flocculated systems it builds up to much lower packing densities. At the highest polymer concentration (Figure 8h), the cream density is almost uniform at $\varphi < 40$ vol% before the cream exhibits slow compaction with time, again uniformly.

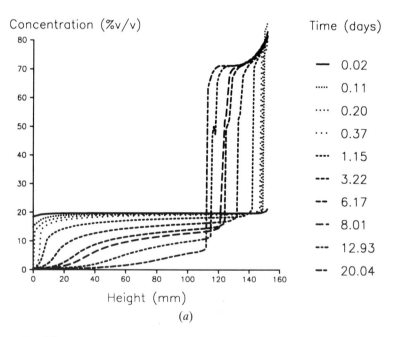

Figure 8 *Oil concentration profiles during creaming of 20 vol% mixed-n-alkane emulsions containing polymer:* (a) 0.02 wt% *polymer,* (b) 0.03 wt%, (c) 0.04 wt%, (d) 0.07 wt%, (e) 0.13 wt%, (f) 0.50 wt%, (g) 0.70 wt%, (h) 1.0 wt%

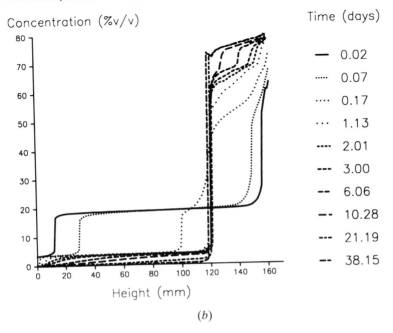

(b)

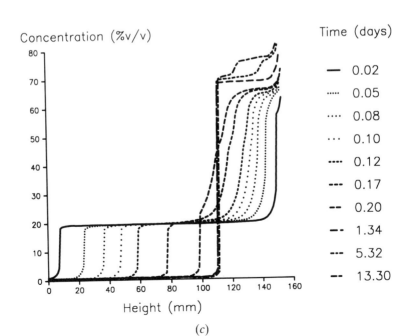

(c)

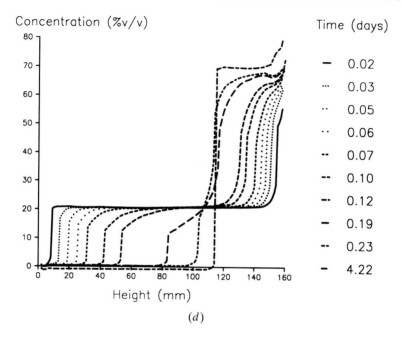

(*d*)

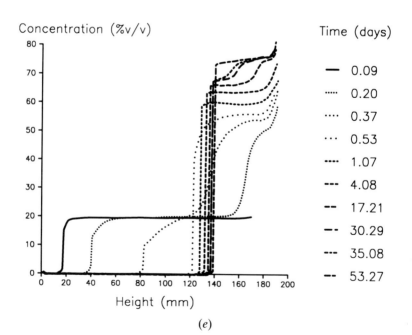

(*e*)

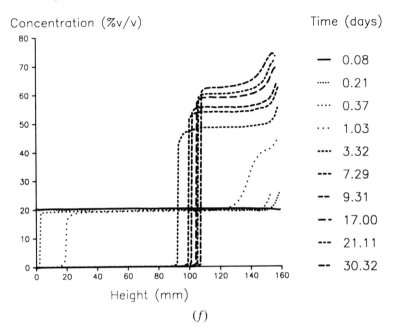

(*f*)

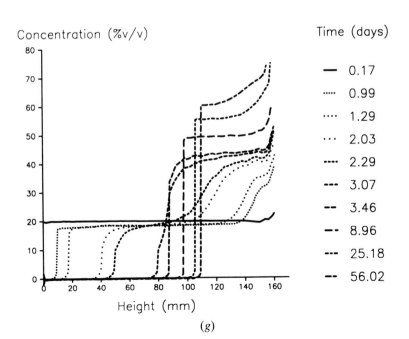

(*g*)

Concentration (%v/v)   Time (days)

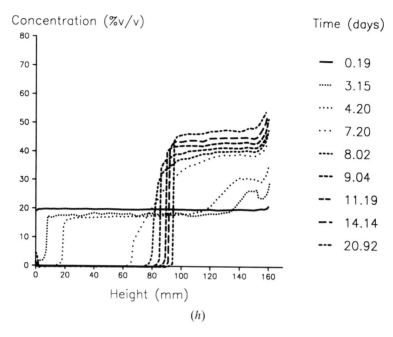

Height (mm)

(*h*)

Flocculation—Depletion or Bridging? The existence of a critical concentration (0.03 wt%) of polymer at which flocculation is initiated is consistent with a depletion rather than a bridging mechanism.[10,11] At the critical concentration, co-existence is observed between a flocculated phase and individual droplets. If one visualizes depletion flocculation as being driven by a need to reduce the volume of continuous phase from which the polymer is excluded, the occurrence of flocculation at the critical concentration should depend on the volume fraction of droplets. This idea is supported by an experiment performed using 0.03 wt% polymer, but with only 5 vol% oil. Figure 9 shows the concentration profiles obtained. No flocculated fraction is observed, all the droplets cream individually as in the systems with less polymer. Another consequence of depletion flocculation is that the final continuous phase, contained in the sub-cream layer, should contain slightly more polymer than the overall concentration, because the cream layer contains continuous phase depleted of polymer. For this reason, we analysed the sub-cream layers for polymer using density measurements and spectrophotometry. Figure 10 shows the estimated polymer concentration in the sub-cream layers. (The density results, which depend on surfactant concentration in addition to polymer concentration, have been corrected to allow for the surfactant present on the surface of the droplets.) There is clearly an enrichment over the initial, overall continuous phase concentration of polymer. Taken as a whole, the evidence is strongly in favour of depletion flocculation. The weakness of the flocs, and the improbability that the polymer could adsorb on to the surfactant-coated droplets, are consistent with the depletion model.

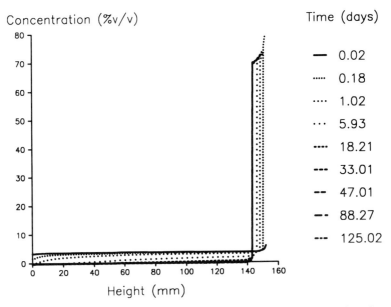

Figure 9 *Oil concentration profiles during creaming of* 5 vol% *mixed-n-alkane emulsion containing* 0.03 wt% *polymer*

The concentration profiles for the emulsions containing increasing amounts of polymer in Figure 8 show that the delay before the onset of creaming is highly dependent on the polymer concentration. At concentrations up to *ca.* 0.1 wt%, there is no measurable delay. At higher concentrations, however, the delay increases dramatically to nearly three days in the case of 1 wt% polymer. The delay may be interpreted as the time required for, firstly, flocculation, and, secondly, the formation of the most favourable channels for the ensuing creaming process. It would be interesting to study this delay in more detail, in order to obtain the dependence on ϕ_0, ϕ_m, *etc.* Clearly the background viscosity η_c will also affect the rate of aggregation of the droplets.

The shape of the concentration profiles shows a considerable amount of detail, particularly at the early stages of creaming with higher concentrations of polymer. The form of the profiles is similar to that observed by Furstenau (unpublished data, cited in ref. 4) and predicted by Russel.[6] However, at present we have not enough information on the physical properties of the flocculated system to reconcile our data with the theoretical treatment of the latter.

4 Conclusions

There is a considerable amount of information to be gained from the measurement of concentration profiles during the creaming of emulsions. Our results imply that the surfactant layer on the oil droplets significantly increases their effective density, which has implications for hydrodynamic sizing methods. The

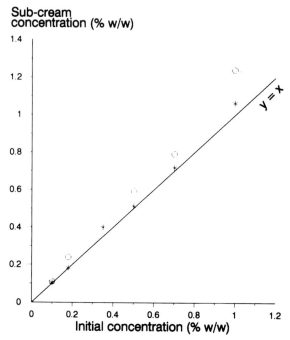

Figure 10 *Polymer concentration in sub-cream layer as a function of initial polymer concentration from density measurement (∗) and spectrophotometry (○)*

primary droplet size appears to be of secondary importance to the creaming rate once flocculation has occurred. The rate of rise of the meniscus in flocculated systems yields information on the strength of the flocculation, although it is difficult to check the inferred pore diameters by other methods. The critical polymer concentration required to flocculate the droplets, the co-existent phases observed at that concentration, and the absence of flocculation at lower ϕ_0, all indicate that the mechanism of flocculation is depletion rather than bridging. The polymer enrichment of the sub-cream layers provides direct evidence for its partial expulsion from the interstices between the close-packed droplets.

Acknowledgements. We are grateful to Dr Andrew Howe, now at Kodak Research Laboratories, Harrow, for his involvement in the early stages of the work, and for helpful discussions since his departure from the Institute. Technical assistance from Chris Carter, and financial support from the Ministry of Agriculture, Fisheries, and Food are also gratefully acknowledged.

References

1. E. Dickinson and G. Stainsby, 'Colloids in Food', Applied Science, London, 1982.
2. P. A. Gunning, M. S. R. Hennock, A. M. Howe, A. R. Mackie, P. Richmond, and M. M. Robins, *Colloid Surf.*, 1986, **20**, 65.

3. P. A. Gunning, D. J. Hibberd, A. M. Howe, and M. M. Robins, *Food Hydrocolloids*,
      1988, **2**, 119.
4. A. S. Michaels and J. C. Bolger, *Ind. Eng. Chem. Fund.*, 1962, **1**, 24.
5. J. F. Richardson and W. N. Zaki, *Chem. Eng. Sci.*, 1954, **3**, 65.
6. F. M. Auzerais, R. Jackson, and W. B. Russel, *J. Fluid Mech.*, 1988, **195**, 437.
7. M. Dubois, K. A. Gilles, J. K. Hamilton, P. A. Rebers, and F. Smith, *Anal. Chem.*,
      1956, **28**, 30.
8. A. M. Howe, A. R. Mackie, and M. M. Robins, *J. Disp. Sci. Technol.*, 1986, **7**, 231.
9. A. M. Howe and M. M. Robins, in 'Food Colloids', ed. R. D. Bee, P. Richmond, and
      J. Mingins, Royal Society of Chemistry, Cambridge, 1989, p. 382.
10. P. R. Sperry, H. B. Hopfenburg, and N. L. Thomas, *J. Colloid Interface Sci.*, 1981, **82**,
      62.
11. P. R. Sperry, *J. Colloid Interface Sci.*, 1982, **87**, 375.

Computer Simulations of the Flow of Deformable Particles

By G. C. Barker and M. J. Grimson

AFRC INSTITUTE OF FOOD RESEARCH, NORWICH LABORATORY, COLNEY LANE, NORWICH NR4 7UA

1 Introduction

The mechanical properties and the flow behaviour of colloidal materials are widely varying and, in many ways, unpredictable. The rheological data, in particular, are difficult to interpret, being most commonly described by complex empirical relationships which tend to be specific to one material. Generally the behaviour is non-linear; that is, there are no simple relationships between the driving forces and the fluxes. Particular examples are termed dilatancy, thixotropy, *etc.* The well dispersed materials which can behave in this way include common suspensions and emulsions, but more complex examples may variously be described as paste, sludge, or even 'pudding'! An illustration of the wide range of detailed rheological behaviour in the subject area of food colloids can be found in specialist volumes,[1,2] but the fundamental question of how to understand these phenomena, in terms of the microscopic character of the materials, remains largely unanswered.

One approach towards the interpretation of this complex behaviour introduces the concept of fluctuations of the internal structure of the droplets, flocs, or particles which form the fundamental constituents of the dispersion. The droplet shape, the droplet orientation, and the floc connectivity—all are perpetually changing, and they are able to respond to external driving forces through their interaction with the suspending fluid. The evolution of particle shapes is therefore an essential part of understanding complex suspension rheology. However, even for rigid particles, a detailed description of suspension dynamics is beyond the scope of rigorous theoretical analysis. This is primarily because of problems encountered with the accurate description of the interactions between moving particles. Some of these interactions are indirect, transmitted by the suspending fluid, and usually called 'hydrodynamic interactions'. Clearly these interaction problems are compounded by the additional motions associated with particle-shape fluctuations. Even the most advanced numerical scheme for computing many-body hydrodynamic interactions, Stokesian dynamics,[3] is only manageable for small systems of rigid particles, and cannot easily be extended to the composite structures considered here. Thus, as a first step towards understanding the dynamics of suspensions of deformable droplets, we have adopted a computational method in which the deformational and hydrodynamical problems are partially decoupled. We have used the free draining approximation,[4] commonly employed in polymer physics, for the interaction between the moving particles and the surrounding fluid.

The additional degrees of freedom associated with variable droplet shapes or variable floc connectivities lead to a large increase in the size of the relevant phase space which must be explored by a computer simulation of a dispersed system. The corresponding requirement for increased computational effort and complexity has restricted previous attempts to simulate suspensions and emulsions. Equilibrium simulations employing the Monte Carlo method have been introduced by Dickinson[5] and Barker and Grimson.[6] The lattice chain model[5] has features in common with both conventional colloidal dispersions and with macromolecular solutions. The lattice animal approach[6] has shown that amendments to rigid particle behaviour can arise either directly from changes to the inter-particle interactions and excluded volumes, or indirectly due to additional contributions to the entropy which arise from shape degrees of freedom.

In order to perform non-equilibrium simulations we have constructed model droplets in the form of aggregates of beads. The beads which make up each droplet are constrained to form a connected unit, but they may continuously alter their relative positions and connections in response to the local flow field of the suspending fluid or as a result of collisions with other droplets. Thus, the droplets are free to change their shapes, their orientations, and their internal structure. The motion of a droplet is a superposition of the motions of its constituent beads. The configurational space of droplet shapes examined in the simulation here is continuous, but otherwise the approach is closely related to the catalogue of four-site lattice animals used previously by Barker and Grimson.[6] The collection of beads forming an individual droplet may, at different times, form an extended chain structure at one extreme, or a folded, strongly bonded globule at the other.

The droplets described by our model most clearly resemble the free-flowing floc structures found in many colloidal materials. However, because of the approximations which have been introduced in order to facilitate an efficient implementation, the simulations are not valid as a model of one particular substance, or of one specific phenomenon, but rather they should be viewed as a well-defined representation of a general class of problems in which internal variables impinge on suspension behaviour. Our model overcomes the serious problems associated with the complete parameterization of flexible shapes, whilst remaining sufficiently economic to allow us to follow the detailed dynamics.

The model described here omits effects due to inertia, Brownian motion, and long-time relaxation phenomena. The droplet shape changes which we shall consider are large, and they cannot be counted as perturbations of symmetrical shapes. Therefore, the omission of Brownian motion is not serious if we restrict the droplets to colloidal sizes. Because of the elementary construction of the droplets, shape changes are divorced from physico-chemical properties, and so our interpretation of the results will be restricted to the mechanical properties of the suspension.

2 Sticky Sphere Simulations

We have performed computer simulations of two-dimensional, flowing, deformable droplets. The simulation technique has been adapted from the aggregating
sticky sphere model developed by Doi and Chen.[7] Each droplet is constructed
from n circular beads of radius a, and the droplets are placed in a square cell of
unit volume. Beads belonging to the same droplet see each other as 'sticky', and
beads belonging to different droplets see each other as 'repulsive'.

The suspending fluid has a single viscosity η_0 and is subject to an enforced
velocity field such that the Cartesian components of the fluid velocity are given by

$$V_x = \dot{\gamma}y, \qquad V_y = 0. \tag{1}$$

This velocity distribution is maintained in the finite simulation cell by applying
simple Lees–Edwards boundary conditions.[8]

The force between repulsive beads has a constant magnitude f_x and extends
over a short distance Da. A bead i at position r_i experiences a force $F_{ij} = f_x n_{ij}$
due to a repulsive bead j at position r_j if $|r_i - r_j|/a < (2 + D)$. Here $n_{ij} =
(r_i - r_j)/|r_i - r_j|$ is the unit vector from the centre of bead j to the centre of bead i.
A bead i experiences a constraining force f_{ij} when it contacts a sticky bead j. In the
simulations, a sticky 'contact' between bead i and bead j means that
$|r_i - r_j|/a < (2 + d)$ where $d \ll 1$. The constraining force is variable and acts to
maintain the two beads in contact. Sticky beads are free to roll over each other,
but cannot slip, and therefore the linear velocities v and the angular velocities w of
two sticky beads i and j which are in contact are constrained such that

$$v_i - aw_i \times n_{ij} = v_j - aw_j \times n_{ji}. \tag{2}$$

Our approximation to the bead dynamics neglects the inertia of both the beads
and the fluid, as well as the fluctuating Brownian forces. This represents an
approximation to low Reynolds number hydrodynamics. Within the free draining
approximation,[4] the hydrodynamic force on a bead is directly proportional to the
relative velocity of the bead with respect to the enforced fluid velocity. In the
absence of inertia this drag must balance the inter-particle forces. Thus we may
write

$$z_t(v_i - \dot{\gamma}y_i \mathbf{i}) = \sum_j f_{ij} + \sideset{}{'}\sum_k F_{ik} \tag{3}$$

$$z_r(w_i + (\dot{\gamma}/2)\mathbf{k}) = a \sum_j f_{ij} \times n_{ij} \tag{4}$$

where z_t and z_r are translational and rotational friction coefficients, and the
summations Σ, Σ' extend over sticky and repulsive beads, respectively.

For each droplet, in a configuration of N droplets with known repulsive forces,
equations (2) to (4) may be combined into m simultaneous equations for the
constraining forces f_{ij},

$$\sum_k [f_{ik} - f_{jk} - (a^2 z_t/z_r) \cdot (f_{ik} \times n_{ik} + f_{jk} \times n_{jk}) \times n_{ij}]$$

$$= \sum_k {}' (F_{jk} - F_{ik}) - (az_t\dot{\gamma}) \cdot (n_{ij}^y i + n_{ij}^x j), \quad (5)$$

where $(n - 1) < m < n(n - 1)/2$. Solution of these equations for each droplet allows us to evaluate the bead velocities using equation (3), and hence make a simple update of the particle positions according to the relation

$$r_i(t + \Delta t) = r_i(t) + v_i \Delta t, \quad (6)$$

where Δt is a small time increment.

Clearly, after the position update, there may be additional pairs of sticky beads which contact one another and a new list of sticky pairs must be formed before the next step. Sticky spheres are also able to break their contacts. If, after the solution of equation(s) (5), any of the constraining forces have an attractive normal component which exceeds a critical value, f_c, this pair of particles become unstuck. A new set of $m' < m$ equations is then formed and solved for f_{ij}. In the droplet simulations presented below, sticky bonds are removed (the strongest bond is broken first) subject to the constraint that the beads forming a droplet remain always connected.

The deterministic dynamics described above is characterized by six parameters. These are the forces f_c and f_x, the capture radii d and D, the velocity gradient or strain rate $\dot{\gamma}$, and the volume fraction of the beads $\phi = \pi n N a^2$. In what follows, we write $z_t a^2/z_r = 3/4$, and we express lengths, forces, and strain rates in units of a, f_c, and $\dot{\gamma}_c = f_c/az_t$, respectively.

The complex bead dynamics manifests itself as the flow of weakly interacting deformable objects which have variable internal structure.

3 Simulation Results

We have computed the particle dynamics by integrating equation (5) using a constant time step. To give numerical stability over a wide range of applied velocity gradient, Δt was chosen to give a constant incremental fluid strain, $\dot{\gamma}\Delta t = 0.01$; and to eliminate accidental breaking of sticky bonds, caused by second-order effects in the numerical integration, we have used the correction scheme proposed by Doi and Chen.[7] A capture radius $d = 0.01$ is consistent with this scheme throughout the range of velocity gradients which we have considered.

The unhindered motions of isolated deformable droplets, which are suspended in a fluid which has an imposed uniform velocity gradient, are summarized in Figure 1. An extended droplet experiences a torque due to the flow field, and it responds by rotating about its centre of mass. (In general, this rotational motion is superimposed on a steady translational motion, the stream velocity, which has been taken as zero for the sake of the representation in Figure 1.) In addition, the individual components of the droplets, i.e. the beads, also experience torques.

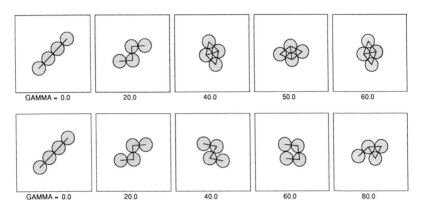

Figure 1   *Configurations during the motion of a chain of sticky beads* (n = 4) *in a fluid flow which has a uniform velocity gradient in the vertical direction. The two rows of pictures correspond to velocity gradients of* 0.5 *(upper) and* 5.0 *(lower).* γ *is the incremental fluid strain*

Relative motions of the beads are resisted partially by the sticky bonds, but these constraints on extreme beads are unbalanced and so they roll over their neighbours. This leads to an overall folding of the droplet. Eventually the extreme beads impinge on other beads within the droplet and new sticky bonds are formed. At a low strain rate (upper row of pictures in Figure 1), the new bonds are permanent, and the droplet forms a tightly structured 'globule' which pursues a steady-state tumbling motion. At a high strain rate (lower row of pictures in Figure 1), the torque imposed by the fluid cannot be constrained by the sticky bonds, and the globule ruptures whilst remaining connected. In this case the steady-state motion involves regular formation and rupture of sticky bonds. For the droplets composed of four beads shown in Figure 1, the steady-state motion is characterized by a mean number of sticky bonds per bead $b = 2.5$ and a radius of gyration $r_g = 1.41$ for strain rate $\dot{\gamma} = 0.5$, and $b \approx 1.5$ and $r_g \approx 1.64$ for $\dot{\gamma} = 5.0$. The isolated motion for droplets with $n = 4$ is typical of that for larger sticky sphere droplets.

We have characterized the repulsive interactions between beads by a range parameter $D = 0.5$ and a strength parameter $f_x = 5.0$. This choice is somewhat arbitrary, and different choices are worthy of investigation over and above that presented in this report, but this set of values is sufficient to prevent regular interpenetration of repulsive beads without swamping the effects due to the changing droplet structures. Without loss of generality, we restrict our discussions here to suspensions in which each droplet is composed of four beads. Simulations have been performed with typically $N = 40$ droplets. We have not noticed any remarkable system size effects.

Each simulation was started from a random sequentially adsorbed configuration of minimally connected—but otherwise arbitrary—droplet shapes. We have investigated volume fractions $\phi < 0.4$ and velocity gradients $0.1 < \dot{\gamma} < 10.0$. There is a transition period in which the suspension relaxes from the initial configuration towards a steady state. This progress is most easily observed by monitoring the average parameter values $\bar{b}$ or $\bar{r}_g$. At low strain rate, this equilibration time is quite small, *i.e.* $\Delta\gamma\,(= \dot{\gamma}t) = 20$; but, for strain rates $\dot{\gamma} \sim 1$, concentrated systems relax very slowly, and so we have $\Delta\gamma \sim 500$. After the transient régime is passed, the system samples states from an ensemble of steady states, and we have evaluated ensemble average properties over intervals which represent several initial relaxation periods.

For each suspension, we have determined the ensemble average of the incremental viscosity η. This is the increase in the viscosity caused by the addition of the droplets and is given by

$$\eta = -(a^2/\dot{\gamma})\left[\sum_{ij} (f_{ij}^x n_{ij}^y + f_{ij}^y n_{ij}^x) + f_x \sum_{ij}{}' r_{ij}^x r_{ij}^y\right], \tag{7}$$

where we have used the standard definition of the stress tensor.[4]

For a constant dispersed phase volume fraction, $\phi = 0.4$, we show in Figure 2 a plot of the incremental viscosity against the logarithm of the strain rate. In the steady state, the simulated viscosity exhibits considerable scatter, and so the points in Figure 2 are subject to 10% error. There are clearly two distinct flow régimes. At low strain rates viscosity decreases rapidly with strain rate and the suspension is shear thinning. At larger strain rates shear thinning is not as

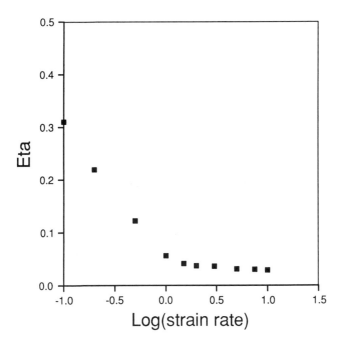

Figure 2 *Incremental viscosity plotted against the logarithm of the strain rate for a suspension with constant dispersed phase volume fraction $\phi = 0.4$*

Figure 3   *Instantaneous configuration of droplets taken from the steady-state régime of a simulation with volume fraction $\phi = 0.4$ and strain rate $\dot{\gamma} = 0.5$*

pronounced. The data in Figure 2 can be approximated by two straight line segments with gradients in *ca.* the ratio 20:1.

The region of rapid shear thinning coincides with a steady decrease of the mean number of sticky bonds per bead. The change of gradient in Figure 2 occurs at a strain rate for which the majority of droplets are minimally connected and $\bar{b} \sim 1.5$. Correspondingly, the mean radius of gyration of the droplets increases from $\bar{r}_g = 1.4$ to $\bar{r}_g = 1.7$ in the shear thinning region, but it increases much less rapidly for $\dot{\gamma} > 2$. These changes in droplet geometry are supported by 'snapshots' of the steady-state configurations. Examples are given in Figures 3 and 4. In Figure 3, the velocity gradient is $\dot{\gamma} = 0.5$, and there are many quadrilateral shapes with five sticky bonds each. In contrast, Figure 4, which corresponds to a velocity gradient of $\dot{\gamma} = 5.0$, contains mainly droplet shapes with chain-like connectivities.

Figure 5 shows a plot of the incremental viscosity against the dispersed phase volume fraction at a constant strain rate $\dot{\gamma} = 2.0$. At small volume fraction, η increases linearly with ϕ, but at $\phi \approx 0.25$ there is a sharp change in slope. This behaviour can be understood in te..ns of a change of position of the break point in Figure 2 as volume fraction is decreased. At $\dot{\gamma} = 2$, dilute suspensions are in the shear thinning régime, and concentrated suspensions are in the region of slowly changing viscosity. For isolated droplets, however, the change in behaviour occurs at $\dot{\gamma} \sim 3$.

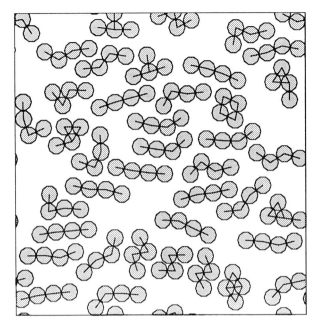

Figure 4 *Instantaneous configuration of droplets taken from the steady-state régime of a simulation with volume fraction $\phi = 0.4$ and strain rate $\dot{\gamma} = 5.0$*

4 Discussion

The simulations show a clear connection between variable droplet structure and complex suspension rheology. The results cannot be compared directly with experiments, however, because of the approximations which have been incorporated in the model. Nevertheless, Figure 2 is in good qualitative agreement with the rheological properties observed for many complex multi-phase colloidal materials, for example, evaporated milk.[9]

The sticky sphere model has been formulated without including aggregation effects or the possibility of droplet break-up, but it still contains features which lead to non-linear rheology. Simulations can play an essential role in highlighting the structural changes which take place within droplets which lead to the shear thinning behaviour. It is clear that these changes would not be distinguished by volumetric measurements of droplet sizes.

The present computations based on the sticky sphere model could easily be extended to three dimensions and to larger aggregates ($n > 4$) where the interplay between droplet structure and suspension behaviour might be more subtle. The shapes of droplets which we have considered here are obviously peculiar to $n = 4$. We make no interpretation of their significance other than their ability to transform easily from globule-like to chain-like configurations. With larger

$*10^{-2}$

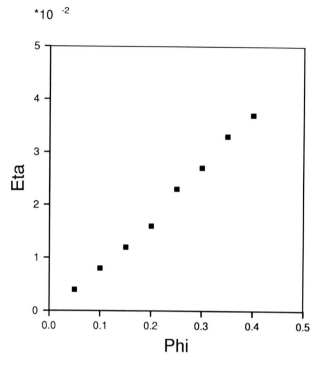

Figure 5   *Incremental viscosity plotted against the volume fraction for a suspension with constant strain rate $\dot{\gamma} = 2.0$*

clusters, the droplet picture is far more realistic, and shapes can be considered to be constrained by an effective surface tension. Recently, single 'drops', modelled as connected chains of hard particles, have been used to study the adsorption of deformable particles at a hard wall and non-linear 'flickering' effects.[10] The dynamic aspects incorporated in the free-draining sticky sphere model may extend the development of these investigations. Realistic direct repulsive interactions between droplets can also be incorporated in our model with only a small increase in computational effort, but accurate account of indirect interactions, *e.g.* using the scheme of Bossis and Brady,[3] awaits significant computational advances. We have made no attempt to investigate the transient régime of the droplet dispersions, but the model is sufficiently sensitive to be prepared in a well defined steady state, and therefore it may be used to follow the relaxation behaviour.

Acknowledgement. This work was supported by the Ministry of Agriculture, Fisheries, and Food.

References

1. 'Food Emulsions and Foams', ed. E. Dickinson, Royal Society of Chemistry, London, 1987.
2. 'Food Colloids', ed. R. D. Bee, J. Mingins, and P. Richmond, Royal Society of Chemistry, Cambridge, 1989.
3. J. F. Brady and G. Bossis, *Annu. Rev. Fluid Mech.*, 1988, **20**, 333.
4. M. Doi and S. F. Edwards, 'The Theory of Polymer Dynamics', Oxford University Press, Oxford, 1986.
5. E. Dickinson, *Phys. Rev. Lett.*, 1984, **53**, 728.
6. G. C. Barker and M. J. Grimson, *Mol. Phys.*, 1987, **62**, 269.
7. M. Doi and D. Chen, *J. Chem. Phys.*, 1989, **90**, 5271.
8. M. P. Allen and D. J. Tildesley, 'Computer Simulation of Liquids', Oxford University Press, Oxford, 1987.
9. P. Walstra, in 'Food Emulsions and Foods', ed. E. Dickinson, Royal Society of Chemistry, London, 1987, p. 242.
10. S. Liebler, R. R. P. Singh, and M. E. Fisher, *Phys. Rev. Lett.*, 1987, **59**, 1989.

only at concentrations above $40\,\mu$M. The conductivity T_{10} increases linearly with Tween 20 concentration although the general shape of the decay changes at concentrations above $80\,\mu$M.[5] The presence of Tween 20 has no effect on the protein foam stability up to a molar ratio of 1:1. However, higher Tween 20 concentrations decrease the amplitude of the conductivity T_{10} to a minimum at a molar ratio of 5:1 ($55\,\mu$M Tween). At still higher concentrations, the amplitude of the conductivity T_{10}, and the shape of the decay curve, is very similar to that obtained with an equivalent concentration of Tween 20 alone, which is consistent with the progressive displacement of the protein from the interface.

The detailed study of air-suspended thin films of mixtures of Tween 20 + β-lactoglobulin has revealed a transition in the drainage properties. Photographs of the 0.3 mm diameter planar films surrounded by their Plateau borders are shown in Figure 2. At Tween 20 concentrations of <0.5:1 ($5\,\mu$M), suspended thin films of the emulsifier + protein mixtures were found to drain like pure protein films.[4] Initially, the films contained static interference fringes or Newton's rings which

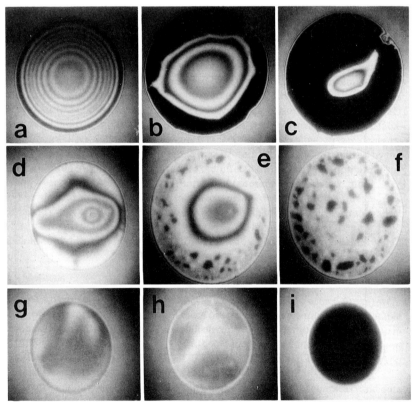

Figure 2 *Drainage patterns of thin films of mixtures of Tween 20 + β-lactoglobulin:* (a)–(c), 0.2:1; (d)–(f), 0.7:1; (g)–(i), 2:1. *See text for further details*

are consistent with regions of equivalent thickness (Figure 2a). The rate of drainage was slow, but the perimeter of the film thinned and became black more rapidly than the central region (Figure 2b and 2c); eventually, the whole film became black. At Tween 20 concentrations of 0.7:1, intermediate drainage behaviour was observed. Transient Newton's rings were present, but were distorted by some 'chaotic' drainage patterns (Figures 2d and 2e). These films did not thin to a uniform appearance, but contained black spots frozen in an otherwise grey film (Figure 2f). At Tween 20 concentrations above 0.8:1, initially the drainage was much more chaotic (Figure 2g). A uniform grey film formed fairly rapidly (Figure 2h) which slowly darkened to black (Figure 2i).

Fluorescence titration showed that Tween 20 was bound tightly by β-lactoglobulin to form a 1:1 complex with a dissociation constant K_d of 0.11 μM. This K_d value was used to calculate the relative concentrations of free Tween 20, free protein and Tween–protein complex as a function of total Tween 20 in the mixture. Up to a 1:1 molar ratio, effectively all the Tween 20 added is bound by the protein, leaving only two species present (*i.e.* free protein and the complex). Thus, a change in the interaction properties of the complex could account for the transition in drainage properties between ratios of 0.5:1 and 0.7:1. At a molar ratio of 2:1, free Tween 20 becomes the predominant component in solution.

A retinol binding site containing a tryptophan[7] has been identified on β-lactoglobulin. We obtain a dissociation constant of 60 nM, which corresponds to weaker binding than that reported previously.[7] Competitive binding studies reveal that the retinol binding site is distinct from the Tween 20 site of β-lactoglobulin.

The dependence of the surface diffusion coefficient of fluorescent β-lactoglobulin on the Tween 20 concentration is given in Table 1. At <0.9:1 the protein appears immobile at the surface, as has been also observed with other protein films. At 0.9:1, however, there is a sharp transition to a mobile protein film. The diffusion coefficient of the protein gradually increases with Tween 20 concentration. Since there is negligible free Tween present in solution at the transition in surface mobility, this phenomenon must arise from a change in interaction of the complex at the surface. This may be explained by steric shielding of the protein interaction sites by the polyoxyethylene chain of the

Table 1 *Surface diffusion of β-lactoglobulin (0.2 mg ml^{-1}) in suspended thin films of mixtures containing Tween 20*

Tween 20 concentration/μM	Molar ratio of Tween 20 to β-lactoglobulin	Diffusion coefficient/ 10^{-7} cm^2 s^{-1}
0	0:1	(immobile)
8.7	0.8:1	(immobile)
9.8	0.9:1	1.67
10.9	1.0:1	1.98
13.1	1.2:1	2.11
21.8	2.0:1	2.55
32.7	3.0:1	3.12
43.6	4.0:1	2.97
54.5	5.0:1	3.50

Tween molecule. This hypothesis is supported by a detectable increase in the hydrodynamic radius of β-lactoglobulin from 3.5 nm to 5.7 nm in the presence of Tween.

4   Conclusions

The three main conclusions of this work are as follows.

(1) Instability in the foam is correlated with the onset of surface mobility.

(2) The change in drainage behaviour and the onset of surface diffusion results from reduced interaction between the Tween–protein complexes at the surface since this species is predominant at the emulsifier concentration corresponding to the transition.

(3) The reduction in interfacial macromolecular interaction may arise from steric interference by the polyoxyethylene chain of the Tween 20 molecule projecting from the complex.

Acknowledgement. The authors would like to thank A. R. Mackie for assistance with the PCS experiments. The work was in part funded by the Ministry of Agriculture, Fisheries, and Food.

References

1. S. Futterman and J. Heller, *J. Biol. Chem.*, 1972, **247**, 5168.
2. M. Enser, G. B. Bloomberg, C. J. Brock, and D. C. Clark, *Int. J. Biol. Macromol.*, 1990, **12**, 118.
3. G. Scatchard, *Ann. N.Y. Acad. Sci.*, 1949, **51**, 660.
4. D. C. Clark, A. R. Mackie, L. J. Smith, and D. R. Wilson, in 'Food Colloids', ed. R. D. Bee, P. Richmond, and J. Mingins, Royal Society of Chemistry, Cambridge, 1989, p. 97.
5. M. Coke, E. J. Russell, P. J. Wilde, and D. C. Clarke, *J. Colloid Interface Sci.*, 1990, **138**, 489.
6. D. C. Clark, R. Dann, A. R. Mackie, J. Mingins, A. C. Pinder, P. W. Purdy, E. J. Russell, L. J. Smith, and D. R. Wilson, *J. Colloid Interface Sci.*, 1990, **138**, 195.
7. M. Z. Papiz, L. Sawyer, E. T. Eliopoulos, A. C. T. North, J. B. C. Findlay, R. Sivaprasadarao, T. A. Jones, M. E. Newcomer, and P. J. Kraulis, *Nature (London)*, 1986, **324**, 383.

Melting and Glass/Rubber Transitions of Starch Polysaccharides

By Mary A. Whittam, Timothy R. Noel, and Stephen G. Ring

AFRC INSTITUTE OF FOOD RESEARCH, NORWICH LABORATORY, COLNEY LANE, NORWICH NR4 7UA

1 Introduction

Starch polysaccharides, both in native granules and in a variety of foodstuffs, can occur with crystallinities ranging from 0 to 30%, the remaining material being amorphous. This amorphous material may be either brittle or rubbery, depending on whether it is below or above the glass transition temperature T_g of the polymer or polymer mixture. In order to understand properly the behaviour of starch during processing and in starch-containing foodstuffs, it is necessary to understand first the thermal behaviour of both the crystalline and amorphous components. The melting temperature T_m and the glass transition temperature T_g are the most important parameters characterizing the behaviour of the starch polymers over a wide temperature range.

Formation of a low molecular weight glass is a result of cooling the melt at a rate too fast for the molecules to relax into their lowest energy state, which below T_m is ideally that of a crystal. A supercooled liquid therefore forms, with a concurrent increase in viscosity, until at a temperature T_g, typically characterized by a viscosity of around 10^{12} Pa s, the liquid-like structure is 'frozen in' to form an amorphous solid. The glass transition is affected by kinetic factors, T_g being influenced strongly by the thermal history of the sample; for example, slower cooling rates give rise to lower values of T_g, providing of course that crystallization does not occur first. The dependence of the glass transition temperature on the heating and cooling rate, q, can be described by the relationship[1]

$$\mathrm{d}\ln|q|/\mathrm{d}(1/T_g) = -\Delta H/R, \qquad (1)$$

where ΔH is the activation enthalpy for the process of structural enthalpy relaxation. Viscosities of glass-forming liquids may be described by the semi-empirical Vogel–Tammann–Fulcher (VTF) equation[2–4]

$$\eta = \eta_0 \exp\left[A/(T - T_0)\right] \qquad (2)$$

where η_0 and T_0 are empirical constants in units of viscosity and temperature, respectively. As T decreases to T_0, the viscosity approaches a limiting value. Thus, T_0 may be thought of as representing a lower limit for the supercooled liquid. Treating viscosity as an activated rate process, a value for E_a, the activation energy for molecular mobility, can be measured from Arrhenius plots of $\ln \eta$ against $1/T$. This activation energy has been found to have similar values to the activation enthalpy as measured by heating rate experiments.[5]

The glass transition is accompanied by a rapid change, or a discontinuity, in thermodynamic properties such as heat capacity C_p and thermal expansivity, which are second derivatives of the Gibbs free energy. This has led to the description of the glass transition as a second-order transition (according to Ehrenfest's classification).[6] Further evidence for the thermodynamic significance of the glass transition comes from the 'entropy catastrophe' first pointed out by Kauzmann.[7] Since the entropy of the supercooled liquid decreases more rapidly with temperature than does that of the crystal, at some temperature T_k the entropy of the liquid will be equal to that of the crystal. Since the liquid cannot have a lower entropy than that of the crystal, this temperature marks the limit of supercooling at which the liquid must undergo freezing to a glass, accompanied by a rapid decrease in heat capacity. By equating the area under the extrapolated C_p curve for the supercooled liquid to the entropy of fusion, ΔS_f, a value for the temperature limit of supercooling can be calculated. This construction indicates that there may be a relationship between T_g and T_m, and this has indeed proved to be the case for a range of materials,[8] for which the semi-empirical rule $T_g/T_m \simeq 2/3$ appears to hold.

Addition of a diluent, or plasticizer, to a glassy material has long been known to lower its glass transition temperature. This process may be explained using a number of theoretical approaches, including free volume theories,[9] statistical theories based on entropic considerations,[10] and classical thermodynamic theories such as that of Couchman and Karasz.[11,12] For a polymer + diluent mixture, a thermodynamic consideration of the entropies of the two components at the glass transition temperature of the mixture leads to the expression[13]

$$T_g = \frac{w_1 \Delta C_{p1} T_{g1} + w_2 \Delta C_{p2} T_{g2}}{w_1 \Delta C_{p1} + w_2 \Delta C_{p2}}, \tag{3}$$

where w is the mass fraction of components 1 and 2, and ΔC_p is their individual heat capacity increments at their respective glass transition temperatures, T_{g1} and T_{g2}.

Polymer crystal melting, too, is affected by the presence of diluents. For polymer + diluent mixtures, the Flory–Huggins theory describes the relationship of T_m, the observed melting temperature at diluent volume fraction v_1, to the melting temperature of the pure crystalline polymer T_m^0:[14]

$$1/T_m = 1/T_m^0 + (R/\Delta H_u)V_u/V_1(v_1 - \chi v_1^2) \tag{4}$$

Here, V_u and V_1 are the molar volumes of the polymer repeating unit and diluent, respectively, ΔH_u is the enthalpy of fusion per repeating unit, and χ is the

Flory–Huggins interaction parameter whose value is determined by the solubilities of polymer and solvent in one another. Since this equation applies only to equilibrium crystals, its application to phase transitions of most starch + water systems, including native starch granules, has met with limited success. However, its application to highly crystalline preparations of fractionated starch molecules is perhaps more justifiable.

The crystalline component of starch granules is almost always in one of the two polymorphic forms, A or B, depending on botanical source. The crystal structures have been widely studied by various experimental techniques, including X-ray and electron diffraction, and by computer simulation, and it is now generally accepted that both forms consist of amylosic chains packed into double helices. The difference between the two polymorphs lies in the spatial arrangement of the double helices themselves. In the B form, six hexagonally arranged helices surround a central cavity containing water molecules, whereas the A form contains less water and is more densely packed, the central cavity being filled by another double helix. Single crystals of fractionated amylose molecules can be grown in either the A form or the B form,[15] although a variety of factors influence which polymorph forms preferentially. The A form is favoured by shorter chain length, higher crystallization temperature, higher polymer concentration, slower crystallization conditions, and by the presence of alcohol. The reverse is true for the B form, which crystallizes readily from pure water.[16,17]

In a previous paper,[18] studies on the effect of water as a diluent on the glass transition of malto-oligomers and starch were reported, and observed behaviour was compared to that predicted from a thermodynamic approach. In this paper, we shall examine thermodynamic and kinetic aspects of the glass transition of both starch and its low molecular weight analogue maltose, including the effect of water as a diluent. We shall also describe the diluent effect of water on the melting of the crystalline forms of amylose, and compare our results with theoretical predictions.

2 Materials and Methods

Materials—Maltose was obtained from Sigma Chemicals. The crystalline maltose was melted at 160 °C for 10 minutes, and then dried in a vacuum oven at 60 °C over P_2O_5. Samples were vitrified by cooling to -30 °C at a rate of 50 °C min^{-1}.

Amylose of approximately DP15 was prepared by acid hydrolysis of potato starch[19] and the product was isolated as described elsewhere.[20] Highly crystalline B spherulites were prepared by heating an aqueous solution of amylose (10 wt%) to 120 °C for 30 minutes, cooling to 80 °C and filtering through a Millipore filter (0.45μm), and then cooling at 5 °C h^{-1} to 5 °C. For preparation of A-type crystals, an equal volume of ethanol was added to the aqueous amylose solution at 80 °C just prior to cooling. Once formed, the crystals were centrifuged and rinsed three times with distilled water to remove any remaining soluble material, and then re-suspended in a little distilled water and stored at 1 °C. X-Ray diffraction patterns of the crystals showed line spacings representative of A and B forms,[21] whilst the sharpness of the lines was similar to that of previously reported preparations with high degree of crystallinity.[22]

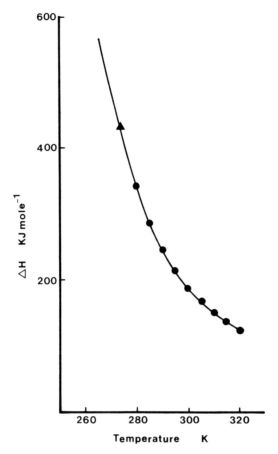

Figure 5   *Plot of activation enthalpy* ΔH *versus temperature for maltose + water mixture containing* 12.4 wt% *water:* ●, *viscometry;* ▲, *calorimetry*

appears to confirm that the two techniques are probing the same process in a complementary manner.

Crystal Melting Behaviour—The effect of heating rate on B crystal melting temperature, measured at the peak of the endotherm observed by DSC, is shown in Table 1. Each value represents the average for at least three determinations. Instrument sensitivity limited the experimental heating rate to 2.5 K min^{-1} on the low side, whilst the highest heating rate attempted was 20 K min^{-1} so as to ensure that there were no temperature gradients present in the samples. A difference of 3 °C in melting temperature was observed when the heating rate was varied from 2.5 K min^{-1} (T_m 74 °C) to 20 K min^{-1} (T_m 77 °C), indicating that there was a small degree of superheating of crystals at higher heating rates. The enthalpy of the melting transition, calculated by computation of the area under the endotherm

Table 1   *Effect of heating rate on melting temperature of B form amylose crystals*

Heating rate $(K\,min^{-1})$	Mass fraction water	T_m (K)	Enthalpy of melting, ΔH_m $(J\,g^{-1})$
2.5	0.93	347	36.5
5	0.92	348	34.7
10	0.92	349	35.4
20	0.92	350	35.5

with a fitted straight base line, shows no progressive increase or decrease with changing heating rate. Within experimental error, an average value for ΔH_m of $35.4 \pm 1.8\,J\,g^{-1}$ was obtained, a value considerably higher than the enthalpy of gelatinization of potato starch of $15.5\,J\,g^{-1}$, which takes place at 67 °C. Potato starch is chosen here for comparison because the length of its short amylopectin branches, generally considered responsible for granule crystallinity, is approximately the same as that for the samples used in these experiments. This indicates the higher degree of crystallinity of the prepared spherulites compared to that of native starch granules, which also contain a large proportion of amorphous material. This amorphous material may itself have an influence on the melting temperature of the crystalline regions.

Drying the spherulites had the expected effect on melting temperature, *i.e.* the crystals melt at higher temperatures as water content is decreased. The melting temperatures for both A- and B-type crystals are shown in Figure 6, plotted as a function of mass fraction of water. Crystals in the A form melted at significantly higher temperatures than B-type crystals at the same water content, even allowing for differences in water of crystallization between the two types. The similar shapes of the two curves shows that water has a similar effect on the melting behaviour of both crystal forms. That the A form might melt at a higher temperature than the B form had previously been postulated,[24] but not demonstrated directly. In theory, the more closely packed A form might be expected to possess a more stable structure than the less densely packed B polymorph, and this is supported by the evidence that A crystals do indeed melt at a higher temperature for any given water content above a mass fraction of water of *ca.* 0.4. Although a difference in melting temperature exists between the two polymorphs, their ΔH_m values for the melting transition appear, perhaps surprisingly, to be similar. Within experimental error, a value of approximately $35\,J\,g^{-1}$ was found for both A- and B-type crystal melting over the whole range of water contents tested. This is not in accordance with observed behaviour of different polymorphs of mono-acid triglycerides, where the more stable, higher melting polymorph also has the greater enthalpy (and entropy) of fusion. However, the presence of water in the amylose systems may lead to additional entropy effects other than those attributable to the melting of the dry crystals alone.

On cooling the samples back to 280 K after melting in the DSC, recrystallization was invariably found to occur. This was apparently always in the B form regardless of the original form of the crystals prior to DSC analysis, since, on rescanning, a melting temperature corresponding to that of the B polymorph was

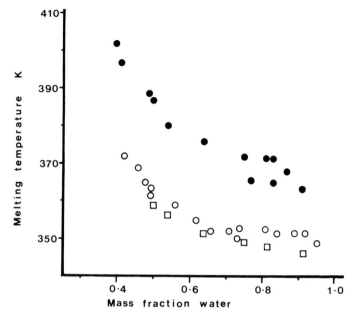

Figure 6   *Plot of melting temperature* versus *mass fraction of water for* A-*type*
(●), B-*type* (○), *and 'rescanned' (□) spherulites of amylose* (DP15)

observed. This melting temperature varied with sample water content in the same
way as that of the crystals initially prepared in the B form, and the results were
superimposable on the B crystal melting curve shown in Figure 6. The enthalpy of
melting for the recrystallized material was lower than that of the original crystal
preparations, and depended more on the period spent at 280 K than on the rate at
which the melted samples were cooled in the DSC.

Fitting the data to the Flory–Huggins equation, equation (4), gives values for
T_m^0, the 'ideal' melting temperature, of around 530 K for both polymorphs, and a
value for the Flory–Huggins interaction parameter χ of approximately 0.5. The
absolute accuracy of the extrapolated T_m^0 values is debatable, since the value of
$1/T_m$ varies rapidly with water volume fraction at the lower water contents, which
were not experimentally accessible. Nevertheless, these values are more realistic
than previously published results from work on whole starch, which gave
calculated T_m values as low as 441 K for potato starch crystallites, which is lower
even than the glass transition temperature of maltohexaose at 445 K shown in
Figure 1. This represents a physically untenable situation, since T_m must always
exceed T_g for any given material, and the glass transition temperature for the
amylose used for these experiments must exceed that measured for maltohexaose. Indeed, by extrapolation of the data in Figure 1, the value of T_g for
amylose of DP15 would be around 473 K, indicating that even our estimate of T_m^0
using the Flory–Huggins equation may be somewhat low. An explanation for the
very low T_m values found using whole starch may be that the amorphous granular

material has a depressing effect on the crystallite perfection and melting temperature; if this is indeed the case, then use of the Flory–Huggins equation is not well justified.

4 Conclusion

It has been shown that a thermodynamic description of the glass transition of binary polymer + diluent mixtures can lead to reasonable predictions about the behaviour of starch polysaccharides. It has also proved very useful to extrapolate from the behaviour of starch oligomers to gain new insight into the behaviour of polymeric material which is difficult to obtain directly. In this paper, we have investigated kinetic aspects of the glass transition using maltose + water mixtures as a convenient system to begin with. Further studies will enable the effects of increasing molecular weight and diluent content to be determined. From the calorimetry and viscometry experiments described, we have found that the activation energies for the relaxation processes observed by the two techniques are mutually consistent, indicating that the techniques are complementary.

The crystal melting experiments confirm that, as has been predicted, the A form of crystalline starch is more stable than the B form. And, as is observed for glass transition temperatures of amorphous polymers, the melting temperature of crystalline starch polysaccharides is depressed by the presence of a diluent, in this case water. Fitting the data to the Flory–Huggins equation gives an extrapolated value for the 'ideal' melting temperature of the crystals; further studies are needed to give insight into the melting and gelatinization behaviour of different starches with varying amylopectin chain lengths.

Starch itself is a complex combination of crystalline and amorphous material. By understanding and predicting the thermal behaviour of the two types of component, it will eventually be possible to build up a picture of the part played by each in determining functional behaviour of starch during food processing and on storage.

References

1. C. T. Moynihan, A. J. Easteal, J. Wilder, and J. Tucker, *J. Phys. Chem.*, 1974, **78**, 2673.
2. H. Vogel, *Phys. Z.*, 1921, **22**, 645.
3. G. S. Fulcher, *J. Am. Ceram. Soc.*, 1925, **77**, 3701.
4. G. Tammann and W. Hesse, *Z. Anorg. Allg. Chem.*, 1926, **156**, 245.
5. C. A. Angell, R. C. Stell and W. Sichina, *J. Phys. Chem.*, 1974, **78**, 2673.
6. P. Ehrenfest, *Proc. Amsterdam Acad.*, 1933, **36**, 153.
7. W. Kauzmann, *Chem. Rev.*, 1948, **43**, 219.
8. S. Sakka and J. D. Mackenzie, *J. Non Crystalline Solids*, 1971, **6**, 145.
9. F. Beuche, 'Physical Properties of Polymers', Interscience, New York, 1962.
10. J. H. Gibbs and E. A. Dimarzio, *J. Chem. Phys.*, 1958, **28**, 373.
11. P. R. Couchman and F. E. Karasz, *Macromolecules*, 1978, **11**, 117.
12. P. R. Couchman, *Macromolecules*, 1978, **11**, 1156.
13. G. ten Brinke, F. E. Karasz, and T. S. Ellis, *Macromolecules*, 1983, **13**, 244.
14. P. J. Flory, 'Principles of Polymer Chemistry', Cornell University Press, Ithaca, 1953, p. 563.
15. A. Buleon, F. Duprat, F. P. Booy, and H. Chanzy, *Carbohydr. Polym.*, 1984, **4**, 161.

16.   M. J. Gidley and P. V. Bulpin, *Carbohydr. Res.*, 1987, **161**, 291.
17.   B. Pfannemuller, *Int. J. Biol. Macromol.*, 1987, **9**, 105.
18.   P. D. Orford, R. Parker, S. G. Ring, and A. C. Smith, *Int. J. Biol. Macromol.*, 1989, **11**, 91.
19.   P. Colonna, A. Buleon, and C. Mercier, *J. Food Sci.*, 1981, **46**, 88.
20.   S. G. Ring, P. Colonna, K. J. I'Anson, M. T. Kalichevsky, M. J. Miles, V. J. Morris, and P. D. Orford, *Carbohydr. Res.*, 1987, **162**, 277.
21.   H. F. Zobel, in 'Methods in Carbohydrate Chemistry', ed. R. L. Whistler, Academic Press, New York, 1964, Vol. 4, p. 109.
22.   S. G. Ring, M. J. Miles, V. J. Morris, R. Turner, and P. Colonna, *Int. J. Biol. Macromol.*, 1987, **9**, 158.
23.   A. Sarko and H.-C. H. Wu, *Starke*, 1978, **18**, 263.
24.   M. J. Gidley, *Carbohydr. Res.*, 1987, **161**, 301.
25.   J. W. Donovan, *Biopolymers*, 1979, **18**, 263.

Calorimetric Study of the Glass Transition Occurring in Sucrose Solutions

By M. J. Izzard, S. Ablett, and P. J. Lillford

UNILEVER RESEARCH, COLWORTH LABORATORY, COLWORTH HOUSE, SHARNBROOK, BEDFORD MK44 1LQ

1 Introduction

Understanding the physical processes that occur, and the state of the non-equilibrium phases that are produced, during the freezing of concentrated saccharide solutions has fascinated research scientists for many years. In 1939, Luyet examined frozen polyhydric alcohol and saccharide solutions by cold-stage microscopy in order to establish the eruptive ice recrystallization temperature.[1] Following this early study on the behaviour of the ice phase, significantly more effort has focused on the physical state that the solute adopts during cooling and rewarming.[2–6]

It is now generally accepted that rapidly frozen dilute aqueous solutions of most saccharides exist as ice crystals embedded in an amorphous metastable glass containing a significant amount of unfrozen water.[7] Eutectic formation rarely occurs and the properties of the system are dominated by kinetics rather than by thermodynamics. This kinetic approach has been developed extensively by the pioneering work of Levine and Slade.[8–12] Their extensive studies on a wide range of saccharides, maltodextrins, corn syrups, and starches have stimulated renewed interest in this area and they have succeeded in drawing the attention of food scientists to concepts which were first developed in the field of polymer chemistry.[8] The differential scanning calorimetric method[10] (DSC), which is based upon the derivative thermograms of 20 wt% solutions, is used to determine two thermal properties, T'_g and W'_g, which are characteristics of the non-crystallizing solute.[10,13,14] T'_g represents the glass transition temperature T_g of the maximally freeze-concentrated solute. This corresponds to the intersection of an extension of the equilibrium liquidus curve and the glass curve. W'_g (expressed as grams of unfrozen water per gram of solute) or C'_g (in wt% solute) is defined as the composition of the glass having a glass transition temperature T'_g.

Sucrose solutions are selected here as a suitable model to study the effect of initial solute concentration on the behaviour of frozen saccharide solutions. For sucrose solutions, values of T'_g and C'_g have been quoted[9,15,16] as $-32\,°C$ and 64.1 wt%, respectively. This study will describe how DSC can be used, on dilute and concentrated sucrose solutions, to investigate the nature of the glass transition in frozen systems. New information on the physical significance of T_g' will be given, and it will be demonstrated that the recently proposed value[15,17] of C_g' is clearly incorrect.

2 Materials and Methods

Sucrose (analytical reagent grade) was obtained from FSA Laboratory Supplies (Loughborough) and BDH (Poole), and samples were used as supplied. Sucrose solutions were prepared for calorimetric studies by gently heating weighed dispersions of sucrose + water, in either a water bath or a domestic microwave oven, until a clear solution was produced. For highly concentrated sucrose solutions, the temperature of the mixture was raised until the crystals would just dissolve with vigorous stirring. These supersaturated solutions were used after they had cooled to room temperature. It was found that 65–70 wt% solutions could be kept at room temperature for several days before solute crystallization occurred. Above 70 wt%, crystallization occurred more rapidly, and so the solutions had to be used immediately. For long-term storage, solutions were stored deep frozen at *ca.* −20 °C, and it was found that none of the dilute or concentrated solutions showed any signs of *solute* crystallization after 1 year. Dilute sucrose solutions containing ice were easily returned to homogenous solutions by warming up to room temperature. Concentrated sucrose solutions remained very viscous and *solute* crystallization only occurred on warming to room temperature.

The concentration of the sucrose solutions was confirmed by refractive index measurements. These were carried out at 20 °C using an Abbe 60/95 refractometer (Bellingham Stanley) with a sodium lamp as the illuminating source; the concentrations were determined using previously published values.[18]

Differential Scanning Calorimetry—Low-temperature calorimetry was carried out on a Perkin–Elmer DSC7 differential scanning calorimeter equipped with either an Intracooler II or a liquid nitrogen driver-controlled cooling accessory. This permitted measurements to be carried out starting from either −55 °C, or at least −100 °C, respectively. The purge gases used were dry nitrogen with the Intracooler II, and dry helium with the controlled cooling accessory.

Temperature calibration was carried out at a scan rate of 5 °C min^{-1} using GPR grades of n-dodecane and n-octane (BDH, Poole). Calibration of the heat flow was carried out by reference to the known melting enthalpy of indium metal (99.99% pure, Goodfellows Metals). Sucrose solutions were accurately weighed into aluminium pans using a Mettler ME30 six-place electronic balance, and were then sealed hermetically. Sample weights of 1–16 mg were used for solutions that formed ice on cooling, and larger sample weights of 10–35 mg were used for sucrose glasses in order to improve the signal-to-noise ratio. The glass transition temperature T_g was obtained from peak analysis of the derivatized thermograms (Figure 1). The increase in heat capacity, ΔC_p, was calculated from the change in heat flow, ΔY, the scanning rate, and the sample weight:

$$\Delta C_p / \text{J g}^{-1}\,°\text{C}^{-1} = \frac{(\Delta Y/\text{mW})}{(\text{sample wt/mg}) \times (\text{scan rate}/°\text{C s}^{-1})}. \qquad (1)$$

Reproducible ΔC_p values (± 0.05 J g^{-1} °C^{-1}) were obtained providing there was at least 10 °C of baseline on either side of the transition. The baseline of most of the thermograms presented here has been corrected using the slope function. No smoothing of the data has been carried out. The initial value of the heat flow has no special significance; it merely assists in the presentation of the experimental data.

Optical Microscopy—Sucrose solution was placed between two circular microscope coverslips and placed in a Linkam THM 600 programmable stage powered by a Linkam TMS90 temperature controller. Fast cooling at 99 °C min^{-1} was achieved by coupling the stage to a Linkam CS196 liquid nitrogen cooling unit. Calibration was carried out using n-octane, n-dodecane, and water.

Good agreement between the quoted literature melting points and the displayed temperature was found at heating rates up to 10 °C min^{-1}. Micrographs were obtained by utilizing the transmission imaging mode of a BioRad MRC-500 confocal microscope. The images were subsequently improved using the image scaling and contrast enhancement package of the system.

3 Results and Discussion

Sucrose Glasses—Sucrose + water glasses were formed when a sucrose solution of concentration greater than 65 wt% was rapidly cooled in the DSC. The effects of heating rate (Figure 2) and cooling rate (Figure 3) were investigated on a

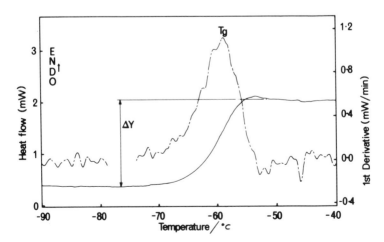

Figure 1 *Typical glass transition (——) and first derivative trace (——-) of a 73 wt% sucrose + water glass*

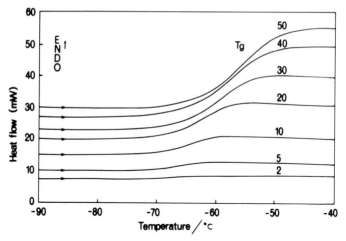

Figure 2   *The effect of heating rate (measured in °C min^{-1}) on the T_g value of a 70.1 wt% sucrose + water glass*

70.1 wt% sucrose solution. The T_g value obtained was highly dependent upon the rate (Figure 4), but at slow scanning speeds, *i.e.* <10 °C min^{-1}, the T_g value obtained was close to the extrapolated zero scanning rate value. A heating rate of 5 °C min^{-1} was used throughout the course of these studies unless stated otherwise. The T_g values of sucrose glasses determined over the concentration range 65–82 wt% are given in Table 1. Also included are data over the higher concentration range 93.9–96.7 wt%.[19]

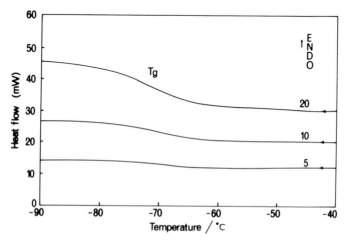

Figure 3   *The effect of cooling rate (measured in °C min^{-1}) on the T_g value of a 70.1 wt% sucrose solution*

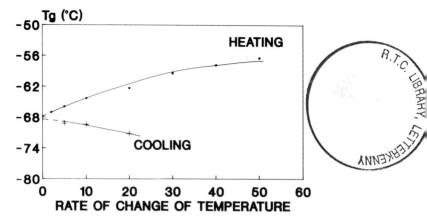

Figure 4   *Glass transition temperatures* T_g *obtained during the heating (■) and cooling (+) of a* 70.1 wt% *sucrose solution*

The magnitude of the increase in heat capacity occurring at the glass transition was measured for 65–82 wt% sucrose solutions at scan rates of 10, 20, 30, and 40 °C min^{-1}.

The results in Table 2 would seem to suggest that higher scan rates tend to give lower values, but further experimentation would be required to confirm this. However, it should be noted that when a 72.4 wt% glass was scanned at rates between 5 and 12 °C min^{-1}, almost identical values were obtained.

Table 1   *Effect of sucrose concentration* C_s *on the glass transition temperature* T_g *of water + sucrose*

C_s/wt%	T_g/°C	C_s/wt%	T_g/°C
65.0	−75	80.4	−39
68.5	−68	82.0	−33
70.1	−64	93.9	23
73.0	−60	95.1	30
75.1	−53	96.5	38
78.8	−43	96.7	42

Table 2   *Effect of DSC scan rate on the* ΔC_p *values of sucrose glasses*

Concentration/%	ΔC_p/J g^{-1} °C^{-1}			
	10 °C min^{-1}	20 °C min^{-1}	30 °C min^{-1}	40 °C min^{-1}
65.0	0.67	0.72	0.68	0.63
68.1	0.804	0.748	0.73	0.70
70.1	0.755	0.712	0.66	0.67
77.4	0.733	0.737	0.742	0.724
81.7	0.828	0.786	0.731	0.715

Glasses Showing Devitrification—Sucrose glasses above 70 wt% never showed the presence of ice in the DSC experiments, but, when the concentration fell below 70 wt%, ice often formed on heating as a result of devitrification. A 68.1 wt% solution, which had been rapidly cooled, was found to have formed a totally vitrified glass. When examined in the heating mode, the amount of ice formation was found to be dependent on the scan rate used (Figure 5). At very high scan rates, almost no ice was formed, and the sample essentially transformed directly from a glass to a viscous solution. On the other hand, for a 65 wt% solution, significant devitrification (ice formation) occurred for all the heating rates employed (5–40 °C min^{-1}). The results demonstrate that the behaviour of these solutions is dominated by the kinetics. The large exotherm present in the DSC thermograms (Figures 5 and 6) was confirmed to be due to devitrification by optical microscopy (Figure 7). It can be seen in Figure 7a that ice nucleation occurs at −55 °C, and that this is followed by ice crystal growth whilst holding at −35 °C (Figure 7c–e). Finally, ice melting occurs at higher temperatures (Figure 7f).

It is possible to observe the changes that occur in T_g as a result of devitrification (Figure 6). When a 65 wt% solution is rapidly cooled in the DSC, no ice is initially formed as the solution vitrifies into a glass. Ice formation does, however, occur on heating. The effect of progressively allowing ice to form was investigated by raising the temperature of this rapidly cooled sample to a temperature above the glass transition for varying periods of time. This was achieved by raising the temperature of a vitrified 65 wt% sucrose solution to −35 °C for a set period of

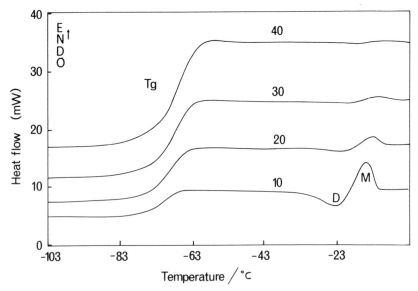

Figure 5   *The effect of heating rate (measured in °C min^{-1}) on the heat flow thermogram of a rapidly cooled 68.5 wt% sucrose solution (T_g = glass transition, D = devitrification, M = melting)*

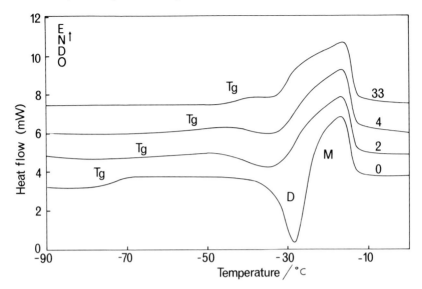

Figure 6   *The effect of holding time in minutes at −35 °C on a rapidly cooled 65 wt% sucrose solution (T$_g$ = glass transition, D = devitrification, M = melting)*

time, followed by rapid cooling to −95 °C. The sample was then scanned in the normal manner to record the glass transition and the ice peak. The T_g values are given in Table 3. These values demonstrate that, as more ice is formed in the sample, so the T_g value is progressively moved to a higher temperature, as a result of freeze concentration. After 15 minutes annealing, the T_g value had increased by 30 °C (Figure 6). This represents an increase in the sucrose glass concentration from 65 to 79 wt% (Table 1). As the T_g value increased, less devitrification was noted, until it was virtually absent after 33 minutes annealing. Broadening of the heat-flow curve over the glass transition temperature range during intermediate annealing times was thought to be due to concentration gradients present in the devitrifying mixture. Annealing for very long periods at these temperatures

Table 3   *Effect of annealing time at −35 °C on the glass transition temperature T$_g$ of a devitrifying 65 wt% sucrose solution*

Time/min	T_g/°C
0	−73.1
0.1	−71.7
1	−71.3
2	−65.0
4	−52.5
15	−43.7
33	−42.1

Figure 7 *Optical micrographs obtained on heating a rapidly cooled* 65 wt%
sucrose solution to (a) −55 °C, (b) −45 °C, (c) −35 °C, (d) −35 °C
(1 *minute*), (e) −35 °C (6 *minutes*), (f) −22 °C

allowed the glass transition temperature very slowly to increase, and, occasionally, it was possible to bring T_g to within one degree of the annealing temperature (Figure 8).

Concentration of the Freeze-concentrated Glasses—Since the 65 wt% and 68 wt% sucrose solutions were able to form significant quantities of ice, the previously reported[15-17] value of $C_g' = 64.1$ wt% is clearly too low. The reported method used to estimate C_g' is one that is often used in the food industry[20,21] for the measurement of ice content, from which the concentration of the freeze concentrated solution is calculated. The amount of ice present is calculated from the total amount of heat (*i.e.* the integral of the DSC ice-melting peak) required to melt the ice in a known weight of sample. The amount of ice present is then calculated from

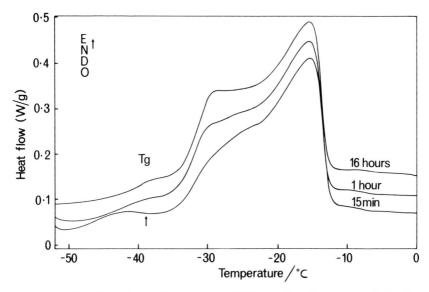

Figure 8   *The effect of annealing time at −38 °C on a 65 wt% sucrose solution (↑ indicates the annealing temperature)*

the known latent heat of ice[22] (334.15 J g^{-1}). The reported[15–17] value of C_g' for sucrose of 64.1 wt% had been obtained from a sample with an initial solute concentration of 20 wt%. Using this same method, we obtain a very similar value (*i.e.* $C_g' = 63.9$ wt%) for the same initial concentration of sucrose. Therefore, this suggests that there is a problem with the method, rather than with its application. This is also highlighted in Table 4, which demonstrates that the calculated C_g' value is dependent on the initial concentration of the sucrose solution. Table 4 clearly demonstrates the inaccuracies of calculating C_g' by this method. Similar reservations about the above method have also been recently expressed following DSC studies on sucrose[23] and galactose[24] solutions.

During our study on sucrose solutions, ice crystal formation and very slow growth was noted in a 73 wt% sucrose solution that had been stored in a deep freeze (−25 °C) for one year. The crystals were shown to be ice because they

Table 4   *Values of C_g' as a function of the initial sucrose solution concentration C_s^0*

C_s^0/wt%	C_g'/wt%
2.6	35.1
10.7	61.9
20.2	63.9
40.0	71.7
60.1*	75.8
65.8*	77.7

* Annealed at −30 °C for 30 minutes.

dissolved when the mixture was allowed to warm up, and completely disappeared when the temperature reached 0 °C. This observation demonstrates that C_g' must be greater than 73 wt%.

Glass and Ice Mixtures—For more dilute sucrose solutions, it was possible to eliminate most of the complexities in the DSC thermograms caused by partial or total vitrification by annealing at −30 °C. This annealing was carried out on solutions having concentrations from 10 to 65 wt% for a period of 30 minutes.

The thermograms obtained are given in Figure 9, together with those for similarly annealed sucrose glasses in the concentration range 75–80 wt% sucrose. The traces have been normalized to take into account differences in sample weight. The solutions had freeze concentrated to produce a glass with a T_g at −45 °C, corresponding to a composition of *ca.* 78 wt% (from Table 1), which was immediately followed by a small amount of devitrification. These transitions in the thermograms were then followed by a much larger step increase in ΔC_p immediately prior to the ice-melting peak. It should be noted that this larger step increase is the so-called T_g' temperature which is what Levine and Slade routinely measured.[8–12]

The increases in heat capacity, ΔC_p, occurring at the glass transition and at this second transition are plotted in Figure 10. These results show that the second transition is significantly sharper than the glass transition observed from concentrated samples when there is no ice present. There is also a much larger increase in ΔC_p than would be expected for a glass transition. Finally, it would seem not to be a second-order transition because there is no step increase in the base line on either side of the ice melting peak. In the light of all this evidence, we propose that

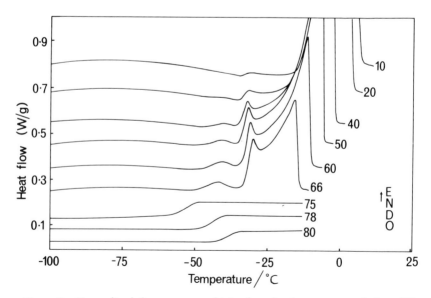

Figure 9   *Normalized thermograms obtained on heating sucrose solutions (10–80 wt%) after annealing at −30 °C for 30 minutes*

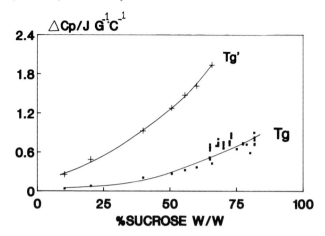

Figure 10   *The change in heat capacity, ΔC_p, occurring at the glass transition and the Levine and Slade transition for annealed sucrose solutions and glasses: ■, glass transition, T_g; +, Levine and Slade transition, T'_g*

the second large transition observed in the DSC thermogram, which has been called the T'_g glass transition, is not a glass transition at all, but is due to the onset of ice melting, which starts to occur soon after the sample has gone through the glass transition, and represents the temperature at which the sample changes from kinetic to thermodynamic control.

Supplemented Phase Diagram—These results, together with the liquidus curve,[18,25–27] the T_g of glassy water,[28] and sucrose,[29,30] can be used to construct a supplemented phase diagram (Figure 11). This is similar to the one proposed by Mackenzie,[5] and Luyet and Rasmussen,[3] but very different to the one recently

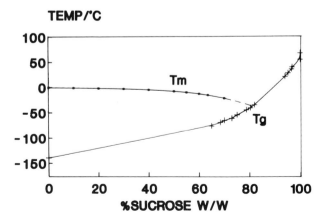

Figure 11   *Supplemented phase diagram for sucrose + water system (T_m = liquidus curve, T_g = glass transition curve)*

proposed by Levine and Slade.[17] It can be seen that the intersection point of the T_m and T_g curves is at a sucrose concentration (*i.e.* C'_g value) of *ca.* 80 wt%, and at a temperature (*i.e.* T'_g value) of *ca.* $-35\,°C$. This is further evidence that the C'_g concentration of 64.1 wt% determined by the method proposed by Levine and Slade[8-12] is wrong.

Acknowledgements. The authors would like to thank Mrs S. J. Lane for invaluable technical assistance, Mr D. P. Ferdinando for the micrographs of the devitrifying sucrose solution, and Drs A. Lips and A. H. Clark for fruitful discussion on aspects of the C'_g measurement. Thanks are finally due to H. Levine and L. Slade, who originally lit the flame to restimulate interest in this area, and who very kindly provided us with fruitful discussion during the early stages of this study.

References

1. B. J. Luyet, *J. Phys. Chem.*, 1939, **43**, 881.
2. F. E. Young, F. T. Jones, and H. J. Lewis, *J. Phys. Chem.*, 1952, **56**, 1093.
3. B. Luyet and D. Rasmussen, *Biodynamica*, 1968, **10**, 167.
4. D. Rasmussen and B. Luyet, *Biodynamica*, 1969, **10**, 319.
5. A. P. MacKenzie, *Phil. Trans. Roy. Soc. (London)*, 1977, **B278**, 167.
6. F. Franks, 'Biophysics and Biochemistry at Low Temperatures', Cambridge University Press, 1985, Chap. 3, p. 37.
7. F. Franks, *J. Microscopy*, 1986, **141**, 243.
8. H. Levine and L. Slade, *CryoLetters*, 1988, **9**, 21.
9. L. Slade and H. Levine, *Pure Appl. Chem.*, 1988, **60**, 1841.
10. H. Levine and L. Slade, *Carbohydr. Polym.*, 1986, **6**, 213.
11. H. Levine and L. Slade, in 'Food Structure—Its Creation and Evaluation', ed. J. M. V. Blanshard and J. R. Mitchell, Butterworths, London, 1988, p. 149.
12. H. Levine and L. Slade, *J. Chem. Soc., Faraday Trans. 1*, 1988, **84**, 2619.
13. F. Franks, in 'Water: A Comprehensive Treatise', ed. F. Franks, Plenum Press, New York, 1982, p. 215.
14. F. Franks, 'Properties of Water in Foods', Martinus Nijhoff, Dordrecht, 1985, p. 497.
15. H. Levine and L. Slade, *CryoLetters*, 1989, **10**, 347.
16. F. Franks, *Process Biochem.*, 1989, **Feb** iii.
17. H. Levine and L. Slade, *Comments Agric. Food Chem.*, 1989, **1**, 315.
18. Handbook of Chemistry and Physics, 58th edn, CRC Press, Cleveland, 1977, p. D-261.
19. J. M. V. Blanshard, A. Gough, and M. T. Kalichevsky, personal communication.
20. Y. H. Roos, *J. Food Sci.*, 1987, **52**, 146.
21. M. Ruegg, M. Luscher, and B. Blanc, *J. Dairy Sci.*, 1974, **57**, 387.
22. J. G. Stark and H. G. Wallace, 'Chemistry Data Book', John Murray, London, 1980.
23. A. B. Biswas, C. A. Kumsah, G. Pass, and G. C. Philipps, *J. Solution Chem.*, 1975, **4**, 581.
24. G. Blond, *CryoLetters*, 1989, **10**, 299.
25. F. E. Young and F. T. Jones, *J. Phys. Chem.*, 1949, **53**, 1335.
26. A. Leighton, *J. Dairy Sci.*, 1927, **10**, 300.
27. H. M. Pancoast and W. R. Junk, 'Handbook of Sugars', AVI Publishing, West Port, Connecticut, 1973.
28. M. Sugisaki, H. Suga, and S. Seki, *Bull. Chem. Soc. Jpn.*, 1968, **41**, 2591.
29. P. D. Orford, R. Parker, and S. G. Ring, *Carbohydr. Res.*, 1990, **196**, 11.
30. L. Finegold, F. Franks, and R. H. M. Hatley, *J. Chem. Soc., Faraday Trans. 1*, 1989, **85**, 2945.

Small-angle X-Ray Scattering and Differential Scanning Calorimetry from Starch and Retrograded Starch

By R. E. Cameron and A. M. Donald

CAVENDISH LABORATORY, UNIVERSITY OF CAMBRIDGE, MADINGLEY ROAD, CAMBRIDGE CB3 0HE

1 Introduction

Starch is laid down in higher plants in the form of insoluble birefringent granules. The size and shape of these spherulitic structures vary with botanical source, the granules of wheat starch being approximately spherical with a diameter of around 10 μm. The granules are composed of two polymers containing glucose units: the highly branched amylopectin and the essentially linear amylose. The physical arrangement of these polymers is thought to be in the form of radially oriented crystallites consisting of alternating layers of amorphous and crystalline regions, the crystalline component being mainly due to the amylopectin.[1,2] A variety of physical techniques has provided information on the nature of these crystallites. Wide-angle X-ray diffraction studies have revealed three types of polymer crystallinity in starches: the A, B, and C forms, which reflect the packing arrangements of the amylopectin helices.[3-6] Line broadening measurements suggest a crystallite width of about 15 nm,[7] a figure which is confirmed by contrast matching experiments with small-angle neutron diffraction.[8] Experiments using small-angle neutron scattering,[8] small-angle X-ray scattering (SAXS),[9-11] and electron microscopy[11] have shown a periodicity occurring at 10 nm (see Figure 1) which is thought to represent the repeat distance between amorphous and crystalline regions of amylopectin. These findings are summarized in the model in Figure 2 for the starch crystallite proposed by Blanshard.[12]

When the granules are heated above a certain temperature range (50–60 °C for wheat starch) in excess water, gelatinization occurs. The granules swell, as water is absorbed and amylose is leached out. The structure is disrupted and the periodicities observed by small- and wide-angle diffraction disappear. This is the process which occurs when starch is cooked. Once cooled, the gels begin to retrograde, regaining some of their structural order. Wide-angle diffraction shows a slow development of the B form of crystallinity with time.[13-14] This is closely related to an increase in the shear modulus of the gels on storage.[15] The

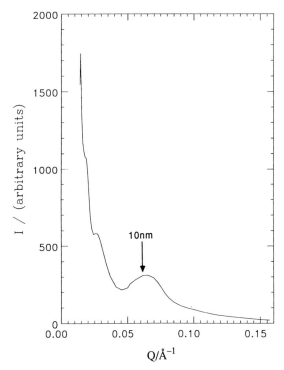

Figure 1   *SAXS intensity profile of a wheat flour suspension in water. Intensity in arbitrary units is plotted against wave number Q. The Bragg peak at 10 nm is indicated*

development of crystallinity on ageing is also indicated by the appearance of an endothermic transition at 50–60 °C as seen by differential scanning calorimetry (DSC). This endotherm has been shown to be due to the amylopectin component.[16] Extensive studies have been carried out on the interpretation of this endotherm,[15,17–23] showing that the moisture content of the gels,[15,20,22,23] temperature[17,18,20,21,23] and time[15,17,18,20,21,23] of storage, play an important part in the retrogradation process.

Although a regular periodicity has been observed in the native starch using SAXS, this technique has not until now been applied to the retrograded material. The present work explores the retrogradation process of wheat starch by using SAXS and by extending the DSC experiments. The results from the two techniques are compared.

2   Experimental

Small-angle X-Ray Scattering—Slurries of 50 wt% wheat starch in water were placed in glass capillary tubes of 0.7 mm diameter. These were sealed with nail varnish and gelatinized by heating to 100 °C for 30 minutes. The sealed tubes were then stored at an ageing temperature of either room temperature or 4 °C.

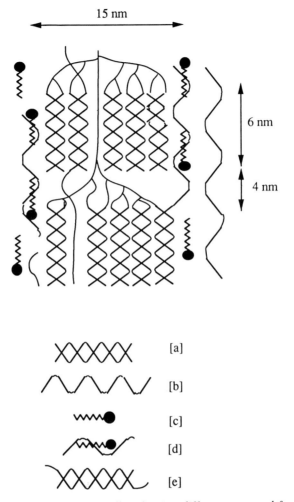

15 nm

6 nm

4 nm

[a]

[b]

[c]

[d]

[e]

Figure 2 *Model of a starch crystallite showing different structural features*: (a) *amylopectin helix*, (b) *free amylose*, (c) *free lipid*, (d) *V-amylose helix*, *and* (e) *hybrid amylose/amylopectin helix*.
[Adapted from Figure 2.6 in 'Starch granule structure and function: a physicochemical approach' by J. M. V. Blanshard, in 'Starch: Properties and Potential', ed. T. Galliard.]

The SAXS experiments were performed on a Kratky camera, using Cu K_α radiation, with the intensity of the scattered beam being recorded on Kodak 'direct exposure' X-ray film. The films were developed simultaneously to eliminate changes in film intensity due to different processing conditions. The developed films were then characterized using a 'Joyce Model Mark III' microdensitometer, and the resulting ink traces transferred to disc using a manually operated digipad. Background scattering was then subtracted using data from a blank run

A Debye–Bueche plot of $I^{-1/2}$ *versus* Q^2 yields the correlation length:

$$l_c = \left(\frac{\text{slope}}{\text{intercept}}\right)^{1/2}. \tag{5}$$

Figure 4 shows the development of correlation length with time for gels stored at room temperature and at 4 °C. The size of the scattering regions increases with time of storage, the increase being greater with a storage temperature of 4 °C. The shape of the curves can be interpreted in terms of a picture of nucleation and growth of amylopectin crystallites within the starch gel. At 4 °C the driving force for nucleation is greater than at room temperature. This results in a faster increase in correlation length with storage time. Guinier analysis, assuming three dimensional scatterers, yields radii of gyration of the order of 3–4 nm which qualitatively show similar trends with time and temperature of storage. The data do not fit an analysis procedure which assumes scatterers of lower dimension. The scatterers can therefore be assumed to extend in three dimensions. The value of the correlation length returns to the unstored value when a stored gel is reheated to 100 °C.

The values of the correlation lengths obtained here are very small, and they are difficult to interpret, although they do provide a crude estimate of the size of the scattering regions within the gel. Likewise, the R_g values obtained by Guinier analysis for gelatinized, unstored gels are considerably lower than the value of about 6 nm reported by Blanshard, based on neutron experiments.[8] Although the trends seen in the values of R_g and l_c are almost certainly qualitatively correct, there are various problems with the present analysis. The use of film as a recording medium may lead to errors in absolute values and less reliable data at very low angles, which means that the effective low-angle cut-off is rather large.

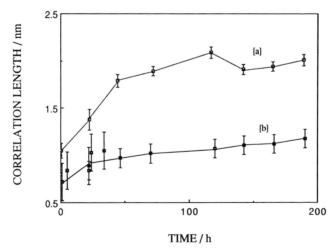

TIME / h

Figure 4 *Correlation length plotted against time for different starch storage conditions*: (a) *aged at* 4 °C, (b) *aged at room temperature*

Another distorting effect may be the relatively high concentration of polymer in the gels. Further experiments are planned using the Synchrotron X-ray Source at Daresbury, making use of proportional detectors. It is hoped that such experiments will yield more quantitatively reliable data, thus enabling us to make more confident estimates of the absolute values of R_g and l_c.

The results from the SAXS can be compared with those from the DSC experiments for samples subjected to the same treatment. Immediately after cooking, the gels register a flat line on the DSC trace, indicating complete gelatinization of the starch, total loss of crystallinity, and an enthalpy change, ΔH, of zero. As the gels age, the amylopectin recrystallizes and the peak corresponding to the melting of these crystallites in the DSC trace grows. The behaviour of ΔH with time under the different storage conditions is shown in Figure 5.

Like the curves obtained from SAXS (Figure 4), the curves in Figure 5 can be interpreted in terms of a picture of nucleation and growth of amylopectin crystallites within the starch gel. The driving force for nucleation at 4 °C (curve a) is greater than at room temperature (curve b). This results in a faster increase in enthalpy. However, the polymers have less freedom of movement at the lower temperature, and the degree of perfection of the resulting crystallites is therefore lower. This is reflected in their melting temperature of 52.3 ± 0.6 °C as compared with 60.9 ± 0.3 °C for crystallites formed at room temperature. In gels stored at −18 °C (curve c), development of the crystallites is severely restricted. However,

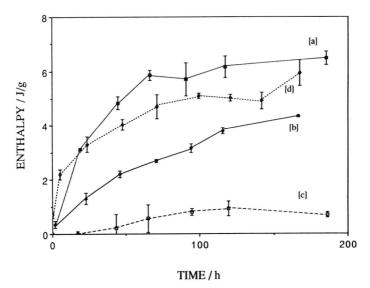

TIME / h

Figure 5 *Enthalpy plotted against time for different starch storage conditions. Sample (a) was aged at 4 °C, and had a DSC peak temperature of 52.3 ± 0.6 °C; sample (b) was aged at room temperature, and had a peak temperature of 60.9 ± 0.3 °C; sample (c) was aged at −18 °C, and gave a broad shallow peak in the DSC; and sample (d) was aged at −18 °C, followed by ageing at room temperature, and had a peak temperature of 60.4 ± 0.5 °C*

if the frozen gel is subsequently aged at room temperature (curve d), the endotherm develops rapidly—more rapidly than for the unfrozen gel at room temperature—although the peak temperature is unchanged. This suggests that nucleation has been encouraged by the freezing process, enabling subsequently faster growth of relatively perfect crystals during storage at room temperature. If any of these samples is immediately run through the DSC a second time, the endotherm is not observed, which implies that the crystallites do not immediately reform. This is the basis for the practice of 'refreshing' stale bread by heating it above the endotherm temperature.

It can be seen that the shape of the curves in Figure 5 is similar to that for the development of the correlation length as observed by SAXS. The value of the correlation length returns to the unstored value if a stored gel is reheated to 100 °C, a similar effect to the loss of the endotherm observed in the DSC after reheating. This indicates that the process observed by SAXS is probably due to the same effect of recrystallizing amylopectin.

4 Conclusions

The results presented here explore the retrogradation of starch by relating information provided by DSC to data given by SAXS. The long-range configuration of retrograded starch is shown to differ from that of the native starch. Rates of recrystallization as a function of storage temperature are reported, and they are interpreted in terms of a picture of nucleation and growth of amylopectin crystallites.

The retrogradation of starch is one of the major processes occurring during the staling of bread and other starch based products. DSC experiments performed on bread crumb yield similar results to those on starch gels.[30–33] Thus, information concerning the rates of recrystallization of amylopectin in starch gels clearly has implications for the optimization of storage conditions for bread. Although storage in the refrigerator slows bacterial spoilage, the staling process under these conditions is dramatically accelerated. Similarly, frozen bread, once defrosted, stales at a faster rate than bread stored at room temperature.

Acknowledgements. The authors wish to thank the Agriculture and Food Research Council and Dalgety plc for financial support.

References

1. K. H. Meyer, *Adv. Colloid Sci.*, 1942, **1**, 183.
2. E. M. Montgomery and F. R. Senti, *J. Polym. Sci.*, 1958, **18**, 1.
3. J. R. Katz, in 'A Comprehensive Survey of Starch Chemistry', ed. R. P. Walton, The Chemical Catalogue Co., New York, 1928, Vol. 1, p. 68.
4. J. R. Katz and T. B. van Itallie, *Z. Phys. Chem.*, 1930, **A150**, 90.
5. H. C. Wu and A. Sarko, *Carbohydr. Res.*, 1978, **61**, 7.
6. H. C. Wu and A. Sarko, *Carbohydr. Res.*, 1978, **61**, 27.
7. S. Hizukiri and Z. Nikuni, *Nature (London)*, 1957, **180**, 436.
8. J. M. V. Blanshard, D. R. Bates, A. H. Muhr, D. L. Worcester, and J. S. Higgins, *Carbohydr. Polym.*, 1984, **4**, 427.
9. C. Stirling, *J. Polym. Sci.*, 1962, **56**, S10.

10. A. H. Muhr, J. M. V. Blanshard, and D. R. Bates, *Carbohydr. Polym.*, 1984, **4**, 399.
11. G. T. Oostergetel and E. F. J. Van Bruggen, *Stärke*, 1989, **41**, 331.
12. J. M. V. Blanshard, in 'Chemistry and Physics of Baking', ed. J. M. V. Blanshard, P. J. Frazier, and T. Galliard, Royal Society of Chemistry, London, 1986, p. 1.
13. J. R. Katz, in 'A Comprehensive Survey of Starch Chemistry', ed. R. P. Walton, The Chemical Catalogue Co., New York, 1928, Vol. 1, p. 100.
14. J. R. Katz, *Bakers' Weekly*, 1934, **81**, 34.
15. M. J. Miles, V. J. Morris, P. D. Orford, and S. G. Ring, *Carbohydr. Res.*, 1985, **135**, 271.
16. M. J. Miles, V. J. Morris, P. D. Orford, and S. G. Ring, in 'New Approaches to Research on Cereal Carbohydrates', ed. R. D. Hill and L. Munck, Elsevier Applied Science, Amsterdam, 1985, p. 109.
17. R. G. McIver, D. W. E. Axford, K. H. Colwell, and G. A. H. Elton, *J. Sci. Food Agric.*, 1968, **19**, 560.
18. K. H. Colwell, D. W. E. Elton, N. Chamberlain, and G. A. H. Elton, *J. Sci. Food Agric.*, 1969, **20**, 550.
19. D. J. Stevens and G. A. H. Elton, *Stärke*, 1971, **23**, 8.
20. L. Longton and G. A. LeGrys, *Stärke*, 1981, **33**, 410.
21. T. Jankowski and C. K. Rha, *Stärke*, 1986, **38**, 6.
22. K. J. Zeleznak and R. C. Hoseney, *Cereal Chem.*, 1986, **63**, 407.
23. Ph. Roulet, A. Raemy, and P. Wuersch, *Food Hydrocolloids*, 1987, **1**, 575.
24. C. G. Vonk, *J. Appl. Cryst.*, 1971, **4**, 340.
25. A. Guinier and G. Fournet, 'Small Angle Scattering of *X*-rays', John Wiley, New York, 1955, p. 126.
26. P. Debye and A. M. Bueche, *J. Appl. Phys.*, 1949, **20**, 518.
27. P. Debye, H. R. Anderson, Jr., and H. Brumberger, *J. Appl. Phys.*, 1957, **28**, 679.
28. F. B. Khambatta, F. Warner, T. Russell, and R. S. Stein, *J. Polym. Sci.*, 1976, **14**, 1391.
29. K. J. I'Anson, M. J. Miles, V. J. Morris, S. G. Ring, and C. Nave, *Carbohydr. Polym.*, 1988, **8**, 45.
30. D. W. E. Axford and K. H. Colwell, *Chem. Ind.*, 1967, 467.
31. W. H. Knightley, *The Bakers' Digest*, 1977, **51**, 52.
32. T. Fearn and P. L. Russell, *J. Sci. Food Agric.*, 1982, **33**, 537.
33. K. J. Zeleznak and R. C. Hoseney, *Stärke*, 1987, **39**, 231.

Weak and Strong Polysaccharide Gels

By V. J. Morris

AFRC INSTITUTE OF FOOD RESEARCH, NORWICH LABORATORY, COLNEY LANE, NORWICH NR4 7UA

1 Introduction

Studies at the Institute of Food Research (IFR) are currently concerned with trying to understand the molecular mechanisms by which polysaccharides associate to form weak or strong gels. The aim of such studies is to develop scientific, rather than empirical, methods for suggesting modifications to processes, or for selecting new polysaccharides, in order to optimize the use of food biopolymers. Investigations fall into three broad areas or themes. Firstly, there is an interest in the investigation of biopolymers which show potential as future food ingredients or *additives*. Such studies are illustrated here by work carried out at Norwich on gellan gum. Secondly, we have an interest in how individual polysaccharides behave when mixed with other biopolymers and cosolutes in complex *multicomponent mixtures*. Mixed systems form more realistic models for foods. Such studies are illustrated by work on starch and binary synergistic gelling polysaccharide blends. Thirdly, we have an interest in a relatively new area aimed at developing *self-textured* or *naturally textured foods* as an alternative to the use of additives. These studies will be illustrated by discussion of extracellular polysaccharides produced by *Acetobacter* and lactic acid bacteria.

2 Gellan Gum

Gellan gum is the extracellular polysaccharide produced by the aerobic fermentation of *Pseudomonas elodea* in batch culture.[1] Toxicity trials have been successful, and gellan gained food approval in Japan in 1988. Food approval is being sought for use in the USA, UK, and Europe as a broad spectrum gelling agent.[2]

Studies at IFR have been concerned with the mechanism of gelation. Gellan is a linear anionic heteropolysaccharide with a tetrasaccharide repeat unit[3,4] as shown in Figure 1. The native product is esterified,[5] but, during production,[6] the broth is allowed to become alkaline resulting in a de-esterified product (called Gelrite). Gelation is dependent on the type and concentration of associated cations.[7,8] In the presence of gel-promoting cations, the X-ray diffraction patterns obtained

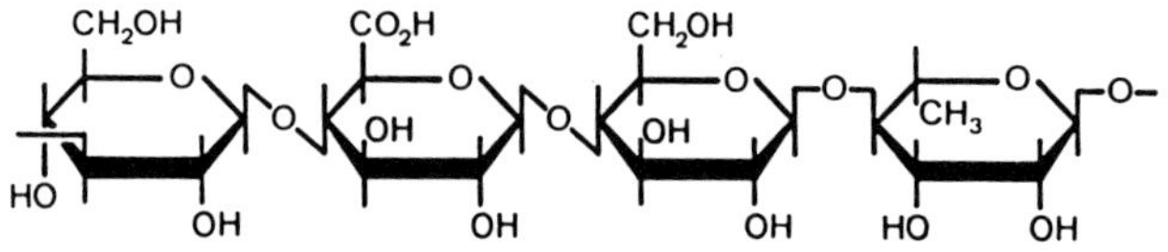

Figure 1 *The tetrasaccharide chemical repeat unit of gellan gum*

from stretched gels are highly crystalline.[7,9-12] The three-fold gellan helix shows an axial rise per chemical repeat equal to one-half the repeat length; it is suggestive of a double helix. The latest modelling studies[13] suggest left-handed three-fold double helices packed into an orthorhombic unit cell. Thus, the mode of association of gellan helices within the gels is well defined and involves the formation of small crystallites containing bound counter-ions.

Bulky tetramethylammonium (TMA) cations inhibit gelation. Physicochemical studies of the viscoelastic fluids suggests an order–disorder transition upon heating and cooling.[14-16] On the basis of the interpretation of optical rotation data, this has been attributed to a helix–coil transition.[14-16] Light scattering studies on TMA gellan in TMA Cl, under conditions favouring the suggested helical structure, are consistent with a stiff double-helical structure.[17] Light scattering studies on 'ordered' TMA gellan in TMA Cl carried out at Norwich has revealed that the measured molecular weight is sensitive to the method of sample clarification. Table 1 compares light scattering data obtained for 'ordered' TMA gellan solutions after clarification by filtration through either $3\,\mu$m or $0.45\,\mu$m filters. Carbohydrate analysis[18] of the TMA gellan samples has yielded a carbohydrate composition of 96%. Analysis of the sample concentrations before and after filtration, using a Chromatix differential refractometer, reveals negligible loss of sample during filtration (Table 2). These studies suggest that filtration breaks down the aggregated gellan molecules. In order to investigate the nature of these aggregates, preliminary light scattering data are displayed as Holtzer plots (Figure 2). The data for the $0.45\,\mu$m filtered sample are consistent with a polydisperse solution of short rod-like molecules. With increasing pore size (for the filters), the shape of the curves suggests progressively elongated and more

Table 1 *Parameters derived from light scattering studies of tetramethylammonium gellan as a function of pore size of filtration membrane used for clarification. Data were measured using a FICA 50 at a wavelength of 436 nm: Rayleigh ratio for benzene $(R_B) = 45.6 \times 10^{-6}\,cm^{-1}$, (dn/dc) = 0.156 cm^3 g^{-1}, depolarization ratio = 0.016*

Method of clarification	Molecular weight (daltons)	Radius of gyration (nm)	Viral coefficient (cm^3 mol g^{-2})
$3\,\mu$m	4.5×10^6	159	0.6×10^{-4}
$0.45\,\mu$m	1.06×10^5	42	8.6×10^{-4}
$0.45\,\mu$m[a]	4.3×10^5	159	22×10^{-4}

[a] From ref. 17.

Table 2 *Polymer concentration* C_p *after filtering tetra-methylammonium gellan solutions through different pore-size filters. Concentrations were measured using a differential refractometer at 25 °C*

Filter size (μm)	C_p (mg ml^{-1})
—	0.82
3.0	0.81
1.2	0.81
0.45	0.78

flexible stiff coils. The plateau value at high scattering vectors, for the 3 μm filtered sample, is clearly of a similar order to that of the 0.45 μm filtered sample. The similar mass per unit length of the aggregates suggests a fibrillar structure involving end-to-end association.

A model for gellan gelation based on the above observations is shown in

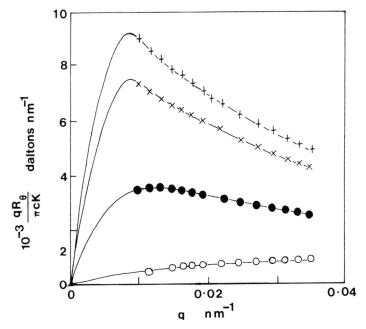

Figure 2 *Holtzer plots for TMA gellan in 0.075 M TMA Cl (1 mg ml^{-1}). Different curves represent clarification using different size filters:* $-\bigcirc-$, *0.45 μm;* $-\bullet-$, *0.6 μm;* $-\times-$, *3 μm;* $-+-$, *5 μm. R_θ is the Rayleigh ratio, K is the scattering constant, c is the polymer concentration, and q is the modulus of the scattering vector. Data were measured using a FICA 50 at a wavelength of 436 nm*

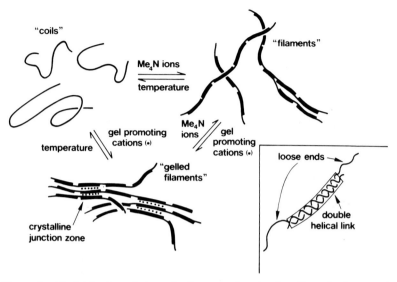

Figure 3   *Model for the gelation of gellan gum showing the effect of gel-promoting cations and of heating and cooling. The inset shows the method envisaged for end-to-end association of the polymers by double-helix formation*

Figure 3. Inter-chain crystallization is inhibited for TMA gellan. Helix formation upon cooling is considered to promote end-to-end association, via double helix formation, into fibrils. These fibrils can thicken into fibres, or bifurcate, by means of a chain end-linking to the middle of a separate chain. Such association would account for the observed high viscosity and shear thinning behaviour of TMA gellan samples. Gel promoting cations promote inter-fibril or intra-fibre crystallization yielding a permanent network. Melting of the double helices or crystallites will determine the setting and melting properties of the gels. Manipulation of the cation content can be used to induce cold-setting or thermosetting gels and to control thermal reversibility or irreversibility.

3   Multicomponent Gels

A binary gel is the simplest form of multi-component gel. Various structures can arise when two polysaccharides are mixed and gelled.[19,20] Schematic models for four basic types of binary gel are shown in Figure 4. In Figure 4a, only one of the polysaccharides associates to form a network. The situation in Figure 4b arises when both polysaccharides form independent networks. If some degree of polymer demixing arises before gelation, then a phase-separated network is formed (Figure 4c). Finally, if one polysaccharide binds to the other, then a coupled network is formed (Figure 4d).

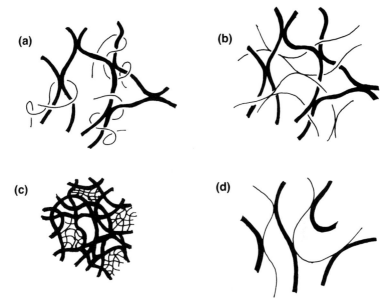

Figure 4   *Schematic models for binary polymer gel networks: (a) network formed
by one polysaccharide alone; (b) interpenetrating network formed by
independent gelation of each polysaccharide; (c) phase-separated
network formed by demixing and subsequent gelation of the two
polysaccharides; (d) coupled network formed by intermolecular binding between the two polysaccharides*

Starch and Staling—Starch gels may be regarded as phase-separated or composite
gels. Recent studies have provided a molecular description for the gelation and
retrogradation of starch.[21–25] Provided the gelatinized starch granules remain
intact, then a starch gel can be modelled as swollen granules, composed of an
amylopectin skeleton, suspended in an interpenetrating gel matrix formed by the
solubilized amylose. Gelation is dominated by gelation of the amylose matrix,
and this step determines the opacity of the starch gel.[22–24] Amylose gelation
involves the phase separation of an amylose-rich amorphous network within
which limited amylose crystallization occurs.[22,23,26] The amylose crystals melt at
temperatures above 100 °C, and thus the amylose gel is normally considered
irreversible. Long-term increases in stiffness of starch gels on storage are thermoreversible and result from amylopectin crystallization.[24] At high starch concentrations, restricted swelling of granules and reduced solubilization of amylose
make the role played by amylopectin increasingly important.[25] Recently, Slade
and Levine[28,29] have stressed the non-equilibrium nature of such crystallization.
The rate of crystallization at a given temperature is controlled by the glass
transition temperature T_g of the amorphous material, which in turn depends on
the water content of the starch gel.[28,29] Baked foods contain co-solutes such as
salts, fats, and sugars. Sugars have been shown[30,31] to affect the staling of starch

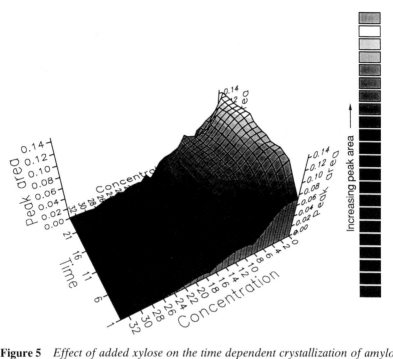

Figure 5 *Effect of added xylose on the time dependent crystallization of amylo-pectin upon storage of wheat starch gels*

gels. This arises by suppression of amylopectin crystallization[32] and Figure 5 shows this effect for added xylose. Sugars are considered to act by raising T_g relative to the value in water, suppressing polymer motion, and inhibiting crystal growth.[28,29] Such studies, coupled with studies of the effect of sugars on the texture of starch gels, provide a route towards replacing sugars in baked products by nutritionally more desirable ingredients with similar function. Amylose crystallization in starch gels is important in determining the digestibility of starch.[33] The ability to vary the level or rate of amylose crystallization, and hence the resistant starch content, offers a route to manipulating the physiological effects of starch-based foods.

Synergisms and Intermolecular Binding—Until recently, synergisms between polysaccharides had often been regarded as synonymous with intermolecular binding. Thus, the synergistic gelation observed for galactomannans or gluco-mannans mixed with either xanthan or certain algal polysaccharides was attribu-ted to the binding of the galactomannan (or glucomannan) backbone to the helix of the other polysaccharide[34–38] (Figure 6). Recent studies,[39–44] however, have suggested that these synergistic gels can be classified into two types (I and II). Type I binary gels are believed to form structures like that in Figure 4a. They comprise mixtures of a gelling and a non-gelling polysaccharide (Table 3).

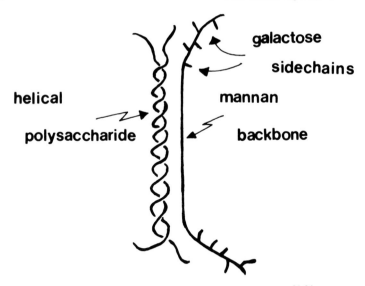

Figure 6   *Original model proposed by Dea and co-workers*[34–36] *to explain gelation in synergistic mixed polysaccharide gels*

Preliminary rheological studies[45] suggest that, at least for a limited range of total polymer concentrations, the fracture stress and low deformation modulus of the mixed gels may be normalized with respect to the equivalent properties of the gelling polysaccharide gel having the same total polymer concentration, resulting in master curves characteristic of the non-gelling component, but independent of total polymer concentration. The immobilized, but soluble, non-gelling component may serve to swell the network opposing synergesis.

Type II synergistic mixtures (Table 3) are composed of non-gelling components. For these systems, evidence has been found for intermolecular binding, and the gels are believed to be examples of coupled networks (Figure 4d). Fibres prepared from stretched carob + xanthan and tara + xanthan mixed gels yield new X-ray diffraction patterns providing direct experimental evidence for xanthan–galactomannan binding.[44,46] The original model[35,36] (Figure 6) proposed to account for intermolecular binding suggested that unsubstituted regions of the galactomannan backbone bound to the ordered xanthan helix. The

Table 3   *Examples of type* I *and type* II *binary synergistic gels*

Type I gels	Type II gels
κ-Carrageenan + carob	Xanthan + carob
κ-Carrageenan + tara	Xanthan + tara
Furcellaran + carob	Xanthan + konjac mannan
Furcellaran + tara	
κ-Carrageenan + konjac mannan	
Furcellaran + konjac mannan	

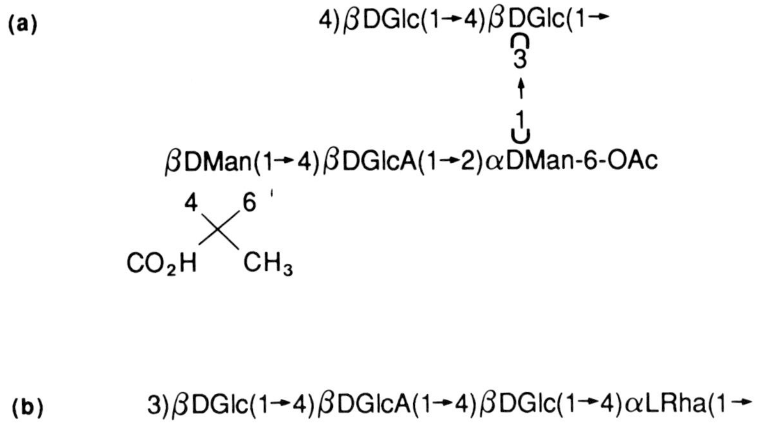

(a)

$$4)\beta DGlc(1\rightarrow4)\beta DGlc(1\rightarrow$$

$$\overset{\Large\cap}{3}$$

$$\uparrow$$

$$\overset{1}{\Large\cup}$$

$$\beta DMan(1\rightarrow4)\beta DGlcA(1\rightarrow2)\alpha DMan\text{-}6\text{-}OAc$$

$$\overset{4}{}\overset{6\ '}{}$$

$$\times$$

$$CO_2H \quad CH_3$$

(b) $3)\beta DGlc(1\rightarrow4)\beta DGlcA(1\rightarrow4)\beta DGlc(1\rightarrow4)\alpha LRha(1\rightarrow$

Figure 7   *Chemical repeat units of the two polysaccharides* (a) *xanthan and* (b) *acetan*

stoichiometry of the interaction is undefined and there is no obvious stereo-chemical compatibility between the galactomannan backbone and the xanthan helix. By varying the conformation of the xanthan molecule, and noting the influence of conformation upon gelation, it has been possible to show that denaturation of the xanthan helix is essential if intermolecular binding and gelation is to occur.[43,44,46–48] The X-ray patterns obtained for the carob + xanthan mixed gels suggest that co-crystallization of the two polysaccharides provides junction points within the gel network. It is likely that such co-crystallization introduces permanent linkages into the otherwise weakly connected xanthan network. Xanthan consists of a $\beta(1\rightarrow4)$ linked D-glucose (cellulosic) backbone substituted on alternate glucose residues with a charged trisaccharide sidechain[49] (Figure 7). Galactomannans consist of a $\beta(1\rightarrow4)$ linked D-mannose backbone in which the primary alcohol group is randomly substituted with galactose. The level and distribution of galactose residues is characteristic of the source of the galactomannan.[37] Galactomannans partially crystallize in an orthorhombic unit cell in which the b and c dimensions are essentially constant, but the a dimension varies depending on the galactose content and hydration.[50] The galactomannan backbones sit in a planar structure with the sidechains projecting into an interplanar spacing (the a dimension) containing water molecules. X-Ray diffraction patterns of carob + xanthan gels resemble carob patterns in the b and c dimensions but all reflections involving the interplanar spacing (the a dimension) are absent. This can be attributed to a co-crystallization in which the xanthan backbone mimics a mannan backbone and the xanthan side-chain is accommodated in the interplanar spacing. Glucose and mannose differ only in the spatial orientation of the hydroxyl residue at C-2, and thus the cellulosic and mannan backbone are equivalent. Molecular modelling studies have shown that substitution of the xanthan backbone does not prevent it adopting a two-fold cellulosic conformation compatible with the measured 'c' dimension of the mixed gel X-ray

diffraction pattern.[51] The presence of xanthan as a defect in the mannan lattice, and especially the influence of the side-chain, is considered to randomize the '*a*' spacing. To accommodate the xanthan side-chains, it is necessary for some of the galactose substituents of the galactomannan to be absent. Too large regions depleted of galactose would be insoluble. Thus, there is a window of galactose/mannose ratios for which xanthan–galactomannan binding and gelation can occur.[37] Highly substituted galactomannans such as guar need to be enzymically modified,[52] reducing their galactose content, in order to induce gelation.

The stereochemistry of glucomannans[53] suggests that they also might bind to denatured xanthan. Konjac mannan, in its native non-gelling form, will gel when mixed with xanthan, and experimental evidence has been obtained for xanthan–glucomannan binding.[54]

In order to replace xanthan, it would be necessary to find polysaccharides for which a stereochemically suitable backbone is disguised under normal solution conditions. A potential candidate is acetan; this is the extracellular polysaccharide secreted by *Acetobacter xylinum*.[55] Acetan has a similar chemical structure to that of xanthan.[55] Acetan possesses a cellulosic backbone substituted on alternate glucose residues with a pentasaccharide side-chain (Figure 7b). Recent *X*-ray studies of mutant xanthan structures[56] have demonstrated the crucial features of the xanthan structure. The three-linked mannose is essential to convert the two-fold cellulosic backbone into a five-fold helical structure. The galacturonic acid serves to solubilize the polymer and permit the helical structure to be denatured in solution. Clearly acetan contains the essential features of the xanthan structure. *X*-Ray fibre diffraction studies[57] show that acetan adopts the five-fold xanthan helix and optical rotation studies[57] show that the helix is maintained in solution and can be melted on heating. However, acetan does not gel with carob or tara because of the extended side-chain. Present studies are concerned with chemically and enzymically modifying the acetan side-chain and exploring the effects of these changes upon synergistic gelation with galactomannans and glucomannans.

4   Self-Textured Foods

Understanding the behaviour of complex food systems offers the ability to select new polymers or to modify processing. One aim of such an approach is to try to develop self-textured foods in which the texture develops naturally during processing rather than being imposed by additives.

One approach hardly used at all at present with polysaccharides is to use food approved enzymes, or natural extracts containing such enzymes, to modify and optimize the rheology of a sample during processing. Such an approach has been demonstrated[58] for the gelation of citrus or sugar beet pectins.

Acetan offers another route to a self-textured product. Xanthan is used as a thickening and suspending agent in many acid-based foods. In such foods vinegar is often used as the acidifier. This prompted a search of the chemical structures of extracellular polysaccharides produced by *Acetobacter* with a view to identifying polymers which should show xanthan-like rheology. Acetan has been shown to adopt the xanthan helical structure in solution.[57] Figure 8 shows that, in the ordered conformation, acetan solutions exhibit the high viscosities and shear-

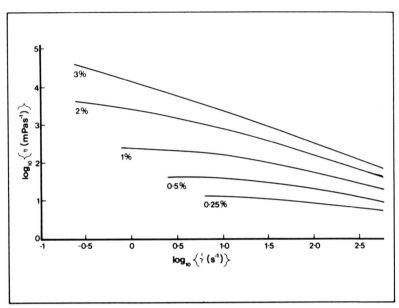

Figure 8   *Viscosity–shear-rate dependence for aqueous acetan solutions. Logarithm of viscosity η is plotted against logarithm of shear-rate γ̇ for various polymer concentrations*

thinning behaviour characteristic of xanthan. Such studies demonstrate the feasibility of fermenting or blending self-thickened vinegars. What would be the status of such a product? Is it a novel food? Does it require food clearance, and would such clearance be eased by selecting natural or genetic mutants which produce xanthan, xanthan derivatives, or acetan derivatives more closely related to xanthan? Would such a product be perceived by the consumer as more natural? Within a few years it will be possible to produce such products. Are they desirable or preferable?

More obvious targets are the fermentation products produced by lactic acid bacteria. Many such products contain polysaccharides which improve their texture. These polysaccharides have been consumed safely for centuries yet little is known about their structure or function. Such knowledge could be used to upgrade the product. For example, the structure of sugary kefir is determined by a gelling polysaccharide produced by *Lactobacillus hilgardi*, one of the bacteria present in the starter culture.[59] Gelation arises because the bacteria modify the normal dextran structure by inserting blocks of $\alpha(1 \rightarrow 3)$ linked D-glucose.[59,60] Such blocks are insoluble, and act to fix the polysaccharides together into a network. Clearly, screening for the production of $(1 \rightarrow 3)$ D-glucose, or the enzyme responsible for inserting such linkages, could be used to improve yield or improve functionality of the polymer produced. New starter cultures could be formulated which include gelation of the final product. Similarly, yoghurts which

are thickened by polysaccharides could be screened for bacteria, which could then be incorporated into new improved starter cultures. Simple extracts from such products could be used as natural ingredients for the thickening of other products.

Acknowledgements. The author wishes to thank P. A. Gunning and P. Cairns for providing unpublished experimental data. This study was funded in part by the Ministry of Agriculture, Fisheries, and Food.

References

1. K. S. Kang, G. T. Veeder, P. J. Mirrasoul, T. Kanecko, and I. W. Cottrell, *Appl. Environ. Microbiol.*, 1982, **43**, 1086.
2. G. R. Sanderson and R. C. Clark, *Food Technol.*, 1983, **37**(4), 63.
3. M. A. O'Neill, R. R. Selvendran, and V. J. Morris, *Carbohydr. Res.*, 1983, **124**, 123.
4. P. E. Jansson, B. Lindberg, and P. A. Sandford, *Carbohydr. Res.*, 1983, **124**, 135.
5. M. S. Kuo, A. J. Mort, and A. Dell, *Carbohydr. Res.*, 1986, **156**, 173.
6. R. Moorehouse, G. T. Colegrave, P. A. Sandford, J. K. Baird, and K. S. Kang, *ACS Symp. Ser.*, 1981, **150**, 111.
7. P. T. Attwool, Ph.D Thesis, University of Bristol, 1987.
8. V. J. Morris, in 'Food Biotechnology', ed. R. D. King and P. S. J. Cheetham, Elsevier Applied Science, London, 1987, Vol. 1, p. 193.
9. M. J. Miles, V. J. Morris, and M. A. O'Neill, in 'Gums and Stabilisers for the Food Industry', ed. G. O. Phillips, D. J. Wedlock, and P. A. Williams, Pergamon, Oxford, 1984, Vol. 2, p. 485.
10. V. Carroll, G. R. Chilvers, D. Franklin, M. J. Miles, V. J. Morris, and S. G. Ring, *Carbohydr. Res.*, 1983, **114**, 181.
11. V. Carroll, M. J. Miles, and V. J. Morris, *Int. J. Biol. Macromol.*, 1982, **4**, 432.
12. P. T. Attwool, E. D. T. Atkins, C. Upstill, M. J. Miles, and V. J. Morris, in 'Gums and Stabilisers for the Food Industry', ed. G. O. Phillips, D. J. Wedlock, and P. A. Williams, Elsevier Applied Science, London, 1986, Vol. 3, p. 135.
13. R. Chandrasekaran, L. C. Puigjaner, K. L. Joyce, and S. Arnott, *Carbohydr. Res.*, 1988, **181**, 23.
14. H. Grasdalen and O. Smidsrød, *Carbohydr. Polym.*, 1987, **7**, 371.
15. V. Crescenzi, M. Dentini, T. Coviello, and R. Rizzo, *Carbohydr. Res.*, 1986, **149**, 425.
16. V. Crescenzi, M. Dentini, and I. C. M. Dea, *Carbohydr. Res.*, 1987, **160**, 283.
17. M. Dentini, T. Coviello, W. Burchard, and V. Crescenzi, *Macromolecules*, 1988, **21**, 3312.
18. M. Dubois, K. A. Gilles, J. K. Hamilton, P. A. Rebers, and F. Smith, *Anal. Chem.*, 1956, **28**, 350.
19. V. J. Morris, *Chem. Ind. (London)*, 1985, 159.
20. V. J. Morris, in 'Gums and Stabilisers for the Food Industry', ed. G. O. Phillips, D. J. Wedlock, and P. A. Williams, Elsevier Applied Science, London, 1986, Vol. 3, p. 87.
21. S. G. Ring and G. Stainsby, *Prog. Food Nutr. Sci.*, 1982, **6**, 323.
22. M. J. Miles, V. J. Morris, and S. G. Ring, *Carbohydr. Polym.*, 1984, **4**, 73.
23. M. J. Miles, V. J. Morris, and S. G. Ring, *Carbohydr. Res.*, 1985, **135**, 257.
24. M. J. Miles, V. J. Morris, and S. G. Ring, *Carbohydr. Res.*, 1985, **135**, 271.
25. P. D. Orford, S. G. Ring, V. Carroll, M. J. Miles, and V. J. Morris, *J. Sci. Food Agric.*, 1987, **39**, 169.
26. K. J. I'Anson, M. J. Miles, V. J. Morris, C. Nave, and S. G. Ring, *Carbohydr. Polym.*, 1988, **8**, 45.
27. C. Mestres, Ph.D. Thesis, University of Nantes, 1986.
28. L. Slade and H. Levine, in 'Recent Developments in Industrial Polysaccharides', ed. S. S. Stivala, V. Crescenzi, and I. C. M. Dea, Gordon and Breach, New York, 1987, p. 387.
29. L. Slade and H. Levine, in 'Food Structure—Its Creation and Evaluation', ed. J. M. V. Blanshard and J. R. Mitchell, Butterworth, London, 1988, p. 115.

30. J. L. Maxwell and H. F. Zobel, *Cereal Food World*, 1978, **23**, 124.
31. R. Germani, C. F. Ciacco, and D. B. Rodriguez-Amaya, *Stärke*, 1983, **35**, 377.
32. K. l'Anson, M. J. Miles, V. J. Morris, L. S. Besford, D. A. Jarvis, and R. A. Marsh, *J. Cereal Sci.*, 1990, **11**, 243.
33. C. S. Berry, K. l'Anson, M. J. Miles, V. J. Morris, and P. L. Russell, *J. Cereal Sci.*, 1988, **8**, 203.
34. I. C. M. Dea, A. A. McKinnon, and D. A. Rees, *J. Mol. Biol.*, 1972, **68**, 153.
35. I. C. M. Dea, E. R. Morris, D. A. Rees, E. J. Welsh, H. A. Barnes, and J. Price, *Carbohydr. Res.*, 1977, **57**, 249.
36. E. R. Morris, D. A. Rees, G. Young, M. D. Walkinshaw, and A. Darke, *J. Mol. Biol.*, 1977, **110**, 1.
37. I. C. M. Dea and A. Morrison, *Adv. Carbohydr. Chem. Biochem.*, 1975, **31**, 241.
38. I. C. M. Dea, *ACS Symp. Ser.*, 1981, **150**, 439.
39. V. Carroll, M. J. Miles, and V. J. Morris, in 'Gums and Stabilisers for the Food Industry', ed. G. O. Phillips, D. J. Wedlock, and P. A. Williams, Pergamon, Oxford, 1984, Vol. 2, p. 501.
40. M. J. Miles, V. J. Morris, and V. Carroll, *Macromolecules*, 1984, **17**, 2443.
41. P. Cairns, V. J. Morris, M. J. Miles, and G. J. Brownsey, in 'Gums and Stabilisers for the Food Industry', ed. G. O. Phillips, D. J. Wedlock, and P. A. Williams, Elsevier Applied Science, London, 1986, Vol. 3, p. 87.
42. P. Cairns, M. J. Miles, and V. J. Morris, *Int. J. Biol. Macromol.*, 1986, **8**, 124.
43. P. Cairns, M. J. Miles, V. J. Morris, and G. J. Brownsey, *Carbohydr. Res.*, 1987, **160**, 411.
44. P. Cairns, M. J. Miles, and V. J. Morris, *Carbohydr. Polym.*, 1988, **8**, 89.
45. P. Cairns, V. J. Morris, M. J. Miles, and G. J. Brownsey, *Food Hydrocolloids*, 1986, **1**, 89.
46. P. Cairns, M. J. Miles, and V. J. Morris, *Nature (London)*, 1986, **322**, 89.
47. N. W. H. Cheetham and E. N. M. Mashimba, *Carbohydr. Polym.*, 1988, **9**, 195.
48. N. W. H. Cheetham and A. Punruckvong, *Carbohydr. Polym.*, 1989, **10**, 129.
49. P.-E. Jansson, L. Kenne, and B. Lindberg, *Carbohydr. Res.*, 1976, **46**, 245.
50. B. K. Song, W. T. Winter, and F. R. Taravel, *Macromolecules*, 1989, **22**, 2641.
51. R. P. Millane, personal communication.
52. P. V. Bulpin, M. J. Gidley, R. Jeffcoat, and D. R. Underwood, *Carbohydr. Polym.*, 1990, **12**, 155.
53. R. Takahashi, I. Kushakabe, S. Kusama, Y. Sakurai, K. Murakami, A. Mackawa, and T. Suzuki, *Agric. Biol. Chem.*, 1984, **48**, 2943.
54. G. J. Brownsey, P. Cairns, M. J. Miles, and V. J. Morris, *Carbohydr. Res.*, 1988, **176**, 329.
55. R. O. Couso, L. Ielpi, and M. A. Dankart, *J. Gen. Microbiol.*, 1987, **133**, 2133.
56. R. P. Millane and T. V. Narasaiah, *Carbohydr. Polym.*, 1990, **12**, 315.
57. V. J. Morris, G. J. Brownsey, P. Cairns, G. R. Chilvers, and M. J. Miles, *Int. J. Biol. Macromol.*, 1989, **11**, 326.
58. G. Williamson, C. B. Faulds, J. A. Matthews, V. J. Morris, and G. J. Brownsey, *Carbohydr. Polym.*, 1990, **13**, 387.
59. M. Pidoux, J. M. Brillouet, and B. Quemener, *Biotechnol. Lett.*, 1988, **10**, 415.
60. M. Pidoux, G. A. de Ruiter, B. E. Brooker, I. J. Colquhoun, and V. J. Morris, *Carbohydr. Polym.*, 1990, **13**, 351.

Structural and Mechanical Properties of Biopolymer Gels

By A. H. Clark

UNILEVER RESEARCH, COLWORTH LABORATORY, COLWORTH HOUSE,
SHARNBROOK, BEDFORD MK44 1LQ

1 Introduction

Gels are notoriously difficult to define.[1] Provided, however, that the objective of obtaining a quantitative definition is abandoned, it is not difficult to agree about what, at least qualitatively, are the principal attributes of materials described as gels. Thus, it is usually recognized that, in some sense or other, gels are solids. The words 'in some sense or other' simply mean that the observer may have to qualify his conclusion about solid character as the mechanical tests he applies widen in terms of time-scales and degrees of external disturbance. Over certain practical conditions of observation, however, gels are able to store the work employed in their deformation, and to recover their original shape, and it this property of 'elasticity' that is, of course, the essence of a solid.

A further mechanical characteristic is that gels are quite often soft solids. That is, where the solid character is measured by a stress–strain relationship, the modulus (ratio of stress to strain) is comparatively low, perhaps of the order of a fraction of a pascal up to several thousands of pascals. Harder gels, however, can be made.

Another important mechanical property of gels is their deformability. A common feature of such materials is that they can respond as solids over a substantial strain interval without rupture, and so, on occasions, they can exhibit a tendency towards so-called 'rubber elasticity'. By no means all gels are rubbery, however, and many are quite brittle (*e.g.* agar gels), with some showing a plastic, rather than rubber-elastic, response. These varying responses simply serve to re-emphasize the point that it is extremely difficult to define a gel in precise mechanical terms, and that the setting up of any precise boundaries is a subjective exercise.

Apart from the mechanical response—or, more technically, the constitutive behaviour—there are other important features of the class of materials known as gels which tend to set them apart. Thus, in terms of chemical constituents, gels usually contain as a major component a substance of comparatively low molecular weight, which in pure form is usually a simple liquid (*e.g.* water). The elastic

character of the gel is then ascribed to the presence of a second much higher molecular weight component, which is often present in much smaller relative amount than the 'solvent' constituent. This higher molecular weight substance may be a polymer, or a colloid particle, *etc.*; but, whatever its nature, the elasticity of the gel usually implies that at least part of this material has become assembled into a full three-dimensional network spanning throughout the entire gel system. Storage of mechanical energy during deformation is thus directly attributable to the properties of this network, particularly to the fact that, upon deformation of the gel, the network strands become altered in both energetic and entropic terms, and so contribute to an increase in free energy. Relaxation of this condition is normally inhibited as long as the bonds of the network remain intact, but, of course, over substantial periods of time, such permanence is by no means guaranteed.

This last point is extremely important, since, in reality, much of the difficulty of defining gels as a class of materials arises as a result of variations in the essential permanence of the underlying networks. In particular, where physical gels are concerned—and these are the main subjects for discussion here—great variations are possible in the lifetimes of cross-links, and in the abilities of such networks to suffer large deformations without breaking down. Nonetheless, it is generally recognized that gels (physical or otherwise) involve a network 'component' (or 'phase') and a solvent component, with the solvent often present in great excess. The solvent phase (or 'sol fraction') may, of course, contain other chemical components, including some polymers or particles which have not become attached to the network. The presence of so much low molecular weight material, however, is one of the most (if not *the* most) important aspects of the gel condition, for it always appears that a large amount of liquid has been 'solidified' by the gelation process. Of course, this is not really the case, as the solvent molecules are usually quite free to move rapidly through the pores of the network.

2 Biopolymer Gels and their Formation

So far, the discussion of gels has been very general. Neither the network-forming material, nor the solvent, has been specified very closely, nor have the forms of interaction between the network-forming particles. In what follows, however, the nature of the gels of interest is specified more exactly. In fact, these systems will be restricted to essentially physical gels formed from biopolymeric material in an aqueous environment. In practical terms, this means that we will principally be concerned with networks formed from proteins or polysaccharides; and the networks will develop in an aqueous medium in which other constituents will be limited to simple salts, or low molecular weight water-soluble solutes. In general, the interactions between polymers responsible for network formation will vary in strength between the limits of simple chain entanglements at one end of the bond strength scale, to covalent bonds at the other. This, of course, implies quite a wide range of rheological response, a point emphasized in the introduction above.

A first issue concerning biopolymer gels is the matter of their generation. This is quite a broad topic, there being a quite a number of procedures available. Thus, if we start from a reasonably stable solution of the biopolymer in water, one very

obvious approach is to substantially raise or lower the solution temperature. Such a temperature jump may lead to a change in the polymer conformational state, and this in turn is often followed by, or directly involves, an association process. Ultimately (if concentrations are high enough) gelation may ensue. Where temperatures are lowered, the likely conformational change is one of ordering, and in this case gelation may sometimes be regarded as a frustrated polymer crystallization event. Here, the network arises kinetically as a metastable (or unstable) intermediate stage, as the system attempts to precipitate from solution. In other cases, such as the heat-setting of globular proteins, disordering of the polymer may be involved, and the process of network formation derives from a complex range of intermolecular interactions including hydrophobic and electrostatic interactions.

In addition to thermally-driven gelation events, there are other approaches to making biopolymer gels. The adding of a new component to the starting solution is an important example. Often a biopolymer solution can be gelled simply by adding an appropriate salt solution, or by changing pH, through addition of acid or alkali. Other neutral small molecules may have an effect, such as alcohols or urea, and in some specialized cases, addition of a higher molecular weight species, such as an enzyme, may induce aggregation (casein renneting, for example, which involves chemical modification of the original polymer). In some cases, the added species simply drives the conformation change, and the association process, in something like the same way as a temperature change would do by, for example, altering the strength of interactions within and between molecules (*e.g.* electrostatic screening of charge, or influence on water structure). In other situations, however, it is known that the additive becomes directly involved in the association process (*e.g.* calcium ions in certain forms of polysaccharide gelation). Of course, it is always quite possible that a combination of the mechanisms just mentioned is involved.

Another way to prepare gels is from existing gels. A gel made by one or other of the above approaches may be further treated by swelling (or de-swelling) it in an appropriate solvent medium. Alternatively, the gel may be made at one temperature (or by the application of a more or less complex heating or cooling profile), and used in practice under another set of conditions. Because of the usually kinetically-controlled nature of the gelation process, gels with different physical properties (hardness, opacity, *etc.*) may exist under apparently similar conditions if the thermal or chemical routes applied during their formation have been different. This emphasizes the difficult question of the proper thermodynamic status of gels (stable *versus* metastable *versus* unstable), as questions of thermal/ mechanical/chemical history would not arise, of course, if such materials really corresponded to a global free energy minumum. In this paper, where the concept of a drive towards equilibrium is referred to, it should be assumed that in most cases a metastable equilibrium state is what is being referred to.

3 The Structure of Biopolymer Gels

The previous section has described a simple classification of gelling biopolymer systems into what amounts to the almost random aggregation of disordered molecules (the denatured protein case), on the one hand, and the association of

ordered polymer chains, on the other. We now consider these two situations in more detail.

Where heat-set globular protein gelation is concerned, there is good evidence[2] that the aggregating molecules behave more as 'sticky' reactive particles, than as totally unfolded random coils. A picture emerges of approximately globular entities, with hydrophobic regions on their surfaces, entering into a branching process. The bonding process may also involve electrostatic interactions mediated by ions, and it is not impossible that co-operative hydrogen bonding of sections of peptide chains (such as β-sheet formation) is also a source of cross-linking. The overall effect is that, on heating, the individual protein molecules (or, in some cases, small groups of these) aggregate into filaments (where double-layer repulsion is high), or dense branched clusters, where long-range repulsion is absent. Depending on such details of inter-particle interaction, gels of varying degrees of homogeneity are able to form.

The rheological character of heat-set globular protein gels is usually quite simple,[2–4] with their constitutive behaviour tending towards a pure elastic response (though with some evidence of viscoelasticity); the gels are usually reasonably extensible prior to rupture (*e.g.* up to 20% strain). In most cases such systems are thermally irreversible; 'melting out' of the network structure below 100 °C is rare (though not unknown). Lowering of the modulus with increasing temperature is a common observation, however, and this may indicate a loss of cross-linking on heating, or a change in network strand character, or both.

Turning to what is perhaps a broader and more common class of gelling biopolymer systems, we now consider cold-set polysaccharide gels,[5–7] and the related polysaccharide gels which form on addition of the appropriate counter-ions.[5–7] Common examples of such materials are gels formed from the marine polysaccharides agar, carrageenan, and alginate, and from the plant cell-wall polysaccharide pectin. Under changing solvent conditions, or just simply changing temperature, such polysaccharides are believed to form ordered helical conformations, including, in some cases, extended ribbons. Where multiple helices are produced (*e.g.* the double helices of agar and the carrageenans), a cross-linking mechanism is already implied. Where the ordered conformation is more extended, however, aggregation may involve either chain dimerization (*i.e.* a degenerate version of the double helix) or sheet formation.

In the non-extended (and non-degenerate) single or multiple helical situation, further aggregation of helices may also be involved, as has been proposed for agar gels,[5–7] and for some carrageenan gels.[8] In many of these situations, the counter-ions involved (and the background electrolyte) are of great importance in determining the outcome of the cold-set network-forming event. Calcium ions, for example, are believed to be actively involved in building ordered conformations during both pectin and alginate gelation, and specific counter-ion binding may be involved in some carrageenan gel situations as well. The overall result is that fibrous networks form from the polysaccharide chains, with a large part of the polymer being involved in the ordered part of the structure. In such a case, the average fibre diameter (and diameter distribution) is mainly determined by conditions of polymer charge, the counter-ions present, and the nature of the supporting electrolyte.

Other polysaccharide systems which show gelling behaviour related to an ordering of chains include the components of starch (amylose and amylopectin), and a number of polymers produced by bacteria. The reader is referred to specialized works[9-11] for a description of these, but it is worth noting that one of the most celebrated of the bacterial polysaccharides, xanthan,[12] provides us with an example of a network-forming polymer in which the bonds between ordered regions are not at all as permanent as those underlying, say, agar or carrageenan gels. The result is that, while xanthan gels appear solid-like at small applied strains, they yield and appear fluid at larger strains. This is a property of xanthan gels which turns out to be of considerable practical significance—for example, when such gels are used as pourable suspending agents for food particles or abrasive particles. Gels of this latter type have sometimes been referred to[7] as 'weak gels' in contrast to most of the materials so far discussed, which are examples of 'strong gels'.

Before ending this brief review of the structural aspects of biopolymer gels, and the mechanisms of their formation, mention must be made of a protein which, without doubt, is one of the most commonly known, and commonly used, gelling biopolymers, *i.e.* gelatin. Gelatin[7,13] is a product of collagen denaturation and hydrolysis, and sets into transparent gels on cooling warm solutions below 40 °C. As a gelling material, it has more in common with the class of cold-setting polysaccharides than the heat-set globular proteins, since cross-linking occurs on cooling, via a disorder-to-order transformation, as the random coil gelatin molecules seek to return to the ordered triple-helical collagen conformational state. The formation of the triple helix is evidently the prime cross-linking mechanism in gelatin gels, as there is little evidence for helix-helix association. A further major difference between gelatin and polysaccharide gels is that the gelatin-based materials show evidence of a more rubber-like (*i.e.* extensible elastic) character, and this is attributable, at least in part, to the far less complete conversion of gelatin to the ordered form when networks are generated. This means that quite a proportion of the network is in a single chain, extensible, form. Like many polysaccharide gels, gelatin gels are thermally reversible, the ordered structure being capable of 'melting out' on heating.

Lastly, in this section, attention is briefly drawn to the nature of the experimental approaches which have been used to derive such detailed descriptions of biopolymer gelling processes. These include spectroscopic techniques such as UV absorption, optical rotary dispersion (ORD), circular dichroism (CD), infrared and Raman spectroscopy, and NMR spectroscopy. Scattering methods have been useful, such as small-angle X-ray scattering, neutron scattering, and light scattering (static and dynamic). Light and electron microscopy have also been applied, as have thermodynamic techniques such as differential scanning calorimetry (DSC), equilibrium dialysis, and potentiometry. These techniques have been discussed individually in numerous specialized textbooks and reviews, but for an overall summary of their value in the study of biopolymer gelation, and the information they can potentially generate, the reader is referred to a recent review.[7] We note that such approaches are often used to study both the approach to the network structure (*i.e.* in a time-resolved dynamic mode), or as methods of studying the properties of the finished structure.

4 Biopolymer Gels—The Growth of the Shear Modulus

The process of polymer aggregation underlying the gelation of a biopolymer solution can be followed by a variety of techniques (*e.g.* X-ray scattering and light scattering), but very few of these are capable of directly detecting the development of a full three-dimensional polymer network spanning the system, or of measuring the properties of this network as it matures. The generation of such a network is, of course, the essence of the gelation process, and it is the origin of the dramatic change in the mechanical behaviour of the system which the sol-to-gel transformation represents. Since a dramatic change in the constitutive behaviour of a biopolymer solution is what is actually to be measured, and since it is a 'solid' or 'solid-like' material that is generated by the gelation process, it is natural that the course of this transformation should be monitored by mechanical testing and, in particular, by testing methods most suitable for the non-destructive study of solids. The methods of oscillatory-shear, dynamic mechanical spectroscopy are ideally suited to this task, and are routinely applied.[14]

The application of dynamic mechanical spectroscopy to study a gelling biopolymer solution proceeds via a so-called 'cure experiment'. This may then be followed by a frequency sweep to test the dependence on frequency (or, more accurately, the expected lack of such dependence) of the gel shear modulus, and a strain sweep to examine the extent of the so-called linear response range (the amplitude of strain over which the modulus is independent of strain). At this stage the gel may yield, or rupture, depending on its network properties. If the gel is not damaged by this treatment, the mechanical spectrometer can then be used to study the temperature dependence of the modulus and to look for 'gel melting' phenomena.

The preliminary cure experiment is the kinetic study of network development. A suitable frequency of measurement is chosen (*e.g.* 1 Hz), and a suitable small amount of strain, and the sample is heated or cooled to the gelling temperature. The real and complex parts of the shear modulus (*i.e.* storage G' and loss G'') are then monitored as a function of time as aggregation and gelling proceed. A typical result is indicated in Figure 1. There are two essential features of the behaviour.

(1) Although not accurately measured, or displayed, in Figure 1, both G' and G'' are expected to be very small quantities to begin with, with G'' greater than G', as expected for a liquid. Eventually there is a cross-over of these values (also not shown in Figure 1, but capable of measurement as indicated, for example, in ref. 15), and, as the network develops, both G' and G'' rise rapidly. At this stage, the ratio G'/G'', which is a measure of elastic character in the system, increases as the solid forms. If available, the cross-over point is sometimes used as a measure of the time elapsed before the network first appears, *i.e.* it is regarded as a measure of the 'gelation time'. But, in practice, other criteria for measuring this quantity are often adopted, such as a simple extrapolation of the rapidly rising modulus back to the time axis. However it is defined, the gelation time is found to depend greatly on sample conditions, such as the concentration of polymer, the gelling temperature, and the solvent conditions; and for work done in which the last two factors are kept constant (the usual situation), polymer concentration is clearly the most important factor. As might be expected, the gelation time increases (sometimes dramatically) as the concentration decreases.

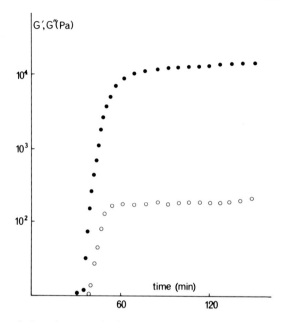

Figure 1   *Typical plot of storage (G') and loss (G'') components of the dynamic shear modulus as a function of time during gelation of a biopolymer solution. Frequency of measurement is ca. 1 Hz, and strain is <0.01*

(2) A second feature of the cure curve is the plateau region which develops some time after the gel point. Although the modulus components G' and G'', never become entirely constant, they become approximately so after a period of time, and generally only experimental results plotted in logarithmic time make it clear that changes are still taking place. It is found that the plateau value of the modulus (*e.g.* the value of G') increases very much with concentration, but that at certain low concentrations the plateau region may not be reached in practice unless very long cure times are employed.

From these essential features (1) and (2), we see that the characteristics of the kinetics of modulus development are a lag time, a period of rapid growth in modulus, and a period of levelling off during which the components of the modulus either become constant or change very slowly. The question may now be asked: what is the molecular explanation for such observations, and can this explanation also rationalize quantitatively, or at least semi-quantitatively, the dependence of the lag period, and the limiting modulus value reached, on polymer concentration?

5   A Kinetic Model for the Growth Process

As a starting point in the search for a rationalization of the observations just described, it is natural to consider the kinetic scheme for the polycondensation of simple monomers into gels proposed by Flory and Stockmayer[16,17] many years ago. Before describing this approach, however, it is worth considering the problem of particle aggregation in more general terms. The complex problem of describing the aggregation of monomeric species to form aggregates is, of course, a very broad one, and cannot be given a general solution. It involves writing down all possible association reaction equations for the aggregating species (including back reactions if these are applicable), and solving the appropriate kinetic equations. These can be set up in all sorts of ways depending on how species of various sizes are believed to react in a particular situation. In practice, therefore, simple models are considered, with certain underlying physical assumptions, and the equations for particular models are solved, and their consequences examined. A familiar example of such a particular solution is Smoluchowksi's approach,[18] in which the rate constant is assumed to be independent of the nature of the reacting aggregates. Such a model allows the molecular weight to grow with time, but, as can easily be shown, it does not provide a rationalization of the form of gelation kinetics described above. To achieve this, other assumptions about the rate constants must be made, such as those originally set out and discussed by Flory and Stockmayer.[16,17] In this latter model, it is assumed that reaction is initially between monomers, each having a number of bonding sites f per monomer, sometimes referred to as 'functionalities'. The assumption is then made that these sites react independently during the polycondensation event, and that the rate constant appropriate to the reaction of an n-mer and p-mer is proportional to the product of the functionalities of these species. In this way, an intrinsic rate constant for the interaction of individual functionalities is defined. This scheme predicts that large species will react very much more rapidly with other large species, than with other monomers, or small aggregates, and solution of the equations of the system reveals that, at a particular stage in the progress of the aggregation process, the weight-average molecular weight of the system will diverge logarithmically to infinity as a 'gel molecule' emerges. In subsequent stages of aggregation, the remaining sol fraction diminishes in amount, and becomes 'cured' into the network. The point of divergence is established to occur when the fraction of functionalities which have reacted reaches the value $1/(f-1)$.

To illustrate how data of the type shown in Figure 1 for biopolymer gelation can be reproduced (at least qualitatively) by a scheme similar to that proposed by Flory and Stockmayer, the following calculation is performed. A monomer with f functionalities (*e.g.* an unfolded globular protein molecule) is assumed to enter into the Flory–Stockmayer reaction scheme, and to have intrinsic rate constants for the forward reaction, K_f, and for the reverse reaction, K_b. The degree of reaction of functionalities, (*i.e.* the fraction of these which have reacted at any time) is denoted by the symbol α, and allowed to be a function of time, $\alpha(t)$. If the reaction of functionalities is described by a second-order forward reaction, coupled to a first-order back reaction (consistent with the Flory–Stockmayer model), a solution for $\alpha(t)$ in terms of the initial monomeric species concentration, the functionality f, and the rate constants, is readily found by the methods of elementary chemical kinetics.

To relate $\alpha(t)$ to the time-dependent shear modulus $G(t)$, methods described by Gordon and co-workers[19,20] using cascade theory, and the concept of an extinction probability, are used to count elastically active chain (EANC) densities; and these, in turn, allow calculation of the shear modulus, a contribution of akT per EANC being assumed.[7,21] Here, the non-ideality parameter a is introduced to account for probable non-ideality of network chains in a real biopolymer network. (For an ideal rubber network a is unity and independent of temperature.) In this way, the function $\alpha(t)$ is transformed into $G(t)$, where in this case no distinction is made between $G = \sqrt{(G' \cdot G' + G'' \cdot G'')}$, and G' itself. This, of course, implies that we have $G'' = 0$ throughout, an assumption which clearly neglects viscoelasticity present in both the gelling polymer solution and the gel. This is unfortunately an inevitable limitation of the model, and means that the modulus data to which such a model can be applied must be data either from a highly elastic gel system or from a gel examined at an experimental frequency high enough to make all of the important network cross-links seem permanent.

A typical set of modulus growth curves is shown in Figure 2 for somewhat

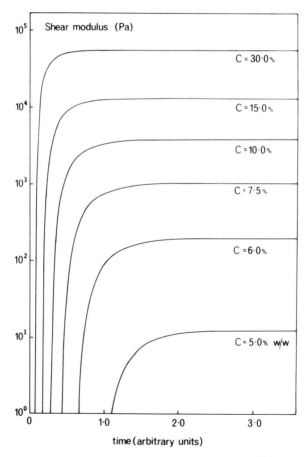

Figure 2   *Simulated shear modulus versus time cure curves for a series of solution concentrations C with the parameter values given in the heading to Table 1 on p. 334*

arbitrary values of the monomer molecular weight (6.65×10^4 daltons), the monomer functionality (3), the forward and back rate constants (1000 and 1, respectively), and the non-ideality factor (10.0). Results for several starting polymer concentrations are shown, and it is clear that the prime features of the kinetics of modulus development are reproduced by the simple cross-linking scheme adopted. The time-scale is in arbitrary units.

Turning to quantitative predictions of the model, it is interesting to examine the concentration dependence of the gelation time implied by the data in Figure 2. Since no G'/G'' crossover exists (G'' is ignored in the calculation), this particular criterion is not available to determine lag times from the theoretical data. If, however, we simply locate the earliest time at which the modulus takes on some limitingly small value (*e.g.* 0.01 Pa), and call that the gelation time, we probably have as reliable an estimate as is ever likely to be available experimentally. Using this procedure, the plot of logarithm of gelation time against concentration which appears in Figure 3 can be constructed. This plot clearly indicates that the gel time is expected to diverge very rapidly to infinity as the concentration of the gelling species approaches some limiting value C_0. This quantity may be called the critical concentration for the gelling system, and, indeed, in practice, the existence of such a threshold seems borne out for many real biopolymer systems.

A second point of interest is the relationship between the terminal (or limiting) modulus and the concentration. A plot of this quantity against concentration for

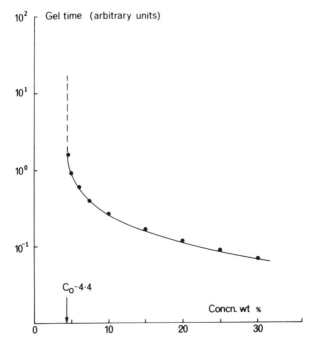

Figure 3 *Gelation times (see text) obtained from the data in Figure 2 are plotted against solution concentration. Note the divergence of log (gel time) as concentration tends to 4.4 wt% from above*

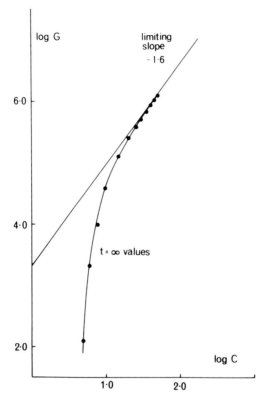

Figure 4   *Plot of logarithm of shear modulus G against logarithm of concentration C derived from the* t = ∞ *modulus values obtained by simulation. The straight line indicates the limiting power law* (G ~ $C^{1.6}$) *behaviour at high concentrations*

the simulated situation is shown in Figure 4. Here a log–log display is adopted, and the results immediately show that no single power-law relationship (*i.e.* $G \sim C^n$) is implied. Instead, the apparent power n in such a relationship varies from a very high value near the critical concentration threshold to a limiting value of less than 2 at high concentrations. It should be noted, however, that at intermediate concentrations, a value for n of *ca.* 5 could be extracted by fitting a straight line to a partial set of data, and that this is in fact the power law quite often quoted in the literature for gelling systems.

An alternative representation of the limiting modulus data in the form of log G *versus* C appears in Figure 5. This is a plot widely used in the literature for the presentation of experimental results. Also shown in Figure 5 are the sorts of limiting log G *versus* C relationships which would have been obtained if the simulated experiments had been stopped at some chosen time before all the moduli had reached plateau values. In the graph, such 'iso-chronal modulus values', as these might well be called, have been plotted for three arbitrary cut-off

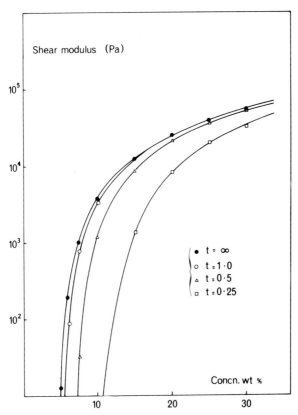

Figure 5   *Data of Figure 4 plotted as log G versus C (●). Corresponding plots for isochronal truncated modulus data (see text) are also included for cut-off times of* t = 0.25, 0.5, *and* 1.0. *Time is measured in arbitrary units. The continuous lines indicate best-fit theoretical curves obtained by least-squares calculations using the* t = ∞ *theoretical function given in ref. 21*

times. It is clear that the equilibrium curve ($t = \infty$) is approached by the others as a limit.

Interestingly, the $t = \infty$ limiting curve in Figure 5 can be used to check the correctness and consistency (with the Flory–Stockmayer approach) of a previous method employed by the author [21] to fit such log G *versus* C results (experimental data). A mathematical formula for the $t = \infty$ result, expressed in terms of the quantities f, a, $K = K_f K_b$, and the polymer molecular weight M, was in fact derived some years ago using much the same arguments and assumptions as those underlying the recent more general kinetic approach. At that time, it was also shown that the critical concentration is given by the expression $C_0 = M(f-1)/Kf(f-2)(f-2)$. Accordingly, as part of the present calculations, the earlier $t = \infty$ G *versus* C formula[21] was applied to the limiting data of Figure 5 using least-squares minimization to determine f, a, K, and the related C_0. The

Table 1   *Results of fits to simulated isochronal shear modulus versus concentration data. Values assumed in simulation are*: $f = 3.0$, $a = 10.0$, $K = 1000.0$ ($K_f = 1000$, $K_b = 1.0$), $M = 6.65 \times 10^4$

	Truncation time (arbitrary units)			
Parameter	0.25	0.50	1.0	∞
f	5.0	5.0	5.0	3.0
a	8.5	6.9	5.5	10.0
K	56	90	121	1000
C_0	10.6	6.6	4.9	4.4

results obtained appear in Table 1 (see column $t = \infty$), and, as would be hoped, the various parameters are exactly equal to the input parameters of the current kinetic simulation. Also, as Figure 5 shows (continuous lines), the fit to the 'experimental' $t = \infty$ data points is essentially perfect.

In summary of the above, therefore, the reversible Flory–Stockmayer approach, coupled with a few other assumptions and some network statistics, seems capable of explaining most of the features observed in practice for gelling biopolymer systems. A qualitative explanation has certainly been achieved, and, though not quoted here, analytical formulae may also be produced for a quantitative description. The questions which still remain are: how good is this quantitative description, and are there any other alternative, kinetic explanations which might be as good or better? These issues are addressed in the next section.

6   Limitations of the Kinetic Model and its Relation to Experiment

While its success in the context of biopolymer gelation has been considerable, it is clear that the kinetic model just described contains a number of drastic simplifications. Accordingly, it must be treated with some caution when considered in the context of real gelling systems, and hence it should be subjected to as many experimental tests as possible.

Before turning to these tests, and examining their conclusiveness or otherwise, it is useful to review some of the assumptions which underly the kinetic model, so that if inconsistencies do arise in application, we may be better positioned to understand their origin. First, and most obviously, the kinetic scheme adopted is simplistic and somewhat rigid. In practice, for the case of real gelling systems, the rate constants will not be truely independent of time, but will change considerably as aggregation proceeds, particularly after the gel point is reached. In addition, bond formation will never be as totally uncorrelated as the model demands, elements of co-operativity (or indeed anti-co-operativity) being possible. Models involving high levels of cross-linking co-operativity, which represent generalizations of the current kinetic scheme, have been proposed in the literature,[22] though their validity is currently under dispute.[23] In addition, other kinetic events not included in the model, such as intramolecular bonding (cyclization) will, in reality, be present. It will also be rare for a complex gelling system to have available to it one single cross-linking mechanism only, and, even if it did, it is unlikely that the aggregating particles would be essentially monodisperse in relation to numbers of bonding sites, bond strengths, *etc.* Thus, we see that the detailed functional form of $\alpha(t)$ is unlikely to be accurately predicted for a real system, even though we may be successful in getting its main features correct.

In addition to the form of $\alpha(t)$, there are further problems with the mapping assumed between $\alpha(t)$ and $G(t)$. As in previous analyses[7,21] in which this same mapping has been adopted, certain simplifying assumptions are implied, such as the neglect of cyclization, and the neglect of other features of real networks such as trapped entanglements.

Returning now to the question of the extent of agreement with experiment, it should be clear that the most rigorous examination of the effectiveness of the model would be to least-squares fit a full collection of experimental gel cure curves, and to examine the quality of the fit achieved. Unfortunately, this has not yet been attempted, and practical testing of the model has so far been restricted to two much simpler (and less conclusive) exercises. The first is to examine the properties of gelation time to see if they do behave as suggested in Figure 3, and diverge at some critical concentration; the second is the fitting of limiting modulus *versus* concentration data. In general, the results of such tests are encouraging, divergence of the gelation time being found in those few cases (e.g. ref. 3) where it has been rigorously searched for, and the $t = \infty$ analytical formula fitting real limiting modulus data very well on the now quite numerous occasions[7,21,23–26] on which this test has been applied. In these last cases also, log G *versus* C plots have been found to have very much the form, and limiting slope, predicted in Figure 4. In addition, where examined,[3,4,24] critical concentrations derived from such fits seem to agree reasonably well with corresponding estimates from gelation times, and values of the non-ideality parameter a usually make sense, being quite close to unity for the more flexible biopolymer networks (e.g. gelatin gels) and much higher for less flexible cases (most polysaccharide gels).

Literal physical interpretation of the parameters derived by fitting modulus–concentration data is not always justified, however, even when the reversible cross-link model is rigorously applicable. A simple computer test, using the simulated data of the last section, readily shows this. This test involves using the $t = \infty$ formula to fit modulus data corresponding to some finite, but quite short, cut-off time, such as the simulated, truncated, isochronal modulus values included in Figure 5. A straightforward least-squares fitting exercise to these three data sets is found to achieve excellent fits (solid lines in Figure 5) and to provide the parameter values included in Table 1. Whilst the parameters f and a are not drastically altered by the time cut-off, it is clear that the extracted equilibrium constants, and the C_0 values, become progressively (and substantially) less accurate, and meaningful, as truncation proceeds. It seems that good fits can be achieved by the formula in situations where it ought not to apply, and that at least some of the parameters likely to emerge in these circumstances will therefore be meaningless. In situations, then, where it is difficult to decide whether modulus values have become constant (and, in practice, this is likely to be quite common), fits may lack any fundamental meaning, however useful they may be as means of interpolating or extrapolating data.

So far, we have found no serious contradiction between experiment and the predictions of the reversible cross-link model, but a little thought soon shows that one of the major successes, *i.e.* the prediction of a limiting critical gel concentration, might have an alternative explanation within the framework of an essentially irreversible cross-linking scheme. The explanation for the existence of C_0 offered by the reversible cross-linking model is that, because of reversibility, certain solutions, at certain concentrations, can never become critically branched. According to the reversible model, this will be true for all gelling systems, but, of course, as the strength of bonding grows, the critical concentration must fall, and so the system becomes more and more completely bonded at low concentrations. This may all seem perfectly plausible, but some thought shows that in the limit of very high cross-linking affinity, there is a fundamental contradiction between the need for a high level of bonding at low concentrations implied by this driving force, and the need for the network to fill the entire volume of the system. At this point the issue becomes one of gel inhomogeneity, an issue not directly addressed by the current model at all, since it is a 'mean field' approach based on ideas of connectivity only. For a highly inhomogeneous gel, the concept of critical concentration may in fact be somewhat different from that discussed above, and its prediction is almost certainly outside the scope of the model. In this case, the critical concentration of practical importance is the concentration below which the gelling system breaks up into microgel fragments, each of which may be assumed to have already passed through the gel point. In such an irreversible situation, an apparent critical concentration is found which is very much higher than would be predicted by rigorous application of the reversible cross-link approach.

Finite critical concentration behaviour could be exhibited by an irreversibly gelling system for yet another reason, *i.e.* as a result of intramolecular reaction, or cyclization. Such reaction in dilute solution would compete with the branching reaction, and lower the effective functionality of the system at the lower concentrations. Thus, for an irreversibly aggregating system, a set of cure curves could be obtained, at a series of concentrations, which have all the appearance of data for a reversibly gelling biopolymer solution moving towards equilibrium.

The conclusions which may be drawn from the preceding paragraphs are as follows. The reversible cross-linking model is most likely to be appropriate to gelling systems where the strength of bonds formed is not high, and where a level of reversibility is to be expected. The model will fail more and more as bonds become permanent. In this last situation, there is a danger that superficial success when using the model to fit modulus data, or to explain gelation times, may obscure the true kinetic situation. In other words, as a result of finding a 'normal' critical concentration, and being able to fit limiting concentration data satisfactorily (as described for the truncated data case earlier), a level of reversibility might be imposed on a system that is in fact unjustified. A proper treatment of irreversible gel formation will probably require inclusion of cyclization in the model, and some recognition of the implications of phase separation.

7 Other Aspects of Biopolymer Gel Formation

The present paper has reviewed the subject of biopolymer gelation from a number of points of view, and has particularly concentrated on the development of elasticity in such systems and its rationalization in terms of the build-up of network structure. The treatment of mechanical properties has been at the level of infinitesimal displacement only. No attempt has been made to discuss or catalogue the various types of response which can arise from such materials when they are subjected to much larger strains.

Another issue which has only been touched upon lightly is the important question of gel network inhomogeneity, and the influence of solvent on this property. In the classical aggregation theory applied above, solvent effects are only considered by implication, that is, in terms of any influence they might have on cross-linking affinity. The Flory–Stockmayer approach does not formally distinguish between homogeneous and inhomogeneous network structures, but, in reality, of course, many gelling systems produce turbid, or highly opaque, gels. A more specific treatment of this issue has been attempted by Coniglio, Stanley, and Klein,[27,28] in which they have combined network connectivity arguments with the ideas of polymer solution theory. To do this they used a lattice model, and by this means were able to calculate 'phase diagrams' for gels. While the approach adopted by Coniglio *et al.* may be regarded as a more genuinely statistical-mechanical attack on the gel problem than that discussed here, it does not appear to deal with the kinetic issues of gelation, being essentially a reversible equilibrium model.

8 Conclusion

The model presented here should be regarded as applying rigorously only to some unrealizable 'ideal gel system'. Like all such ideal descriptions, it provides a reference point with which to relate the actual behaviour of real systems. In this respect, it is hoped that the present paper will encourage workers in the area of food gels to locate and measure such discrepancies, for by this route it is likely that our knowledge about the mechanisms of biopolymer gelation will substantially increase.

Acknowledgements. I would like to thank my colleagues Dr S. B. Ross-Murphy and Mr R. K. Richardson for many useful discussions of polymer gelation and for the provision of an example of their experimental data. I also thank Mrs L.A. Linger for technical assistance in the preparation of diagrams.

References

1. P. J. Flory, *Faraday Discuss. Chem. Soc.*, 1974, **57**, 7.
2. A. H. Clark and C. D. Lee-Tuffnell, in 'Functional Properties of Food Macromolecules', ed. J. R. Mitchell and D. A. Ledward, Elsevier Applied Science, London, 1986, p. 203.
3. R. K. Richardson and S. B. Ross-Murphy, *Int. J. Biol. Macromol.*, 1981, **3**, 315.
4. R. K. Richardson and S. B. Ross-Murphy, *Brit. Polym. J.*, 1981, **13**, 11.
5. E. R. Morris, D. A. Rees, D. Thom, and E. J. Welsh, *J. Supramol. Struct.*, 1977, **6**, 259.
6. D. A. Rees and E. J. Welsh, *Angew. Chem. Int. Ed. Engl.*, 1977, **16**, 214.
7. A. H. Clark and S. B. Ross-Murphy, *Adv. Polym. Sci.*, 1987, **83**, 57.
8. E. R. Morris, D. A. Rees, and G. Robinson, *J. Mol. Biol.*, 1980, **138**, 349.
9. M. J. Miles, V. Morris, and S. G. Ring, *Carbohydr. Res.*, 1985, **135**, 257.

10. S. G. Ring, P. Colonna, K. J. I'Anson, M. T. Kalichevsky, M. J. Miles, V. J. Morris, and P. D. Ordford, *Carbohydr. Res.*, 1987, **162**, 277.
11. P. A. Sandford, *Adv. Carbohydr. Chem. Biochem.*, 1979, **36**, 265.
12. S. A. Frangou, E. R. Morris, D. A. Rees, R. K. Richardson, and S. B. Ross-Murphy, *J. Polym. Sci., Polym. Lett. Ed.*, 1982, **20**, 531.
13. D. A. Ledward, in 'Functional Properties of Food Macromolecules', ed. J. R. Mitchell and D. A. Ledward, Elsevier Applied Science, London, 1986, p. 171.
14. J. D. Ferry, 'Viscoelastic Properties of Polymers', 3rd edn, John Wiley, New York, 1980.
15. D. Durand, C. Bertrand, A. H. Clark, and A. Lips, *Int. J. Biol. Macromol.*, 1990, **12**, 14.
16. P. J. Flory, *J. Am. Chem. Soc.*, 1941, **63**, 3083, 3091, 3096.
17. W. H. Stockmayer, *J. Chem. Phys.*, 1943, **11**, 45.
18. M. von Smoluchowski, *Z. Phys. Chem.*, 1917, **92**, 129.
19. M. Gordon, *Proc. Roy. Soc. (London)*, 1962, **A 268**, 240.
20. M. Gordon and S. B. Ross-Murphy, *Pure Appl. Chem.*, 1975, **43**, 1.
21. A. H. Clark, in 'Food Structure and Behaviour', ed. J. M. V. Blanshard and P. J. Lillford, Academic Press, London, 1987, p. 13.
22. D. Oakenfull, *J. Food Sci.*, 1984, **49**, 1103.
23. A. H. Clark, S. B. Ross-Murphy, K. Nishinari, and M. Watase, in 'Physical Networks', ed. W. Burchard and S. B. Ross-Murphy, Elsevier Applied Science, London, 1990, p. 209.
24. A. H. Clark and S. B. Ross-Murphy, *Brit. Polymer J.*, 1985, **17**, 164.
25. A. H. Clark, M. J. Gidley, R. K. Richardson, and S. B. Ross-Murphy, *Macromolecules*, 1989, **22**, 346.
26. M. Watase, K. Nishinari, A. H. Clark, and S. B. Ross-Murphy, *Macromolecules*, 1989, **22**, 1196.
27. A. Coniglio, H. E. Stanley, and W. Klein, *Phys. Rev. Lett.*, 1979, **42**, 518.
28. A. Coniglio, H. E. Stanley, and W. Klein, *Phys. Rev.*, 1982, **B25**, 6805.

Mixed Gels Formed with Konjac Mannan and Xanthan Gum

By P. A. Williams, S. M. Clegg, D. H. Day, G. O. Phillips

FACULTY OF SCIENCE AND INNOVATION, NORTH EAST WALES INSTITUTE, CONNAH'S QUAY, DEESIDE, CLWYD CH5 4BR

and K. Nishinari

NATIONAL FOOD RESEARCH INSTITUTE, MINISTRY OF AGRICULTURE, FORESTRY AND FISHERIES, TSUKUBA, IBARAKI 305, JAPAN

1 Introduction

Konjac Mannan (KM) is a β-D-(1–4) linked glucomannan and contains 1–3 linked branches occurring at C-3 of glucose and mannose residues.[1-3] The D-mannose to D-glucose ratio has been shown to be 1.6:1 and approximately 1 in 19 sugar units are acetylated.[4] Whereas native KM is soluble in water, deacetylation results in the formation of a thermally irreversible gel.[4] Xanthan gum has a β-D-(1–4) linked glucan backbone with short trisaccharide side-chains consisting of α-D-mannose, β-D-glucuronic acid, and β-D-mannose on alternating glucose residues.[5,6] The mannose residue linking the side-chain is acetylated and the terminal mannose contains pyruvate groups approximately every other side-chain. Xanthan gum does not form true gels in aqueous solution, but has been shown to undergo a thermally induced conformational change which is sensitive to the presence of electrolyte, and there is still considerable debate as to whether the ordered structure adopted involves single or double helices.[7-14]

Mixtures of xanthan + KM have been reported to form thermally reversible gels, and the gels produced are much stronger and have higher melting points than those formed between xanthan and galactomannans.[15-20] Early work suggested that a specific interaction occurred between the ordered xanthan molecule and the galactomannan or glucomannan chain.[15-17] More recently, Tako et al.[21] concluded that intermolecular interaction occurs between the xanthan side chains and the galactomannan backbone, and that the molecules adopt the same ordered conformation as in the solid state. Brownsey et al.[19] have suggested that intermolecular binding involves co-crystallization of sections of the disordered xanthan chain with the structurally similar segments of the galactomannan or glucomannan chain. They showed that gelation of xanthan and KM only occurred if the solutions were mixed at temperatures above the xanthan order–disorder transition. In solutions containing 0.5 mol dm^{-3} CaCl$_2$, where the xanthan conformational change is shifted to above 100 °C, they found that gelation did not occur on cooling. They also carried out *X*-ray fibre diffraction studies on xanthan + KM mixed gels, and the diffraction pattern obtained provided evidence for intermolecular binding. The patterns were related to that for xanthan alone, unlike the case for xanthan + galactomannan gels, where the pattern was related to that for the galactomannan alone.

The purpose of the present paper has been to provide further evidence of intermolecular binding between xanthan and KM, and to shed further light on the gelation mechanism.

2 Materials

Konjac mannan was supplied by FMC Marine Colloids Div. The sample was found to have a molecular mass of 454 000 daltons as determined by light scattering measurements, and was shown to contain 1 acetyl group for every 17 sugar units as determined by the method of Maekaji.[4] A portion of the KM was spin-labelled by modifying the procedure of Cafe and Robb.[22] A sample of 4 g of KM was dissolved in 1.8 dm^3 of distilled water containing 0.02 wt% sodium azide by mixing at 85 °C for 30 minutes. After cooling, the solution was adjusted to pH 7.0 and cyanogen bromide (0.4 g/10 cm^3 H$_2$O) was added and the solution stirred for 4 hours. Some 4-amino Tempo spin label (0.2 g/5 cm^3 H$_2$O) was then added, and the solution mixed for a further 18 hours when the reaction was considered complete. The solution was dialysed against 0.02 wt% sodium azide until free of unreacted spin label, and then the solution with solids content *ca.* 0.2 wt% was stored at 4 °C. The KM was not freeze dried since it had been previously found that freeze-dried material was very difficult to redissolve.

Xanthan was supplied by Kelco International Ltd. and was coded Keltrol T. Analysis revealed that the xanthan contained one acetyl group every pentasaccharide repeat unit, and the number of pyruvate groups determined by the method of Sloneker and Orentas[23] was found to be 0.34 per repeat unit. The intrinsic viscosity was 30 dl g^{-1} in 0.5 mol dm^{-3} NaCl at 25 °C as determined by capillary flow viscometry.

3 Methods

Gel Strength—Xanthan and KM were dissolved together in water at varying ratios, but at a total polysaccharide content of 1.0 wt%, by heating at 95 °C for 30 minutes with vigorous stirring. The solutions were then equilibrated in a water bath for 2 hours at 25 °C and the 'breakforce' measured using a Stevens LFRA Texture Analyser fitted with a 0.4 cm diameter probe.

Electron Spin Resonance Spectroscopy—KM and xanthan + KM solutions were prepared at 80 °C and were placed in a flat quartz cell suitable for aqueous solutions and the ESR spectra were recorded as a function of temperature using a

Jeol JES ME IX ESR spectrometer. Spectra were obtained between 10 °C and 80 °C, cooling and heating over a 6 hour period and allowing 10 minutes equilibration before recording each spectrum.

Differential Scanning Calorimetry Measurements—DSC measurements were made on KM, xanthan and xanthan + KM mixtures using a Setaram Micro DSC equipped with 1 cm^3 batch vessels. The polysaccharides were dissolved individually at 95 °C and were then mixed thoroughly in varying proportions to give a total polysaccharide concentration of 1.2 wt% in water or electrolyte. The DSC heating and cooling thermograms were monitored at a scan rate of 0.2 °C min^{-1} and samples were subjected to an initial cycle of heating and cooling prior to recording the curves in order to ensure the same thermal history.

Rheological Properties—The storage modulus G' of xanthan + KM mixtures was recorded as a function of temperature using a CarriMed Constant Stress Rheometer fitted with a 4 cm parallel plate and solvent trap. Solutions were prepared as for the DSC experiments, and G' was determined on cooling from 85 °C at *ca.* 0.2 °C min^{-1}. Measurements were made using an amplitude of 6 milli-radians and at a frequency of 0.5 Hz.

4 Results

The breakforce values of gels prepared from blends of xanthan + KM at a total polysaccharide concentration of 1.0 wt% are given in Figure 1. They show that,

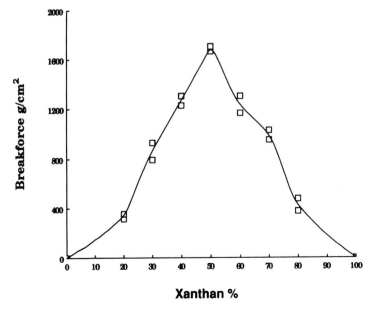

Figure 1 *Breakforce of* 1.0 wt% *xanthan* + KM *mixed gels*

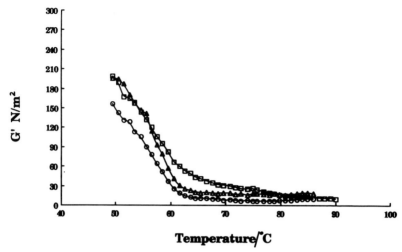

Figure 2   *Storage modulus G' on cooling xanthan + KM solutions* (1.2 wt%): ○,
25/75; □, 50/50; and △, 75/25 xanthan/KM

whereas neither KM nor xanthan are able to form gels alone, in admixture they
give rise to very strong gels. The strongest gels are produced at xanthan/KM ratios
of *ca.* 1:1 by weight, which on a molar basis would be equivalent to approximately
five KM molecules for every one xanthan molecule. The gels formed are
thermoreversible, and the onset of gelation as monitored by the increase in G' on
cooling xanthan + KM solutions at 1.2 wt% total polysaccharide concentration
and at varying ratios is illustrated in Figure 2.

The samples show a sharp transition at about 61–63 °C corresponding to gel
formation which appears to be independent of the polysaccharide ratio. This
agrees closely with the gel melting temperature of 63 °C reported by Dea *et al.*[16]
for a 0.25 wt% xanthan + 0.25 wt% KM mixture and is much higher than the
melting temperatures reported for xanthan + locust bean gum mixtures
(≈41 °C).

The ESR spectra obtained on cooling a spin-labelled KM solution alone and a
0.3 wt% xanthan + 0.2 wt% KM mixture are given in Figures 3A and 3B.
Whereas, in the absence of xanthan, the spectra remain narrowed indicating a
high degree of segmental motion, in the presence of xanthan the spectra possess
an anisotropic component which is indicative of a reduction in segmental motion
caused by chain aggregation as discussed in more detail previously.[24] Below
40 °C, the anisotropic component represents >40% of the signal as resolved by
computer analysis, indicating, therefore that a significant proportion of the KM
segments are involved in specific interaction with xanthan. It is possible, although
unlikely, however, that the anisotropic component could arise from aggregation
of KM molecules themselves and further work is required to fully clarify this
point. The anisotropic component is first detectable on cooling to about 62 °C and
thus corresponds to the onset of gelation. On heating the solutions, the anisotropic component disappears showing that the aggregation is reversible.

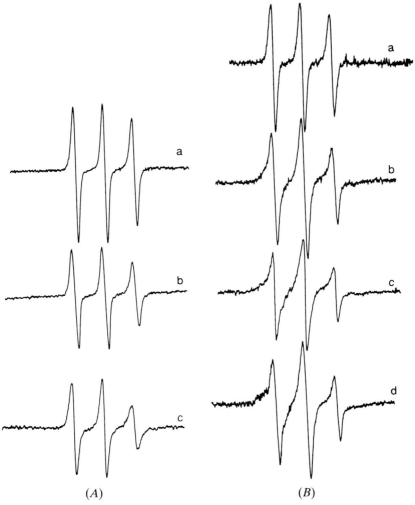

Figure 3   (*A*) *ESR spectra for KM alone at* (a) 75 °C, (b) 42 °C, *and* (c) 10 °C. (B)
ESR spectra for xanthan + KM mixtures at (a) 74 °C, (b) 40 °C, (c) 8 °C
and (d) 3 °C

The DSC cooling and heating curves for xanthan + KM solutions at 1.2 wt%
total polysaccharide concentration and at varying ratios are given in Figures 4A
and 4B. Xanthan alone is seen to undergo a thermally reversible enthalpy change
which has been assigned by other workers to a conformational order–disorder
transition.[10,11,25] The transition does not display hysteresis and the mid-point
temperature T_m is 51 °C which is close to the expected value as reported by
others.[8,10] ΔH was calculated to be 2.5 J g^{-1}, which is of the same order as that
indicated by Holzwarth and Ogletree,[26] but less than the values quoted by other
workers.[10,11,25] For the xanthan + KM mixtures, an increased enthalpy change

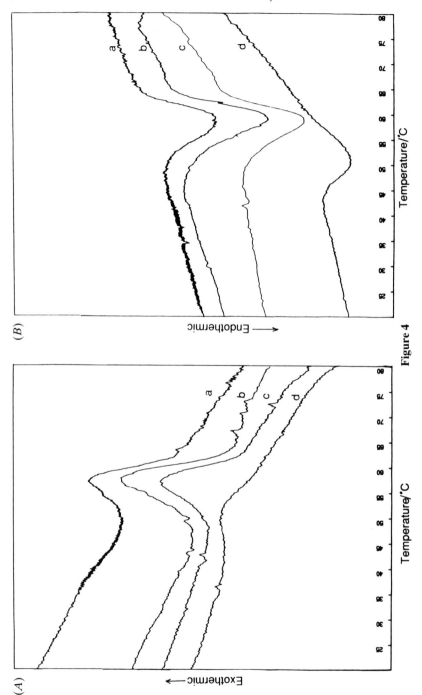

Figure 4

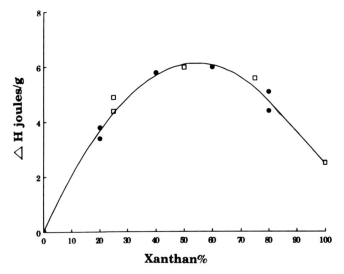

Figure 5 *Enthalpy of gelation, ΔH, of solutions of xanthan + KM at varying ratios:* □, *1.2 wt% total polymer;* ●, *0.5 wt% total polymer*

was noted with a T_m for the cooling curve of 57 °C and for the heating curve of 59 °C. The start of the transition on the cooling curve occurs at about 62–66 °C which corresponds closely to the gelation temperature obtained by rheological measurements and to the onset of chain aggregation as monitored by ESR spectroscopy. The DSC data thus provide further evidence of intermolecular binding. The enthalpies of the transitions of the various mixtures for total polysaccharide concentrations of 1.2 wt% are given in Figure 5. The maximum enthalpy change is obtained at xanthan/KM ratios between 1.5:1 and 1:1.5 by weight, which corresponds to the compositions giving the maximum gel strength. KM alone does not give rise to an enthalpy change.

DSC cooling curves for xanthan alone and a 1:1 mixture of xanthan + KM both in 0.04 mol dm^{-3} NaCl and both at a total polysaccharide concentration of 1.2 wt% are given in Figure 6. The T_m value for the xanthan alone had increased to 84 °C as expected from the observations of other workers.[8,10,25] The curve for the mixture shows two peaks, one at about 82.5 °C, which corresponds to the xanthan conformational transition, and another at 42–44 ° C. This latter peak was found to correspond to gelation as indicated by the sharp increase in G' shown in Figure 7.

Figure 4   (A) *DSC cooling curves for* (a) 25/75, (b) 50/50, (c) 75/25 *and* (d) 100/0 *xanthan + KM solutions at* 1.2 wt% *total polymer concentration*
(B) *DSC heating curves for* (a) 25/75, (b) 50/50, (c) 75/25 *and* (d) 100/0 *xanthan + KM solutions at* 1.2 wt% *total polymer concentration*

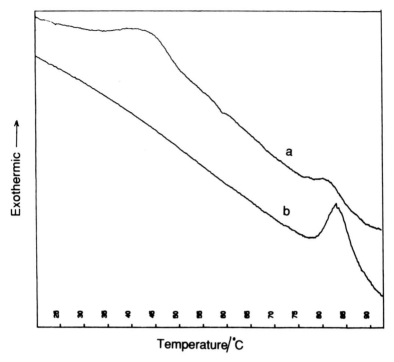

Figure 6 *DSC cooling curve for* (a) 50/50 *xanthan* + *KM solution* (b) *xanthan alone at* 1.2 wt% *total polymer in* 0.04 mol dm^{-3} NaCl

5 Discussion

The techniques of ESR spectroscopy and DSC measurement both indicate that gelation of xanthan + KM solutions arises from specific binding between the two polysaccharides. The current debate on the gelation mechanism of xanthan with glucomannans or galactomannans hinges on whether the interaction involves specific binding between the ordered or disordered form of xanthan and the KM or galactomannan chain. Dea *et al.* have shown[16] by optical rotation that xanthan undergoes a conformational change during gelation of xanthan/galactomannan blends and also that the transition temperature occurs 10 °C higher. In our studies in water, the DSC curves for xanthan + KM mixtures exhibit only one peak, the T_m for the enthalpy change occurring about 6 °C above the expected T_m for the xanthan conformational transition (see Figure 4). It is reasonable to suggest, therefore, that the enthalpy associated with gelation arises from xanthan–KM molecular interaction as well as ordering of xanthan chains.

The DSC cooling curves obtained in the presence of 0.04 mol dm^{-3} NaCl give different results. At this ionic strength, two enthalpy transitions are observed, one at 82.5 °C, which corresponds to the conformational transition of xanthan alone, and one on further cooling at *ca.* 42 °C, which corresponds to gelation. It is

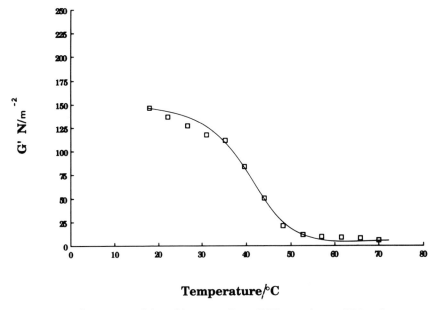

Figure 7 *Storage modulus G' on cooling 50/50 xanthan + KM solution at*
1.2 wt% total polymer in 0.04 mol dm^{-3} *NaCl*

apparent that, at temperatures above *ca.* 82.5 °C, disordered xanthan molecules
do not interact with KM molecules, presumably as a consequence of their high
thermal energy. On cooling, the xanthan molecules initially form their ordered
structure, and then at much lower temperatures interact with KM molecules to
form the gel network.

The fact that there is a reproducible difference in the gelation temperatures and
of the enthalpy of gelation in 0.04 mol dm^{-3} NaCl as compared to those in water
suggests a different mode of molecular aggregation at the two different ionic
strengths. In 0.04 mol dm^{-3} NaCl interaction must occur between KM and
ordered xanthan chains, but it is believed that in water the KM molecules interact
predominantly with disordered xanthan chains. Brownsey *et al.*[19] ruled out the
possibility of interaction between KM molecules and ordered xanthan molecules
since they found that solutions had to be heated above the xanthan order–
disorder transition temperature before gelation could occur. We have found that
mixing KM and xanthan solutions at low temperatures, where the xanthan is in
the ordered form, leads to the formation of very weak gels which increase in
strength on ageing, and that, although solutions must be heated to produce firm
gels, the temperature required does not have to be as high as the temperature for
the xanthan conformational change. Thus, whereas a solution of 1:1 xanthan + KM at a total polymer concentration of 1.2 wt% in 0.04 mol dm^{-3} NaCl
does not gel significantly when mixed thoroughly together at 25 °C, heating the
mixture to 70 °C (*i.e.* well below the xanthan order–disorder transition at 82.5 °C)

results in gel formation on cooling to *ca.* 42 °C. Heating, therefore, is believed to promote gelation by disrupting aggregated xanthan chains, and so enabling interaction to occur with KM molecules. Heating the same solution to 95 °C, and then cooling, still leads to gelation at *ca.* 42 °C (compare Figure 6).

6 Conclusions

It is concluded, therefore, that KM interacts with xanthan molecules in two different conformational forms. In the presence of 0.04 mol dm^{-3} NaCl, the interaction is between KM and the ordered xanthan chains giving rise to gelation at 42 °C. In water, the interaction is believed to be predominantly between KM and disordered chains leading to a more stable gel with a setting temperature of 57 °C.

References

1. F. Smith and C. Srivasta, *J. Chem. Soc.*, 1959, **81**, 1715.
2. K. Kato, T. Watanabe, and T. Matsuda, *Agric. Biol. Chem.*, 1970, **34**, 532.
3. H. Shimahara, H. Suzuki, N. Sugiyama and K. Nisizawa, *Agric. Biol. Chem.*, 1975, **39**, 293.
4. K. Maekaji, *Agric. Biol. Chem.*, 1974, **38**, 315.
5. P. E. Jansson, L. Keene, and B. Lindberg, *Carbohydr. Res.*, 1975, **45**, 275.
6. L. D. Melton, L. Mindt, D. A. Rees, and G. R. Sanderson, *Carbohydr. Res.*, 1976, **46**, 245.
7. G. Holzwarth, *Carbohydr. Res.*, 1978, **66**, 173.
8. M. Milas and M. Rinaudo, *Carbohydr. Res.*, 1979, **76**, 189.
9. D. A. Brant and G. Paradossi, *Macromolecules*, 1982, **15**, 874.
10. S. Paoletti, A. Cesaro, and F. Delben, *Carbohydr. Res.*, 1983, **123**, 173.
11. I. T. Norton, D. M. Goodall, S. A. Frangou, E. R. Morris, and D. A. Rees, *J. Mol. Biol.*, 1984, **175**, 371.
12. S. A. Jones, D. M. Goodall, A. N. Cutler, and I. T. Norton, *Eur. Biophys. J.*, 1987, **15**, 185.
13. W. Liu and T. Norisuye, *Int. J. Biol. Macromol.*, 1988, **10**, 44.
14. W. Liu and T. Norisuye, *Biopolymers*, 1988, **27**, 1641.
15. I. C. M. Dea and A. Morrison, *Adv. Carbohydr. Chem. Biochem.*, 1975, **31**, 241.
16. I. C. M. Dea, E. R. Morris, D. A. Rees, E. J. Welsh, H. A. Barnes, and J. Price, *Carbohydr. Res.*, 1977, **57**, 249.
17. C. T. Prest and K. Buckley, Eur. Patent, EP 185, 511 (1986).
18. S. Okonogi, H. Yagachi, and Y. Mugazaki, Eur. Patent, EP 208 313 (1987).
19. G. J. Brownsey, P. Cairns, M. J. Miles, and V. J. Morris, *Carbohydr. Res.*, 1988, **176**, 329.
20. B. V. McCleary, *Carbohydr. Res.*, 1979, **71**, 205.
21. M. Tako, A. Asato, and S. Nakamura, *Agric. Biol. Chem.*, 1984, **48**, 2995.
22. M. Cafe and I. D. Robb, *Polymer*, 1976, **17**, 91.
23. J. H. Sloneker and D. G. Orentas, *Nature (London)*, 1962, **194**, 4827.
24. D. H. Day, G. O. Phillips, and P. A. Williams, *Food Hydrocolloids*, 1988, **2**, 19.
25. S. Kitamura, T. Kuge, and B. T. Stokke in 'Gums and Stabilisers for the Food Industry', ed. G. O. Phillips, D. J. Wedlock, and P. A. Williams, Oxford University Press, 1990, Vol. 5, p. 329.
26. G. Holzwarth and J. Ogletree, *Carbohydr. Res.*, 1979, **76**, 277.

Flow and Viscoelastic Properties of Mixed Xanthan Gum + Galactomannan Systems

By Jean-Louis Doublier and Geneviève Llamas

LPCM-INRA, CENTRE DE RECHERCHE DE NANTES, RUE DE LA GÉRAUDIÈRE,
B.P. 527, 44026 NANTES CEDEX 03, FRANCE

1 Introduction

Galactomannans (guar gum and locust bean gum) and xanthan gum are two types of polysaccharides widely used in the food industry. Galactomannans are neutral polymers with a main chain of mannose units linked by β-(1–4) bonds, and this main chain bears galactose units irregularly distributed along the mannan backbone. Locust bean gum and guar gum differ primarily in their galactose to mannose ratio which is of the order of 1:2 for guar gum and 1:4 for locust bean gum. Xanthan gum is the extracellular polysaccharide produced by *Xanthomonas campestris*. The repeat-unit of this polymer is a pentasaccharide, the backbone of which is a chain of β-(1–4) glucose units as in cellulose with a trisaccharide side-chain linked at C-3 of alternate glucosyl residues. The terminal unit of this side-chain may contain pyruvic acid at a variable level of content.

These two types of polysaccharides display widely different rheological properties, which have been extensively investigated. Galactomannans show the typical behaviour of macromolecular solutions with topological entanglements where there is no interaction between macromolecules.[1,2] This behaviour is related to the conformation of the macromolecules which is known to be coil-like. In contrast, the behaviour of xanthan gum in aqueous medium is more complex, and is still a matter of some debate.[3] Overall, the flow and viscoelastic properties seem to correspond to a weak, gel-like network of associated xanthan molecules giving rise to supramolecular aggregates which are easily broken down on shearing.[4] The high low-shear viscosity of xanthan aqueous dispersions may be ascribed to these superaggregates and the pronounced shear-thinning behaviour can be explained by their rupture.[4,5] An alternative explanation is related to the rigid worm-like helix conformation of the macromolecules in the aqueous medium.[6]

When mixed together, xanthan gum and locust bean gum display spectacular synergistic properties, yielding strong gels while each polysaccharide separately is unable to produce a gel. This phenomenon has been extensively investigated for its rheological aspects, as well as for the molecular mechanisms underlying the properties.[7–11] In contrast, guar gum and xanthan gum do not yield a gel when mixed. However, increases in viscosity have been reported, which suggests that some kind of molecular interaction may arise from the mixing of the two polysaccharides.[7,12]

The molecular mechanisms underlying these rheological phenomena are still a matter of debate. Several authors have reported a relation between the proportion of 'smooth regions' in the galactomannan backbone and the interaction strength with xanthan gum. This interaction has been ascribed to the existence of specific associations between xanthan helices and unsubstituted regions of the galactomannan. This is the basis of the different models proposed for the interpretation of xanthan + locust bean gum synergy, although the models differ in the detailed description of the mode of association.[7–10] Viscosity increases in xanthan + guar gum mixtures cannot, however, be explained on a similar molecular basis. Although the galactose distribution along the mannan main chain is known to be irregular, as in locust bean gum,[13] the amount of galactose is too high to yield a sufficient amount of unsubstituted galactose-free regions to form junctions zones as in locust bean gum. The mechanisms involved in these interactions have been ascribed to regularly substituted regions in the mannan backbone which would be able to interact with xanthan helices.[8] The rheological aspects of these phenomena have been mainly investigated in order to evaluate the conditions of optimum synergism.[10–12] It was implicitly assumed that the rheology of the mixed gels thus obtained does not differ from that of classical polysaccharide gels as described by Clark and Ross-Murphy,[3] with the storage modulus being constant within the frequency range 0.01–10 Hz and $G' > 10G''$. Recent investigations, however, have given evidence that the detailed rheology of xanthan + locust bean gum systems is more complex and can not be described so simply.[11] It has thus been shown that two plateau regions exist in the mechanical spectrum (G' and G'' varying as a function of frequency), suggesting that two molecular mechanisms are involved in the gelation process. One is ascribed to mannan–mannan associations, and the other would seem to arise from xanthan–mannan interactions.[11]

It is thus clear that rheological techniques can be useful for studying the mechanisms of xanthan–galactomannan interactions. The object of the present work was to use rheological methods to study xanthan + galactomannan (locust bean gum or guar gum) mixtures at a very low level of xanthan gum. Our aim is to obtain a detailed description of the way by which the rheology of a galactomannan solution is modified by the presence of a small amount of xanthan gum. An additional objective is the development of model liquid systems whose flow properties would be close to those of the galactomannans whilst displaying much more pronounced viscoelastic properties.

2 Experimental

Materials—The galactomannans used were obtained from Meyhall Chemical (Switzerland). They were pure samples (<0.2 wt% protein). Intrinsic viscosity values $[\eta]$ of these samples were 1130 and 1450 ml g^{-1} for guar gum and locust bean gum, respectively. The locust bean gum sample was therefore of higher molecular weight than the guar gum sample, and in order to compare their rheological properties we chose concentrations, c, giving a similar reduced concentration ($c[\eta]$), which allowed us to make comparisons at a similar degree of volume occupancy.[3,14] These concentrations are 0.5 wt% ($c[\eta] = 7.25$) and 0.7 wt% ($c[\eta] = 7.9$) for locust bean gum and guar gum, respectively. Xanthan gum was provided from Sanofi Bio Industry (France).

Preparation of Solutions and Mixtures—Guar gum and locust bean gum solutions were prepared by dispersion in water at 25 °C, and then heating at 80 °C for 30 minutes in order to achieve a total solubilization. The xanthan gum solutions (concentration 0.7 wt% or 0.5 wt%) were prepared by dispersion in KCl (0.2 M) at 25 °C with strong stirring for 2 hours, and then centrifuging in order to remove air bubbles.

Blends with a variable xanthan:galactomannan ratio and a constant total polysaccharide content were obtained by mixing appropriate amounts of each polysaccharide at 80 °C under magnetic stirring. The hot mixtures were then poured into the rheometer, and cooled immediately to 25 °C.

Oscillatory Shear Measurements—These experiments were performed using a CarriMed Rheometer with a cone–plate device (cone angle 4°, diameter 6 cm). Immediately after quenching at 25 °C, gel cure experiments were performed at 1 Hz for 2 hours at a constant deformation amplitude (0.04). At the end of this ageing period, measurements were made as a function of frequency within the range 0.01–10 Hz at the same deformation amplitude.

Viscosity Measurements—These were performed at 25 °C using a low-shear viscometer for measurements within the range 0.017–128.5 s^{-1} and the CarriMed Rheometer for the shear rate range 1–600 s^{-1}.

Creep Experiments—These were carried out with the CarriMed Rheometer using the same experimental device as above. A low shear stress was applied (0.18 Pa) and the shear deformation was recorded for 2 minutes.

3   Results and Discussion

Figure 1 shows the variation of G' as a function of time for xanthan + locust bean gum blends at a ratio (X:L) ranging from 3:97 to 10:90. The curves are compared to locust bean gum alone at the same concentration (0.5 wt%). It is clearly seen that G' values are much higher for the blends whatever the X:L ratio, and that the mixed systems reach an equilibrium value beyond around 30 minutes. This shows that the properties of the mixed systems are different from that of locust bean gum even at a low X:L ratio, and that the mixed systems reach an equilibrium state quite slowly.

Figure 2 shows the mechanical spectrum of a xanthan + locust bean gum system (X:L = 4:96) which is compared to that of the locust bean gum system at the same total concentration. The spectrum obtained with locust bean gum is

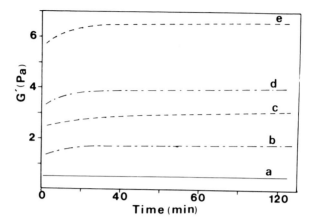

Figure 1   *Variation in storage modulus G' at 1 Hz as a function of time for xanthan + locust bean gum blends (0.5 wt%) at 25 °C: (a) locust bean gum, (b) X:L = 3:97, (c) X:L = 5:95, (d) X:L = 6:94, (e) X:L = 10:90*

typical of that of a macromolecular solution with $G'' > G'$ over all the range of frequencies, and there is a trend towards a crossover of the G' and G'' curves beyond 10 Hz. These data are consistent with what has been reported for galactomannan solutions.[1] On the other hand, it is clear that the viscoelastic behaviour of the mixture is significantly different. We see that at low frequency G' is much higher than G'' and is almost constant. Also the G'–G'' curves are superimposable at high frequency. Moreover, it is noteworthy that the G'' curve

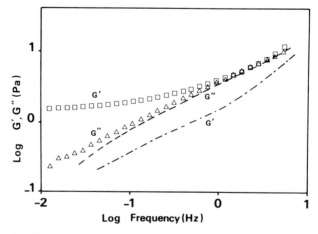

Figure 2   *Mechanical spectrum of a 0.5 wt% xanthan + locust bean gum mixture (X:L = 4:96): □, G'; △, G'; –·–·–, G' (pure locust bean gum); – – –, G'' (pure locust bean gum)*

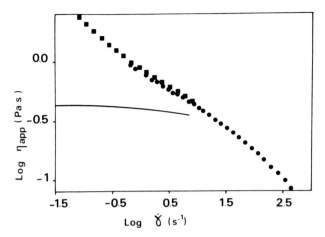

Figure 3 *Flow curves of the same blend as in Figure 2. Logarithm of apparent viscosity η_{app} is plotted against logarithm of shear-rate $\dot{\gamma}$. ■, data from low-shear rheometer; ●, data from CarriMed rheometer; ——, pure locust bean gum*

of the mixture does not differ much from that of the locust bean gum solution with the exception of the low frequency range. Figure 3 shows the flow curves displayed for the same mixture as in Figure 2. The flow behaviour of the locust bean gum is typical of a macromolecular solution with a shear-thinning behaviour and a Newtonian region at low shear-rate. The flow curve of the mixture is noticeably different. The behaviour exhibited is still shear-thinning, but the limiting low-shear viscosity is absent. This trend at low shear-rate is classically ascribed to the existence of a yield stress. It is to be emphasized that the xanthan content of this sample was very low, and accounted for only 4 wt% of the total polysaccharide. Nevertheless, the apparent viscosity at low shear-rate was found to be significantly higher than for galactomannan alone, and this synergy became more and more pronounced as the X:L ratio increased. The same type of behaviour was found for X:L ratios as low as 1:99.

Figure 4 shows a set of results obtained for a xanthan + guar gum mixture. The example given here shows the mechanical spectrum of a mixture at an X:G ratio of 5:95. This is compared to guar gum alone at the same polysaccharide concentration. As discussed in the experimental part, this concentration was chosen at 0.7 wt% in order to account for the lower molecular weight of the guar gum sample in comparison to the locust bean gum sample. The reduced concentration ($c[\eta] = 7.9$) was thus slightly higher for guar gum than for the 0.5 wt% locust bean gum systems ($c[\eta] = 7.25$). This explains why the mechanical spectrum of guar gum is slightly different from that of locust bean gum, particularly with the G'–G'' curves intercepting at high frequency. It is, however, clear that the result obtained for this mixture is qualitatively similar to that from mixtures with locust bean gum: G' is higher than G'' at low frequency and levels out, the G'–G'' curves are superimposed over the high frequency range, and the G'' curve of the

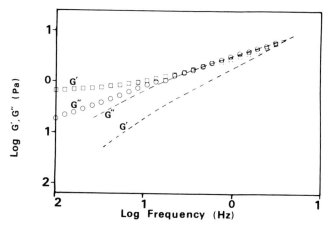

Figure 4   *Mechanical spectrum of a 0.7 wt% xanthan + guar gum mixture (X:G = 5:95): □, G′; ○, G″; ---, G′ and G″ for 0.7 wt% pure guar gum*

mixture is close to that of the locust bean gum. The differences from the pure gum were found to become more and more pronounced as the xanthan content increased. Here, also, it is clear that the rheological behaviour of the guar gum is strongly influenced by the presence of even very small amounts of xanthan gum.

Figure 5 shows the creep curves over the first 30 seconds for the experiment with xanthan + guar gum mixtures at different ratios. The curve displayed by the guar gum solution (curve a) is typical of a viscoelastic solution, the creep

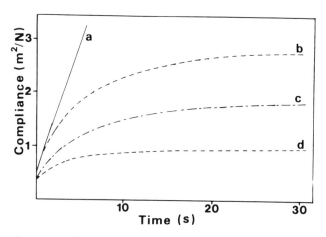

Figure 5   *Creep compliance response obtained from 0.7 wt% xanthan + guar gum mixtures. Applied shear stress is 0.18 Pa, except for pure guar gum (0.09 Pa): (a) guar gum, (b) X:G = 3:97, (c) X:G = 4:96, (d) X:G = 6:94*

compliance varying linearly with time beyond a few seconds. This means that a steady flow régime is rapidly attained. The creep curves exhibited by the mixtures are spectacularly different: the creep compliance reaches a constant value, and no steady flow is observed. This is typical of a viscoelastic solid-like system. Here, also, it is clear that the behaviour of the guar gum, which is the main polysaccharide component of the system, is strongly modified by the presence of very small amounts of xanthan gum. It is to be noted that in most of the experiments a flow was experienced only when the applied shear stress was above 2–3 Pa. Similar results were obtained for the xanthan + locust bean gum mixtures. Comparable creep curves have also been described for xanthan + glucomannan mixtures.[15] However, they were obtained at a higher xanthan content (X : G = 1 : 1).

4 Conclusions

The three rheological techniques employed in the present study show that xanthan gum plays a major role in the rheology of xanthan + galactomannan mixed systems even if it is present at very low amount. Galactomannans are known to exhibit a rheological behaviour which is typical of macromolecular solutions. The presence of xanthan gum at a ratio as low as 1 : 99 induces, in effect, a transition of the system from a macromolecular solution to a structured system characterized by a yield stress and pronounced viscoelastic behaviour. An important result is that the system flows at high shear rate, and that the flow properties are close to those of the galactomannan in the high shear-rate range. Conversely, the rheology is dramatically modified at low shear-rate with the appearance of a yield stress which we are able to estimate from creep experiments to be of the order of 2–3 Pa.

The fact that there is no difference in general behaviour between guar gum and locust bean gum suggests that the same mechanism is involved in each case whatever the fine chemical structure of the galactomannan. This indeed means that the molecular mechanisms which have hitherto been proposed to explain gelation of xanthan + locust bean gum systems are unable to explain the present rheological results. Since the properties described here display several similarities with the behaviour of xanthan gum alone, a tentative explanation is suggested which is based on the mixing of non-interacting polysaccharides, each type of macromolecule being excluded from the volume occupied by the other. This volume exclusion would yield two separate phases, each being enriched by one polysaccharide. The large proportion of galactomannan makes this polysaccharide occupy the main part of the available volume. However, since it is the xanthan molecules that play the major role in the properties, it is assumed that a second continuous phase is formed. If the xanthan gum is sufficiently concentrated within this phase, a weak network can be obtained as in xanthan suspensions. This description assumes that a bi-continuous system is formed, one of the phases being a macromolecular solution, and the other a weak xanthan network. This provides an explanation for the fact that the behaviour is close to that of the galactomannans at high frequency, or high shear-rate, and similar to that of xanthan, although less pronounced, at low frequency, or low shear-rate.

References

1. G. Robinson, S. B. Ross-Murphy, and E. R. Morris, *Carbohydr. Res.*, 1982, **107**, 17.
2. B. Launay, J. L. Doublier, and G. Cuvelier, in 'Functional Properties of Food Macromolecules', ed. J. R. Mitchell and D. A. Ledward, Elsevier Applied Science, London, 1986, p.1.
3. A. H. Clark and S. B. Ross-Murphy, *Adv. Polym. Sci.*, 1987, **83**, 57.
4. V. J. Morris, D. Franklin, and K. l'Anson, *Carbohydr. Res.*, 1983, **121**, 13.
5. S. B. Ross-Murphy, V. J. Morris, and E. R. Morris, *Faraday Symp. Chem. Soc.*, 1983, **18**, 115.
6. G. Cuvelier and B. Launay, *Carbohydr. Polymers*, 1986, **6**, 321.
7. I. C. M. Dea, E. R. Morris, D. A. Rees, E. J. Welsh, H. A. Barnes, and J. Price, *Carbohydr. Res.*, 1977, **57**, 249.
8. I. C. M. Dea, A. H. Clark, and B. V. MacCleary, *Carbohydr. Res.*, 1986, **147**, 275.
9. P. Cairns, M. J. Miles, V. J. Morris, and G. F. Brownsey, *Carbohydr. Res.*, 1987, **160**, 411.
10. M. Tako, A. Asato, and S. Nakamura, *Agric. Biol. Chem.*, 1984, **48**, 2995.
11. G. Cuvelier, Ph.D.Thesis, Université de Paris XI, 1988.
12. M. Tako and S. Nakamura, *Carbohydr. Res.*, 1985, **138**, 207.
13. B. V. MacCleary, R. Amado, R. Waibel, and H. Neukom, *Carbohydr. Res.*, 1981, **92**, 269.
14. E. R. Morris, A. N. Cutler, S. B. Ross-Murphy, D. A. Rees, and J. Price, *Carbohydr. Polym.*, 1981, **1**, 5.
15. G. J. Brownsey, P. Cairns, M. J. Miles, and V. J. Morris, *Carbohydr. Res.*, 1988, **176**, 329.

Concentration Dependence of Gelation Time

By Simon B. Ross-Murphy

CAVENDISH LABORATORY, UNIVERSITY OF CAMBRIDGE, CAMBRIDGE CB3 0HE

1 Introduction

Many gels of both synthetic and biological macromolecules are known to be formed from non-covalent cross-links involving intermolecular junction zones, the cross-links being of small but finite energy, and/or of finite lifetime; such materials are called physical gels,* and include both biological and synthetic systems.[1] Only very recently has much effort been employed in trying to elucidate structure–property relations for this class of materials.

The presence of non-covalent cross-links complicates any physical description of the network properties enormously, because their number and position can, and do, fluctuate with time and temperature. In many cases the nature of the cross-links themselves is not known unambiguously, and often involves such disparate intermolecular forces as Coulombic, dipole–dipole, van der Waals, charge transfer, and hydrophobic and hydrogen bonding interactions.[3] In many cases there is a subsequent lateral aggregation of chains, after the initial contact. These factors must influence the *actual* number of physical cross-links and, consequently, the modulus of the final gel.

Physical gels formed from synthetic polymers include certain isotactic and syndiotactic polymers in organic solvents (*i*- and *a*-polystyrene in decalin), ionomer systems in solvents of low dielectric constant, and a number of A–B–A type block copolymers—aggregation of heterostructural chains is often a common feature of such materials.[4] In the case when block A is compatible with a solvent and block B is incompatible, the B–B interaction between units on adjacent chains in solution forms the physical cross-link. As far as the biopolymers, in particular, are concerned, non-covalent cross-links are formed by one or more of the mechanisms listed above, usually combined with more specific and complex mechanisms involving junction zones of known, ordered secondary structure, for example multiple helices.[5] Double helices occur in carrageenans and agarose, and triple helices in gelatin and curdlan.

---

* The term 'physical gel' appears to have been introduced by de Gennes.[2] However, the class of materials covered by his description implies thermoreversibility, which is not the case with every system included here, or more generally in the literature. This article uses a more general definition, and includes all non-covalently cross-linked systems.

A series of recent papers by Oakenfull and co-workers[6-8] has proposed a simple method for the determination of the molecularity parameter n (*i.e.* 2 for a simple double helix) from measurements of the concentration dependence of the gelation time for physical gels. In a more recent paper[9] we have shown that the actual exponent n' extracted from the plot proposed by Oakenfull is not, in general, equal to n, nor is it constant; rather it is suggested that n' should increase as the critical gel concentration C_0 is approached, and become infinite at C_0. In the present paper, both models are discussed, together with some related experimental and theoretical topics.

2   The Measurement of Gelation Time

There are potentially a large number of ways of determining the gelation time of a gelling system, but the most precise methods are only applicable to covalently cross-linked networks, since close to the gel point the incipient network is very weak. Contrary to common expectation, this does not imply that the gelation time itself is altered by mechanical perturbation (in fact there is good evidence that this does *not* happen), but what it does mean is that the true gelation time is more difficult to determine, and is apparently delayed.

The most rigorous method of directly measuring the gelation time (t_c) employs the application of a low-frequency, small-amplitude oscillatory strain to the system, and many experiments of this nature have been reported. Even so, to obtain the precise t_c may require measurements to be made at a range of frequencies and strains, and then double extrapolated to zero amplitude and frequency (strain and strain rate), respectively.[5]

In the study of thermoset resins, it is common to regard the gelation time to be that when the real and imaginary parts of the complex modulus are equal (*i.e.* when the loss tangent is unity). Recent work by Winter and co-workers[10,11] has concentrated on analysis of the time when $G' = G''$ over a wide range of frequencies (a more testing criterion than that above), *i.e.* both moduli are proportional to ω^x, and the exponent x is constant over a wide range of frequencies. (In the original work x was thought to be close to 0.5, but more recently it has been seen to lie in the range 0.2–0.8, depending upon the precise nature of the gelling system.)[12] Te Nijenhuis and Winter have investigated the applicability of the method to physical gels.[13]

Other methods of determination of t_c involve measurement on the pre-gel state, perhaps of the viscosity or the (weight average) molecular weight and determining t_c by extrapolation of these values to infinity. The former method must be carefully employed, since, although the strain rate may be low, the absolute strain is implicitly high. Later we will describe some of our own experiments in which both small-deformation oscillatory and viscosity measurements were made on the same system.[14]

All of these methods are rather complicated, and the method employed by Oakenfull in his original work,[6] although perhaps not of great absolute accuracy, does allow a number of systems to be studied simultaneously, and has the advantage of requiring little in the way of apparatus. The gelation time is assessed by pouring a fixed amount of pre-gel solution into a series of small vials held in a thermostatted waterbath. These vials are then inverted sequentially, and the time, t_c, required to form a gel just strong enough to remain held in position is recorded, for each of a series of concentrations.

3 Estimation of Molecularity from Gelation Time/Concentration Data

The details of the Oakenfull gelation kinetics method[6-8] are as follows: it is proposed that the time required to form a gel of small, but precisely determined rigidity is a measure of the rate of gelation, and that the setting time (gelation time) is inversely proportional to the initial gelation rate. By confining measurements to the very early stages of the gelation process, it is then proposed that since each potential cross-linking locus (L) along a chain acts as an independent species in solution, when n of these form a junction zone (J) then

$$nL \rightarrow J,$$

and the rate of gelation v is given by

$$v = d[J]/dt = kC^n \tag{1}$$

where k is a rate constant, [J] is the concentration of J units, and C is the polymer concentration. Following this argument, the slope of log (rate of gelation) *versus* log (concentration) gives a 'reaction order', which is assumed to be the number of polymer chains participating in the formation of a junction zone. The rate of gelation is then assumed to be proportional to $1/t_c$, a procedure equivalent to estimating the rate constant from the initial rate method of chemical kinetics.

Analysing data for ι and κ-carrageenan by this method, Oakenfull and Scott found n to be 4.5 ± 0.5 for ι-carrageenan, and 12.5 ± 0.9 for the κ sample. The values obtained for the exponent n noted above were argued to be reasonable, in view of the 'domain model' of carrageenan gelation proposed by Morris *et al.*,[15] in which ion-mediated gelation is assumed to involve the stacking of double helices into a fringed micellar type structure. In other words, from the above plot it was suggested that ι-carrageenan junction zones involve 2–3, and κ-carrageenan around 6, such associated double helices. The model was subsequently applied to other systems, including gelatin (where n was close to 3—consistent with the expectation of a mechanism involving reversion to collagen-like triple helices), pectin, and furcellaran.

4 Critique of the Oakenfull Kinetic Model

Although the Oakenfull model is attractively simple, and appears to give reasonable results, consistent with current 'molecular' understanding, the present author has criticized the model on a number of points. The rest of this section follows his previous argument quite closely; the three major problems with the simple gelation time model are now discussed in more detail.

The Number of Chains Involved in a Junction Zone—The first, and perhaps most direct, argument against the kinetic approach is that it requires that n-fold collisions of the macromolecules must occur, and that these n polymer chains are simultaneously involved in the nucleation of the junction zones. Binary collisions are, of course, very common, and even ternary collisions can occur (although with a much lower probability than binary), but values of $n > 3$ seem to be of progressively much decreased likelihood.

Even for gelatin, where n is expected to be 3, the evidence is equivocal. According to Busnel *et al.*, the reaction order of renaturation, monitored by the usual optical rotation technique, is 1 at low concentrations, increasing to 2 at higher concentrations.[16,17] The first-order reaction rate is associated with intramolecular helix formation, which 'wastes' cross-links, so that not all helix formation results in elastically effective junction zones. (In the absence of 'trapped entanglements',[18] these should only involve intermolecular helices.) The second-order process occurring at higher concentrations is assumed to involve an anti-parallel triple stranded double chain helix. Conversely, static light scattering measurements by ter Meer and co-workers[19] on the same sample (molecular weight increase monitored as a function of time) conclude that n is indeed 3. This contradiction has yet to be completely resolved. (Of course, subsequent side-by-side helix aggregation could then occur, but the neutron scattering data of Djabourov *et al.*[20] seem to rule out this possibility.)

The Absence of a Critical Gel Concentration—The kinetic scheme suggested by equation (1) requires that, at any concentration, gelation will occur, provided that a long enough time is allowed to elapse. This follows because [J] is a monotonically increasing function of time for $C > 0$ and $n > 0$. However, a critical concentration must exist below which gelation can never occur. [There are arguments in the literature which suggest that this concentration, here denoted C_0, may be related to the C^* overlap concentration, in particular suggesting that $C_0 = C^*$, although there are certainly many counter examples (more usually $C_0 < C^*$)—discussion of this issue is given in our recent review.[5]]

Nevertheless, the presence of such a concentration is not in dispute and may be related to the Flory gel point requirement,[21,22] *i.e.* that a critical number of cross-links (junction zones) per primary polymer chain are required to produce a continuous gel network. Further aspects of this requirement are discussed below, but the crucial point is that v, the gelation rate defined in equation (1), must always be zero for concentrations less than C_0.

In reality this requires that there must be (at least) one other term in the kinetic scheme of equation (1), either a back reaction (equilibration) of the form of the Ostwald dilution law, as assumed by other workers[23,24] and explicitly stated in both the Oakenfull–Scott[25] and Clark–Ross-Murphy[26] calculations of the gel modulus as a function of concentration, or a wastage term, involving intramolecular reaction steps,[27,28] sometimes known as cyclization. Both of these lead naturally to a more physical model of the gelation rate.

The Relationship between Gel Modulus and [J]—The assumption that a gel is self-supporting when the modulus becomes greater than a certain value (say G_c) is probably quite realistic. Of course, inverting a vial containing a gel sample is not a simple rheological experiment, since both shear and tensile components may be present, and it is by definition a large deformation experiment (at least for the samples which are judged not to have gelled), and the deformation rate is unknown (it also requires that the adhesion of the gel is not itself a strong function of concentration).

However, even accepting that there is a sharp distinction between ungelled (with $G < G_c$) and gelled ($G > G_c$) samples, the assumption that G_c is proportional to [J], the concentration of junction zones, is a very poor one. In particular, following the argument above, at and below C_0 the equilibrium modulus G is zero, and becomes finite only for $C > C_0$. There is a corresponding critical value of [J], here denoted [J_c]. The modulus is therefore not dependent simply upon [J] but upon the proportion of junction zones above this critical amount. In the notation of the physics of critical exponents we have written this dependence in the form[29]

$$G \sim (([J]/[J_c]) - 1)^p, \qquad (2)$$

which is valid for values of $[J]/[J_c] \geqslant 1$. For the case of gelation (elasticity), the 'critical exponent' p is 3 in the classical theory of percolation on a treelike or Bethe lattice[30] and around 1.8 for percolation on a cubic lattice.[29] (The quoted exponents are only valid extremely close to the gel point.) For the Bethe lattice, and most probably also for other lattice graphs, as the degree of cross-linking is increased, the 'wastage' effect results in a lowering of the exponent below this limit value, as the difference ($[J] - [J_c]$) is increased.

In other words, if we are to apply such a formula when the ratio $[J]/[J_c] \gg 1$, the measured exponent will be somewhat less than this 'critical value', and will, in general, depend upon this difference. For the moment, however, the precise value of the exponent is not important; what is significant is that for low-modulus gels [J] is still only a little greater than [J_c], so that the term $(([J]/[J_c]) - 1)$ is less than 1.

The slope of a log G versus log [J] plot is proportional to $[J_c]/([J] - [J_c])$ and for this the slope increases, becoming infinite as $[J] \to [J_c]$ from above. In our previous work we have asserted that p must be around 2; this is quite close to the non-classical percolation exponent, but sufficiently below the classical exponent to implicitly include some wastage effects.

5 Features of Our Alternative Model

Any model suggested as an alternative to that of Oakenfull and co-workers must still predict very high slopes in the plot of log (gelation time) [or log (gelation rate)] *versus* log (concentration) since these have repeatedly been reported for a wide range of systems. For example, Bisschops[31] noted that the slope of modulus growth, another measure of gelation rate, dG/dt, was proportional to C^{22} for physical gels of poly(acrylonitrile) in DMF, and we have reported that for heat set globular protein gels this slope is $\propto C^{27}$ at low concentrations, and even at higher concentrations ($>3C_0$) it is $\propto C^6$.[14]

The first assumption of our approach[9] is that $d[J]/dt$ is proportional to C^2—a second-order mechanism implying a binary intermolecular complex in the activated complex, *i.e.*

$$d[J]/dt = kC^2 \text{ (or more generally } = k_n C^n, \textit{cf.} \text{ equation 1).} \qquad (3)$$

Note that there is no specific mechanism to give a critical concentration; this is incorporated by choosing an arbitrary $[J_c]$. The second assumption is that the gelation time corresponds to a fixed low-modulus system, and that the relationship between modulus and $[J]$ is calculated using the exponent $p = 2$. The gel point can now be defined in terms of $[J_c]$, and so the gelation time t_c will be given as the time when $[J] = [J_c]$.

The gelation time could, in principle, be calculated from any kinetic scheme, but the simplest way to examine its dependence on concentration is to say that the measured gel point corresponds to the sample having a small, but finite modulus, G_c. The modulus will then correspond to a small, but constant ratio (J_r) of $[J]/[J_c]$ a little greater than one. As explained earlier, to avoid gelation at concentrations less than the critical concentration C_0, the ratio J_r must always be less than 1, however long is the time. If we assume that J_r is very small, the ratio $[J]/[J_c]$ is very well approximated by the differential ratio $d[J]/d[J_c]$. Since the concentration corresponding to $[J_c]$ is the critical concentration C_0, we can use equation (3) and say

$$J_r \approx d[J]/d[J_c] \approx (d[J]/dt)/(d[J_c]/dt) = kC^2/kC_0^2 \qquad (4)$$

[and more generally we have the result that $J_r \approx (C/C_0)^n$]. Substituting equation (4) into equation (2), we can then say $G \approx [(C/C_0)^n - 1)^p]$ and, since as $C \rightarrow C_0$, then $G \rightarrow 0$ and $t_c \rightarrow \infty$, we now assume that $G \propto 1/t_c$ (in practice this is quite a reasonable assumption—see below). The gelation time is given by

$$t_c \approx K/[((C/C_0)^n - 1)^p]. \qquad (5)$$

Further implications of the calculation are discussed in the original paper. The values of K and p (assuming that $n = 2$) are now of particular interest; it is easy to establish that the limiting value of the measured exponent n' at high C/C_0 is equal to np, and as expected n' becomes infinite at the critical concentration C_0.

6   Application of the Model

Carrageenan Gels—We were able to apply equation (5) to the Oakenfull–Scott data for ι- and κ-carrageenan.[6] Fitting was carried by minimizing the sum of $\{(\log (1/t_c)_{\text{measured}} - \log (1/t_c)_{\text{calculated}}\}^2$ values. The advantages of the weighting implied by minimizing the sum of logarithm values rather than the values themselves has been discussed recently,[32] and we have retained this weighting scheme in the present work.

For the ι-carrageenan a satisfactory fit was obtained (we were fitting three parameters to five points and so this is not too surprising), but the value of p

obtained (2.28) was close to 2, and the quality of the fit was not really impaired when p was so constrained. The fit to the κ-carrageenan data was not so convincing; the curvature in the data was almost the opposite to that expected from the model, but in a direction that is not physically reasonable—this is due to difficulties in measuring the very short gelation times accurately (D. Oakenfull, personal communication). Despite this, the best overall fit corresponds to $p = 1.75$, again not so far from our estimate of 2.

Bovine Serum Albumin Gels—In view of the small number of data points for each of the two carrageenan samples mentioned above, in our previous publication we also reproduced some earlier data for heat set protein gels (BSA) at 65 °C.[14] In this data set, there were a total of 13 data points, but even more usefully these were collected using two completely different rheological techniques, *viz.* small-shear-strain oscillatory measurements, and pre-gel viscosity measurements.

The oscillatory data were measured with a self-sustaining torsion pendulum (at a nearly constant frequency of around 10 rad s^{-1}), *i.e.* under conditions of small strain, but large strain rate. By contrast, the viscosity data were measured using a simple falling shear viscometer, so that the strain rate is low (and becomes zero at the gel point), but the absolute strain is high. From the earlier discussion, it should be clear that the former method would slightly underestimate the true gelation time (due to the finite frequency), whereas the latter might contribute an over estimate, because of the effect of the non-negligible strain. Figure 1 illustrates the experimental data; the former data are given as circles, the latter as squares. It is indeed clear that, at the same concentrations, the apparent gelation time from the

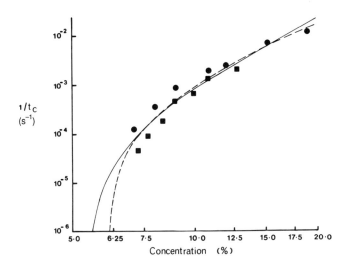

Figure 1 *Log* $1/t_c$ *plotted against log* C *for bovine serum albumin gels. Data correspond to measurements with a torsion pendulum [low strain, high strain rate], (●), and a falling sphere viscometer [high strain, low strain rate], (■); —, p = 2 constrained fit; - - - -, full model*

torsion pendulum is less, and $1/t_c$ is greater, than that measured by the falling sphere viscometer.

Figure 1 also shows the fits obtained, first by constraining p to be 2 and secondly by allowing the three parameters K, p, and C_0 of equation (5) to float independently. The best fit is actually obtained for $p \approx 1.70$, and $C_0 = 6.04 \, \text{wt}\%$ (a value rather close to that estimated from the concentration dependence of the gel modulus for these samples), but the $p = 2$ constrained fit is still fairly reasonable (the sums of squares are 0.425 and 0.453 respectively), and in this case C_0 is found to be 5.36 wt%. The exponent determined for the unconstrained fit is actually rather close (1.70 as opposed to ca. 1.8) to that for non-classical percolation, but in view of the comments earlier about the extent of the critical region, this must be regarded as largely coincidental. Overall, on the basis of these measurements, equation (5) seems to be very successful, and our qualitative argument about the value of the exponent p appears to be borne out in practice.

Gelatin Gels—In view of the apparent success of the simple model described above, it is important to test its more general applicability. Unfortunately, there are not many published sets of data of the dependence of gelation time upon concentration, other than our own, and that of Oakenfull himself. It was therefore decided to try and apply the method to data by te Nijenhuis for gelatin gels.[33–35] His study, carried out for a range of concentrations and ageing temperatures, using a very high precision, home-constructed, oscillatory rheometer, is one of the most complete studies of gelatin gelation. For this reason, the apparent gelation time for a range of his measurements was estimated from his data of modulus growth *versus* time. The results are illustrated in Figures 2a and 2b. Regrettably, despite the breadth of his measurements, they do not really provide an extensive test of the model. In particular, for any given temperature

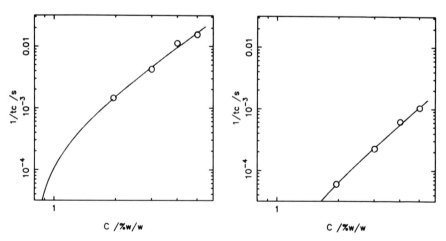

Figure 2   As Figure 1 for gelatin gels at (a) −1.2 °C and (b) 25.2 °C. Solid line from unconstrained model

there seem to be data available for only a limited number of concentrations—for example, in the present case only four points for each of the $-1.2\,°C$ and $25.2\,°C$ runs. Whilst good fits to each of these can be obtained, the parameter p is equal to 1.21 and 1.41, and the calculated C_0 values are 0.79 and 0.83 wt%, respectively. The latter values agree with te Nijenhuis's own data, since no gels were prepared at concentrations below 1 wt%, whereas the former reflect the range of C/C_0 values which are being measured, *viz.* > 2.4. Unfortunately in this region the model should not really be expected to be valid, and this is also indicated by the very low values of p. Using the original Oakenfull argument, the limiting slopes $(= np)$, *i.e.* 2.42 and 2.82, appear to be consistent with a mixture of double and triple helices for gelatin, but we regard this as coincidental. Nonetheless, we require more such data measured closer to C_0, *i.e.* at very long gelation times. Such measurements are difficult, but nevertheless it is to be hoped that many more results will become available in the future.

7 Conclusions

The Analogy between Concentration Dependence of Modulus and of Reciprocal Gelation Time—In the previous sections we have several times commented that the C dependence of $1/t_c$ appears quite similar to that of the gel modulus (G). On a double logarithm plot, both functions have a singularity as $C \to C_0$ from above, and the slopes decrease monotonically to a limit ($\sim np$ in the former case and ~ 2 in the latter case). In previous work we have commented on this,[14] and suggested that a G *versus* C model may be used empirically to fit the data for $1/t_c$ *versus* C.

In Table 1 we have compared the sum of squares for the BSA data, when fitted to such a G *versus* C function,[5,26] with that obtained from equation 5. For the same number of independent parameters (3), the fit is only marginally less good. The three parameters in this case are an equilibrium constant K', a scale factor, and the apparent network functionality, f.* For large f, we can even use the relationship $C_0 = 1/K'f^2$, so that C_0 is again the parameter of interest. In view of the similarity of the fit in this case we have not illustrated it. On the scale of Figure 1, it falls almost exactly between the two curves.

Table 1 *Fit of gelation data*

System	Model	p	C_0	Σ^2
BSA	Equation (5)	1.70	6.04	0.425
	Equation (5)	2.00	5.36	0.453
	$G \propto 1/t_c$	$\sim1000^a$	5.61	0.431
Gelatin	Equation (5)	1.41	0.83	0.002
25.2 °C	Equation (5)	2.00	1.06	0.248
Gelatin	Equation (5)	1.21	0.79	0.007
−1.2 °C	Equation (5)	2.00	0.97	0.035

^{a} In this case the apparent network functionality f is quoted.

* The equation is not given here since, as it involves a recursive relation, it cannot be written in any simple analytic form.

At this point a cautious note must be sounded. Although the two functions can be made to have similar curvatures, they actually involve very different assumptions. In the latter, the making and breaking is governed by an equilibrium (with constant K') between cross-linking loci and junctions, whereas the model of equation (5) is essentially kinetic—the reversibility is 'fudged' by asserting that a critical value, $[J_c] > 0$, exists. Nevertheless a kinetic equivalent of the gel modulus formulation has been derived,[36] where the K' is just the ratio of forward and back reaction rates, which itself leads to predictions of gelation times.

A Model for the Temperature Dependence of Gelation Time—Many years ago, Eldridge and Ferry[24] showed that, for thermoreversible (gelatin) gels, the concentration and molecular weight dependence of the gel melting temperature could be described by a simple model, essentially that of Flory for the crystalline melting of a polymer. In earlier work[5] we sketched how this could lead, for thermoreversible gels close to their melting temperature, to a dependence $G \propto (1/T)^e$, where T is absolute temperature.

In view of the discussion above, it is quite appropriate to consider how t_c will depend upon T. (Our ultimate target must be to develop a sensible but unified model to describe the dependence of both G and t_c on T and C). At the moment we present a purely empirical description, which is to say that, at the gel melting

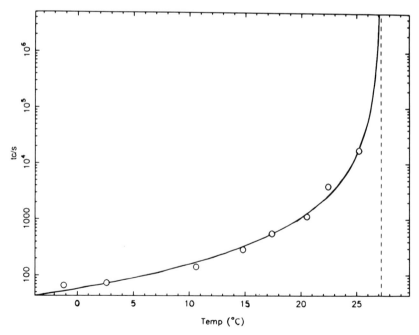

Figure 3 *Log* t_c *versus* T *for 1.95 wt% gelatin gels:* —, *equation* (6); ----, $T_c = 27.24\,°C$

temperature (assuming complete thermoreversibility), the gelation time becomes infinite. By comparison with equation (5), we can then write

$$t_c = K_T(1 - T/T_c)^q \qquad (q < 0) \qquad (6)$$

in which the analogy with the critical behaviour near a thermal phase transition should also be clear.* Equation (6) was applied to the data of te Nijenhuis for 1.95 wt% gelatin gels,[33] and a good fit (Figure 3) was obtained with an exponent q of -2.22, and with $T_c = 27.24\,°C$. In his thesis[33] te Nijenhuis has also fitted the gelation time (in his terminology, the 'induction time') to a simple model, by plotting $t_c^{1/2}$ against T. This plot, which corresponds to equation (6) with $q = -2$, was quite linear, with an intercept on the T axis at 26.3 °C; the difference between the two values of T_c merely reflects the effect of constraining the q exponent. In his thesis te Nijenhuis comments: 'why the relation [between $t_c^{-1/2}$ and T] is linear, is still unclear to the author'. As far as the present work is concerned, it is at least reasonable to assert that the correlation must follow the form of equation (6). Nevertheless, the exact significance of the exponents 2.22 (2.0) is still not clear, and must remain subject to further investigation.

Acknowledgements. I would like to thank Dr Allen Clark, Unilever Colworth, Prof Ed Morris, Silsoe/Cranfield Institute of Technology, and Dr David Oakenfull, CSIRO, North Ryde, NSW, Australia, for invaluable comments and discussions. I am also indebted to Dr Ir. Klaas te Nijenhuis, Delft Institute of Technology, The Netherlands, for sending me a copy of his thesis.

References

1. 'Physical Networks—Polymers and Gels', eds. W. Burchard and S. B. Ross-Murphy, Elsevier Applied Science, London, 1990.
2. P.-G. de Gennes, 'Scaling Concepts in Polymer Physics', Cornell University Press, Ithaca, N.Y., 1979.
3. E. Tsuchida and K. Abe, *Adv. Polym. Sci.*, 1982, **45**, 1.
4. P. J. Flory, *Faraday Discuss. Chem. Soc.*, 1974, **57**, 7.
5. A. H. Clark and S. B. Ross-Murphy, *Adv. Polym. Sci.*, 1987, **83**, 57.
6. D. G. Oakenfull and A. Scott, in 'Gums and Stabilisers for the Food Industry', ed. G. O. Phillips, D. J. Wedlock, and P. A. Williams, Elsevier Applied Science, London, 1986, Vol. 3, p. 465.
7. D. G. Oakenfull and V. J. Morris, *Chem. Ind.*, 1987, 201.
8. D. G. Oakenfull and A. Scott, in 'Gums and Stabilisers for the Food Industry', ed. G. O. Phillips, D. J. Wedlock, and P. A. Williams, IRL Press, Oxford, 1988, Vol. 4, p. 127.
9. S. B. Ross-Murphy, *Carbohydr. Polym.*, 1990, *in press*.
10. H. H. Winter and F. Chambon, *J. Rheol.*, 1986, **30**, 367.
11. F. Chambon and H. H. Winter, *J. Rheol.*, 1987, **31**, 683.
12. M. Muthukumar, *Macromolecules*, 1989, **22**, 4656.
13. K. te Nijenhuis and H. H. Winter, *Macromolecules*, 1989, **22**, 411.
14. R. K. Richardson and S. B. Ross-Murphy, *Int. J. Biol. Macromol.*, 1981, **3**, 315.
15. E. R. Morris, D. A. Rees, and G. Robinson, *J. Mol. Biol.*, 1980, **138**, 349.

* This is not to say that we expect the behaviour to be 'universal' or to extract a meaningful 'critical' exponent.

16. J.-P. Busnel, S. M. Clegg, and E. R. Morris, in 'Gums and Stabilisers for the Food Industry', ed. G. O. Phillips, D. J. Wedlock, and P. A. Williams, IRL Press, Oxford, 1988, Vol. 4, p. 105.
17. J.-P. Busnel, E. R. Morris, and S. B. Ross-Murphy, *Int. J. Biol. Macromol.*, 1989, **11**, 119.
18. N. R. Langley, *Macromolecules*, 1968, **1**, 348.
19. H.-U. ter Meer, A. Lips, and J.-P. Busnel, in 'Physical Networks—Polymers and Gels', eds. W. Burchard and S. B. Ross-Murphy, Elsevier Applied Science, London, 1990, p. 253.
20. I. Pezron, T. Herning, M. Djabourov, and J. Leblond, in 'Physical Networks—Polymers and Gels', eds. W. Burchard and S. B. Ross-Murphy, Elsevier Applied Science, London, 1990, p. 231.
21. P. J. Flory, *J. Am. Chem. Soc.*, 1941, **63**, 3083.
22. P. J. Flory, *J. Am. Chem. Soc.*, 1941, **63**, 3091.
23. J. R. Hermans, *J. Polym. Sci.*, *Part A.*, 1965, **3**, 1859.
24. J. E. Eldridge and J. D. Ferry, *J. Phys. Chem.*, 1954, **58**, 1034.
25. D. Oakenfull, *J. Food Sci.*, 1984, **49**, 1103.
26. A. H. Clark and S. B. Ross-Murphy, *Brit. Polymer J.*, 1985, **17**, 164.
27. J. Y. Chatellier, D. Durand, and J. R. Emery, *Int. J. Biol. Macromol.*, 1985, **7**, 311.
28. D. Durand, J. R. Emery, and J. Y. Chatellier, *Int. J. Biol. Macromol.*, 1985, **7**, 315.
29. D. Stauffer, A. Coniglio, and M. Adam, *Adv. Polym. Sci.*, 1982, **44**, 103.
30. M. Gordon and S. B. Ross-Murphy, *Pure Appl. Chem.*, 1975, **43**, 1.
31. J. Bisschops, *J. Polym. Sci.*, 1955, **17**, 89.
32. A. H. Clark, S. B. Ross-Murphy, K. Nishinari, and M. Watase, in 'Physical Networks—Polymers and Gels' eds. W. Burchard and S. B. Ross-Murphy, Elsevier Applied Science, London, 1990, p. 209.
33. K. te Nijenhuis, Ph.D. Thesis, University of Delft, 1979.
34. K. te Nijenhuis, *Colloid Polym. Sci.*, 1981, **259**, 522.
35. K. te Nijenhuis, *Colloid Polym. Sci.*, 1981, **259**, 1017.
36. A. H. Clark, this volume, p. 322.

On the Fractal Nature of Particle Gels

By Pieter Walstra, Ton van Vliet

DEPARTMENT OF FOOD SCIENCE, WAGENINGEN AGRICULTURAL UNIVERSITY,
P.O. BOX 8129, 6700 EV WAGENINGEN, THE NETHERLANDS

and Leon G. B. Bremer

DEPARTMENT OF PHYSICAL AND COLLOID CHEMISTRY, WAGENINGEN
AGRICULTURAL UNIVERSITY, DREIJENPLEIN 6, 6703 HB WAGENINGEN,
THE NETHERLANDS

1 Introduction

A gel is a continuous three-dimensional network of connected molecules or particles in a continuous liquid phase. This structural definition may be supplemented by a mechanical one, stating that a gel is characterized as having a yield stress, and thus has the properties of a solid, while its modulus is fairly low (an instantaneous shear modulus below, say, 10^6 N m^{-2}). A gel generally has viscous as well as elastic properties, but the preponderance of elastic over viscous properties varies widely.

Most researchers think primarily of a gel network as consisting of long flexible macromolecules that are cross-linked at some places, be it by covalent bonds, micro-crystalline domains, entanglements or other linkages. But there are other types. Fairly small amphiphilic molecules may associate to give a three-dimensional structure throughout the solution. Suspended particles may aggregate and thereby form a rather irregular continuous network, and this latter type of gel is that which will be considered here. Examples are plastic fats (fat crystals in oil), several heat-set protein gels, soy curd, and renneted or acidified skim milk. The latter are essentially casein gels, and although casein consists of macromolecules, these associate into more-or-less spherical particles which, in turn, may aggregate to form a gel. Most of our results are on casein gels; experimental details have been given elsewhere.[1-6] The voluminosity of the casein particles studied here is *ca.* 2.9 ml g^{-1}.

2 Some Characteristics

Compared to macromolecular gels, particle gels show the following differences in properties.

(1) The elastic properties are caused by enthalpic rather than entropic changes. In a typical macromolecular gel the strands between cross-links can—due to heat movement—assume numerous conformations; any deformation of the material lowers the entropy of the strands, and this is the main cause of resistance against permanent deformation. This effect is rarely important in particle gels, because the strands here are much shorter relative to their thickness and they have very little freedom for Brownian motion. Deformation of the material now causes deformation of bonds, and the enthalpy change involved is the primary cause of the elastic properties. This implies that Flory-type theories (rubber-like elasticity theories) are not applicable.[7]

(2) Due to the enthalpic nature of the 'springs' (*i.e.* the elastically effective strands between cross-links), the so called linear region (*i.e.* the deformation up to which the ratio of stress to deformation is constant) is small. If we define the deformation as the Hencky strain

$$\varepsilon \equiv \left|\ln\left(H/H_0\right)\right| \tag{1}$$

where H is the length of a piece of the material after deformation and H_0 is the original length, we have roughly that the deformation is linear for: rubber up to $\varepsilon = 3$, for gelatin up to $\varepsilon = 1.5$, for alginate up to $\varepsilon = 0.2$, for acidified skim milk up to $\varepsilon = 0.03$, and for margarine fat up to $\varepsilon = 0.0005$. Rubber and gelatin are typical 'entropic' materials, casein gels and plastic fats are 'enthalpic' gels, and alginate is intermediate. Moreover, a particle gel is mostly much 'shorter', *i.e.* it fractures at a far smaller deformation. These aspects are illustrated in Figure 1.

(3) Because the building blocks of a particle gel are large compared to molecules, particle gels are rather coarse. This is best expressed in the permeability coefficient B of the Darcy equation,[8]

$$v = Bp/l\eta, \tag{2}$$

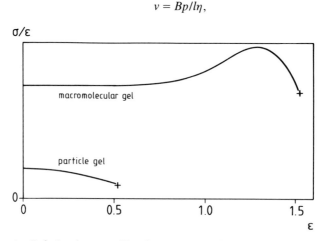

Figure 1   *Relation between Hencky strain ε and stress σ divided by strain ε for macromolecular and particle gels. At point (+) the gel fractures. Curves are highly schematic and are meant only to illustrate trends*

where v is the superficial velocity of a liquid of viscosity η that moves through a porous material over a distance l due to a pressure (difference) p. For protein gels of about 2% dry protein in water, we have as an order of magnitude $B = 10^{-12}$ and 10^{-17} m^2 for particle and macromolecular gels, respectively. The coarseness may also be reflected in a low modulus: for the same examples we may have shear moduli of, say, 50 and 500 N m^{-2}, respectively.

(4) Most particle gels have a fractal nature, or more precisely, they are built of fractal aggregates.[6] This is discussed below. One of the implications is that some physical and mechanical properties of the gel very steeply depend on the concentration of aggregating material.

3 Theory of Fractal Flocculation

The random aggregation of equal sized spherical particles in translational Brownian motion, taking cluster–cluster aggregation into account, can be simulated with a computer. It then turns out that the number of particles (of radius a) in an aggregate (of radius R) is given by

$$N_p \sim (R/a)^D \tag{3}$$

where D is a constant, called the fractal dimensionality.[9] The radius of a floc or aggregate is rather ill-defined, but any type of radius taken (*e.g.* the radius of gyration) leads to a remarkably good fit to equation (3) over a very wide range of R (with a numerical proportionality constant near unity), as soon as R is appreciably larger than a. The flocs generated in this way have a fractal nature, *i.e.* their structure is scale invariant, again at scales appreciably larger than a. The flocs show a limited number of fairly long strands, mostly of a thickness of only one particle; a few thicker nodes are also present. If the aggregating particles and clusters are allowed to stick immediately after they touch and cannot alter their relative position afterwards, it turns out that the fractal dimensionality is very close to 1.78.[9] If the aggregation is not purely diffusion-limited, as above, and refinements like a limited chance of sticking after an encounter (for which the chance may be size dependent), or a certain probability for deflocculation, are introduced into the model, higher values of D are obtained, but they are usually not higher than 2.1.[10] In many cases, good agreement between the calculated and experimentally determined scaling results for floc geometry have been obtained.[9,10]

The number of particles that could be present in a floc, if it were to be close-packed, is given (apart from a numerical constant) by

$$N_a = (R/a)^3, \tag{4}$$

This implies that the volume fraction of particles in a floc is given by

$$\Phi_a \equiv N_p/N_a = (R/a)^{D-3}. \tag{5}$$

The floc thus becomes ever less dense as R increases, and as soon as the average

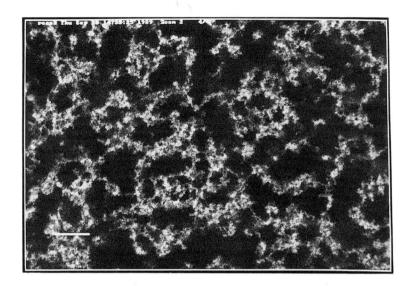

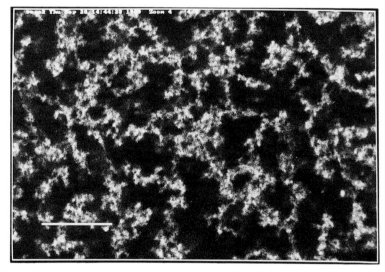

Figure 2   *Micrographs made by confocal scanning laser microscopy (CSLM) in a fluorescent mode of two acid casein gels (Type 2) of dry casein concentrations of 16.2 and 25.5 g kg^{-1}, optical thickness 4 and 2 μm, respectively, and differing by a factor 2 in magnification. The bars denote 10 μm*

Φ_a of the flocs equals the total volume fraction Φ of the particles in the system, the flocs completely fill it and we have obtained a gel.[6] The critical (average) floc radius at which a gel is formed is thus

$$R_c = a\Phi^{1/(D-3)}. \tag{6}$$

This implies that R_c strongly depends on Φ. This can be seen in Figure 2, which shows micrographs of two casein gels of different concentrations (and thus different values of Φ), but at differing relative magnifications such that the size of the flocs making up each gel looks the same. From the relation between magnification and concentration, we obtain from equation (6) that $D \approx 2.35$, and this same value of D was also found for different concentration ranges of several casein gels. This value of D is thus larger than model calculations predict, but definitely less than 3, and equation (6) is found to hold very well.

The gel formed is thus inhomogeneous, and the more so as Φ becomes smaller. The dimensionality, which characterizes the relation between the amount of aggregated material and the length scale, varies according to the scale. At distances not greater than a, we have $D = 3$, and this is also the case for distances much greater than R_c: if one takes a cube of gel, the amount of material in it is, of course, proportional to its edge size cubed. But at scales in between, D is less than 3, and for the most part, in the gels considered here equal to *ca*. 2.35. It remains to be studied, however, for instance by three-dimensional correlation analysis, precisely how D depends on scale.

Several properties of the gel will thus greatly depend on Φ. For instance, the permeability coefficient B defined in equation (2) can be expressed by[6]

$$B = (a^2/k)\Phi^{2/(D-3)} \tag{7}$$

where k is a dimensionless constant. This relation is indeed observed experimentally, and an example is given in Figure 3.[4] We obtained $D = 2.3 \pm 0.1$ for the various casein gels studied. The dependence of mechanical properties on Φ is somewhat more complicated, and is briefly discussed later on. The fractal dimensionality of a gel, *i.e.* the D value of equation (6), can also be derived from the relation between the turbidity and the wavelength of the light.[6,11] Although the interpretation is not quite unequivocal, comparable results are found. The values of D obtained for different casein gels by various methods are compiled in Table 1.[6,12]

4 Some Consequences

The occurrence of 'fractal' flocculation has several important consequences.

(1) Irreversible flocculation of particles leads to a gel, not a precipitate, unless the flocculation is disturbed by external forces. We shall consider two cases.

Velocity gradients in the liquid as caused, for instance, by stirring may break up the flocs or at least bend the strands in the flocs, making them far more dense. This happens very often in practice when a reagent is added to a

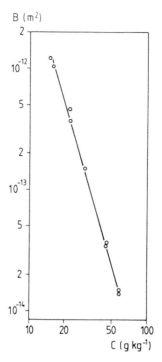

Figure 3   *The permeability coefficient* B *as a function of casein concentration* C *of acid casein gels (Type 1). (After Roefs[4])*

dispersion to induce flocculation: flocculation mostly is so rapid that it occurs while stirring, and a precipitate is formed. Such a disturbance will occur more readily if Φ is smaller (since this implies less dense and thus weaker flocs), and if a is larger (since shearing forces acting on a particle are proportional to a^2 while

Table 1   *The fractal dimensionality of casein gels made by rennet-ing and by acidification, where Type 1 refers to gels made by acidification in the cold followed by quiescent warming and Type 2 to slow acidification by means of glucono δ-lactone, as determined in various ways*

Method	Rennet	Acid Type 1	Acid Type 2
Cluster size *versus* Φ		2.35	2.35
Permeability *versus* Φ	2.23	2.39	2.36
Modulus *versus* Φ	2.17[a]	2.24[a]	2.36[b]
Fracture stress *versus* Φ		2.2[a]	2.2[a]
Turbidity *versus* wavelength	≈2.4		2.27

[a] Exponent = $2/(3 - D)$.
[b] Exponent = $3/(3 - D)$.

colloidal interaction forces are mostly proportional to *a*). In extreme cases, convection currents caused by temperature fluctuations may disturb gel formation. In our experiments, we have deliberately tried to avoid these complications.

The particles, and the flocs formed, may sediment before they have formed a gel. As a first approximation, we may state that the average time needed for two particles to flocculate, t_f, must be less than the time needed for a particle to sediment over a distance equal to its own diameter, t_s. From Smoluchowski's theory for perikinetic flocculation,[13] we derive

$$t_f = \pi\eta a^3 W/8\Phi kT \tag{8}$$

where kT has its usual meaning, and W is the ratio of particle encounters over those encounters leading to lasting contact, the so-called stability factor. A value of $W > 1$ mostly is due to a repulsive interaction energy between pairs of particles, which is, in turn, due to colloidal forces. From the Stokes equation for unhindered sedimentation, we derive

$$t_s = 9\eta/ga\Delta\rho, \tag{9}$$

where g is the acceleration due to gravity and $\Delta\rho$ is the difference in density between particle and continuous phase. We thus obtain a dimensionless number Q which is given by

$$Q = t_f/t_s = \pi a^4 gW\Delta\rho/72\Phi kT \approx (10^{20} \text{ kg}^{-1} \text{ m}^{-1})a^4 W\Delta\rho/\Phi, \tag{10}$$

where Q should be much smaller than unity. Thus, for small Φ, and for large $\Delta\rho$, W and especially a, disturbance can occur. We have observed for flocculating emulsion droplets (covered by casein) with $a \approx 10\,\mu m$, $\Phi = 0.1$ and presumably $W \approx 10$, a gel to be formed, but then we had set $\Delta\rho \approx 0$ by using an appropriate mixture of triglyceride oil and brominated triglyceride oil for the droplets.

Presumably, anisometric particles can withstand somewhat greater disturbances during flocculation and still form a gel.

(2)   There is no critical volume fraction of particles below which no gel can be formed. As can be seen from Figure 4, we have observed a gel being formed at values of Φ as low as 0.001 and even smaller, where no modulus could be measured but the existence of a gel could be visually observed; the bend in the curve will be discussed later on. There is, however, a geometrical constraint imposed by equation (6): the vessel in which the dispersion is held must be larger than $2R_c$. It is somewhat uncertain which value for D should be taken in this case, and it may be argued (see p. 378) that it should be less than 2.35. Assuming that $D \approx 2$ as long as the flocs have not yet formed a gel, we have roughly for the minimum volume fraction the value

$$\Phi_{min} \approx 2a/x \tag{11}$$

where x is the vessel size. For example, for $a = 1\,\mu m$ and $x = 1$ mm, we find

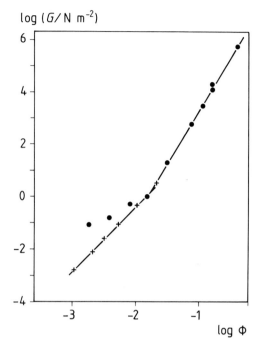

Figure 4 *The shear modulus G as a function of particle volume fraction* Φ *of gels of polystyrene latex particles (a = 35 nm) coated with palmitate and brought to gelation by adding glucono δ-lactone. Moduli were partly measured in a mechanical spectrometer at a frequency of 1 Hz (●), which is unreliable for G < 1 N m^{-2}; and partly in a frictionless parallel plate viscometer (+)*

$\Phi_{min} \approx 0.002$. Whether a gel is observed for low Φ in a rheometer may thus depend on the measuring geometry. Moreover, at very low Φ, the gel may be so weak that the rheometer cannot sense it: taking the results in Figure 4, the stress exerted at Φ = 0.001 and at a relative deformation of 1%, would only be 10^{-5} Pa.

It also follows that explanations based on percolation theories,[14] which have also been tried to interpret the formation of casein gels,[15] cannot properly hold for particle gels. The simplest site–bond percolation model leads to a value of $\Phi_{min} = 0.312$, and although various refinements lead to a lower value of Φ_{min}, it always remains far above the observed values.

(3) The coagulation time of a dispersion, *i.e.* the time which elapsed before coagulation can be visually observed, will often be close to the time needed for gelation. Combining Smoluchowski's theory for perikinetic flocculation with equation (6), we obtain for small Φ:[16]

$$t_{gel} \approx (\pi a^3 \eta / kT)\Phi^{3/(D-3)}W. \tag{12}$$

This equation predicts a very strong dependence on Φ. The theory has been applied to the heat coagulation of casein micelles, and it is found to hold rather well, with $D \approx 2.2$ to 2.4.[17]

(4)   As has been mentioned already, the gels are very inhomogeneous, containing thin strands, as well as thick lumps of particles, and the holes in between vary widely in size[4] (see Figure 2). A macromolecular gel always contains so-called 'spoiled' strands, *i.e.* parts of macromolecules that form at most one bond with the network and thus do not contribute to its elastic properties. In a particle gel we must for similar reasons have 'spoiled particles'. Presumably, a rather greater part of the network material is 'spoiled' as compared to most macromolecular gels.

5   Some Complications

The theory as given above is rather idealized and several factors may, to a greater or lesser extent, interfere with it. We shall briefly discuss some of the more obvious complications.

(1)   The computer simulations of flocculation are over simplified. For instance, rotational diffusion of the aggregating clusters is generally not taken into account. In real systems, if the flocs become large, *i.e.* if Φ is low, the strands in them may become very pliable, and their thermal motion relative to the rest of the floc cannot be neglected any more. This would be expected to lead to denser flocs, hence a higher value of D. We presume that this is indeed what occurs in casein gels, and that it causes many of the strands in the network to be thicker than would follow from the simplified simulations. Nevertheless, the proportionality between $\log \Phi$ and, say, $\log$ (modulus) holds good over a remarkably wide range of Φ, albeit with a somewhat higher fractal dimensionality.

(2)   The theory predicts a sharp gel point in time. But we must assume that there are flocs, and individual strands within a floc, that can still form contacts with the network and will do so after the first continuously connected network has been formed. The stiffness should thus increase with time. However, we know of no study in which the reactivity of the particles, and other factors affecting the stiffness (mentioned under points 6–8, below) were kept constant.

(3)   It has until now been tentatively assumed that the original particles and, at any time, the clusters have equal size; this is, of course, unrealistic. One may take the appropriate average size[6] and assume that the shape of the size distribution does not alter too much. However, in practice, it turns out that the size of the clusters making up the gel (*i.e.* R_c) is remarkably constant. Figure 5 shows the results of a two-dimensional correlation analysis on a casein gel, and it can be seen that clear minima are found at $r = 5.5$, 16.5, and $27.5\,\mu m$, all consistent with a value of R_c of $5.5\,\mu m$. The explanation for this monodispersity is presumably that the smallest clusters diffuse fastest and thus have the highest probability of flocculating. It may be noted in passing that in orthokinetic flocculation, *i.e.* in a velocity gradient, the cluster size distribution will tend to become ever more *poly*disperse.

(4)   The clusters are bound to inter-penetrate to some extent near the moment

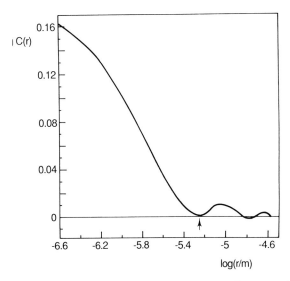

Figure 5   *Two-dimensional correlation function* C *as a function of distance* r *from a central point, derived from a micrograph* (CSLM) *of an acid casein gel* (Type 2) *of a concentration of* 16.2 g *casein* kg^{-1}

of gel formation. This would tend to imply that the size of the clusters in the gel will be (somewhat) smaller than predicted by equation (6), when taking the dimensionality of the clusters before gelation occurs.

Cluster inter-penetration will cause the gel to become somewhat more homogeneous, since the outer layer of a floc is the least dense, and we presume this to be a cause for the determined fractal dimensionality to be higher than the 2.0, or maybe 2.1, predicted by computer simulations. Unfortunately, we do not yet have a quantitative theory for this effect.

(5)   Particles may also deflocculate, or, in other words, the bonds between particles may (occasionally) spontaneously break. This process depends primarily on the activation free energy for bond breakage, *i.e.* on the depth of the free energy minimum of a particle pair. The various situations are illustrated in Table 2. If defloculation is possible and the number of particles is so large (owing to their size being very small) that mixing entropy is considerable, one should expect a weak reversible gel to be formed, as described earlier.[18] Presumably, the gel becomes somewhat more homogeneous in this way. If the particles are larger, the possibility of defloculation favours syneresis.[19] The aggregating particles are reactive over their total surface, and many more bonds between particles could be formed were it not that they are unable to reach each other; this is why the gel formed can be regarded as 'permanent'. But if some defloculation occurs, reorientation is possible, permitting a denser packing, and syneresis can occur. If such syneresis occurs already in the flocs before a gel is formed, it will tend to increase *D*; in extreme cases it may lead to the formation of a precipitate rather than a gel.

Table 2 *The situation occurring in a dispersion of fairly isometric particles at a not very low volume fraction if there is no disturbance, as a function of the free energy of activation F relative to kT for flocculation of the particles (subscript f) and for deflocculation (subscript d), and particle diameter d. (See text for further details)*

F_f/kT	F_d/kT	d	Situation
high	—	—	dispersion
low	high	—	permanent gel
low	low	small	weak reversible gel
low	low	large	syneresing gel

(6) After a gel is formed, any tendency to exhibit syneresis may not be manifest on a macroscopic scale, because the gel is geometrically confined and/or the liquid cannot be expelled because of the large distance over which it would have to move. In such a situation, microsyneresis occurs, *i.e.* in some regions of the gel larger holes are formed, which leaves other regions more dense; this can be observed as a gradual increase in permeability.[1,3] One result may be a decrease in D, but this may be off-set by the above mentioned increase before the gel is formed. Another clear change is that those strands in the gel, that were originally curved and twisted, become straightened.[12]

(7) Another change that can occur with some materials is that the particles shrink after the gel is formed: this also causes the strands to straighten. We have observed this in acid casein gels that were made by acidifying the dispersion in the cold and subsequently warming it; this causes the particles to become reactive as well as to shrink.[4,12] Gels with straightened strands we call Type 1 gels (they were the first we studied), and the other, more 'normal' ones, we call Type 2. Both types do not differ greatly in permeability, but they do differ greatly in mechanical properties (see below). It should be realized that any particle gel when just formed is of Type 2 character, and that it may change into a Type 1 gel only later.

(8) After gelation, the bonds between any two particles may (slowly) become stronger or increase in number. This was observed for casein particles,[4,19] which are fairly soft, and it occurs in many (if not most) systems, if the particles are not very small. It implies that stiffness increases, but also that any tendency for deflocculation, and thereby for microsyneresis, gradually disappears.

6 Mechanical Properties

The shear modulus G of a fractal gel can in principle be calculated. Since the clusters making up the gel are scale invariant, the number of contact points between clusters is independent of their size. Consequently, the number of contacts (bonds) between clusters is inversely proportional to their individual surface area, and hence proportional to R_c^{-2}, and the same relation holds for the number of stress-carrying strands per unit area of cross section in the gel. If the strands in the gel are straightened (Type 1), the modulus of the gel must be proportional to the (elongational) modulus E of the casein particles, assuming their deformation, if small, to be much easier than moving their surfaces away from each other. It can now be derived[12] that the Type 1 gel modulus is

$$G_1 = c_1 E \Phi^{2/(3-D)} \tag{13}$$

where c_1 is a numerical constant (~ 1) that can be approximately calculated.

In Type 2 gels, the strands are not straight and any deformation leads to a certain straightening or further bending. Here the modulus of the gel will be proportional to the bending modulus of the strands, which is inversely proportional to their length, and hence to R_c; the bending modulus is clearly lower than E. It can be derived[12] that

$$G_2 = c_2 E \Phi^{3/(3-D)} \tag{14}$$

where the numerical constant c_2 is much smaller than c_1, but cannot be easily calculated.

The results, like for instance those given in Figure 6, reasonably agree with this theory. For the Type 2 gel we find $D = 2.36$, in agreement with the other results. For the Type 1 gel we find $D = 2.25$, *i.e.* somewhat lower; it can, however, not be expected that the changes occurring during the transition of a Type 2 to a Type 1 gel are fully independent of cluster size. It is also seen that both gels, for which the nature and the number of bonds between particles must be the same, would

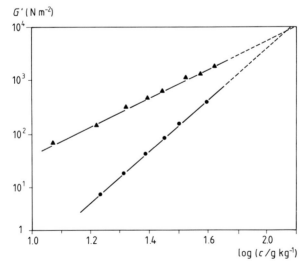

Figure 6 *Dynamic shear modulus G' at 1 Hz of acid casein gels as a function of casein concentration* c: *Type 1 (▲) and Type 2 gels (●)*

indeed give the same modulus at a concentration corresponding to about $\Phi = 0.35$; at such a high Φ, it becomes impossible to distinguish between bent and straightened strands, because $R_c/2a$ is only *ca*. 2.5.

The results in Figure 4 concern gels of fairly small hard particles. It is seen that a bend in the curve occurs at $\Phi \approx 0.015$. Below the bend, the slope is 2.46, and equation (13) yields $D = 2.19$; above the bend, the slope is 4.12, and equation (14) yields $D = 2.27$. Presumably, at very low values of Φ, syneresis occurs, converting the Type 2 gel into a Type 1 gel; this is in agreement with the idea that longer and thinner strands have a greater tendency to break, and thus to lead to syneresis.[20]

Figure 7 shows stress–strain curves for the two casein gel types at large deformation until fracture. There are clear differences in the strain values at which fracture occurs; the data in Figure 7 and other results show a difference of about 0.6. In a Type 2 gel, the strands are curved, and they will be straightened first before they become stretched, as will occur directly on deformation of the straight strands in a Type 1 gel. It is also seen that the stress at fracture is not greatly different, which agrees with the ideas given above.

One would expect the dependence of the fracture stress on Φ to be as in equation (13), since the strands are straightened at fracture. It turns out then that we have $D \approx 2.2$,[12] a value a little lower than that found for the scaling of *G*. It is, however, not unreasonable to expect limited cluster-size-dependent changes upon large deformation.

It may, finally, be remarked that some of the above conclusions only hold for gels made of soft particles. Also, our theories on the mechanical properties show independence of the size of the primary particles, whereas other theories predict a dependence; experimental results vary as well.[12] We have chosen to leave out these complicating aspects here.

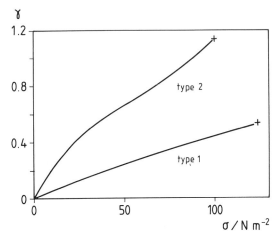

Figure 7 *Stress (σ) versus shear strain (γ) curves for acid casein gels of Types 1 and 2, as determined in a concentric cylinder rheometer over a time scale of 3 min*

382   *On the Fractal Nature of Particle Gels*

Acknowledgements. We are indebted to Professors B. H. Bijsterbosch and M. H. Ernst for useful discussions, to Mr R. Schrijvers for performing and evaluating several of the experiments, and to the Technische en Fysische Dienst voor de Landbouw (Wageningen), and Mr S. Henstra, for the making of the CSLM micrographs.

References

1. H. J. M. van Dijk, 'Syneresis of curd', Ph.D. Thesis, Wageningen Agricultural University, 1982.
2. S. P. F. M. Roefs and T. van Vliet, in 'Advances in Rheology', ed. B. Mena, A. Garcia-Rejon, and C. Rangel-Nafaile, Universidad Nacional Autonoma de Mexico, Mexico, 1989, Vol. 4, p. 249.
3. H. J. M. van Dijk and P. Walstra, *Neth. Milk Dairy J.*, 1986, **40**, 3.
4. S. P. F. M. Roefs, 'Structure of acid casein gels', Ph.D. Thesis, Wageningen Agricultural University, 1986.
5. P. Zoon, T. van Vliet, and P. Walstra, *Neth. Milk Dairy J.*, 1988, **42**, 249.
6. L. G. B. Bremer, T. van Vliet, and P. Walstra, *J. Chem. Soc., Faraday Trans. 1*, 1989, **85**, 3359.
7. T. van Vliet and P. Walstra, *Neth. Milk Dairy J.*, 1985, **39**, 115.
8. A. E. Scheidegger, 'The Physics of Flow through Porous Media', Oxford Univ. Press, London, 1960.
9. 'Kinetics of Aggregation and Gelation', ed. F. Family and D. P. Landau, North-Holland, Amsterdam, 1984.
10. P. Meakin, *Adv. Colloid Interface Sci.*, 1988, **28**, 249.
11. D. S. Horne, *Faraday Discuss. Chem. Soc.*, 1987, **83**, 259.
12. L. G. B. Bremer, B. H. Bijsterbosch, R. Schrijvers, T. van Vliet, and P. Walstra, *Colloids Surf.*, 1990, in press.
13. J. T. G. Overbeek, in 'Colloid Science', ed. H. R. Kruyt, Elsevier, Amsterdam, 1952, Vol. 1, p. 278.
14. D. Stauffer, A. Coniglio, and M. Adam, *Adv. Polym. Sci.*, 1982, **44**, 103.
15. M. Tokita, *Food Hydrocolloids*, 1989, **3**, 263.
16. P. Walstra, *J. Dairy Sci.*, 1990, in press.
17. J. A. Nieuwenhuijse and P. Walstra, this volume, p. 523.
18. T. van Vliet and P. Walstra, in 'Food Colloids', ed. R. D. Bee, P. Richmond, and J. Mingins, Royal Society of Chemistry, Cambridge, 1989, p. 206.
19. P. Walstra, H. J. M. van Dijk, and T. J. Geurts, *Neth. Milk Dairy J.*, 1985, **39**, 209.
20. T. van Vliet, H. J. M. van Dijk, P. Zoon, and P. Walstra, *Colloid Polym. Sci.*, submitted for publication.

Mechanical Properties and Structure of Model Composite Foods

By Keith R. Langley, Margaret L. Green, and Brian E. Brooker

AFRC INSTITUTE OF FOOD RESEARCH, READING LABORATORY, SHINFIELD, READING, BERKSHIRE RG2 9AT

1 Introduction

Most food systems can be considered as composites whose composition can be very complex, *e.g.* pâté, pastes, fruit-and-nut-filled chocolates, and cakes. As a first step in understanding how mechanical properties depend on structure, such composites may be described in terms of a simple filler–matrix system. A simple model of a food composite would be one where the filler particles are well defined structures, *e.g.* fibres or rigid spheres. Particulate composites are of great interest in the construction, automobile, and aircraft industries, where the inclusion of particles, particularly if they adhere to the matrix, has a desirable strengthening effect. Food composite requirements are quite different, however, since the desirable properties are 'mouthfeel' and 'texture', the perception of which involves a complex combination of shearing, compression, and fracture.

The mechanical properties of the food composite depends on the distribution, composition, and volume of each phase, and the nature of the interaction between the particles and the matrix. This paper is concerned with the measurement of the extent of the particle–matrix interaction, and how it is affected by the (surface) chemistry of the particles and the type of failure mechanisms.

Composite theories, previously applied to polymers and plastics,[1–4] have been applied successfully to model food systems containing glass particles of differing sizes[5] and Sephadex particles of differing degrees of cross-linkage.[6] Previously, we applied composite theory to a model food system consisting of hard glass spherical particles in a relatively soft protein matrix,[7] and we found that the data could be explained by the modification of an existing theory.[5] The present work expands on our previous study, and it includes new data for soft non-porous and semi-hard porous particles in the same protein matrix.

2 Materials and Methods

Materials—Lead glass ballotini spheres of different sizes were obtained from Jencons Scientific Ltd. (Leighton Buzzard, Bedfordshire, UK). Sephadex cross-linked dextran spheres of different sizes, porosity and chemical composition were obtained from Pharmacia Ltd. (Milton Keynes, Buckinghamshire, UK). Gel filtration particles of various sizes, graded 'superfine' (mean diam. 35 μm), 'fine' (50 μm), 'medium' (100 μm), and 'coarse' (200 μm), with porosities G10 to G200 (porosity increasing with increase in G number), and LH-20, a lipophilic particle prepared by hydroxypropylation, were used. Protein rich powder (Lactein 75) was obtained from Dairy Crest Foods (Surbiton, Surrey, UK) and pure corn oil (Mazola) was purchased locally.

Preparation of Composites—Composites containing glass and corn oil in a heat-denatured whey protein matrix were prepared as described previously.[7] The final average diameter of the oil droplets was controlled by adjusting the power output of the ultrasonic homogenizer and the emulsification time. Sephadex particles were preswollen in the gelling solution prior to gelation.

Mechanical Tests—Most of the tests have been described elsewhere.[7] The stress at failure was determined by loading 15 mm diam. × 15 mm long cylinders in compression at 0.83 mm s^{-1}, and by 3-point bending of 50 mm long × 15 mm diam. cylinders over a 25 mm span at 0.17 mm s^{-1}.

Absorption of Protein—The amount of protein absorbed (*not ad*sorbed) by Sephadex particles was determined by fast protein liquid chromatography (FPLC).[8] The amount of protein absorbed by the Sephadex was determined from the amount of protein remaining in a 100 ml solution of 25 wt% protein at pH 7 when the solution was mixed with 1 g of dry Sephadex powder.

Scanning Electron Microscopy—Surfaces of the composites, fractured at ambient temperature, and at −170 °C to −180 °C, were examined after dehydration and critical point drying. Details of the method have been described previously.[7]

Theory—The relative stress at failure of a composite containing particulates depends on the nature of the particle–matrix interaction.[5] For strong interaction between the particle and the matrix, we have

$$\frac{F}{F_o} = \frac{(1 - \phi)^{5/2}}{(1 - \phi)^{1/3}};$$ (1)

but, where particle–matrix interaction is weak, we have

$$\frac{F}{F_o} = (1 - \phi)^{3/2}.$$ (2)

In equations (1) and (2), F and F_o are the stress values at failure of the composite and the matrix, respectively, and ϕ is the phase volume of particles. Equation (1) predicts that F/F_o decreases with increase in ϕ between $\phi = 0$ and $\phi = 0.11$, where F/F_o is a minimum, and then increases with ϕ so that at $\phi = 1.0$ we have $F/F_o \rightarrow \infty$. Equation (2) predicts that F/F_o increases continuously, with no minimum value, with increase in ϕ up to $\phi = 1.0$ when again we have $F/F_o \rightarrow \infty$.

For a composite of a whey protein matrix containing spherical glass particles, where there is a strong interaction between the matrix and the particles, we have shown[7] that the term in ϕ has to be modified to include a term ϕ_g, which can be considered as a maximum packing fraction. Equation (1) can then be rewritten as:

$$\frac{F}{F_o} = \frac{(1 - \phi/\phi_g)^{5/2}}{(1 - \phi/\phi_g)^{1/3}}. \tag{3}$$

The parameter ϕ_g is a constant for each size of particle, but it increases with increase in particle diameter. This can be explained by assuming that a layer of protein of constant thickness adheres to the surface of each glass sphere.

3 Results and Discussion

Hard, Non-porous Particles—The thickness of the protein layer adhering to the hydrophilic (clean or protein-coated) glass particles was estimated to be $9 \pm 4 \mu m$.[7] Failure probably occurs at the surface of the particle-bound protein, within the protein matrix (Figure 1a). At the maximum packing of the particles, almost all the protein was found to be adhered to the glass beads, and it seems likely that in this case failure occurs at the boundary between adjacent particle-bound protein surfaces.

Composites containing glass particles with a hydrophobic (alkylated or silicone) coating were found to exhibit mechanical failure adjacent to the particle surface (Figure 1b). The mechanical properties, however, are consistent with a strong particle–matrix interaction composite model, but with little or no adhered protein layer, although a composite would normally be expected to obey a weak particle–matrix interaction composite model. Examination of the fractured surface indicated that small amounts of protein were attached to the glass surfaces, presumably at areas that did not have the hydrophobic coating. The total amount of protein adhered to the surface was small, and so the average thickness of this protein layer could not be sensibly calculated.

Soft, Non-porous Particles—It was possible to prepare composites of corn oil droplets in heat-denatured whey protein gels provided that the mean droplet diameter and oil volume fraction were lower than $1 \mu m$ and 0.4, respectively. Outside this range, *i.e.* with large droplets or high phase volumes, oil coalescence was observed. This results in very weak unstable gels as shown in Figure 2.

Mechanically stable oil-droplet–protein gel composites are consistent with a strong particle–matrix interaction model, similar to that found for the hydrophilic glass–protein gel composites discussed above; the calculated thickness of the protein layer is $0.04 \pm 0.02 \mu m$.

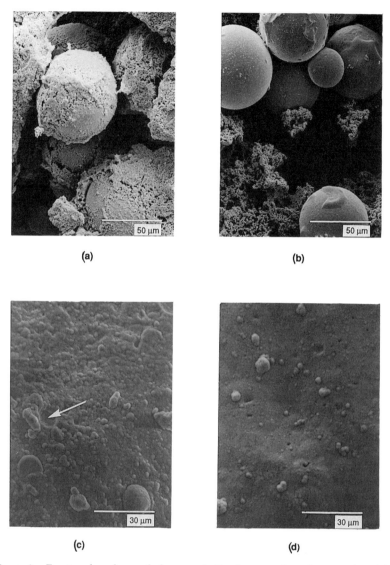

(a)   (b)

(c)   (d)

Figure 1   *Fractured surfaces of glass– and oil–whey protein gel composites*: (a)
20 vol%, 30 μm *diam. hydrophilic coated glass*; (b) 20 vol%, 30 μm
diam. hydrophobic coated glass; (c) 30 vol%, 3 μm *diam. oil droplets,*
fractured frozen (→ *indicates particle removed from surface*); (d)
30 vol%, 3 μm *diam. oil droplets fractured at ambient temperature*

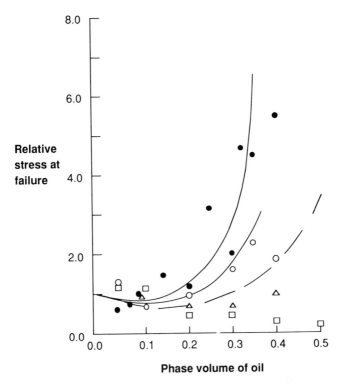

Figure 2   *The effect of corn oil volume fraction on the stress at failure in
compression of an oil–whey protein gel composite:* ●, *0.5 μm diam.
droplets;* ○, *1 μm diam. droplets;* △, *3 μm diam. droplets;* □, *5 μm
diam. droplets;* – –, *theoretical line for maximum packing. Solid lines
are drawn through experimental points*

Frozen oil-droplet–protein composites exhibited fracture at the oil-bound
protein–protein interface, in some cases the protein-coated oil droplets being
removed from the matrix (Figure 1c). When similar composites were fractured at
ambient temperature, the oil droplets ruptured and coalesced. This resulted in oil
being spread over the surface of the protein matrix, giving a smooth, almost
featureless, appearance (Figure 1d).

Semi-hard, Porous Particles—The mechanical strength of Sephadex particle–
whey protein gel composites increases with increase in particle volume, the
increase being largest for the lowest porosity particles (see Figure 3). For a given
particle porosity, no effect of particle size on the mechanical strength could be
observed. However, when the particles were hydrated with the protein solution
prior to gelation, they all swelled, and the amount of swelling increased with
increase in particle porosity, the particle diameter increasing by a factor of 1.8 for

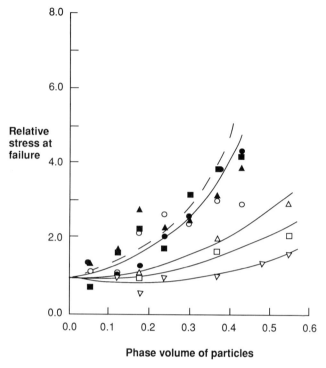

Figure 3   *The effect of Sephadex volume fraction on the stress at failure in compression of a Sephadex–whey protein gel composite:* ○, *G50 superfine;* ■, *G50 fine;* ▲, *G50 medium;* ●, *G50 coarse;* △, *G100 superfine;* □, *G200 superfine;* ▽, *LH20 lipophilic;* – – –, *theoretical line for maximum packing. Solid lines are drawn through experimental points*

G25 particles up to a factor of 3.9 for G200 particles. This increase in swelling could be associated with an increase in the amount of protein that was absorbed by the particles, sample G200 absorbing the greatest amount of protein of the particles assessed.

The general shape of the strength *versus* particle volume fraction curve closely follows the theoretical equation for a composite model assuming little or no matrix–particle interaction, *i.e.* equation (2). The term ϕ is modified, in a similar manner to the modified high interaction model, to include the term ϕ_g, the maximum packing fraction. The modified equation is

$$\frac{F}{F_o} = \frac{1}{(1 - \phi/\phi_g)^{3/2}}. \tag{4}$$

The term ϕ_g is the summation of the volume fraction of protein absorbed by the

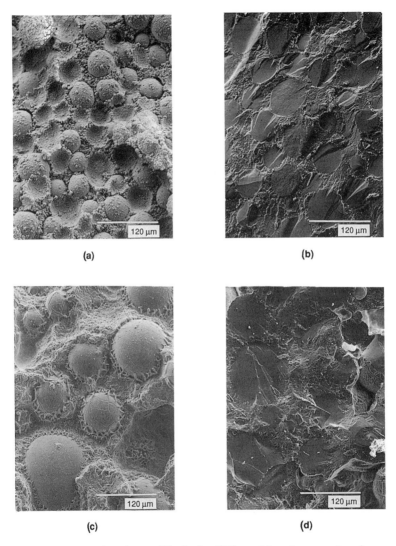

(a)   (b)

(c)   (d)

Figure 4   *Fractured surfaces of Sephadex G25 particle–whey protein gel composites*: (a) 60 vol% superfine particles, fractured at ambient temperature;
(b) 60 vol% superfine particles, fractured frozen; (c) 60 vol% fine
lipophilic particles, fractured at ambient temperature; (d) 60 vol% fine
lipophilic particles, fractured frozen

particle, as determined by FPLC, and the maximum packing fraction of random hard spheres, $\phi = 0.637$.[9]

At ambient temperature, failure was observed to occur at or near the particle–protein matrix interface, in some cases the particle being completely removed from the surface (Figure 4a). When similar composites were frozen, fracture occurred through the particles, the fracture surface being smooth (Figure 4b). It would be expected that the fine porous internal structure of the particle would not allow the protein to aggregate on heating and hence the protein would behave as in solution. On freezing failure would occur through the frozen solute, ice.

The strengths of composites containing lipophilic LH20 particles were found to be much lower than their G25 counterparts. These systems were the weakest of any of the Sephadex particle composites tested (Figure 3). The lipophilic coating appears to influence the structure of the protein matrix. Adjacent to the particle surface, the protein structure has a 'string bag' appearance (Figure 4c), with fracture occurring within this localized structure at or near the particle surface. This localized structure may have been influenced by the amount of the individual whey protein absorbed by the particles. Lipophilic LH20 particles absorbed less β-lactoglobulin and more proteose peptones than the equivalent G25 particle, the amount of α-lactalbumin absorbed being similar. This localized structure was still apparent in composites that were fractured in the frozen state (Figure 4d). Here, fracture occurs at both the particle surface and through the centre of the particles, presumably through ice surfaces in a similar manner to that observed for uncoated Sephadex particles.

4 Conclusions

We have shown that the mechanical properties of composites containing three types of particles (*i.e.* soft and hard non-porous, and semi-hard porous) in a relatively soft matrix behave as predicted by a simple composite theory, with modifications being made for protein absorption or adhesion to the particle.

The mechanical properties are strongly dependent on the protein–particle interaction and the distribution of the protein within the matrix. Clean glass beads, and those with the same whey protein mixture as used for the matrix, adsorb protein so strongly that they behave as if permanently coated with a protein layer. This reduces the maximum particle packing density from the theoretical value and also dictates the fail mechanism. Failure occurs within the protein matrix rather than at the particle–matrix interface. Alkylated or silicone-coated glass particles were found clearly to adhere to the protein matrix much less strongly, with failure occurring at the particle–matrix interface. However, protein was observed to adhere to small areas of the particles, and this may explain why a strong interaction model fits the data.

Oil particles behave as if they are coated with a protein layer, reducing the maximum packing volume below the theoretical value. The physical state of the oil dictates the fail mechanism. The adsorbed protein layer is thin, and, at room temperature, above the melting point of the oil, it is easily ruptured, releasing the contained oil, and resulting in oil coalescence at the fracture surface. When the oil in the droplets is solid, below the melting point, failure occurs within the protein matrix, coated oil droplets remaining intact.

The Sephadex particle–matrix interaction is weak, and the associated composite strength is dictated by the interaction between matrix protein and protein absorbed by the particles, failure occurring at this interface. The protein absorbed into the particle could not aggregate when heated, and it would seem likely that protein at or near the particle surface would be involved in the composite strengthening matrix protein–particle protein interaction. The inhibiting effect of the particle on protein aggregation is illustrated by the failure of frozen Sephadex particle composites, the protein behaving as in solution, and the fracture occurring within the ice layers.

The presence of particles influences the matrix structure. When there is strong interaction between particle and matrix (hydrophilic glass beads or oil droplets), the protein matrix is coarse, becoming finer at or near the particle surface. Lipophilic particles (LH20 Sephadex) influence the protein structure to such an extent that, adjacent to the particle, the structure has a 'string-bag' appearance.

Acknowledgement. The authors wish to thank Mr A. Martin, Mr J. Taylor, and Miss A. Lucas for their skilled technical assistance.

References

1. L. E. Nielsen, 'Mechanical Properties of Polymers and Composites', Marcel Dekker, New York, 1974, Vol. 2.
2. J. A. Mason and L. H. Sperling, 'Polymer Blends and Composites', Plenum, New York, 1976.
3. C. B. Bucknall, 'Toughened Plastics', Applied Science, London, 1977.
4. T. S. Chow, *J. Mater. Sci.*, 1980, **15**, 1873.
5. S. B. Ross-Murphy and S. Todd, *Polymer*, 1983, **24**, 481.
6. G. J. Brownsey, H. S. Ellis, M. J. Ridout, and S. G. Ring, *J. Rheol.*, 1987, **31**, 635.
7. K. R. Langley and M. L. Green, *J. Texture Studies*, 1989, **20**, 191.
8. A. T. Andrews, M. D. Taylor, and A. J. Owen, *J. Chromatogr.*, 1985, **384**, 177.
9. C. Rha and P. Pradipasena, in 'Functional Properties of Food Macromolecules', ed. J. R. Mitchell and D. A. Ledward, Elsevier Applied Science, London, 1986, p. 79.

Fracture and Yielding of Gels

By T. van Vliet, H. Luyten and P. Walstra

DEPARTMENT OF FOOD SCIENCE, WAGENINGEN AGRICULTURAL UNIVERSITY,
P.O. BOX 8129, 6700 EV WAGENINGEN, THE NETHERLANDS

1 Introduction

A gel consists of a continuous three-dimensional network structure, which can resist a stress—over the time-scale considered, *i.e.* the time during which a certain load is applied—in a continuous liquid phase. On the one hand, it may be distinguished from an ordinary solid by the presence of a continuous liquid phase, which makes it less stiff. Shear or Young's moduli are typically of the order of $10^6 \, \mathrm{N \, m^{-2}}$ or (much) lower, whereas for most solids the moduli are *ca.* $10^{10} \, \mathrm{N \, m^{-2}}$. On the other hand, gels may be distinguished from liquid dispersions or macromolecular solutions by their ability to resist a stress over the time-scale considered. The latter condition implies that materials may behave as an elastic solid over relatively short time-scales and as a viscous liquid over long time-scales. In practice, such behaviour is certainly observed, *e.g.* for the inner part of a mature Camembert cheese. More generally, it is often observed that the reaction of a gel to a stress is, simultaneously, partly elastic and partly viscous (Figure 1); the system is said to be viscoelastic. The ratio of the elastic and the viscous components will depend on the time-scale of the measurement.

In essence, two types of network structures may be distinguished in food gels, *viz.*, those built of particles (*e.g.* crystals, emulsion droplets, or protein aggregates) and those built of (flexible) macromolecules (*e.g.* gelatin or many polysaccharides). In food products, a less important type of gel is a network of ordered lamellar phases (*e.g.* soap gel or phospholipids). Because of their network structure, gels are always inhomogeneous. The scale of these inhomogeneities may vary greatly, not only between different gels, but also, although to a lesser extent, within a gel. For instance, in gels formed by aggregation of the casein in milk, inhomogeneities may be distinguished at various levels: that of the fractal clusters,[1] that of the casein strands, and that of the casein particles.[2] In addition, filler particles may induce inhomogeneities. These inhomogeneities may strongly influence the fracture properties of the gel.[3–7]

The fracture and yielding behaviour of food gels is an important mark of quality, affecting aspects such as (i) eating quality; (ii) usage properties *e.g.* ease of cutting and spreading; (iii) handling properties during storage and/or further

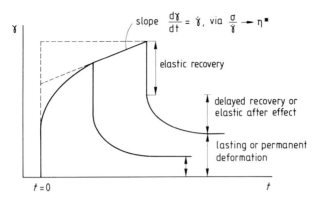

Figure 1 *The strain γ of a viscoelastic material as a function of time t elapsed after exerting a constant stress σ and after a subsequent release of the stress. The lasting or permanent deformation is due to viscous flow and increases with the time-scale of the experiment*

processing in connection with *e.g.* shape retention and pumping characteristics. These aspects pertain to the fracture and yielding behaviour of gels under widely varying conditions and time-scales. During eating, for example, a gel is quickly deformed until it fractures; the time-scale of this deformation is less than a second. On the other hand, the time-scale of shape retention during storage may be of the order of days or (much) longer.

Fracture and yielding are properties which are concerned with large deformations. However, most of the fundamental rheological studies have been done at small deformations. Often, there is no direct relation between them.[5,8,9] In spite of its importance, the fracture and yielding behaviour of food has been studied relatively little from a more fundamental point of view, and then mainly for gels that are primarily elastic, *e.g.* gelatin gels.[3,4,10] Fundamental studies of viscoelastic gels have only been performed rarely.[5,6] There even does not exist a conclusive description of what fracture and yielding are in these systems. In fact, there is not a clear distinction between the two.

General characteristics of macroscopic fracture are: (1) simultaneous breaking of bonds between the structural elements forming the network (*e.g.* atoms, molecules, crystals) in a certain macroscopic plane; (2) breakdown of the structure of the material over length-scales clearly larger than the structural elements, resulting in the formation of cracks; and (3) ultimately, the falling apart of the material into pieces. The first characteristic also applies to viscous flow. Yielding does not include characteristic (3); it results in a material which flows. Sometimes yielding precedes fracture as for, say, ductile steel[9] or young Gouda cheese[5] (especially at slow deformation rates). In other cases, it may be unclear whether one has to speak of fracture or of yielding, as during strongly deforming or even cutting of rennet-induced milk gels. Then the protein network clearly fractures in visible macroscopic planes, resulting in curd particles, but the whole remains continuous due to the abundance of solvent. However, the latter stems primarily from fast syneresis of the formed curd particles. When syneresis is nearly absent, as during cutting of milk gels formed by acidification, the whole tends to fall into pieces. Therefore, we choose to speak of fracture in cases like these. In other cases, yielding clearly does occur, as during the spreading of margarine or butter.

Fracture Mechanics—For the fracture and yielding behaviour of construction materials, extensive theory has been developed.[9,11] The starting point is that a material fractures due to the growth of defects (inhomogeneities), because there the stresses are locally higher. The extent of fracture depends on the shape of the irregularities and on material properties. According to Inglis, the stress σ at the tip of a flat elliptical notch of length l and a radius r at the tip in an ideal, isotropic, elastic material loaded perpendicular to the notch is[12]

$$\sigma = \sigma_0[1 + 2\sqrt{(l/r)}], \tag{1}$$

where σ_0 is the average stress remote from the notch.

A disadvantage of equation (1) is that it gives σ only in one point of the sample and does not indicate how it falls off to σ_0. The stress σ_{ij} at a certain place near the crack tip is given by[1]

$$\sigma_{ij} = \frac{K}{\sqrt{(2\pi R)}} f_{ij}(\theta) \tag{2}$$

where K is the so called stress intensity factor, and R and θ are the cylindrical polar coordinates of the point where $\sigma = \sigma_{ij}$ with respect to the crack tip. K is proportional to $\sigma_0\sqrt{l}$, and so the relation between σ_{ij} and l is as in equation (1).

Fracture starts (is initiated) if the local stress in the material exceeds the breaking stress of the bonds between the structural elements giving the solid-like properties to the material. Subsequently, fracture will spontaneously propagate if enough strain energy is available at the point where the crack grows to form new surfaces.[9] The energy that is necessary to create one unit of new surfaces is called the specific fracture energy R_s (J m^{-2}). Strain energy for crack growth becomes available because the stress in the material just around the crack can relax, *i.e.* the strain energy W' (J m^{-3}) stored in a certain volume is released (Figure 2A). This energy can be used for further crack growth if it is transported to the tip of the crack. Fracture propagates spontaneously if the differential energy released during crack growth surpasses the differential energy required[13] (Figure 2B):

$$W'(\tfrac{1}{2})\pi d[(l + \delta l)^2 - l^2] \geqslant R_s d\delta l. \tag{3}$$

Then, the lower limit for spontaneous crack growth is

$$R_s = \pi l W'. \tag{4}$$

Equation (4) clearly shows the importance of the size of inhomogeneities present

A   B

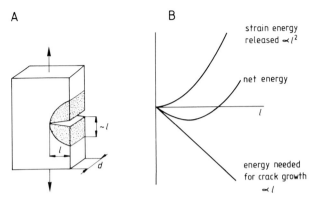

Figure 2   (A) *Volume in which strain energy is released around a growing crack.* (B) *Fracture energy needed and amount of stored energy available for crack growth as a function of crack length* l

for the onset of crack propagation. For instance, in linear–elastic materials, W' equals $(\tfrac{1}{2})\sigma\varepsilon$, and R_s then is given by

$$R_s = \pi l \int_{\varepsilon=0}^{\varepsilon_f} \sigma(\varepsilon)\mathrm{d}\varepsilon = \frac{\pi \sigma_f^2 l}{2E} \qquad (5)$$

where σ_f is the overall stress at the onset of fracture propagation (normally equal to what is called the fracture stress), and E is the Young modulus. Because R_s and E are material constants, σ_f is proportional to $l^{-1/2}$. So a curve which shows σ_f as a function of the crack length at propagation is similar (but not identical) to the crack initiation line (see equation (1)).[11]

2   Materials and Methods

Preparation of the Gels—Skim milk gels were prepared from a skim milk powder dispersion which was stirred at 30 °C for about 16 hours before use. Gelation was induced by addition of calf rennet[14] or by slow acidification in the cold followed by quiescent heating at a rate of 0.5 °C min^{-1}.[15]

Gouda cheese was made as usual from standardized pasteurized milk.[16] Care was taken, as far as possible, to avoid the formation of eyes, slits, *etc.*[5]

For the preparation of starch gels, dispersions of 1.5 wt% potato starch (Avebe, Foxhol, the Netherlands) in water were first gelatinized. After cooling to about 40 °C, the gelatinized starch was mixed with a dispersion of potato starch and glass beads (1–53 μm diam.) in water (room temperature). The final concentration of starch was 10 wt% in the water phase; the final concentration of glass was 0 to 23 vol% in the gel. Long cylindrical moulds were filled with the dispersion and immediately heated further in a microwave oven up to about 95 °C. In this way homogeneous samples could be made without visible defects. Measurements were performed after an ageing time of 25 h at room temperature.

Determination of Fracture and Yielding Properties—The behaviour at large deformation of the skim milk gels was determined in a constant stress apparatus, a Deer PDR 81 Rheometer from Deer Ltd fitted with a stainless steel measuring body of coaxial cylinder geometry.[15,17] The gels were formed in the measuring body. A thin layer of paraffin oil was put on the gel to prevent drying of the upper part of the gel. Experiments were done at 30 ± 0.1 °C.

Fracture and yielding properties of cheese and starch gels were determined using an Overload Dynamics traction–compression apparatus, fitted with a 100 or 2000 N load-cell.[5] Essentially, this apparatus consists of a fixed bottom plate and a crosshead bar containing a load-cell. The moving bar can be raised or lowered at various fixed speeds. The apparatus is mounted in a temperature-controlled box (±0.5 °C). By applying different measuring geometries, uniaxial compression and tension tests could be performed.

Cheese was examined in tension and in uniaxial compression tests at 20 ± 0.5 °C. Samples were cut at ripening temperature (13–14 °C). For the compression tests, cylindrical samples were used with a height of about 30 mm and a diameter of about 20 mm. The samples were compressed between perspex plates. The samples for the tension experiments were cut with a special formed mould (a 'dumb bell' shape). The samples were of thickness *ca.* 10 mm, the central part of the sample having a width of *ca.* 20 mm and a length of *ca.* 90 mm. The actual size was measured with a micrometer just before testing. The influence of stress concentration was determined by applying a notch with a razor blade in the central part of the sample. The size of the notch was determined afterwards.[5] With unnotched samples, the complete central part of the test-piece was made narrower to prevent fracture starting at the grips.

The Hencky strain ε_t was calculated as

$$\varepsilon_t = \ln\left(\frac{h_0 + \Delta h}{h}\right) = \ln\left(\frac{h_t}{h_0}\right), \tag{6}$$

where h_0 is the starting length, and h_t the length at time t. Stresses were calculated using the actual stress-bearing area. The latter was calculated assuming constant volume of the elongated part of the test-piece.

Starch gels were examined in uniaxial compression tests between perspex plates at 21 °C. Cylindrical samples were used with a height of *ca.* 10 mm and a diameter of *ca.* 14 mm. The strain was again calculated as a Hencky strain, with the stress taken as the force divided by the actual stress-bearing area, assuming constant volume and a cylindrical shape of the test-piece.

3 Results and Discussion

General Fracture and Yielding Behaviour—As mentioned in the introduction, the range of fracture and yielding behaviour of gels varies greatly. Fracture stress, strain at fracture, as well as fracture toughness (*i.e.* the total energy required to propagate fracture),[5,9,18] may show wide variation depending on such factors as the concentration of network-forming material, the strain rate, and the tempera- ture. Moreover, the shape of the curves depends on the nature of the network-

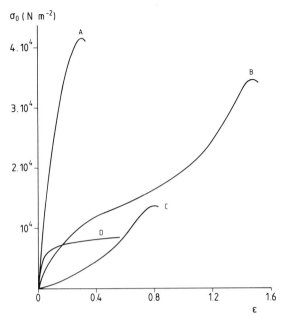

Figure 3   *Stress–strain curves in uniaxial compression for different gel-like materials. Overall stress σ_0 is plotted against strain ε: A, 9-month-old Gouda cheese; B, 2-week-old Gouda cheese; C, 10 wt% potato starch gel; D, 12 wt% glycerol lacto-palmitate gel in a 2 wt% sodium caseinate solution[19]*

forming material (Figure 3), and it may also be influenced by the strain rate.[5,17] Of the materials shown in Figure 3, the potato starch gel behaves elastically,[20] the mature Gouda cheese reacts primarily elastically,[5] but the young Gouda cheese and the glycerol lacto-palmitate gel exhibit strong time-dependent deformation and fracture behaviour.[5,19] Comparing the first two materials, the energy necessary to obtain a certain strain was clearly higher for the mature Gouda cheese than for the potato starch gel. If R_s in the two cases were the same, as it nearly is (*ca.* 1 J m^{-2} for 9 months old Gouda cheese[5] and 1–2 J m^{-2} for the potato starch gel),[20] this would imply, according to equation (3), that the mature Gouda cheese would be much more sensitive to fast fracture propagation after initiation than would the potato starch gel. This is indeed what is observed.

The stress–strain curves for the young Gouda cheese and the glycerol lacto-palmitate gel in 2 wt% sodium caseinate solution have regions where σ_0 increases only slowly with increasing ε, after an initial much steeper rise. The same behaviour has been observed for rennet milk gels.[17] In the region where the slope of the σ_0–ε curve strongly decreases, extensive local 'fracture' probably occurs, resulting in flow processes, but initially not in macroscopic fracture. The material initially yields and fractures only at much larger deformations. The conclusion that the flat part in both curves primarily stems from a flow process after yielding is strongly supported by the observation that the extent of this region increases with decreasing deformation rate.[5,19] Visible fracture occurs just before the end or curves B and D. The strain at which the strong convex curvature occurs in curve D, which is interpreted as a yield point, is roughly independent of the strain rate.[19] This was not found to be the case for the young Gouda cheese, however, where this characteristic strain increased with decreasing strain-rate. This agrees with the strong viscous component in the rheological behaviour at small and at large deformation for this material,[5] while the glycerol lacto-palmitate gels behave mainly elastically before the yield point.

Effects due to Inhomogeneities—The effect of an inhomogeneity (defect) on the fracture properties may be studied by determining the fracture stress and strain as a function of the length (at constant shape) of artificial notches.[5,9,11,18] In notch-insensitive materials (*e.g.* muscle tissue), the overall fracture stress σ_f decreases in proportion to the relative decrease in cross section of the test piece due to the notch. In notch-sensitive materials (*e.g.* biscuits), the decrease is much stronger (equation (1)). These relations may be used to estimate the size of inherent defects present in gels.

In some cases, the size of the inherent defects may be estimated directly from a plot of σ_0 as a function of the notch length l (Figure 4). The defect size of the

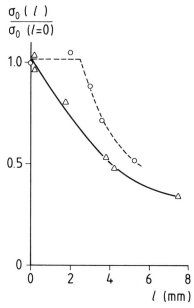

Figure 4   *Overall stress σ_0 (l) at fracture versus notch length l curves for Gouda cheese of different age.[5] The stresses are given relative to the overall stress at zero notch length. The experiments were done in tension at 20 °C with $\dot{\varepsilon}_{initial} = 2.8 \times 10^{-3}$ s^{-1}: ○, age 3 days; △, age 3 weeks*

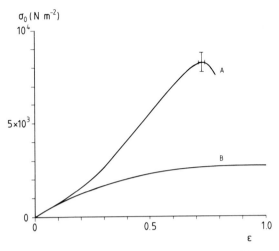

Figure 5   *Stress–strain curves of potato starch gels obtained by uniaxial compression with* $\dot{\varepsilon}_{initial} = 1.7 \times 10^{-2}$ s^{-1}: *A*, 10 wt% *potato starch*; *B*, 10 wt% *potato starch in water containing a volume fraction of 0.23 of small glass beads*

three-day-old cheese was *ca.* 2 mm, which compares well with the size of the curd particles. It indicates as yet insufficient fusion of these particles. At this age, the cheese matrix still fractures between the curd particles. For an older cheese, however, the inherent defect length is much smaller. Using equation (1) (*r* was assumed to be constant), we calculate a value of about 0.1–0.3 mm, probably due to incidentally occurring imperfect fusion of curd particles.[5] In the same way, an inherent defect length for the potato starch gels was obtained of *ca.* 0.2 mm. This value compares well with the size of the swollen starch granules. Fracture probably occurs between these granules.

The presence of many inhomogeneities may also influence the shape of the stress *versus* strain curves. This is illustrated for potato starch gels in Figure 5. The gels containing the glass beads were found, by microscopic inspection, to be inhomogeneous. The glass beads were concentrated in certain regions, with in-between regions of at maximum 0.3 mm without glass beads. Moreover, at a volume fraction of glass beads of 0.23, larger sized regions, predominantly containing liquid, of size 2–3 mm by 1 mm, could be seen by visual inspection. As may be seen, not only is the σ_f decreased, as might be expected, but also the shape of the curve is changed into something like that of material D in Figure 3. The filled potato starch gels apparently show yielding first and fracture later. Just as for the materials B and D in Figure 3, it probably takes time before a defect is formed of size large enough to permit fracture propagation at the stress level considered [see equations (4) and (5)].

Energy Considerations—As indicated above, the fracture properties of gels often depend on the strain-rate, *i.e.* on the time-scale of the experiment. This can be understood better by considering the energy balance of the fracture or yielding process. Moreover, fracture propagation can *only* be understood using energy balances. In general, during deformation, an amount of energy W is supplied to a material equal to

$$W = \int_{\varepsilon=0}^{\varepsilon_f} \sigma(\varepsilon,\dot{\varepsilon},t)\, d\varepsilon \tag{7}$$

where σ may depend both on the strain ε, the strain rate $\dot{\varepsilon}$, and the deformation history denoted by the time t. At small deformations of elastic materials, this energy is stored, whilst for viscoelastic materials—purely viscous materials, in which no energy can be stored over the relevant time scales, are not considered—part of the energy is stored (W') and part is dissipated (W''_m). Moreover, at large deformation, energy dissipation due to friction between components caused by inhomogeneous deformation of the gel (W''_c) and due to fracture (W_f) may occur.[5,9,21] So the following energy balance may be written:

$$W = W' + W''_m + W''_c + W_f \tag{8}$$

The energy for fracture during fracture propagation must originate from W', implying that energy dissipation makes a material tougher.

Normally, energy dissipation processes are strain-rate dependent, and thus cause W and σ_f [equation (5)] and sometimes ε_f to depend on $\dot{\varepsilon}$.[3,5,17,19,20,22] The latter can be qualitatively understood for viscoelastic materials, for which at lower $\dot{\varepsilon}$ the relative importance of the viscous properties is greater, implying the energy dissipation to be relatively higher and thus W' to be lower at a certain strain. Then the material must be deformed further before W' is high enough for fracture propagation to occur. Such behaviour is shown, for instance, by rennet-induced skim milk gels[17] (Figure 6A) and young Gouda cheese with a pH > 5.15.[5] For materials like acid milk gels (Figure 6A), young Gouda cheese with pH < 5.15, and mature Gouda cheese, of which the ratio of the viscous properties over the elastic ones is lower and does not significantly change over the time-scale considered, the strain at fracture does not depend on the time-scale of the experiment. In all cases, σ_f increases with increasing strain-rates, but this dependence is much stronger for the rennet-induced skim milk gels than for the less viscoelastic, acid ones (Figure 6B). For viscoelastic materials, W' will always increase with increasing $\dot{\varepsilon}$, while W''_m may increase, remain the same, or even may decrease with increasing $\dot{\varepsilon}$. The latter may be the case if the ratio of the viscous over the elastic component drastically decreases with increasing $\dot{\varepsilon}$ over the considered range of time-scales.

Besides flow of the matrix material (viscoelastic behaviour), friction between components due to inhomogeneous deformation may cause energy dissipation. In gels this may be friction between the liquid component and the matrix-forming material, between components forming the matrix (*e.g.* after local yielding or fracture), or between filler material and the gel matrix. The latter processes probably play a part in the large deformation and fracture behaviour of the glass-filled potato starch gels reported in Figure 5, of mature Gouda cheese,[5] and of

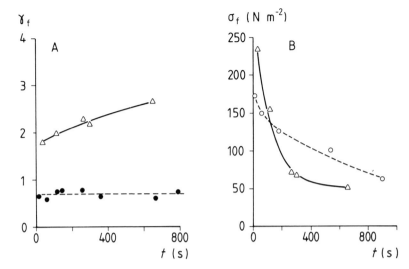

Figure 6   *The fracture strain γ_f (A) and fracture stress σ_f (B) at 30 °C as a function of time t up to fracture:[17] ●, ○, acid skim milk gels (aged for 16 h at 30 °C, pH = 4.6); △, rennet-induced skim milk gels (aged for 3.5 h at 30 °C, pH 6.65)*

frozen meat.[22] Normally friction processes dissipate more energy if they take place at higher speeds, thereby causing W_c'' and consequently W and σ_f to increase with increasing $\dot{\varepsilon}$.

Fracture Propagation—The critical overall stress for crack propagation is directly related to energy dissipation; a certain amount of strain energy W' must be present and be transported to the crack tip for a crack to propagate spontaneously. Also during the latter process, energy may be dissipated. Because near a crack tip the stresses are highest [equation (2)], there also the strain is highest as well as the changes in it (*i.e.* the strain-rate). This normally causes the energy dissipation to be highest there too. Moreover, for a viscoelastic material it may cause the tip of a crack (notch) to become blunted as can nicely be seen during fracture of a young, non-acid, Gouda cheese in tension.[5] These processes will retard fracture propagation, as may be seen, say, by comparing fracture initiation and propagation of young and more mature Gouda cheese (Figure 7). Moreover, extensive (local) energy dissipation, which increases with increasing strain-rate, will slow down the cracking speed. For a young Gouda cheese, a crack speed is observed of about 2×10^{-4} m s^{-1} (1 cm min^{-1}), while for a mature cheese the speed is much faster.

Another point determining fracture propagation and crack speed in a given material is the presence of weak spots. Weak spots met by an advancing crack may result in additional energy dissipation due to extensive local yielding or formation of side cracks. Moreover, it may cause the stress concentration [equation (1)] to

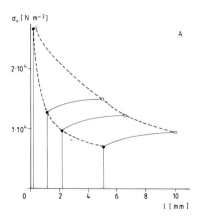

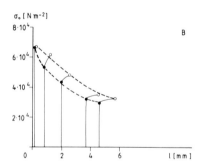

Figure 7   *Overall stress σ_0 versus notch length l for* (A) *a young and* (B) *a more mature cheese. The experiments were done in tension. Results are given for crack initiation (●) and crack propagation (○)*

be less. Probably these phenomena also play a part in the deformation and fracture behaviour of the glass-filled potato starch gels (Figure 5). The regions containing liquid, and maybe partly also those around the glass beads, will probably act as weak spots and thereby impede crack propagation.

Fracture or Yielding—The relative importance of the different mechanisms described above will determine if a material shows a clear fracture point or more of a yielding behaviour. In general, we expect yielding behaviour if (i) the material contains many, more-or-less equal, weak points, so that fracture is initiated at many points at roughly the same time, and (ii) strong energy dissipation occurs after fracture initiation so that not enough energy can be stored to allow fracture propagation Presumably, the latter condition is essential, and the former favours yielding. The main cause of energy dissipation may be friction between the different components due to the inhomogeneous deformation.

References

1. L. G. B. Bremer, T. van Vliet, and P. Walstra, *J. Chem. Soc., Faraday Trans. 1.*, 1989, **85**, 3359.
2. S. P. F. M. Roefs, A. E. A. de Groot-Mostert, and T. van Vliet, *Colloids Surf.*, 1990, in press.
3. S. B. Ross-Murphy and S. Todd, *Polymer*, 1983, **24**, 481.
4. G. J. Brownsey, H. S. Ellis, M. J. Ridout, and S. G. Ring, *J. Rheol.*, 1987, **31**, 635.
5. H. Luyten. 'The rheological and fracture behaviour of Gouda cheese', Ph.D. Thesis, Wageningen Agricultural University, Netherlands, 1988.
6. H. Luyten and T. van Vliet, Proceedings of Conference on Rheology of Food, Pharmaceutical and Biological Materials, Coventry, 1989, in press.
7. M. L. Green, R. J. Marshall, and K. R. Langley, Proceedings of Conference on Rheology of Food, Pharmaceutical and Biological Materials, Coventry, 1989, in press.
8. L. G. B. Bremer, B. H. Bijsterbosch, R. Schrijvers, T. van Vliet, and P. Walstra, *Colloids Surf.*, in press.
9. A. G. Atkins and Y-M. Mai, 'Elastic and Plastic Fracture', Ellis Horwood, Chichester, 1985.
10. H. McEvoy, S. B. Ross-Murphy, and A. H. Clark, *Polymer*, 1985, **26**, 1483.
11. H. L. Ewalds and R. J. H. Wanhill, 'Fracture Mechanics', Delftse Uitgevers Maatschappij, Delft, 1984.
12. J. E. Gordon. 'The New Science of Strong Materials', Penguin Books, Middlesex, 1968, Chap. 4.
13. A. A. Griffith, *Phil. Trans. Roy. Soc. (London)*, 1920, **A221**, 169.
14. P. Zoon, T. van Vliet, and P. Walstra, *Neth. Milk Dairy J.*, 1988, **42**, 249.
15. S. P. F. M. Roefs, 'Structure of acid casein gels', Ph.D. Thesis, Wageningen Agricultural University, Netherlands, 1986.
16. P. Walstra, A. Noomen, and T. J. Geurts, in 'Cheese: Chemistry, Physics and Microbiology', ed. P. F. Fox, Elsevier Applied Science, London, 1987, Vol. 2, Chap. 2.
17. P. Zoon, T. van Vliet, and P. Walstra, *Neth. Milk Dairy J.*, 1989, **43**, 35.
18. P. Purslow, in 'Food Colloids', eds. R. D. Bee, P. Richmond, and J. Mingins, Royal Society of Chemistry, Cambridge, 1989, p. 246.
19. J. M. M. Westerbeek, 'Contribution of the α-gel phase to the stability of whippable emulsions' Ph.D. Thesis, Wageningen Agricultural University, Netherlands, 1989.
20. H. Luyten and T. van Vliet, 'Gums and Stabilisers for the Food Industry', eds. G. O. Phillips, P. A. Williams, and D. J. Wedlock, IRL Press, Oxford, 1990, Vol. 5, p. 117.
21. T. van Vliet, H. Luyten, and P. Walstra, to be published.
22. B. J. Dobraszczyk, A. G. Atkins, G. Jeronimidis, and P. P. Purslow, *Meat Sci.*, 1987, **21**, 25.

Light Scattering Studies of Milk Gel Systems

By David S. Horne

HANNAH RESEARCH INSTITUTE, AYR, SCOTLAND KA6 5HL

1 Introduction

Some of the most important properties of colloids are those which owe their origin or nature to the size of the colloidal particles. These particulate properties can be classified under three headings: optical, rheological, and statistical. Common examples of optical effects arising from dispersed particles are the colours of the rainbow, the blue of the sky, the crimson of sunsets, and the occurrence of natural whites in clouds, milk, paper, teeth, and the plumage of swans. Rheological effects due to dispersed particles are illustrated by tomato ketchup, paint, inks, dough, and in such processes as the manufacture of cheese. The most common example of a statistical property is the Brownian motion of the suspended particles. All three of these groups of properties are merged in this paper, in that we describe the use of a new optical (*i.e.* light scattering) method to monitor the dynamic properties of a suspension as gelation progresses and the rheological properties develop.

Whilst clouds and milk undoubtedly owe their whiteness to the particulate scattering of light, the use of light scattering as a practical research tool has hitherto been confined to studies of dilute suspensions. From such measurements, it is relatively simple to extract information on the molecular weight of the scatterers from angular variation of the static intensity, or on the size of the particles from the time-dependent fluctuations in the scattered light intensity. As the concentration of scatterers increases, however, the system becomes effectively opaque, and physically a multiple scattering régime is rapidly approached. Significant practical and theoretical deficiencies have so far prevented meaningful interpretation of data taken under such circumstances. In the majority of cases available, previous theories[1,2] are only applicable to systems where relatively weak multiple scattering prevails, and so effectively only single scattering with minor corrections could be considered. Recently, however, in optically thick media exhibiting a very high level of multiple scattering, this impediment has been turned to advantage by exploiting the diffusive nature of light transport in such systems.[3-5] Practically, of course, there is also the very important requirement that light must pass through the medium from source to detector. In our work, this is accomplished by using a bifurcated fibre-optic bundle as the light-guide in a back-scattering geometry.

The new light scattering technique reported here has been given the name Diffusing Wave Spectroscopy (DWS). After verifying its applicability to our experimental configuration using monodisperse polystyrene latices,[5] the study concentrates on its practical usage in a more mundane but nevertheless important system—the archetypal white liquid, milk, or more correctly skim milk (from which the fat has been removed by centrifugation). This system is no longer describable as an emulsion, but rather as a particulate suspension which owes its light scattering properties to the so-called casein micelles. These aggregates of caseins with calcium phosphate form particles of diameter in the range 50–500 nm,[6] containing thousands of individual protein molecules. It is on the properties of the caseins and the casein micelles that the industrial properties of milk largely depend. We first of all consider the interaction properties of these micelles as evidenced by the concentration dependence of their mobility before applying the diffusing wave technique to monitoring curd formation in milk following renneting and acidification.

2 Verification of the Applicability of Diffusing Wave Spectroscopy

All our light scattering measurements on latices and milks were made using the apparatus shown in Figure 1. An extensive description of the equipment has been given previously.[5] Based on a Malvern autocorrelator, the instrument operates in back-scattering mode and incorporates a bifurcated fibre-optic bundle as light-guide. Half the fibres, distributed randomly over the face of the common leg, carry light from the laser source. When dipped into a scattering medium, the other half of the fibres carry back-scattered light to the photo-multiplier, where masking by slit and pinhole ensures that light from only a small area falls on the detector.

The bundle therefore approximates to an extended plane-wave light source incident on the front face of a slab of thickness L and infinite extent with the scattered light collected at a point on the same front face. Applying the diffusional approximation for light transport to such an experimental configuration, Pine et al.[4] have shown that the electric field correlation function for the multiply scattered light can be written

$$g^{(1)}(t) = \frac{1}{1 - \gamma l^*/L} \cdot \frac{\sinh\left[(L/l^*)(6t/\tau)^{1/2}(1 - \gamma l^*/L)\right]}{\sinh\left[(L/l^*)(6t/\tau)^{1/2}\right]} \tag{1}$$

$$\approx \exp\left[-\gamma(6t/\tau)^{1/2}\right] \quad \text{(for } L \gg l^*\text{)} \tag{2}$$

where l^* is the diffusive mean-free-path for light in the system. This is also the length scale over which the direction of light propagation is randomized. In practice, this means that the diffusion approximation will provide an accurate description of the temporal fluctuations in the scattered light, only when the sample dimensions are greater than l^* and the distance the photons travel through the system is very much greater than l^*. The parameter γ is introduced as a multiplicative factor to locate the conversion point within the sample at which the passage of light can be considered diffusive, providing a physically meaningful

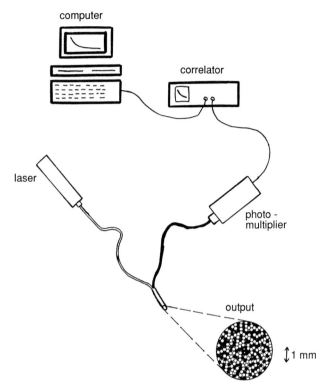

Figure 1 *Schematic diagram of apparatus configuration. Enlargement shows output face of common leg of fibre-optic bundle, illustrating random distribution of fibres (black) from the light source leg. The white fibres all carry scattered light to the photomultiplier but masking ensures that only a small area of this leg is active*

geometrical boundary condition for the solution of the diffusion equation by traditional techniques.[7]

Figure 2 compares the behaviour of the normalized autocorrelation function obtained for a 330 nm diameter polystyrene latex under dilute solution single-scattering conditions (volume fraction $\phi \sim 10^{-5}$) with that typically obtained from a concentrated solution ($\phi \sim 0.1$), when measured in back-scattering mode using the fibre-optics apparatus. In dilute solution, as expected from mono-disperse particles, the decay observed is a single exponential, and the semi-log plot *versus* delay time shows a straight line behaviour. In contrast, the correlation function obtained from the concentrated solution in the multiply-scattering régime, though monotonically decreasing, is markedly non-exponential, showing a more rapid initial decay followed by a slower decreasing tail. The wide range of decay times present in the multiply scattered function reflects the contributions from diffusing light paths over a wide range of length-scales. The longer paths

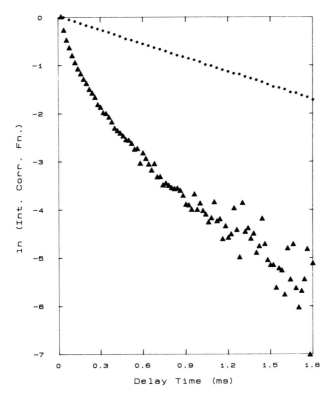

Figure 2 *Comparison of the semi-log plots of the normalized intensity correlation functions, obtained for* 330 nm *diameter polystyrene spheres in dilute* $(\phi \sim 10^{-5})$ (●) *and concentrated* $(\phi \sim 0.1)$ (▲) *solution conditions. The dilute solution measurements were made at 90° using conventional optics; the data for the concentrated solution were obtained in the back-scattering mode using fibre optics*

include a larger number of scattering events and so decay more rapidly, probing the motion of individual particles over shorter length- and time-scales. Conversely, the shorter paths consist of a smaller number of scattering events, and thus decay more slowly, probing in turn the motion of individual particles over longer length-scales and times.

The success of this approach in fitting data collected using our apparatus is demonstrated in Figure 3 for various depths of solution, L. For a sufficient depth of solution, the theory predicts that in back-scattering the logarithm of the intensity autocorrelation function should be linear in the square root of the delay time. This is effectively achieved for the latex suspension used at a depth of 2 mm (Figure 3). Note that under these conditions the scattering is independent of l^*. Reducing the depth of solution introduces curvature into the plots at shorter delay times due to the loss of the longer light paths. The solid lines in Figure 3 are as

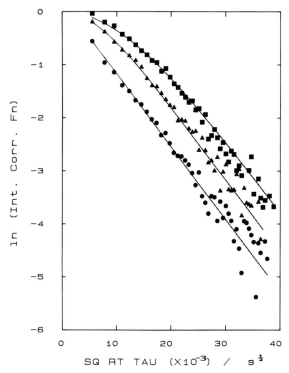

Figure 3 *Effect of slab thickness* L *at* $\phi \approx 0.01$ *on the semi-log plots of intensity correlation function against the square root of the delay time* τ: ●, L = 2 mm; ▲, L = 0.5 mm; ■, L = 0.25 mm. *Full curves are fits of equation* (1) *to the data using* $l^* = 0.217$, $\gamma = 0.87$, *and* $\tau = 9.641 \times 10^{-4}$ *throughout*

calculated from equation (1) and provide a very good fit to the data. Note that this equation contains no adjustable parameters, since l^* and γ can be obtained independently from transmission data whilst τ is confirmed as the infinite dilution value by the infinite thickness measurements.

The use of γl^* as the onset distance for diffusive light transmission is a device to permit easy solution of the diffusion equation. When more rigorous account is taken of the transition between ballistic and diffusive light transport, the value of γ is predicted to depend on both polarization and the anisotropy of the scattering.[8] Experiment confirms this polarization memory effect together with a particle-size dependence reflecting single-scattering anisotropy.[9] Parallel polarization shows γ increasing with size whilst perpendicular polarization shows the opposite, both asymptotically approaching a value of *ca.* 2.0 as size increases. Because our apparatus employs a bundle of multimode fibres as the light-guide, all polarization memory is lost, the parallel and the perpendicular effects cancel, and a constant value of γ is obtained. Equation (2) predicts linear behaviour when the logarithm of the correlation function is plotted against the square root of the delay time, the slope of this line being given by $\gamma\sqrt{6/\tau}$. The relaxation time τ is

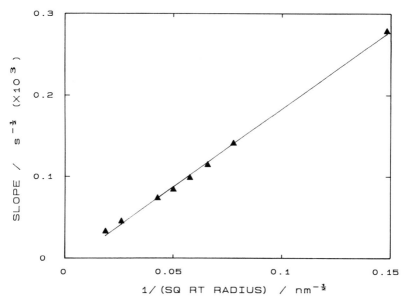

Figure 4   *Gradient of plots similar to 2 mm slab plot of Figure 3 as a function of the square root of latex particle nominal radius. All suspensions were measured individually at $\phi \sim 0.1$. The straight line is the linear least-squares fit to the data points*

directly proportional to particle radius, and so a constant γ suggests that the slope should be linear in the reciprocal of the square root of particle size. The linearity of the plot in Figure 4 confirms γ to be independent of particle size and demonstrates the usefulness of our apparatus as a size-measuring device in concentrated opaque dispersions. In passing, we note that the value of γ obtained from the gradient of this plot in Figure 4 is in excellent agreement with previous independent measurement.[4]

3   Liquid Milk

With undiluted skim milk as the light scattering medium, a preliminary study of light transmission yielded a value of 0.54 ± 0.01 mm for the diffusive mean free path, l^*, and 0.9 ± 0.02 for γ. The value for γ was confirmed by fitting to equation (1) the autocorrelation functions obtained with slabs of milk up to 1 mm thick. As predicted, these correlation functions rolled over at short delay times owing to the finite thickness of the slab cutting out the longest paths of the diffusing light—the paths which decay fastest and on the shortest time-scales. With still thicker slabs, however, the autocorrelation functions were found to continue to roll over, and a more extensive series of fitting operations traced this behaviour to the finite radius of the fibre bundle.[10]

At long delay times, equation (1) behaves asymptotically as the stretched exponential form given by equation (2). With a known value of γ, it is therefore possible to extract a relaxation time τ from the gradient of the long-time tail of the

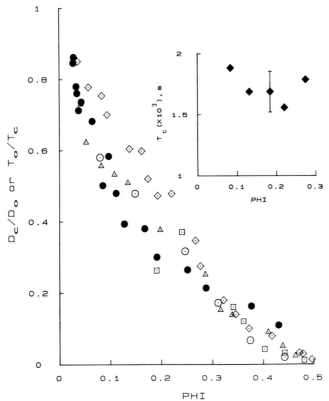

Figure 5 *Normalized relaxation time τ_o/τ_c as a function of the casein micelle volume fraction ϕ ($\bullet$). Tracer diffusion results (D_c/D_o) from references 14 ($\diamond$), 15 ($\square$), 16 ($\triangle$), and 17 ($\bigcirc$) are also shown. The inset shows the behaviour of τ_c as a function of volume fraction ϕ for a monodisperse micellar fraction*

plot obtained when the autocorrelation function is plotted logarithmically against the square root of the delay time, as in the upper plots of Figure 3. With monodisperse polystyrene latices, up to volume fraction $\phi = 0.1$, this relaxation time was found to be independent of concentration and equivalent to the infinite-dilution free-particle value. This lack of sensitivity to the particle-interaction effects, which are expected to have an increasing effect on diffusive motion at these concentration levels, was explained by Maret and Wolf[3] as a counter-balancing of hydrodynamic and structure factor effects. For milk solutions, however, we find an increasing relaxation time as the concentration is increased, indicating decreased micellar mobility (Figure 5).

The polydispersity of the casein micelle system can be reduced by fractionation in a series of centrifugation steps, the pellets being resuspended in permeate prepared by ultrafiltration of the original milk.[11] Though the system is not truly monodisperse, the measured polydispersity and the lack of angular variation in the infinite-dilution micellar diffusion coefficients indicate a marked narrowing of the size distribution.[11] Several such micellar fractions were prepared and their scattering behaviour examined over the concentration range previously covered in the skim milk experiments. Relaxation time data for one such pellet are shown in the inset to Figure 5. The steep concentration dependence encountered with milk and its polydisperse micellar system has been removed by the fractionation and the narrowing of the size distribution. The polydispersity of the original distribution is unmasking a slower mode, dependent on the interactions, whose strength is increasing with concentration of the dispersion. Pusey *et al.*[12] have argued that the intensity autocorrelation function obtained for light scattering from a polydisperse system consists of two independent modes with well separated decay times. The faster-decaying mode they identified with collective compression–dilation motions, *i.e.* collective diffusion, which is present even for monodisperse systems. The second, slower mode describes the exchange of particles of different types and is closely related to long-time self-diffusion. Rigorous separation of these modes is achievable only for optical polydispersity, but experimental evidence on their occurrence in systems polydisperse in size has recently been presented.[13] The relaxation time whose concentration dependence we are observing in milk is derived from the long-time tail of the correlation function, and hence must be the slowest relaxation process in the system, equating with the longer-lived self-diffusion mode.

Normalizing the data to their extrapolated infinite dilution value, the decay of relaxation time with concentration superimposes on the behaviour of the tracer long-time self-diffusion coefficient observed by several other groups[14–17] (Figure 5). DWS is measuring a dynamic structure factor sensitive to the local motions of the particles, which, at higher volume fractions, require the accommodating movements of their neighbours to diffuse an appreciable distance.

4 Milk Gels

If DWS is indeed sensitive to particle interactions in polydisperse systems, as the above results indicate, this opens up possible use of the technique as a monitor of gel formation where particle motion becomes increasingly hindered by network formation. Gel formation in milk is readily induced by acidification or enzyme action, or a combination of both. Though this aspect has not been stressed in the discussion so far, we note that, as well as the dynamic properties we have been highlighting here, our instrument also provides information on the integrated or static light scattering from the dispersion.

With milk as the substrate, changes are seen in both the static and the dynamic light scattering as gel formation progresses. Such data can be utilized at several levels of sophistication. With rennet gels, for example, the onset of these changes denote critical times during the progress of curd formation. Increases in the relaxation time coincide with the onset of visible clot formation in the rotating test-tube assay.[18] The reciprocal of the relaxation time defines a micellar mobility. This declines by some three to four orders of magnitude as gelation ensues, and it has been suggested that its extrapolation to zero might provide a gelation time—an objective measure of the optimum time for curd-cutting.[18]

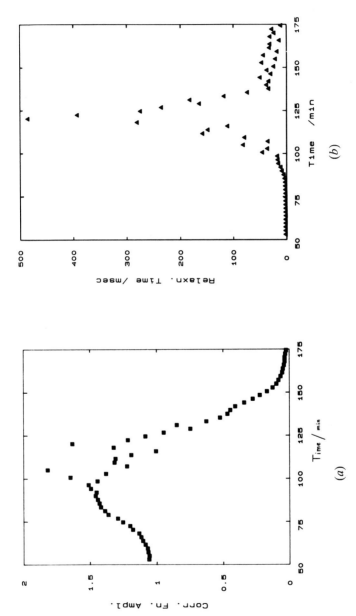

Figure 6 *Changes in* (a) *correlation function amplitude and* (b) *relaxation time recorded during the combined rennet and acid induced coagulation of skim milk. Reagents were added to the milk at time zero*

A major advantage of this technique is the non-destructive nature of the measurement. It therefore provides the continuity of output required for the derivation of fundamental kinetic and mechanistic information from the earliest stages of the gelation process. The information content of DWS alone will be insufficient to answer all questions relating to mechanism, but, as indicated below, it could provide information on elastic and frictional properties of these gels, which, with the help of suitable theoretical models, may provide molecular insight.

Figure 6 provides typical examples of the changes in static and dynamic light scattering of milk during acid gelation. Starter bacteria and a low level of rennet were added at time zero. The relaxation time (Figure 6b) exhibits a sharp maximum. Under other reaction conditions the rise and fall are neither so steep nor so far. If particle motion is considered to be governed by the resultant of viscous and elastic forces, then this relaxation time is simply the ratio of viscosity to elastic modulus, η/E. As aggregation begins, an individual aggregate meets increasing frictional resistance to its motion, and a growing viscous component is apparent. The relaxation time diverges. Later, as gel network formation dominates, the elastic forces come to the fore and the relaxation time decreases.

A similar qualitative interpretation can be extracted from the behaviour of the correlation function amplitude during the gelation process. One component of the scattered intensity equation is the static structure factor, delineating the influence of neighbouring particles on the scattering behaviour. The structure factor and the radial distribution function of the suspension are related *via* a Fourier transform equation, the radial distribution function defining the probability of finding a second particle at a distance r from the chosen reference particle. The changes seen in light-scattering intensity during the gelation reaction could be considered as mirroring the changes in the radial distribution function as the interparticle distance r is decreased. In the earliest stages, we see average liquid-like behaviour with no preferred position for particles within the suspension. As the reaction proceeds, and the intensity is observed to increase, this indicates that the most probable distance between the particles is decreasing, approaching the first shell of the radial distribution function. This could mean that the majority of particles in the system are now being incorporated into aggregates, though its initial rise could imply that the particles are simply able to approach one another more closely, as might occur, say, following the depilation of the 'hairy' casein micelles.[19] Indeed, it has proved possible to relate this initial rise in intensity during rennet curd formation to the proteolysis stage of the reaction.[18] The fall in intensity thereafter would then indicate a locking of particles into a gel matrix where the most probable position would be determined by the interparticle potential energy or, effectively, the gel strength. Hence, the larger the drop, the stronger the gel. Alternatively, the static structure factor can be straightforwardly expressed in terms of the osmotic compressibility of the system, and hence related to the elastic modulus of the gel network. The decline in intensity at longer reaction times may therefore be considered to reflect the growing elasticity of the gel.

In whatever manner the static and dynamic light scattering behaviour is interpreted, by virtue of its relation to the relaxation modes of the system, it must eventually reflect a viscoelastic property. The use of the apparatus described here provides a complete time history of the process. Its major advantage over traditional mechanical techniques is that it is non-perturbing. Particularly with very weak gels, the very act of applying a mechanical probe to the system may distort or destroy the nascent structure, whereas DWS simply measures the effect of the spontaneous thermal fluctuations present in all systems.

Acknowledgement. The author thanks Mrs Celia Davidson for skilled technical assistance in this project.

References

1. G. D. J. Phillies, *J. Chem. Phys.*, 1981, **74**, 260.
2. J. K. G. Dhont and C. G. De Kruif, *J. Chem. Phys.*, 1983, **79**, 1658.
3. G. Maret and P. E. Wolf, *Z. Phys.*, 1987, **B65**, 409.
4. D. J. Pine, D. A. Weitz, P. M. Chaikin, and E. Herbolzheimer, *Phys. Rev. Lett.*, 1988, **60**, 1134.
5. D. S. Horne, *J. Phys. D.*, 1989, **22**, 1257.
6. T. C. A. McGann, W. J. Donnelly, R. D. Kearney, and W. Buchheim, *Biochim. Biophys. Acta*, 1980, **630**, 261.
7. H. S. Carslaw and J. C. Jaeger, 'Conduction of Heat in Solids', 2nd edn, Oxford Univ. Press, 1959.
8. F. C. MacKintosh and S. John, *Phys. Rev.*, 1989, **B40**, 2383.
9. F. C. MacKintosh, J. X. Zhu, D. J. Pine, and D. A. Weitz, *Phys. Rev.*, 1989, **B40**, 9342.
10. D. S. Horne, *J. Chem. Soc., Faraday Trans.*, 1990, **86**, 1149.
11. D. S. Horne and D. G. Dalgleish, *Eur. Biophys. J.*, 1985, **11**, 249.
12. P. N. Pusey, H. M. Fijnaut, and A. Vrij, *J. Chem. Phys.*, 1982, **77**, 4270.
13. A. Van Veluwen, H. N. W. Lekkerkerker, C. G. De Kruif, and A. Vrij, *J. Chem. Phys.*, 1988, **89**, 2810.
14. W. Van Megen and S. M. Underwood, *J. Chem. Phys.*, 1989, **91**, 552.
15. M. M. Kops-Werkhoven and H. M. Fijnaut, *J. Chem. Phys.*, 1982, **77**, 2242.
16. R. H. Ottewill and N. St. J. Williams, *Nature (London)*, 1987, **325**, 232.
17. A. Van Veluwen and H. N. W. Lekkerkerker, *Phys. Rev.*, 1988, **A38**, 3758.
18. D. S. Horne and C. M. Davidson, *Milchwissenschaft*, 1990, **45**, in press.
19. P. Walstra, V. A. Bloomfield, G. J. Wei, and R. Jenness, *Biochim. Biophys. Acta*, 1981, **669**, 258.

Relationships between Sensory Perception of Texture and the Psychophysical Properties of Processed Cheese Analogues

By Richard J. Marshall

AFRC INSTITUTE OF FOOD RESEARCH, READING LABORATORY, SHINFIELD, READING, BERKSHIRE RG2 9AT

1 Introduction

In the instrumental measurement of food texture, a primary objective is to eliminate or reduce the need for sensory assessments. Most attempts to do this have only been partially successful mainly because of the complexity of the sensory process which involves such steps as biting, chewing, and swallowing.[1] Some improvement in the relationships has been achieved by increasing the deformation rate during uni-axial compression of cheese analogues[2] and by using free-choice profiling.[3] However, further improvement in instrumental measurement requires more knowledge of the fracture processes within the food and of the means of describing the fracture processes in psychophysical terms that relate to the sensory processes. Oral viscosity evaluation of fluid foods is closely related to the flow between the surface of the tongue and the hard palate, and the sensory stimulus varies with the viscosity of the food.[4] Kokini[5] showed that the viscous and frictional forces measured instrumentally, and applied to a geometric model of the mouth, compare well with sensory smoothness, slipperiness and thickness of liquid foods. These properties refer, however, to relatively simple systems. Interpretation of results becomes more difficult with solid foods because work has to be done to perceive the texture,[6] and because the texture changes during mastication.

Any comparison of the sensory evaluation of texture with the physical properties of the food must be based on theories describing the physical basis of textural sensations.[7] For example, in chewing food, the subject will perceive in some form the effort required as a force, or a time, or a combination of both.[8] Thus, in physical terms the chewiness may be related to work input and to the solidity, elasticity, or plasticity of the food. Similarly, moistness may be related to the moisture within the food and to the fluidity of any fat that is present.

The objective of this current work was to make a preliminary measurement of the physical and sensory properties of processed cheese analogues containing different types of fat in varying amounts and to search for psychophysical relationships between them. The derived information could then help us to design better experiments in which we could examine the relationships between sensory and instrumental measurements more carefully. Such insights might suggest ways of improving instrumental testing, and tell us how to vary or reduce the fat content of composite foods whilst maintaining normal texture.

2 Experimental

Materials—Sunflower oil (SO) was purchased locally. Hycoa 5 (H5) and Coberine (Co) were obtained from Loaders & Crocklaan Ltd, London, UK. Other materials were as described previously.[2]

Preparation of Model Processed Cheeses—These were made as described previously[2] to the compositions in Table 1 using four different fats (SO, butteroil (BO), H5 and Co), each at three concentrations, but at constant values of moisture in non-fat solids (MNFS).[9] The composition of the cheeses was measured using standard methods as described previously.[2]

Sensory Analysis—This was carried out by free-choice profiling.[3] Eight panelists were able to attend all the sessions, and others were excluded from the analyses. During training, which was spread over six 40-minute sessions, panelists were presented with all possible combinations of composition of the model cheeses in order to develop their individual descriptor profiles. During the experiment, panelists were presented with four cheeses per session at each of three sessions. Order of presentation was fully randomized across the panelists and the sessions. Further presentations were not possible because of the limitations on panelists time.

Mechanical Tests—Rheological parameters were measured as described previously.[2] Maximum stress (MS) and work to maximum stress (WMS) were measured by uni-axial compression of cylindrical samples of the cheeses at 0.083, 0.83, and 8.33 mm s^{-1}, and at 80% deformation. Stiffness was measured by cyclic compression to 2% deformation at 0.033 mm s^{-1}, and work of fracture was determined by cutting samples of the cheeses with tungsten wires of different diameters.

Structural Analysis—The internal structure of the cheeses was examined by cryo-scanning electron microscopy of freeze-fractured samples.[2] The structure of fracture surfaces after deformation by 80% was examined in frozen samples using the cryo-equipment.

Table 1 Moisture in non-fat solids (MNFS), fat content, protein content, and rheological properties of processed cheese analogues. (MS, maximum stress; WMS, work to maximum stress; WF, work of fracture.) Values are means of duplicate measurements for MNFS, fat content, and protein content, means of six measurements for MS and WMS, and means of three measurements for WF and stiffness

MNFS wt%	Fat content wt%	Protein content wt%	MS N m^{-2} Deformation rate (mm s^{-1})			WMS kJ Deformation rate (mm s^{-1})			WF^a J m^{-2}	Stiffness^b N m^{-2}
			0.083	0.83	8.33	0.083	0.83	8.33		
Sunflower oil (SO)										
55.23	6.20	27.82	149.50	238.32	304.05	0.081	0.141	0.165	5.99	1.27
55.64	13.75	25.14	99.31	191.74	198.35	0.065	0.116	0.120	6.38	1.84
54.91	20.75	22.84	92.21	154.33	165.04	0.053	0.088	0.094	5.98	2.41
Butter oil (BO)										
54.60	6.65	28.28	172.81	254.25	349.01	0.097	0.156	0.184	4.59	3.11
55.19	14.00	25.39	140.06	238.04	272.40	0.093	0.158	0.172	6.55	4.60
55.24	21.65	23.29	103.64	162.49	163.60	0.068	0.104	0.105	4.78	2.22
Hycoa 5 (H5)										
54.86	6.80	28.33	189.86	326.42	349.34	0.109	0.186	0.198	2.19	3.61
54.97	14.50	25.84	199.13	287.95	316.81	0.156	0.234	0.251	6.66	13.40
54.34	21.50	23.29	189.13	290.17	282.01	0.178	0.263	0.252	9.91	16.40
Coberine (Co)										
54.24	6.90	28.71	206.90	375.27	390.81	0.121	0.225	0.244	7.20	3.80
53.64	14.80	26.41	194.68	326.81	366.00	0.162	0.257	0.290	6.83	16.71
55.37	21.00	22.90	189.13	284.62	324.92	0.173	0.260	0.319	15.56	22.51

^{a} Deformation rate = 0.83 mm s^{-1}.
^{b} Deformation rate = 0.033 mm s^{-1} (cyclic).

Statistical Analysis—Sensory descriptor scores were analysed by generalized Procrustes analysis* for each individual panelist within fat type, across fat content, within fat content, and across fat type. Individual Procrustes scores were themselves inserted into Procrustes analysis to determine the overall consensus configurations for the cheeses.

Rheological and compositional data were compared by regression of terms up to the quadratic. Cheeses containing the different fats were considered separately. To compare behaviour across the range of fats, arbitrary factors describing the approximate relative 'hardness' of each fat at 22 °C were included in the regressions. These factors are: SO, 1; BO, 4; H5, 12; and Co, 14.

The individual Procrustes scores were compared with the rheological and compositional data by step-wise forward and backward regression of all terms including logarithmic and reciprocal derivatives. Psychophysical derivatives of the compositional parameters were selected on the basis of their apparent relevance to the sensory perception of the texture of the cheese analogues. 'Softness' was defined as protein content/MNFS, 'moistness' as fat content/MNFS, and 'chewiness' as WMS/MNFS. These quantities were examined both as simple functions and as logarithmic and reciprocal transformations.

3   Results

Composition of the Cheeses—The mean MNFS content for all cheeses is 54.85 ± 0.58 wt% (Table 1). The mean fat contents at the three levels are 6.61 ± 0.31, 14.3 ± 0.48, and 21.20 ± 0.42 wt%. Protein content, which varies from 22.84 to 28.70 wt% over the 12 cheeses, is inversely proportional to fat content.

Relationships of Rheological Parameters to Fat Content and Type—As the deformation rates were increased, the values of MS and WMS for the cheeses containing a single fat type were found to increase, but there was a much greater increase between 0.083 and 0.83 mm s^{-1} than between 0.83 and 8.33 mm s^{-1} (Figures 1a,b,c,d and 2a,b,c,d). With the cheeses containing the soft fats SO and BO (Figure 1a,b), the MS decreases as the fat content increases. There is a similar but a smaller decrease as the content of H5 or Co is increased (Figure 1c,d). The WMS decreases with increasing SO and BO content (Figure 2a,b), but increases as the H5 and Co contents increase(Figure 2c, d). Generally, the stiffness of all the cheeses increases as the fat content increases, but this increase is much greater for the hard fats than for the soft ones (Figure 3). The effect of BO was found to be unclear. Work of fracture changes very little as the SO or BO content increases, but it increases as the H5 or Co content increases (Figure 4).

---

* In generalized Procrustes analysis the sensory scores are considered as co-ordinates in multi-dimensional space. These are then transformed by translation, rotation–reflection and scaling to give a configuration that minimizes the residual variation between samples. This new configuration has co-ordinates on a number of axes of which the first accounts for most of the variance. For further details, see Marshall and Kirby,[3] Banfield and Harries,[10] Gower,[11] Langron,[12] and Vuataz.[13]

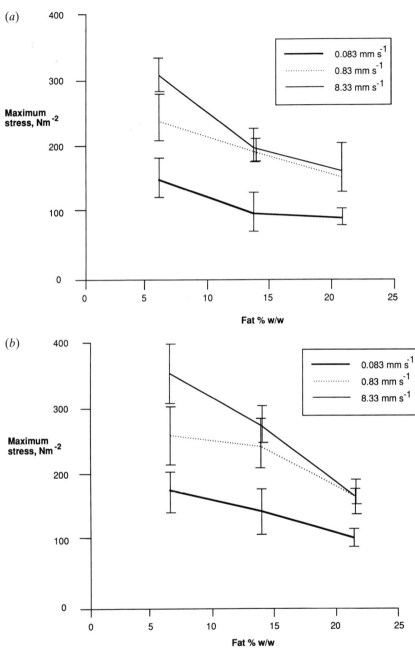

Figure 1 *Relationship of maximum stress to fat content and type in processed cheese analogues*: (a) *sunflower oil*, (b) *butter oil*, (c) *hycoa 5, and* (d) *coberine.* (*Continued*)

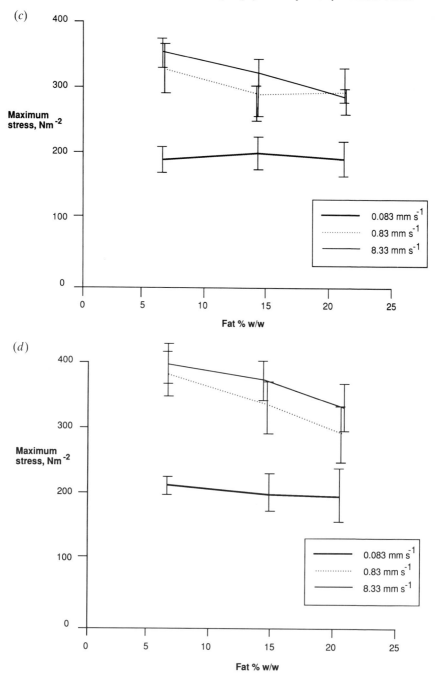

Figure 1 *continued*

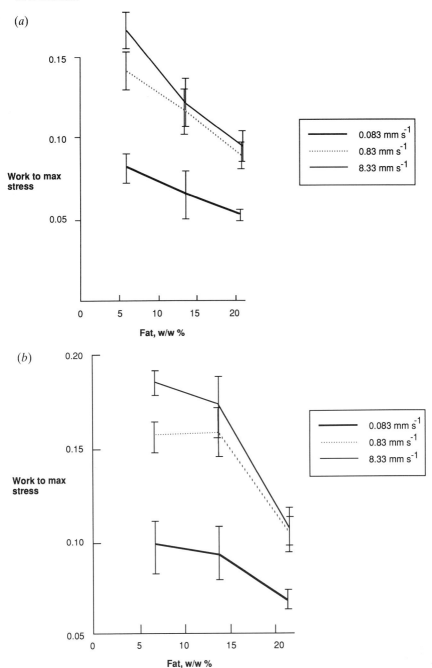

Figure 2 *Relationship of work to maximum stress to fat content and type in processed cheese analogues: (a) sunflower oil, (b) butter oil, (c) hycoa 5, and (d) coberine. (Continued)*

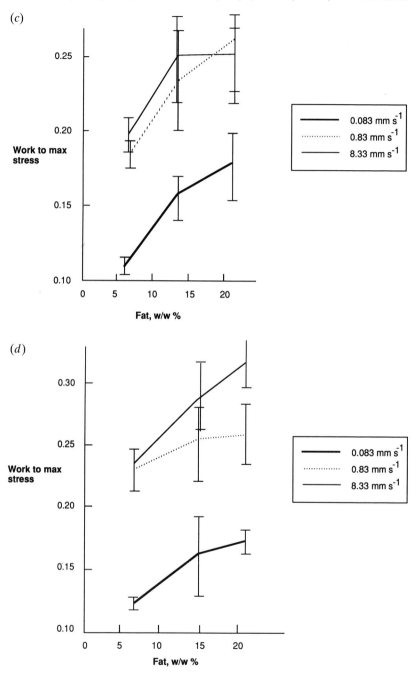

Figure 2 *continued*

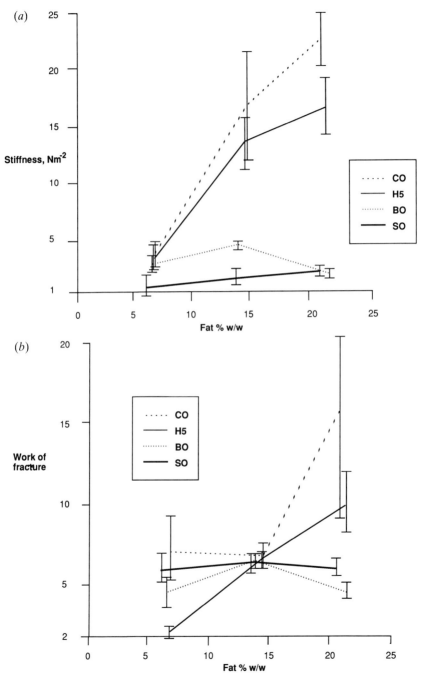

Figure 3 *Relationship of* (a) *stiffness and* (b) *work of fracture to fat content and type in processed cheese analogues*

Internal Structure of the Cheeses—The internal structures of frozen, freeze-fractured samples of the different cheeses are similar (Figure 4a,b). There is some indication that the fat globules are slightly smaller in the cheeses containing H5 and Co than in those with SO and BO.

In frozen samples of the cheeses that had been fractured by uni-axial compression at 8.33 mm s^{-1}, there were observed to be major differences in the structure of the fracture surfaces. In the SO and BO cheeses, the fat appears to be spread

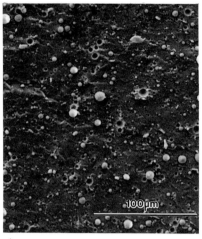

(a) 15% w/w Butter oil cheese, 55% MNFS

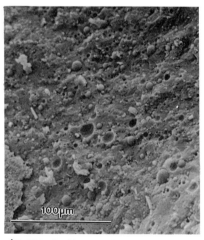

(b) 15% w/w Hycoa 5 cheese, 55% MNFS.

Figure 4   *Internal structure of freeze–fracture surfaces of processed cheese analogues*: (a) *butter oil cheese and* (b) *hycoa 5 cheese*

out over the surfaces, and very few intact fat globules can be seen (Figure 5a). In the H5 and Co cheeses, the fracture surfaces are knobbly with only few fat globules visibly damaged (Figure 5b). The globules appear to be just below the fracture surface.

Sensory Analysis—Generalized Procrustes analysis of an individual panelist's scores within each fat type showed that most panelists discriminate well between

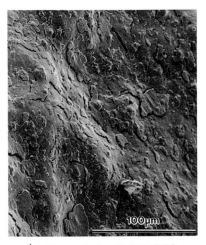

(a) 15% w/w Butter oil, 55% MNFS

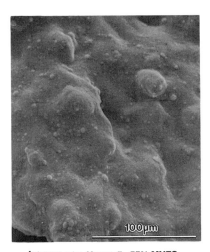

(b) 15% w/w Hycoa 5, 55% MNFS.

Figure 5 *Structure of fracture surfaces of processed cheese analogues produced by uniaxial compression: (a) butter oil, (b) hycoa 5*

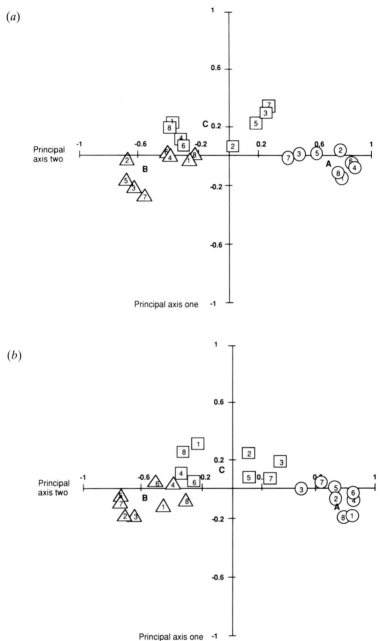

Figure 6 *Scores on first two principal axes from generalized Procrustes analysis of individual panelist's scores within each fat type: (a) sunflower oil, (b) butter oil, (c) hycoa 5, (d) coberine. A, B, and C denote overall*

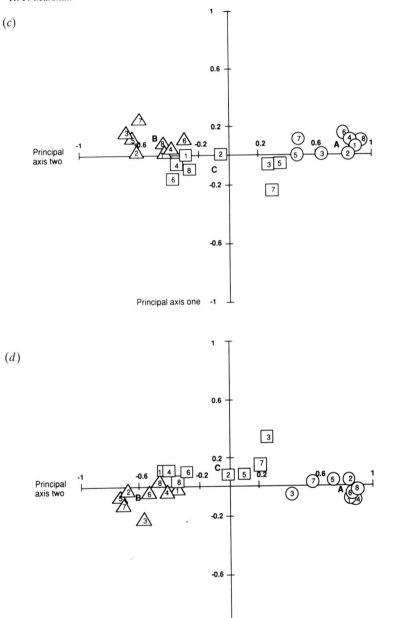

consensus points for fats at 5, 15, *and* 20 wt% (*Nominal concentration*), *respectively. Numbers* 1 *to* 8 *refer to different individual panelists for samples with fat contents*: ○, 5 wt%; △, 15 wt%; □, 20 wt%

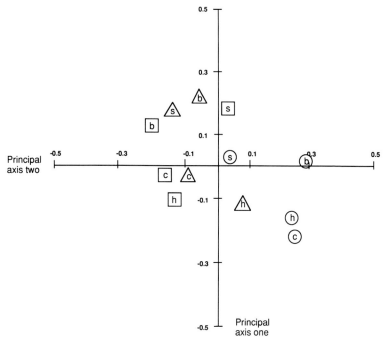

Figure 7   *Examples of scores on first two principal axes from generalized Procrustes analysis of individual panelist's descriptor scores:* s, *sunflower oil*; b, *butter oil*; h, *hycoa 5*; c, *coberine*; ○, 5wt% fat; △, 15 wt% fat; □, 20 wt% fat

cheeses containing one type of fat (Figure 6a,b,c,d). However, three panelists placed the cheeses containing 15 wt% and 20 wt% of both H5 and Co quite close together. The analyses across the fat type showed that panelists tend to place the SO and BO and the H5 and Co cheeses with similar fat contents close together. By way of illustration, Figure 7 shows the data from one panelist.

The first set of principal axes from the Procrustes analyses accounted for over 65% of the variance, the second set of axes about 30%, and the third the remainder. Therefore, only the first two axes were used to compare the sensory data with the physical data.

Relationships of Procrustes Scores to Physical Parameters—In no case was there found to be any significant correlation between the scores on the second set of axes and the physical parameters. When the data are compared over all fat types and contents, the Procrustes scores from four panelists are closely correlated with the logarithm of fat content (Table 2). Scores from two other panelists are best correlated with 'moistness', those from another one with 'fat content', and those from the last one with logarithm of 'hardness'.

Table 2 *Relationship of physical parameters to individual Procrustes scores on the first set of principal axes*

Panelist number	Rheological or compositional parameter	% Variance accounted for	t-Value (df = 10)	Probability
1	Log (fat content)	94.9	−14.40	<0.001
2	Log (fat content)	60.9	4.26	0.002
3	Log (hardness)	53.7	3.71	0.004
4	Moistness	95.4	15.40	<0.001
5	Log (fat content)	95.2	−14.88	<0.001
6	Fat content	95.0	14.55	<0.001
7	Moistness	96.4	17.20	<0.001
8	Log (fat content)	96.0	16.24	<0.001

The correlations between single descriptor scores for each panelist and their Procrustes scores on the first and second sets of principal axes are quite low (Table 3). Those terms with the best correlations tend to describe surface properties rather than any fracture properties.

4 Discussion

The results presented here are from a preliminary investigation, and as such can give only a general indication of relationships between sensory and physical

Table 3 *Best correlations of individual descriptor scores with Procrustes scores on the first set of principal axes*

Panelist	Descriptors	% Variance accounted for
1	Stickiness / Smoothness	11.7
2	Gluiness / Crumbliness	45.6
3	Greasiness / Clean bite	5.5
4	Clamminess / Aggregatability	37.3
5	Smoothness	14.2
6	Viscousness / Plasticine	28.3
7	Springiness	4.3
8	Softness	22.0

parameters. On the basis of their rheological properties, the cheeses could be divided into two groups. When deformed to 80%, those containing SO and BO were found to be weaker and more plastic than those containing H5 and Co, which were more elastic. The increasing stiffness of all the cheeses as the fat content increases shows that, under small deformations, all types of fat strengthen the matrix, but that hard fats are more effective than soft ones. Unlike previous work,[2] increasing the BO content did not result in a decrease in the work of fracture of the cheeses and similarly the SO content had little effect. However, in the previous work, both the BO content and the MNFS of the cheeses were varied giving data over a wider range, whereas the present work examines only one level of MNFS. The SO may have such a great effect that further increase in its concentration does not give additional lubricating effects as seen before.[2]

The large differences in physical properties and behaviour probably account for the poor correlations between the Procrustes scores and the physical parameters, in contrast to previous results with cheese analogues containing BO only.[2,3] It is, however, possible to relate some of the compositional variables to the Procrustes scores. Generally, the individual scores are correlated with a compositional factor related to fat content. On the other hand, the individual descriptors correlating with their respective Procrustes scores tend to describe 'mouthfeel'. Thus, with these particular cheeses, it appears that the panelists were all observing a similar property. Yet the consensus Procrustes scores are not related to any physical parameter to a significant degree. Therefore, the individual responses may have revealed more about perception than the overall response. The panelists tended to divide the cheeses into two groups according to fat type, as shown by Procrustes analyses across the types, where they grouped the SO and BO cheeses together and the H5 and Co cheeses together. This division is similar to that found in the rheological measurements.

Thus, it appears that the mouthfeel of the cheeses is altered by the differences in lubricating properties of the fats (Figure 8). More precise definition is not possible because of the complexity of mastication,[1,6,9] the relative simplicity of the rheological tests, and the way in which the sensory analysis was carried out. The present results suggest that, for a more detailed study of the oral perception of texture, the sensory evaluation should be broken down into a number of steps,

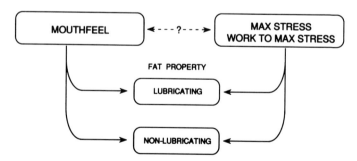

Figure 8 *Relationship of sensory perception of texture to compositional para-meter and rheological properties*

e.g. 'first bite', 'several chews', 'chew until ready to swallow', and 'swallowing'. This approach, together with the use of foods with smaller differences in composition and hence texture, could allow the examination of perception in much greater detail. This could in turn lead to improvements in the design and use of rheological equipment.

Acknowledgments. The author thanks Dr Margaret L. Green for helpful discussions, Mr Brian E. Brooker for the microscopy and for discussions, Misses Catherine Gelaky and Nicola A. Elliman for skilled technical assistance, and the anonymous panelists for their skilled assistance.

References

1. M. C. Bourne, 'Food Texture and Viscosity: Concept and Measurement', Academic Press, New York, 1982, p. 24.
2. R. J. Marshall, *J. Sci. Food Agric.*, 1990, **50**, 237.
3. R. J. Marshall and S. P. J. Kirby, *J. Sensory Studies*, 1988, **3**, 63.
4. P. Sherman, in 'Sensory Properties of Foods', ed. G. C. Birch, J. G. Brennan, and K. J. Parker, Applied Science, London, 1977, p. 303.
5. J. L. Kokini, *J. Food Eng.*, 1987, **6**, 51.
6. C. M. Christensen, *Adv. Food Res.*, 1984, **29**, 159.
7. D. A. Booth, personal communication.
8. J. B. Hutchings and P. J. Lillford, *J. Texture Studies*, 1988, **19**, 103.
9. R. C. Lawrence and J. Gilles, *N.Z. J. Dairy Sci. Technol.*, 1980, **15**, 1.
10. C. F. Banfield and J. M. Harries, *J. Food Technol.*, 1975, **10**, 1.
11. J. C. Gower, *Psychometrika*, 1975, **40**, 33.
12. S. P. Langron, 'The statistical treatment of sensory analysis data', Ph.D. Thesis, University of Bath, 1981.
13. L. Vuataz, 'Some points of methodology in multi-dimensional data analysis as applied to sensory evaluation', Nestlé Research News 1976–77, p. 57.

Discussions

Paper by Lips, Campbell, and Pelan

Dr. D. G. Dalgleish (Ayr): With respect to the temperature dependence of the thickness of the casein layers, can you comment on the possible relationship between your results and the well known temperature-dependent associations of the individual caseins and their mixtures?

Dr. A. Lips (Bedford): Our studies of adsorbed layer thickness by PCS relate to a level of addition of caseinate of 3 mg m^{-2}. On the basis of published isotherms for caseinate on polystyrene latex, this corresponds to *ca.* 75% of the pseudo-plateau adsorption limit. We have assumed, because of high affinity adsorption, that the added caseinate is completely adsorbed in the range of temperature studied, and that the observed decrease in adsorbed layer of thickness with temperature reflects a change in conformation of caseinate on the surface, rather than a transfer of caseinate from surface to bulk solution.

For a constant adsorbed amount, the Scheutjens–Fleer theory predicts a decrease in layer thickness with decreasing solvent quality (*i.e.* increasing χ). It could be inferred, therefore, that water becomes a poorer solvent for caseinate as temperature increases. The greater aggregation of β-casein with increasing temperature is consistent with this; light-scattering studies by Thurn and Burchard indicate both an increase in the molecular weight and a local chain collapse (possibly associated with a coil–globule transition). If the assumption of constant adsorption is correct, the change in adsorbed layer thickness is more likely to be a consequence of contraction of monomer chains than of the concomitant increased association at the interface.

Dr. E. Dickinson (Leeds): To what extent are your results on the bridging flocculation of the model polystyrene latex system applicable to more realistic food colloid systems such as those composed of protein-coated emulsion droplets?

Dr. A. Lips: Our model stability studies on dilute polystyrene latex systems enable food polymers to be ranked according to their capacity to give rise to attractive interactions between hydrophobic surfaces. In principle, this approach could be extended to study mixtures of biopolymers and competitive emulsifier–biopolymer effects. For example, depletion effects with protein-coated particles in the presence of non-adsorbing polysaccharides could be investigated. As we have shown, model studies of temperature effects with caseinate are clearly relevant to the behaviour of model food emulsions.

The stability method which we have described is more sensitive to the long-range details of the polymer–polymer interaction which control such phenomena as storage stability and creaming. Stability to shear and coalescence, on the other hand, depend on steric interactions at close surface-to-surface separations, to which the Brownian coagulation studies respond less sensitively. It is not unreasonable, however, to expect that bridging effects at long range and steric effects at short range might be inversely related, so that Brownian stability measurements may nevertheless provide useful guidelines for assessing the comparative shear and whip behaviour of emulsion samples.

Paper by Robinson, Manning, and Morris

Professor A. Prins (Wageningen): By means of a simple experiment, you demonstrated in your lecture that, immediately after shaking an aqueous gellan solution, a gel structure is formed which exhibits a certain yield strength. When such a system is diluted, what is the effect on the time required to form a gel after shaking?

Professor E. R. Morris (Cranfield): The lecture demonstration showed plastic spheres (*ca.* 5 mm diameter) suspended in a 'weak gel' matrix. When the container was inverted, the system flowed freely; but, as soon as flow ceased, the particle-suspending properties were restored with no detectable sedimentation of the spheres. By this simple visual criterion, the recovery is instantaneous, but a proper examination of concentration dependence would really require a more objective index of 'weak gel' properties, *e.g.* the stress-overshoot measurements described by R. K. Richardson and S. B. Ross-Murphy [*Int. J. Biol. Macromol.*, 1987, **9**, 257]. Unfortunately, no such measurements have yet been made on gellan.

Paper by Griffin and Griffin

Dr. D. C. Clark (Norwich): Can you say whether the photochemically induced grating in the sample cell is two-dimensional or three-dimensional?

Is interaction between BSA and gelatin a possible explanation for the appearance of a second component in the mixtures examined?

Dr. M. C. A. Griffin (Reading): The transient diffraction grating formed in the sample by the crossed writing beams is a thick grating. The approximate thickness, $t \approx 1$ mm, is determined by the crossing angle α and the writing beam diameters d at the region of intersection (see Figure 1). The grating has a sinusoidal concentration profile in *cis*- and *trans*-isomers of the chromophore spanning the region of intersection of the two writing beams.

As discussed in the paper, further experiments are required to establish the nature of any interactions between the gelatin network and the BSA probe molecules. It is expected that the decrease in mobility of the probe molecules comes about partly by hydrodynamic effects, and partly by direct obstruction of the passage of probe molecules by the network. To explain quantitatively the increase in polydispersity with time in terms of a binding between gelatin and BSA, it would be necessary to assume a rapid exchange equilibrium with time-dependent binding constants, or to suppose that some BSA molecules become gradually bound to a mobile, polydisperse fraction of gelatin molecules. Self-

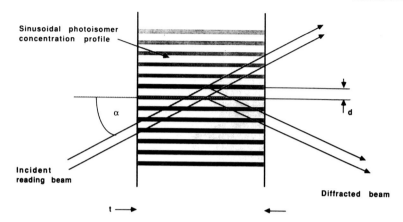

Figure 1   *Schematic diagram of FRS interference fringes*

aggregation of the BSA does not seem a plausible explanation at the probe concentration used.

Paper by Jones and Brass

Dr. E. Dickinson (Leeds): From either the theoretical or the experimental viewpoint, do you have any comment to make concerning the relative extent or nature of surfactant binding to *adsorbed* protein (*e.g.* SDS + lysozyme) as compared with the same binding interactions in bulk solution?

Dr. M. N. Jones (Manchester): I think one must consider two separate possibilities here. Firstly, if the protein is adsorbed at the interface before exposure to surfactant, then at high surfactant concentrations I think it is likely that the binding of surfactant to protein will cause some desorption. On the other hand, if the surfactant–protein complexes are already formed in the bulk phase, they might adsorb at the interface, although they might be less surface-active than the unliganded protein. In either case, the same equilibrium state should be found. In fact, however, we have not studied surfactant binding to a pre-unfolded protein– as might be present in surface denaturation. It should be possible in principle to do this, but there could be kinetic effects if the denatured protein were to become highly aggregated.

Dr. J. Mingins (Norwich): Your molecular graphics shows the bound SDS molecules lying down on the surface of the protein. Have you investigated the effect of surfactant chain length on the nature of the binding?

Dr. M. N. Jones: We have not yet investigated chain length effects by molecular dynamics, but we do have experimental data on this point. On shortening the chain length, we find that the binding affinity is reduced, which means that the alkyl chains form an intrinsic part of the interaction. Complex formation requires both the ionic head-group interaction with a charged site on the protein surface together with the binding of the alkyl group to hydrophobic patches on the protein surface. This is confirmed by the molecular dynamics calculations, which begin with only the head-group interaction; as the energy is minimized, the alkyl chains are observed to 'tuck in' to the protein surface.

Professor Z. U. Haque (Mississippi): When you alter conditions that affect ligand–macromolecule interactions, the self-interactions between the macromolecules may be influenced. If there is macromolecular association due to hydrophobic interactions, then the hydrocarbon–aqueous interface on the macromolecular surface is changed, and the ligand–macromolecule interaction will decrease if that too is hydrophobically driven.

Dr. M. N. Jones: That is true, but under the conditions that we use the concentration of protein is low, and so macromolecular association is minimized.

Professor Z. U. Haque: Yes, but in food systems the protein concentration is much higher.

Dr. M. N. Jones: While our studies all relate to low protein concentrations (0.1–0.2 wt%), there is no reason to suppose that surfactant binding and protein unfolding would not be similar at higher protein concentrations, provided that the weight ratio of surfactant to protein is kept the same. Obviously, complex–complex interactions would become more important at higher protein concentrations, and the binding isotherms would not be identical to those at lower concentrations. Over a limited range of protein concentrations, we have shown that the average number of surfactant molecules bound decreases with lysozyme concentration at constant free SDS concentration in the hydrophobic binding region of the isotherm [M. N. Jones and P. Manley, *J. Chem. Soc., Faraday Trans*.1, 1979, **75**, 1736].

Paper by Roscoe

Dr. D. G. Dalgleish (Ayr): Whether the κ-casein is in polymeric or monomeric states may make a difference to your observations. Since κ-casein forms extensive disulphide-linked polymers, it is possible to control the polymerization by breaking the disulphide bonds. Can you tell us whether the protein was reduced or not?

Dr. S. G. Roscoe (Nova Scotia): The κ-casein used in these experiments was not reduced.

Dr. M. C. A. Griffin (Reading): Further to Dr. Dalgleish's comment about the nature of oxidized κ-casein compared to reduced κ-casein, I should like to point out that, in experiments to study the ability of κ-casein to penetrate lipid monolayers [M. C. A. Griffin *et al.*, *Chem. Phys. Lipids*, 1984, **36**, 91], we showed that reduced κ-casein added to the sub-phase below a phospholipid monolayer causes a much greater increase in surface pressure than that which results from the addition of oxidized κ-casein. It seems likely—especially in view of the results of T. Arnebrant and T. Nylander [*J. Colloid Interface Sci.*, 1986, **111**, 529]—that, if Dr. Roscoe repeats her experiments with purified reduced κ-casein, she will still observe at the electrode stronger adsorption of κ-casein than β-lactoglobulin.

Dr. S. G. Roscoe: Yes, I believe Dr. Griffin is correct with regard to possibly enhanced adsorption for reduced κ-casein at the electrode surface as compared with oxidized κ-casein. This aspect will be investigated in the near future.

It is important to point out that this is a preliminary study using the electro-chemical technique of cyclic voltammetry to measure the blocking efficiency to oxide deposition at a platinum electrode by adsorbed species. This type of analysis, comparing the enhanced anodic (under-potential deposition) charge density resulting from adsorption with the charge density resulting from oxide reduction measured during the cathodic sweep at a platinum electrode, has been used here for the first time with proteins. The technique was used initially with protein supplied by Sigma Chemicals to determine, firstly, whether the technique could be applied to protein adsorption, and, secondly, whether a clear distinction can be made between the adsorption behaviour of two different proteins. The results obtained are in agreement with those of Arnebrant and Nylander even without the use of highly purified κ-casein. The surface concentration of 6 mg m^{-2} obtained for κ-casein at the higher anodic potentials was calculated on the basis of a one-electron transfer process per protein molecule. It is certainly conceivable that denaturation of the protein at the electrode surface could lead to two electrons being transferred per protein molecule, which would be in total agreement with the surface concentration observed by Arnebrant and Nylander. On the other hand, it is interesting to note that J. A. de Feijter *et al.* [*Biopolymers*, 1978, **17**, 1759] have reported surface concentrations of κ-casein of 6 mg m^{-2} at the air–water inter-face, which they have interpreted as being due to multi-layer formation.

The results presented for the adsorption of β-lactoglobulin are in good agreement with those reported in the literature using other techniques, and even the results obtained with the κ-casein that had not undergone rigorous purifi-cation were reasonable. Clearly, then, the technique appears to be suitable for this type of study, and it will be interesting to investigate in detail the reduced *versus* the oxidized forms of κ-casein, as well as the *para*-κ-casein *versus* the larger hydrophilic macromolecular fragment of κ-casein.

Professor T. J. Lewis (Bangor): Presumably the electrical charges on the amino acids and the biopolymers will influence adsorption. These charges will, in turn, depend on the pH. What influence does the fact that the local pH in the double-layer is cyclically different from the bulk solution pH during voltammetry have on the results?

Dr. S. G. Roscoe: In a well mixed solution, and particularly at a rotating electrode, the double-layer region will be small. The bulk solution in these experiments was aqueous phosphate buffer (pH 7). Any change in pH in the double-layer region will result from potential cycling over the restricted range of the under-potential deposition regions for hydrogen and oxygen. Ions formed as a result of this potential cycling will be immediately hydrated in the aqueous solution, and will be repelled from the double-layer region as the sweep pro-gresses (anodically for hydrated protons and cathodically for hydrated hydroxide ions). Since the changes in the hydration free energy are greater for small ions, proteins with a large surface-to-charge ratio, resulting from charged domains, will be less strongly hydrated. Thus, in the double-layer region, the free energy changes associated with interaction between protein with a small hydration shell and the polarized electrode should allow closer approach to the surface by the protein (in comparison with smaller more strongly hydrated ions), resulting in electro-chemical adsorption, as discussed by J. O'M. Bockris and A. K. N. Keddy ['Modern Electrochemistry', Plenum, New York, 1970, Vol. 2, p. 738–745]. If this process is fast, any localized change in pH due to potential cycling should be minimal, and should have a negligible effect on the electrochemical processes.

Paper by Fleer

Dr. D. S. Horne (Ayr): Am I correct in thinking that your calculations of the electrostatic potential are one-dimensional and normal to the surface? If so, how do you take into account radial interactions between charged sites in the same plane, and of a distribution of charged sites along a polyelectrolyte backbone? Does your approach not introduce constraints or preferential conformations for the adsorbing polyelectrolyte, and how does this affect the final segment density distribution?

Professor G. J. Fleer (Wageningen): In the mean-field model which I have described, the potential (and also its electrostatic part) depends only on the distance from the surface, and hence it is the same for any segment in a given layer. In this sense the model is one-dimensional with respect to the direction normal to the surface. Radial variations in the same plane are neglected: the charge is smeared out just as in the well known Stern-layer and Goüy–Chapman models. This approximation probably neglects some preferential conformations, especially when the segment density is low (*e.g.* in the bulk of a dilute solution). It is difficult to predict how the segment density distribution is affected, but the main trends for the adsorbed layer are probably correct as the concentrations are higher near the surface. While the model is, admittedly, rather simple, we console ourselves with the fact that agreement between theory and experiment is quite satisfactory.

Professor T. J. Lewis (Bangor): Your multi-layer model requires that each layer be assigned a dielectric constant. However, the dielectric constant is an 'outer-sphere' macroscopic concept, and it is not really appropriate for a monolayer, for which 'inner-sphere' microscopic polarizabilities must prevail, which will depend on intra- and inter-layer forces. How important are such effects in relation to your model?

Professor G. J. Fleer: The points you raise are, in principle, valid. Microscopic effects are disregarded in our model. Our use of a macroscopic dielectric constant is equivalent to that of the Poisson–Boltzmann approach and that in a simple Stern-layer model. It is known that these approaches work quite well in practice, unless the ionic concentration becomes very high. We trust, therefore, that our theory has some validity for the usual practical situations where the ionic strength is well below 1 M.

Professor A. Prins (Wageningen): In your paper you mention the presence of a 'depletion zone' near the interface when a polyelectrolyte is adsorbed. What is the effect of this depletion zone on the colloid stability when such a polyelectrolyte is used as a stabilizer?

Professor G. J. Fleer: The depletion zone in Figure 4 of my paper, applying to adsorbing polyelectrolyte, has a completely different origin from that occurring with a non-adsorbing polymer. In the latter case, there is a negative surface excess of polymer (due to a loss of conformational entropy not compensated for by an adsorption energy), which may lead to the so-called 'depletion flocculation'. Although Figure 4 shows a region where the polyelectrolyte concentration is lower than in the bulk solution, the overall surface excess is positive because of the affinity of the segments for the surface. Hence, for adsorbing polymer, no depletion flocculation is found.

The free energy of interaction in the presence of adsorbing polyelectrolyte is a complicated interplay of bridging attraction (especially at low coverage), steric repulsion (when the segment density is high), and electrostatic interaction. A detailed analysis for weak polyelectrolytes has been given by M. R. Böhmer *et al.* [*Macromolecules*, 1990, in press]. The main conclusions are as follows. The adsorbed amount is such that the surface charge is (slightly) over-compensated for by the polyelectrolyte charge. As a result, the electrostatic contribution is weakly repulsive, and the overall free energy of interaction is more positive than for an uncharged polymer present in the same amount. However, as this amount is relatively low for polyelectrolytes, bridging may still occur. For this reason, stabilization by polyelectrolytes might be more difficult than by uncharged polymers.

Paper by Mackie, Mingins, Dann, and North

Dr. D. G. Dalgleish (Ayr): Based on your work, can it be taken that end-on adsorption of dimeric β-lactoglobulin definitely does not occur, or is it possible that the act of adsorption causes the dimeric form of β-lactoglobulin to dissociate into monomers, which then absorb separately?

Dr. J. Mingins (Norwich): At pH 7.2, where there are dimers in solution, the contrast profile shows no evidence for end-on dimers at the surface. That is not to say, of course, that end-on dimers cannot adsorb. We cannot distinguish whether the protein adsorbs as monomers, or as dimers in any orientation with subsequent reorientation, dissociation, or conformational change. All we can say is that, over the time-scale of the SAXS measurement (1–2 hours), we have molecules at the surface which are not end-on dimers. The molecules have been conformationally flattened to half the diameter of the native monomer, but even here we cannot say whether the layer is composed of flattened individual monomers, or flattened sideways-on dimers, or a combination of the two.

Comparison of the plateau adsorption value with the two-dimensional parking limit would seem to indicate that a layer of end-on dimers is not initially formed. Our argument is that, for rapid irreversible adsorption with no lateral movement along the surface, the surface concentration is restricted to the two-dimensional parking limit irrespective of what conformation or dissociation changes occur subsequently.

Dr. M. C. A. Griffin (Reading): You indicate that the surface coverage of polystyrene latex by β-lactoglobulin at low pH is such that the protein adsorbed could no longer be associated as octamers. From the studies of Timasheff and

Townend [*Nature* (*London*), 1964, **203**, 517], it is clear that association into octamers is observed only in the cold, *i.e.* well below room temperature, and our own dynamic light scattering studies on β-lactoglobulin A have confirmed this. Did you carry out your experiments under conditions where octamerization is to be expected?

Dr. J. Mingins: While we chose an optimum pH for octamer formation, namely pH 4.65, we did not optimize the temperature. With the set-up on the SAXS station, it was difficult to hold samples at temperatures away from ambient (typically *ca.* 21 °C). The samples were 'equilibrated' for 12 hours in a separate laboratory before carrying out the SAXS measurements, but no thermostatting was done, and so the temperatures would have varied widely throughout the year. In these preliminary measurements, we were primarily interested in whether the SAXS technique would work, and whether we could isolate any protein shell characteristics. The adsorption measurements were also made at ambient temperatures.

In the paper you quote, the authors refer to their earlier work (ref. 8 and 9 of our paper) as having shown that β-lactoglobulin forms 'tetramers' (our octamers) at pH 3.7–5.2 in the cold. But nowhere in their Nature paper does it say that octamers are excluded at other temperatures in this pH range. In that Timasheff and Townend suggest that the A-variant aggregates strongly and reversibly to a 'tetramer' in this pH range in the cold, it is reasonable to expect some aggregates also at higher temperatures. Indeed, in their paper (ref. 7), Townend and co-workers say that the association is strongly temperature dependent, being maximal at low temperatures, but still detectable at room temperature. Their ultra-centrifugation of pooled β-lactoglobulin would seem to suggest a heavy fraction of *ca.* 10% at 20 °C, with nothing significant at 25 °C. It must be noted, however, that the ionic strength here is 0.1 M, and that the protein concentration is very high, so that the presence of any octamers to any significant degree under our conditions is rather uncertain. You are therefore quite right to draw attention to the importance of temperature. All that we can say at present is that the contrast profiles show no octamers at the surface, and that we intend to repeat the measurements at a temperature below 5 °C.

Paper by Dalgleish and Leaver

Dr. D. C. Clark (Norwich): Given that your earlier studies on the addition of a second protein to a protein-stabilized emulsion generally showed an increase in the protein load, can you comment as to why the trypsin does not merely adsorb?

Dr. D. G. Dalgleish (Ayr): There may be a small amount of trypsin adsorption, since there is a concentration below which the trypsin does not work. However, there are two reasons why adsorption may not be important. Firstly, the concentration of trypsin is very low, compared with the other experiments to which you refer. And, secondly, the spaces between the β-casein chains are likely to be sufficiently small that there is little likelihood of the trypsin molecule finding its way through to the interface, rather than finding a susceptible site for enzyme action.

Paper by Bergenståhl

Dr. M. N. Jones (Manchester): Have you found any evidence for lateral phase separation in binary lipid mixtures? I am thinking particularly about the PC + PI system, which, in the presence of divalent ions, separates into a PI-rich phase above the chain-melting temperature of PC.

Dr. B. Bergenståhl (Stockholm): There is no such evidence for the PC + PI system under the conditions used in my experiments (*i.e.* with no added ions beside the sodium ions associated with the PI). I believe that any lateral phase separation within the bi-layers would result in a macroscopic phase separation in the liquid crystalline system. At certain ratios of phospholipids in the PC + PE and PI + PE systems, however, the systems separate into lamellar and reversed hexagonal phases.

Paper by Dickinson and Euston

Dr. D. C. Clark (Norwich): Rapid lowering of interfacial tension leads to good emulsification, and high molecular weight of the adsorbing species leads to good emulsion stability. Can you comment on the relative importance of these two factors when attempting to produce an emulsion where both emulsifying and stabilizing activity are built into a single molecule?

Dr. E. Dickinson (Leeds): It depends very much, I think, on the way in which the emulsion is made, and on the time-scale and nature of the potential instability. Under diffusion-controlled conditions one cannot reconcile a very high molecular weight with fast transport to the interface and an associated rapid lowering of interfacial tension, but under highly turbulent conditions the large surface-active species may be rapidly transported to the new oil–water interface by a non-diffusive mechanism. The best compromise may be to form the stabilizing complex just after emulsification (or in a second homogenization step) by adding a stabilizer which interacts with the adsorbed emulsifier, but does not hinder adsorption of the emulsifier at the nascent interface during the primary homogenization step.

Professor Z. U. Haque (Mississippi): You carried out the creaming experiments for up to seven days. Did you have any bacterial problems?

Dr. E. Dickinson: All the emulsion samples prepared at neutral pH contained a small quantity of sodium azide to inhibit microbiological activity.

Paper by Westerbeek and Prins

Dr. T. van Vliet (Wageningen): You show that glycerol lactopalmitate may form a liquid crystalline phase. Do you think that this is a thermodynamically stable state or just a metastable state? If the latter, how far do you think that this is related to the fact that the glycerol lactopalmitate used is really a mixture of many different components?

Dr. J. M. M. Westerbeek (Veghel): You are pointing towards an important practical property of glycerol lactopalmitate (GLP). Normally, an α-gel phase for amphiphilic lipids like distilled monoglycerides is a metastable state from a thermodynamical point of view. This is probably also the case for GLP. However, the fact that GLP is very complex mixture of hydrophilic components makes the system kinetically stable in the α-modification for at least one year, as was found with wide-angle X-ray diffraction.

Paper by Dickinson, Goller, McClements, and Povey

Dr. D. G. Dalgleish (Ayr): You say that the ultrasonic technique is applicable to the study of crystallization of triacylglycerols. To what extent does the fact that many food oils are mixtures of different triglycerides affect the possibility of using the technique to study emulsions containing such oils? Also, can you indicate how your measurements on such systems relate to DSC results on the melting of these oils?

Dr. M. J. W. Povey (Leeds): Variations in triacylglycerol contents and polymorphic forms do make small differences (<20 m s^{-1}) in the measured velocity of the ultrasound. These differences are, however, relatively small ($\leqslant 7\%$) in comparison to the change in velocity at the solid–liquid transition. So, the differences you refer to do not affect the general conclusion of our work, that ultrasonic velocity measurement is an excellent way of following the crystallization process in emulsions; nevertheless, it is clear such considerations do need to be taken into account quantitatively during the analysis of the experimental results. We can confirm that there is good agreement between the ultrasonic velocity measurements and the DSC results for the melting of these oils.

Paper by Wedzicha, Zeb, and Ahmed

Dr. M. N. Jones (Manchester): Could you comment on the possibility of sulphydryl compounds, such as cysteine, partitioning into the oil phase of the two-phase systems?

Dr. B. L. Wedzicha (Leeds): Our data show that even non-ionic thiols such as undissociated mercaptoethanol or mercaptoacetic acid have little tendency to partition into sunflower oil. We would expect cysteine to remain essentially in the aqueous phase.

Paper by Leeson, Velissariou, and Lyddiatt

Professor Z. U. Haque (Mississippi): Since you are having some problems in interpreting the experiments using the polyclonals from sheep, have you considered raising monoclonals for your antigens?

Dr. A. Lyddiatt (Birmingham): We have not yet attempted to raise monoclonal antibodies against foam antigens—basically for two reasons. Firstly, we need to complete the work which is possible with the polyclonal sera we have available. The complex story that is emerging is just as likely to be due to the heterogeneity of the foam antigens as it is to the polyclonal nature of the antibody preparations. After all, the brewing process is characterized by the complex exploitation of chemical, physical and enzymic reactions, where the only certainty is the complexity of the molecular end product, particularly in respect of the structures of prospective protein antigens. Secondly, as the ultimate goal of this research is to produce a diagnostic test of foam quality, given the antigen complexity just referred to, the use of polyclonal antibodies may well be superior to monoclonals. My gut feeling is that we need to clarify the structure of the foam complex before investing time and effort in the search for a good monoclonal antibody.

Paper by Narsimhan

Dr. M. N. Jones (Manchester): A major factor in the rupture of thin films exposed to air is the loss of water by evaporation. Do you think that this is an important effect in your system, where the top of the foam is open to the atmosphere?

Professor G. Narsimhan (Indiana): It is true that evaporation can play an important role in the rupture of the thin films exposed to air over long periods of time. However, the present experiments on foam drainage were conducted over relatively short time periods (*ca*. 30 minutes) so as to ensure the absence of bubble coalescence during the measurements. This was necessary for comparing the model predictions with the experiments, since the model does not account for coalescence due to rupture of thin films. The observed coalescence at the top of the foam column at large times can be attributed to the rupture of thin films as a result of external perturbations including evaporation.

Dr. D. C. Clark (Norwich): Did you include a contribution from cylinder wall effects in the analysis of your drainage experiments?

Professor G. Narsimhan: Wall effects were not accounted for in the analysis of the foam drainage experiments, which were conducted in a 5 cm diameter cylindrical column. The bubbles generated by sparging nitrogen gas through the sintered glass discs were small enough to keep the column to bubble diameter ratio to 100 or more. Under such conditions, the effect of the wall on the rate of drainage of liquid from the foam is expected to be negligibly small.

Paper by Muschiolik

Dr. R. D. Bee (Bedford): What is the influence of the order of addition of proteins and polysaccharides on the final properties of your spreads?

Being water-continuous, how are your spreads stabilized against microbiological deterioration?

Professor G. Muschiolik (Potsdam): The first step was the preparation of an oil-in-water emulsion by homogenization of the oil in the protein solution. There was no marked difference in the final properties of the spreads if xanthan (not more than 0.2 wt%) was added to the protein solution or to the emulsion system. There is a larger effect, however, with some other polysaccharides, *e.g.* low-methoxy pectin.

The water activity of the spreads is in the range 0.70–0.75. Potassium sorbate (0.3 wt%) was added for microbiological stability.

Paper by Clark, Coke, Wilde, and Wilson

Dr. E. Dickinson (Leeds): I find your results very interesting. Your observation that a low concentration of surfactant increases the surface mobility of the adsorbed protein film is consistent with recent measurements of our own [E. Dickinson *et al.*, *Prog. Colloid Polym. Sci.*, 1990, **82**, in press] for systems containing casein and the water-soluble non-ionic surfactant $C_{12}E_8$ (octaoxyethylene dodecyl ether). At $C_{12}E_8$ concentrations very much below that required for protein displacement, the surface viscosity at the oil–water interface is drastically reduced as compared with the situation in the absence of surfactant. However, our interfacial tension data for the same system show no evidence for any surfactant–protein complex formation, and we also find no interaction with $C_{12}E_8$ + gelatin or $C_{12}E_8$ + lysozyme. On the other hand, we do find evidence for a surface complex with mixtures of the anionic surfactant SDS + casein or gelatin.

Dr. D. C. Clark (Norwich): It is unusual for proteins to bind non-ionic detergents such as Tween 20 with the affinity that we have observed with β-lactoglobulin, but I have no doubt that foam destabilization arises in part from the formation of a surfactant–protein complex. We believe that the initial onset of mobility can be explained by a steric mechanism associated with the polyoxyethylene chain projecting from the molecular complex. A plot of the surface diffusion coefficient of β-lactoglobulin against the Tween 20 concentration reveals that the surface mobility continues to increase as the Tween 20 concentration is raised above that which causes the initial appearance of mobility. We believe that this further increase is due to the appearance of free (non-bound) Tween 20 at the surface. In a non-interacting system, such as the one you refer to, the formation of a mixed component interface at concentrations of detergent below that required to cause protein displacement could account for the decrease in surface viscosity. We have observed this effect with α-lactalbumin + Tween 20 mixtures at Tween 20 concentrations where there is no detectable binding of the detergent. This suggests that small amounts of adsorbed detergent can indeed be sufficient to disrupt interactions between protein molecules at the interface. It is my opinion, however, that different proteins will behave in different ways with different detergents!

Dr. D. G. Dalgleish (Ayr): Our experiments on protein displacment have shown the difficulty of displacing one already adsorbed protein by another. However, you have now managed to partly mobilize the protein, even though it is still adsorbed. It may be the case, therefore, that the treatment with Tween 20 will render the adsorbed β-lactoglobulin more susceptible to displacement by another protein. Is there any evidence that it can be displaced by, say, β-casein?

Dr. D. C. Clark: The foams and films in our experiments were prepared from premixed samples of Tween 20 and β-lactoglobulin. We did not attempt to displace adsorbed β-lactoglobulin from the air–water interface by treatment with another protein or with the Tween 20, although protein displacement by the latter from the oil–water interface has been reported [J. A. de Feijter *et al.*, *Colloids Surf.*, 1987, **27**, 243]. Nevertheless, we have observed that the amount of adsorbed protein decreases with increasing Tween 20 concentration in the premix, especially at surfactant/protein ratios above 5:1, as determined by the decrease in fluorescence from the protein at the interface. The steric interference mechanism, which we propose to explain the onset of mobility in the surface-adsorbed β-lactoglobulin–Tween 20 complex, could facilitate desorption or displacement of β-lactoglobulin by another protein that does not form a high-affinity complex with the Tween 20. The protein β-casein may fall into this category.

Professor Z. U. Haque (Mississippi): Would you like to comment as to why there was such a difference in the plot of surface tension against surfactant concentration on addition of β-lactoglobulin?

Dr. D. C. Clark: Surface tension data for Tween 20 as a function of concentration were obtained in the absence and presence of 0.2 mg ml^{-1} $(11 \, \mu M)$ β-lactoglobulin. The protein alone gives a tension of 54 mN m^{-1} after 30 minutes of adsorption. In the absence of protein, $11 \, \mu M$ Tween 20 produces a tension of less than 40 mM m^{-1}. Therefore, Tween 20 on its own reduces the surface tension to a greater degree than the protein. However, $11 \, \mu M$ Tween 20 in the presence of $11 \, \mu M$ β-lactoglobulin only reduces the tension to 46 mN m^{-1}. This is because the surfactant is not available in its free state, since approximately 90% of that added is tightly bound to the protein. Our interpretation of the observed surface tension is that either the interface is populated purely by the surfactant–protein complex which is slightly more surface-active than the protein alone, or that the surface activity of the complex is indistinguishable from that of the protein alone, and that the small reduction in tension from 54 to 46 mN m^{-1} is due to small amounts of free Tween 20 at a surface that is mainly populated by complex. The data are discussed in detail by M. Coke *et al*. [*J. Colloid Interface Sci.*, 1990, **138**, 489].

Paper by Whittam, Noel, and Ring

Mr. M. Roozen (Wageningen): Are you sure that it is valid to use the Couchman equation for your system? Is it not applicable only if the values T_{g1} and T_{g2} are close together?

How did you determine the water content of your samples?

Dr. M. A. Whittam (Norwich): The Couchman equation has been shown to apply to the depression of the glass transition temperature of polymer networks by diluents [G. ten Brinke *et al*., *Macromolecules*, 1983, **16**, 244]. The problem of widely different T_g values is at least partially solved by considering only small diluent contents. The fact that the equation underestimates the depression of T_g for starch indicates that other factors are also involved—possibly in this case the specific interaction of water and polymer molecules.

The water content was determined by drying over P_2O_5 in a vacuum oven at $60 \, ^\circ C$ for 24 hours. Great care was taken to avoid any inadvertent moisture sorption prior to weighing.

Dr. R. D. Bee (Bedford): The fact that you obtain a change in the activation energy as a function of temperature suggests that there are different processes occurring at high and low temperatures. Can you give some physical interpretation of this?

Dr. M. A. Whittam: It seems likely that the change in activation energy with temperature reflects a structural change within the liquid. It is possible that clusters of molecules are formed, with an increase in the number of molecules involved in each cluster as temperature decreases. Co-operative reorientation of these larger units might then give rise to the larger observed activation energies.

Paper by Cameron and Donald

Professor E. R. Morris (Cranfield): How do the DSC endotherms for your retrograded samples compare in peak size and width (*i.e.* the degree of co-operativity) with the original gelatinization DSC endotherm?

Dr. R. E. Cameron (Cambridge): The size and shape of the endotherm are dependent on the ratio of starch to water in the gelatinizing mixture. At 50 wt% starch, a two-stage process occurs, giving a double peak in the DSC curve. This double peak has a total enthalpy of $7.1 \pm 0.1 \, \mathrm{J \, g^{-1}}$. The temperature of the first (stronger) peak is $61.6 \pm 0.5 \, ^{\circ}\mathrm{C}$. The first peak starts at a temperature of *ca.* 53 °C, and the second ends at *ca.* 91 °C, giving an overall width of *ca.* 40 °C. Retrograded starch gels containing 50 wt% starch in water give a single melting peak of size and peak temperature as indicated in Figure 5 of our paper. These also have a total width of *ca.* 40 °C.

Dr. D. S. Horne (Ayr): Your X-ray scattering spectra for retrograded starch do not show the periodicity of *ca.* 10 nm observed for granular starch, but you have analysed these data using the Guinier and Debye–Bueche approaches, both of which give effective 'sizes' for the scattering regions. Fractal concepts being much in vogue these days, have you considered attempting to derive a fractal dimension from the double logarithmic plot of intensity *versus* wave vector?

Dr. R. E. Cameron: Small-angle neutron work by A. Vallera [unpublished] on amylose gels indicates a fractal dimension of 2.58. A plot of logarithm of intensity *versus* log Q for our small-angle X-ray data, however, does not yield a straight line. This suggests that the scatterers in a starch gel are non-fractal.

Paper by Clark

Professor E. R. Morris (Cranfield): You refer to a final minimum slope of *ca.* 1.6 for the concentration dependence of the modulus in the log–log plot, as opposed to the limiting C^2 dependence that has been proposed previously. Could you comment on this?

Dr. A. H. Clark (Bedford): The apparent limiting power of 1.6 arises because of my choice of $f = 3$ in the model simulation. Had I chosen a much higher value (say, $f = 100$), I would have found a power (at the same C/C_0) much nearer the commonly found experimental result of 2.0. The variation in the limiting power with f is apparently one of the predictions of the classical theory. I have no simple physical explanation for this at the moment; nor am I aware of any experimental data for a gelling system that gives a result as low as 1.6. It should be remembered, however, that the power-law index varies with concentration, and it may be quite rare that gels can be made that are sufficiently concentrated to allow the limiting region to be truly accessed. Also, in practice, network defects will come into play, which may influence the modulus–concentration slope in a way not accounted for by the current theory.

Dr. D. S. Horne (Ayr): I have a couple of queries about the polyfunctional condensation model. Firstly, what functionality did you assume for your monomers, and how are your predictions of lag time or modulus as a function of concentration affected by variations in f? Secondly, the theory assumes that all functionalities remain equally active and accessible throughout the duration of the reaction. When monomers are particles of the size of casein micelles, how do you take into account the burial or shielding of reactive sites during gelation?

Dr. A. H. Clark: A value of $f = 3$ was assumed in the calculations. All other things being equal, the main effect of increasing f is to lower the critical gelation concentration, *i.e.* the concentration at which the gelation time diverges to infinity. The exact relationship between gelation time (or shear modulus) and the ratio of concentration to critical concentration is also affected, as I described in my answer to Professor Morris. In reply to your second query, the model in its present form does not allow for any inhibition of reaction, although the effect is no doubt important in some real situations. An allowance could possibly be introduced into the cascade theory treatment as a so-called 'substitution effect'.

Paper by Ross-Murphy

Professor E. R. Morris (Cranfield): The Oakenfull method for determining the gelation time seems to be a large-deformation measurement, and it might therefore be expected to give very different values from the small-deformation criteria described in your paper. In particular, for systems such as xanthan, it would presumably give infinite times, although the small-deformation storage modulus could be quite large. Work on gelatin in our laboratory has indicated a constant *ratio* between the onset of the rapid increase in G' and the appearance of a network strong enough to resist flow. Does this fit in with your own experience?

Dr. S. B. Ross-Murphy (Cambridge): Certainly, the Oakenfull method would give $t_c \to \infty$ for xanthan or other 'weak gels', although the result might depend on the concentration and the bore of the tube! I have only compared the small-deformation and large-deformation values of t_c very crudely, but it does appear that the latter is slightly larger than the former. The model which I have suggested implies that t_c is proportional to $1/([J]/[J_c] - 1)^p$ [see equation (2) of my paper], and this *could* reflect your experience, because $[J_c]$ would apparently be greater in the large-deformation case. This does not imply, however, that the modulus corresponding to the large deformation t_c is constant, and indeed one might expect it to decrease as C increases, because p is not really constant. It would be valuable to examine your gelatin data in more detail.

Paper by Walstra, van Vliet, and Bremer

Dr. W. G. M. Agterof (Vlaardingen): The fractal dimensions you quote are very high. Your explanation for the high fractal dimension is that the particles may rearrange during the aggregation process. However, rearrangement of particles or clusters is assumed when one goes from a diffusion-limited cluster–cluster aggregation process to a reaction-limited process. Do you believe therefore that the interaction potential is not taken into account adequately in the standard reaction-limited process, and that this is the reason for the higher fractal dimension found in practice as compared with that given by the computer simulations?

Professor P. Walstra (Wageningen): In 'reaction-limited' cluster–cluster aggregation simulations, rearrangements are not taken in account: the clusters become denser because particles and small clusters can, on average, diffuse 'deeper' into an existing cluster before becoming fixed. A different interaction potential does lead to differences on short length-scales, but it does not seem to affect the fractal dimensionality on longer scales, according to simulations [G. C. Ansell and E. Dickinson, *Phys. Rev.*, 1987, **A35**, 2349].

Dr. D. S. Horne (Ayr): I would like to comment on the high value of the fractal dimension found in these gel studies. They are similar to values found in other protein aggregation systems, and they are certainly far higher than the values of 2.0–2.1 found in computer simulations of reaction-limited cluster–cluster aggregation. They are also higher than the values found by Jullien or Meakin in simulations where limited annealing and rearrangement of the clusters are permitted, leading to increases in d_f of the order of 0.1–0.2 above the original value of 2.0. One problem with all these calculations is that they are effectively carried out at infinite dilution. Clusters interact as pairs, with A initially approaching B from beyond the immediate vicinity of B. Lack of reactivity allows inter-penetration, and the associated sampling of an increased number of configurations allows closer packing and a higher fractal dimension, but only up to an upper limit of $d_f = 2.1$ in these simulations.

My work with ethanol-induced aggregation of casein micelles under diffusion-limiting reaction conditions [D. S. Horne, *J. Dairy Res.*, 1989, **56**, 535] reveals a concentration-dependent fractal dimension, which starts low at the diffusion-limited cluster–cluster values of $d_f \approx 1.8$ at low concentrations, and progresses asymptotically to values of 2.4–2.5 at the higher concentrations employed. (This latter value is close to the diffusion-limited *particle*–cluster limit of $d_f = 2.5$ seen in simulation studies.) What I think may be happening in these higher concentration systems—mine with ethanol and yours with rennet—is the interdigitization of clusters as they are in the process of being formed, so that reaction can be concurrently proceeding at more than one position within any one cluster, with one or more other clusters. This model could possibly lend itself to computer simulation, but I know of no such calculations having been attempted.

Dr. E. Dickinson (Leeds): The phenomenon referred to by Dr. Horne of an increase in effective fractal dimension with increasing particle volume fraction has been reproduced in Brownian dynamics simulations of irreversible coagulation of spherical particles with periodic boundary conditions [G. C. Ansell and E. Dickinson, *Faraday Discuss. Chem. Soc.*, 1987, **83**, 167].

Professor P. Walstra: We expect no difference in 'interdigitization' between small and large aggregating clusters because the cluster structures are essentially scale-invariant. In the case of gelation, the effective volume fraction is always near to unity independent of the particle volume fraction—only the scale is different. Our effective fractal dimensionality relates the initial volume fraction to that scale; this is a physically important parameter. When considering individual fractal

aggregates prior to gelation, several complicating factors besides those mentioned in our paper may have to be taken into account (*e.g.* time-dependent
annealing of bonds between particles). We can, however, only speculate about
these aspects at present.

Dr. S. B. Ross-Murphy (Cambridge): You show a plot of modulus *versus* volume
fraction with a break-point down to a very low value of modulus (even with your
frictionless system). Could you explain what correction was made for *sample*
inertia and surface tension effects for the low modulus case?

You claim no percolation threshold in your systems. Can you compare your
analysis with the measurements of M. Tokita *et al.* [J. Chem. Phys., 1985, **83**,
2583] made with a very high precision rheometer, where they were able to
measure a non-classical percolation exponent? Were these earlier experiments
carried out on a different type of casein gel?

Could you comment on the *highest* value of fractal dimension that can be
measured (even for *non*-fractal systems). I understood this to be *ca.* 2.4–2.5, *i.e.*
not very much higher than your values.

Professor P. Walstra: The measurements at volume fractions below 0.01 were
from creep experiments where sample inertia was negligible. Surface tension
cannot affect the results, but a surface shear modulus can; the effect of the latter is
negligible, however, for modulus values above 10^{-3} N m^{-2}.

As may be clear from the paper, we consider the concept of a percolation
threshold to be meaningless for particle gels of low volume fraction. The results of
Tokita *et al.* for rennet gels are unreliable because they used a constant rennet-to-
casein ratio. At low volume fraction, the aggregation is thus extremely slow,
permitting considerable sedimentation to occur before the flocs become large
enough to form a gel. In our experiments with latex particles, the density
difference was essentially zero.

As far as I understand the theory, it is possible for aggregates to have all values
of fractal dimensionality in the range 1–3.

Paper by Langley, Green, and Brooker

Dr. D. G. Dalgleish (Ayr): Your quoted layer thicknesses of protein around the
included particles in your gels are very much greater than the thicknesses of
protein layers formed at air–water or oil–water interfaces. Can you explain this,
or suggest why the presence of an included particle influences the structure of the
gel in its vicinity to the extent which you claim? Also, why should this vary so
much (from 40 nm to 8 μm, if I remember correctly) with different included
materials?

Mr. K. R. Langley (Reading): The thickness of the protein layer attached to the
hydrophilic glass particles (9 ± 3 μm) and to the corn oil droplets (40 ± 20 nm) is
much bigger than that typically found from air–water or oil–water interfacial
measurements. In our study, the proteins were heat gelled in intimate contact
with the particles at 90 °C for 1 hour. The particles then form an integral part of
the matrix, with fracture occurring at the edge of the attached protein layer. From

our calculations, it would seem likely that, at maximum packing, all the available protein has become adhered to the particles.

Paper by Horne

Professor P. Richmond (Norwich): Can you explain how the approach you describe to light scattering relates to that given by the electromagnetic theory? Do the parameters that describe your diffusion process relate to dielectric permittivities in any simple way?

Dr. D. S. Horne (Ayr): Diffusion-wave spectroscopy is no different from any other light-scattering technique in the sense that it relies on differences in refractive index or dielectric permittivity between the particles and the medium to produce the scattering effects that we monitor. The approach embodies an extension of dynamic light scattering into the strong multiple-scattering régime. Whereas the latter conventionally uses correlation techniques to analyse the temporal fluctuations of singly scattered light, the new technique treats the propagation of light in strongly scattering media as a diffusion process, thereby relating the temporal fluctuations of multiply scattered light to the motion of the scatterers. Whilst providing a marvellously simple physical picture of the ongoing process, the approach does contain many simplifying assumptions—most notably that of uncorrelated disorder within the suspension, and the diffusion approximation itself. As often is the case, the justification of the acceptability of the assumptions was initially provided by the agreement of prediction with experiment. More recently, however, F. C. MacKintosh and S. John [*Phys. Rev.*, 1989, **B40**, 2383] have put the theoretical framework on a more rigorous and formal foundation, deriving a solution for the transport equation whose kernel can be written in terms of eigenfunctions describing the spectral content of the specific intensity within the medium. So, yes, the diffusion approach can be related to the wave theory of light, but not in any simple way.

Dr. W. G. M. Agterof (Vlaardingen): Your paper reminds me of investigations of multiple scattering systems by pulsed lasers in the fempto-second régime. There it has been shown that the scattering angle determines the path length of light in the system. This feature seems to drop out from your equations. Can you explain why?

Dr. D. S. Horne: In optically dense random media, multiple scattering randomizes the direction, phase, and polarization of the incident wave. It is this randomization that permits the successful application of a scalar diffusion theory to the description of the light propagation. Though all knowledge of the incident wave is lost, the solution of the diffusion equation is determined by geometrical boundary conditions set by the experimental configuration. The resultant solution for the back-scattering geometry employed by us is given in equation (1). This is still a function of the mean-free-path for light diffusion in the system, l^*. It is only a mathematical artefact which allows fortuitous cancellation of the l^*-dependent terms when the sample thickness L is very much greater than l^*, and so gives the simple stretched exponential form in equation (2).

Food Polymers, Gels, and Colloids—A Teacher's View*

By G. Stainsby

PROCTER DEPARTMENT OF FOOD SCIENCE, UNIVERSITY OF LEEDS,
LEEDS LS2 9JT

1 Introduction

Most chemists do not consider any food to be a worthwhile topic for proper research, since even a fairly simple food, such as milk, is seen to be a complex multi-component, multi-phase, colloidal system. Those who are bold enough to rise to the challenge realize that the cornerstone for a real understanding lies with a thorough grasp, at the molecular level, of the nature of every component and of the changes and mutual interactions that take place when the complete system is processed and stored. Technological progress, of course, has been extensive even without such a sound scientific base, but inevitably product development in these circumstances is bound to involve considerable time wastage. Increasingly, the food industry compares unfavourably with science-led enterprises, such as those involving synthetic polymers.

What, then, should a teacher of food science impart to the next generation of product developers, production manager, *etc.*? In this article, I aim to indicate how we have tried to answer the question, at Leeds, for the specific areas of food chemistry that are surveyed in this particular meeting. Even this restricted area is too broad to cover properly in the space available, and so I shall deliberately select a few topics that continue particularly to interest me. Perhaps not surprisingly, these topics will mostly involve gelatin, since it is the biopolymer with which I am most familiar. In my view, the protein gelatin is a good example for a teacher to use, even though it is quite atypical of most food proteins in being a denatured, partly-degraded, derived product. The words 'gel' and 'gelatin' are generally synonymous to most people; and gelatin is easy to use, is readily obtained, and is a versatile component because of its compatibility with other ingredients. Gelatin is still the only protein used widely to produce clear food gels, although its pre-eminence in this respect seems to be under challenge from the transparent heat-induced gels that can now be formed from proteins extracted from legume seeds.

* Invited lecture given on the occasion of the award to Dr Stainsby of the 1989 Senior Medal of the Food Chemistry Group of the Royal Society of Chemistry.

Against this development, gelatin is somewhat old-fashioned, but not yet an outcast.

In a more general sense, the teacher of food science is fortunate that it is a new subject (when compared with established disciplines like chemistry and physics, or even biochemistry). During the thirty years or so since it was first thought possible to assemble an integrated discipline at University level in the UK, food science has become a major meeting place between the biological and physical sciences. Though it relies on many other subjects also (see the Appendix), the central feature of the scheme at Leeds is food chemistry. This, in turn, is based on the nature and properties of the components, large and small, derived from animals and plants. The way in which these chemical aspects are integrated into the degree scheme is outlined in Table 1 of the Appendix. For the present purpose, the salient point is that the structure, interactions, and functionality of food biopolymers are studied in honours year 3, and their role in particular food commodities is studied in the final year (year 4).

An important objective of the undergraduate scheme is to provide a sound and comprehensive scientific background to the study of the chemistry of food, beginning with well-established concepts and introducing more contentious issues later on. I shall, therefore, comment first on the main scope of our courses on biopolymers, gels, emulsions, and foams, and mention the approaches we use for some of the topics. On the way, I shall draw attention to those places where greater clarity, in a few selected instances, seems to be needed before the topic can be taught satisfactorily, in the hope that this will provoke research. The article ends with a brief discussion of one food product—ice-cream—to illustrate the extent to which colloid and biopolymer science can now provide understanding of the processing of a representative but rather complex commodity.

2 Biopolymers in General

It has to be said that many students find the detailed study of biopolymers rather tedious. For convenience, this course is divided into three sections, one each for proteins and polysaccharides, and one concerned with general macromolecular solutions. Proteins are dealt with first, starting with the properties of the backbone and then introducing the chemistry of the side chains (see Table 2 in the Appendix). Most of the information the student needs here was available before the degree was inaugurated (*e.g.* the classic work of Sanger, Edman, Moore, and Stein on composition and sequence; the main types of secondary structure; titration behaviour; and classical methods—osmosis, ultracentrifuge, total intensity light scattering—for molecular size).

From the teacher's point of view, the most useful developments over the past thirty years have been the improvements in undergraduate texts generally, and the steady accumulation of information on well-characterized food macromolecules. Some textbooks now are written in a precise style with easy and imaginative wording and quite superb illustrations. Articles in popular journals , such as Anfinsen's account[1] of his Nobel Prize lecture, provide excellent reading for students (and teachers!). New concepts, such as hydrophobic bonding, and new techniques, such as dynamic light scattering, chromatographic methods dependent on molecular volume, isoelectric focusing, and SDS–PAGE, have gained prominence, and have had to be included in the course. Overall, though, this aspect of the course has not required any really drastic revision to maintain its relevance.

Experience has shown that students should have the opportunity of handling space-filling scale models of polypeptides (and, eventually, of polysaccharides). The essential features of the α-helix or the β-sheet cannot easily by appreciated from drawings in books. Nor do such drawings reveal how suitable sequences of side chains can permit a sharp change in the direction of the backbone, and hence the development of a compact tertiary structure. The fact that most food proteins are not simply polypeptide chains, but may contain bound carbohydrate or lipid, also requires elaboration.

The ionization of some of the amino-acid side chains is, I believe, best introduced by first dealing with simple buffer systems. The approach used by Michaelis and others early this century is recommended. This centres around the need for every positive charge in a solution to be balanced by a negative charge, and avoids the specious treatment in so many modern chemistry or biochemistry textbooks, where the ionic properties of water are ignored even when one of these ions dominates. At the first degree level, it is in order to consider an ionizable macromolecule as a set of independent weak acids. Gelatin or highly esterified pectin form useful examples. When the charge density along the backbone is intense, as in alginate, then the effective acid strength of a particular residue is moderated by the presence of charged neighbours and differs significantly from the strength of the residue in isolation as a free monomer. A comparison of the titration behaviour before and after extensive hydrolysis clearly reveals that this is so. A more sophisticated theoretical treatment, involving calculating the electrostatic interactions of charges at fixed sites on the surface of a globular protein, for instance, is not justifiable. The simpler approach is adequate for understanding charge distribution at a selected pH, and for grasping the difference between isoionic and isoelectric situations, once the binding of other ions (*e.g.* Ca^{++}, detergents, dyestuffs) has been dealt with.

A combination of the titration curve, the condition of electrical charge neutrality in every phase, and the requirements for equilibrium between phases, allows the ideas behind the Donnan equilibrium to be tackled. Perhaps not surprisingly, in the Procter Department, this is not dismissed too lightly, but is amplified by at least one full example class. Modern hand calculators take the drudgery out of the calculations for ionic compositions, which require substantial precision when inert salts are present at high concentrations. The student soon learns that weighing can be quite inadequate for the preparation of phases for equilibrium, and that dialysis—correctly carried out—is essential. Just what is being measured when the pH of marmalade or meat, for instance, is being determined is another corollary of this section of work. The shrinkage of whole cherries in jelly provides a nice demonstration of the Donnan effect. It can also be used to show that class B gelatins (isoelectric pH *ca.* 5) are very superior to class A gelatins (isoelectric pH *ca.* 9) for this purpose, as the requisite pH for optimum flavour is close to 4. Cherries with damaged skins do not shrivel at any pH, of course.

A more theoretical course on macromolecular solutions, occupying about one half the time, runs parallel with the protein course. Topics covered include the thermodynamic non-ideality of such systems, even when very dilute, the behaviour of more concentrated solutions, phase separations of mixtures of biopolymers, and adsorption theory (especially the distinctions between isotherms for small surfactants and polymers). Although the adsorption of a disordered protein such as β-casein may conveniently be treated like that for a synthetic homopolymer, the likelihood that a rather different situation is found for ordered food proteins, perhaps especially at the oil–water interface, has to be discussed. It is at this stage that the limitations of our present understanding of macromolecular interfacial conformation are brought into focus. The question of reversibility, for macromolecular adsorption, provides another contentious topic, and has the merit of highlighting the distinction between thermodynamics and kinetics.

Polysaccharides are dealt with after proteins. Students first need to gain familiarity with ball-and-stick models to discover the limitations of Haworth projections for monomers and, more especially, polymers. Then, the so-called Reeves projections—though the idea seems first to have been proposed by Sachse in 1890—are introduced, and the restriction on rotation in the backbone considered. Whilst the rather limited chemical reactivity of the groups on the monomers is readily understood, most students have considerable difficulty with the geometrical subtleties of polysaccharide chains. An important section of this course is the extension of earlier ideas on the properties of macromolecular solutions to take into account of the effects of polydispersity, and, for amylopectin and galactomannans, for example, the effects of chain branching. One example is the contrast between the needle-sharp boundaries of polysaccharide solutions, in sedimentation velocity experiments, and the apparently normal peak for commercial gelatin, which, though highly polydisperse, looks like that for a monodisperse polymer.

Gelatin—Two biopolymer systems in particular, starch and gelatin, have been investigated now for over a century, but, as yet, neither is fully understood. Though starch research seems to be in a healthy state, there has been only limited progress in gelatin research in recent years. A novel and interesting suggestion (see p. 455) about the nucleation of the folding of gelatin chains, however, should renew interest in the self-association to form gels, and in interactions in interfacial layers. As proline residues play a dominant role, it would seem to be worthwhile to consider the implications for interfacial caseins also, since these polymers, like gelatin, are rich in proline amino-acid residues. Progress will not be made, though, in my opinion, if researchers continue to use commercial gelatin. This is highly complex, both in chemical and in physical terms; this is because of the nature of the starting material, collagen, and because the hydrolytic procedures that are commonly used to solubilize this extensively cross-linked protein lack selectivity. Ideally, the cross-bonds between the ordered collagen monomers would first be broken, and the monomers then disorganized gently to release intact the component chains. A very small proportion of the collagen monomers in a tissue—those which have been incorporated most recently—are held together by only feeble cross-bonds. These monomers are quite readily extracted in mildly acidic conditions, and provide a source of intact chains on denaturation. Under suitable conditions of pH, ionic strength, and temperature, such chains can reassemble into perfect collagen monomers provided the solution is sufficiently dilute. It is thought that renaturation begins from the C-terminus, where there is a sequence, of some fifteen residues, which is particularly favourable for the specific folding of collagen. Once this localized folding has occurred, the rest then follows quickly. The analogy with a zip fastener is often used to describe the overall process.

The collagen of commerce, however, is held together by mature cross-bonds, and cannot be dissolved so readily. In gelatin making, some cross-bonds and backbone peptide bonds are broken to yield a product having an extremely wide spectrum of molecular weights. A typical sample includes intact single chains (95 000 daltons), a spectrum of fragmented chains, and a range of oligomers of up to at least 15–20 chains and fragments still held together by the native covalent cross-bonds.[2] Not surprisingly, the connection between functional activity and size distribution remains obscure. The problem is compounded by the fact that the polypeptide chains in the mature tissues used commercially are not all chemically identical.[3] Mature collagen monomers consist of two α1-chains and one α2-chain. The two types have considerable homology, as one would expect from their self-assembly to the ordered monomer, but the small differences in composition produce an importance difference in organizational stability. Thus, an artificial monomer of α1-chains denatures at *ca.* 34 °C, whilst one composed entirely of α2-chains denatures at *ca.* 22 °C in the same solvent. It is suspected, but not proved, that α2-chains hydrolyse more readily than α1-chains, after disorganization of a native monomer, so that the fragmented chains in a commercial gelatin could be mostly α2-chain fragments. Whether there are differences in surface activity or gelling ability between intact α1- and α2-chains has still to be studied.

Chain length is important for gelation as the gelatin quality (the potential 'strength') of a sample seems to depend on the proportion of intact chains of both kinds.[5] The function of the oligomers is to speed up the setting rather than to add to the eventual strength. In a typical food gel, at room temperature, some 30% of the gelatin is still free in solution and can be removed by dialysis or ultracentrifugation.[6] This 'sol' fraction is actually incapable of gelling—it is not simply composed of the lower molecular weight chain fragments.

Chemical studies have shown[7] that a good quality gelatin contains many fewer imino end-groups (from proline or hydroxyproline) than would be expected from random chain scission (1 in 25 are found; 2 in 9 are expected). In the sol fraction from such a gelatin, on the other hand, the proportion of chains with terminal proline or hydroxyproline residues is very large. These results are in accord with the idea[8] that nucleation of the collagen structure, which has been shown (by *X*-rays and infra-red dichroism) to be present in a set gel, involves the C-terminal sequence, just as in renaturation in very dilute solutions. In the gel, however, the renaturation is only partial, and provides the junctions in the network. (I am pleased to note the aptness of the expression 'frustrated renaturation' recently introduced[9] to describe the struggle to associate.)

Studies of gel melting, however, have led to an alternative model.[10] This also requires proline residues for the nucleation stage. A meta-stable hairpin bend brings two sections of the same chain into close spatial proximity, and is stabilized by interaction with another chain to form the start of a collagen helix (junction zone). A hairpin bend, which turns the backbone through 180°, consists of four amino-acid residues. Positions 2 or 3, on the outside of the bend, could be occupied by proline. Hydrogen bonding is needed, between the peptide bonds of residues 1 and 4, to give the turn some stability. Although the authors do not discuss it, an important feature of the collagen helix is that the three component chains are each staggered along the helix by one residue, so that glycine is never directly alongside glycine, but is always bonded to another type of residue. This means that glycine could not occupy position 1 (and 4) in the sharp bend, but would have to occupy either position 2, or position 3, to get the intra-chain alignment suitable for subsequent helix formation. The turn, therefore, has to utilize a triplet with only one imino residue, and with suitably flexible neighbouring sequences. (There are several regions like this along the chain.) This type of nucleation, therefore, is very different from the earlier suggestion that adjacent proline-rich triplets are necessary. In order to decide definitively between these views of the mechanism of nucleation in highly entangled systems, it will, in my opinion, be essential to study the behaviour of homogeneous chain fragments of known sequences. Methods for isolating intact α-chains, and for breaking them selectively into fragments which can be readily separated, have been available for a long time. They have had to be reliable, or the complete sequence along the α-chain could not have been determined. Once the role of each region along a chain has been established, the functionality of traditional gelatin will be properly understood. It should then be possible to aim for a 'designer gelatin' with improved gelling power and interfacial activity. And, on the way, it may be possible to overcome the main defect of traditional gelatin—the unacceptably low melting temperature for some uses—without changing the 'mouthfeel' which sharply differentiates gelatin from other denatured proteins and also from polysaccharides.

3 Functional Activity of Food Polymers

The foundation courses on the individual biopolymers having been completed, the teacher now tackles a more challenging problem—how to deal with the basic physico-chemical concepts behind thickening, gelation, emulsification, and foaming. Thickening is the least difficult topic. The salient features for the simplest systems, with one type of biopolymer, are (i) pH stability (we note that the glycosidic bonds between neutral sugar residues, unlike those involving acidic residues, readily undergo acid catalysed hydrolysis) and (ii) the concentration dependence of the rheological properties. Thickening can be followed by gelation, so the interactions that give 'body' in the more complex systems, containing several biopolymers and fillers, can be discussed later. It is salutory to remind the student that Alfred Bird in 1837 patented the replacement of protein by polysaccharide (egg by starch) in order to thicken milk in a more convenient way. What a gap there was in those days between technological innovation and scientific understanding!

4 Food Gels

The simplest kind of food gel—a table jelly, for example—is a protein or polysaccharide network holding a large volume of water containing low molecular weight solutes. Ferry's review[11] on the association of disordered chains to network gels, therefore, forms a good starting point, with gelatin as the example. Emphasis on the interrupted, periodic chain structures of polysaccharides links the topic to the earlier part of the course, and provides a molecular description of the ordered zones in gels, in terms of flat ribbons, buckled ribbons, hollow helices, and so on. Although agar, like gelatin, is versatile and needs only water, other polysaccharides require specific co-solutes for gelation: sugars for high methoxy pectin, divalent ions (usually from milk) for alginate, potassium ions for carrageenan, *etc.* These requirements highlight the types of inter-chain bonding in the ordered zones. The general transparency of the gels relates to the limitation of the size of structured regions. Globular proteins usually give opaque gels (*e.g.*, boiled egg), and are naturally the next to be considered. Weight for weight they are very ineffective gelling agents when compared with polysaccharides. Over the lifetime of the undergraduate course, the earlier simplistic view that fully denatured chains associated through the formation of β-sheet structures must now be modified. It is now held that only partial denaturation of globular proteins is needed. When the β-sheet content is determined, it does not seem fully to correlate[12] with the extent of aggregation, and additional forms of inter-chain association—particularly hydrophobic bonding—need to be considered. The contrasting rheological properties of these types of simple gel, the syneresis, and the behaviour on freezing and thawing, are all discussed to bridge the gap from molecular structure to consumer acceptability. This section concludes with aspects of quality control, both chemical and rheological, particularly for the plant and algal gelling agents as they are the ones that exhibit the greatest variability. It is at this point that the extracellular microbial gelling agents, xanthan and gellan, with their potentially greater consistency and reliability of supply, are considered.

Mixed Networks—Most food gels, however, are not so simple; they may contain more than one type of biopolymer species, as well as deliberate inclusions (bubbles, solid particles, *etc.*) in order to provide the required texture. Following the nomenclature used by Morris,[13] they can be conveniently divided into mixed network gels and filled gels, and the mixed networks further divided into phase-separated and coupled gels, as extreme types.

A mixture of agar + gelatin, which has long been used in meat pies, is an example of a phase-separated gel. Here the two biopolymers are mixed and then gelled, when they compete for the available water and contribute independent networks. The properties of these two networks are very different, particularly the temperature dependence, so that the shear modulus falls sharply on heating, for instance, when gelatin is the main component, but the agar prevents complete melting up to the boiling point. Selective staining for globular protein + agar mixed gels has demonstrated the microscopic thermodynamic phase separation. Most pairs of non-associating biopolymers, at concentrations favouring their thermodynamic incompatability, will give phase-separated gels provided the separation is not inhibited on kinetic considerations.

Favourable interactions betwen biopolymers, however, can lead to coupled network gels; here a single network results with new rheological properties. Perhaps the most common examples have sometimes been ascribed to synergism. There are often savings on costs, and there are usually new textures as well. Examples include the case of alginate + pectate, where the diaxially linked galacturonic residues in pectate are geometrically compatible with the diaxially linked guluronic residues of alginate and which, in the absence of preferred interactions with calcium ions, provide gels in the pH range most favourable for fruit flavour. Synergism between carob gum (a galactomannan) and helix-forming polysaccharides (such as agarose, κ-carrageenan or xanthan) is used in the food industry, though it is not always clear (to me, at least) whether the mixture is phase-separated or coupled. The presence of coupling is in no doubt, of course, when the biopolymers are held together by covalent bonds, but is uncertain when, as in the above examples, molecules interact by secondary forces. Gum arabic is a good example of a coupled natural biopolymer, and no doubt many other extracellular microbial gum exudates are examples of polysaccharide coupled with protein. Though gum arabic does not gel, the covalent bonding between protein and polysaccharide is important in its interfacial behaviour through the linking of an emulsifier (*i.e.* protein, to provide short-term stability) and a stabilizer (*e.g.* polysaccharide, to ensure long-term stability). This aspect is considered in detail elsewhere in this volume (see p. 135).

The main advantage of covalent bonding is that, when a gel is formed, there is a single hybrid network which is thermostable. On prolonged heating, its gel-like character is lost only through the disruption of chemical bonds. What may then look like melting is in fact degradation, and subsequent cooling will not lead to setting again. Only a very limited number of cross-bonds need to be introduced to create a suitable gel. If there is no additional linking by secondary forces, then the gel behaves like a simple rubber. It is not generally worthwhile, however, to try to calculate the elastic shear modulus, which depends on the number average molecular weight between cross-links, as some of the network consists of one biopolymer whilst the rest is composed of another. The fraction of each in the mixed network would also need to be determined. It cannot simply be presumed that the gel is formed from the components in the proportions that were present before cross-linking.

An example which has been studied in some detail,[14] and which has found use in edible products,[15] is the coupling of gelatin to alginates which have been partially, but highly, esterified with propylene glycol. (This idea was conceived originally to prevent the melting and to control the swelling of the 'emulsion' during the development of photographic plates and films.) The reaction proceeds smoothly, when the protein and ester solutions are mixed, at a rate that is very sensitive to pH and temperature. Under the optimum conditions for cross-bonding, the ester is also readily hydrolysed, leaving unreactive alginate in the gel. The main reactive site on the protein is the ε-amino group of lysine. If this is blocked chemically, then gel formation with alginate ester is inhibited. The few lysine residues that are lost through coupling pose no nutritional or digestive problems. Almost all the sites for attack by the digestive enzymes are still readily available.[16] The cross-bonds are very stable to the acidic environment in the stomach, but their stability during the subsequent stages of digestion does not seem to have been investigated. The strongest gels are formed with disordered proteins (*e.g.* gelatin, caseinate), but weak gels can be also made using ordered globular and fibrillar proteins. If these are first partially denatured, then stronger gels form. It seems to me to be a particularly beneficial way of utilizing unattractive protein waste (providing it is safe for human consumption, of course!) to create attractive, shaped, freeze-thaw stable components of food-stuffs. In the absence of non-covalent association, the gel is rather chewy and 'short', *i.e.* it has hardly any plastic flow and breaks apart when the deformation is very small. It is rather like the sort of agar jelly used in microbiological work. In coupled gelatin gels, where hydrogen bonding as well as covalent cross-links co-exist, the system is less chewy at room temperature. Only the cross-bonds remain on heating, however, and so it is rather similar to a gelatin + agar mixed gel.

When alginate esters having different proportions of mannuronic and guluronic residues are used, the strongest gels form from the ester with the highest proportion of mannuronic residues.[17] These are linked to one another, in sections of the backbone, di-equatorially to give a more flexible and extended configur-ation than the guluronic sections, where the glycosidic links are di-axial. Propyl-ene glycol pectate, which has no di-equatorial links, cannot make covalent bonds with proteins in aqueous solution. Since the alginate ester reaction is highly inefficient, through simultaneous de-esterification, it would be useful to know whether the mannuronic ester hydrolyses less readily than the guluronate. Attempts to determine these relative rates have so far not provided an unambig-uous answer, and so it is not certain that the superiority of the mannuronic ester is solely due to geometrical considerations.

Filled Gels—Food gels are even more complicated when they are filled—with bubbles, with solid fat, with crystalline ice, or with cellular material such as swollen starch granules. Model experiments[18,19] have demonstrated how the mechanical behaviour of a gel, or a viscoelastic solution, is affected by the volume fraction, the shape, and the deformability of the inclusion. Gelation and thicken-ing processes with starch granules provide simple examples. The gelation of starch alone, and the subsequent retrogradation when stored, are also usefully considered from this viewpoint. Anisometric fillers are particularly effective for stiffening solutions and gels. Microcrystalline cellulose, with its nutritional inertness, is perhaps the simplest material to give 'body'. The production of anisotropy through the distortion of droplets by flow, prior to gelation, has been suggested[20] as a novel way of making fabricated foods such as simulated 'meats'. When gases or lipids are present, the adsorption of the biopolymer also needs to be considered. Model experiments have shown[21] how loss of gelling agent by adsorption can weaken the gel. The filler then gives negative re-inforcement, which can be overcome if additional gelling agent can be included in the formulation.

Clearly, the preceding material represents more than enough subject matter on gels for a teaching course, and some careful selection is necessary. Since a good student expects topicality, it might now also be appropriate to pay some attention to fibrinogen and blood clotting, in view of the recent publicity* over one type of reformed 'steak'. Whilst there is plenty of information on fibrinogen, there is unfortunately still too little information on the major fibrillar protein in a carnivorous diet, myosin, for this to be a suitable undergraduate topic.

5 Food Colloids

This topic is one where there has been significant progress during recent years. At the start of my career, there were very few fundamental papers of relevance, despite considerable technology. The research by Haydon and colleagues on protein adsorption, or by Cumper on protein foams, for instance, seemed to me to be only peripheral to real food systems. The systematic interfacial studies of Graham and Phillips, on really pure single food proteins, and the pioneering work of Musselwhite on competition between proteins, has catalysed interest in research and has set a landmark for teaching. Interfacial behaviour is now being described in terms of the competition between biopolymers and surfactants to secure a place in the interfacial region, and the co-operation to provide the necessary film stability. Methods for producing emulsions are discussed, and those for determining the droplet-size distribution are assessed. Loss of acceptability in emulsions through creaming, flocculation and coalescence is considered in some detail, as are chemical and physical techniques for determining the composition of the interfacial film and its slow change with time. Parallel considerations apply to foaming behaviour. Here, there are the additional complications of instability involving increase in bubble size through transfer of gas (Ostwald ripening) and instability by rapid drainage of the liquid between the bubbles. Interactions between components in the adsorbed films relate back to protein–protein and protein–polysaccharide interactions in solutions and bulk gels.

It is evident from this conference proceedings that all the above issues still command the attention of researchers. The two earlier volumes[22,23] in this series, together with an edited review volume covering recent advances,[24] give some indication of the level of world-wide interest in the surface functionality of food biopolymers. These books also show how the teacher has been quite overwhelmed with new information in the past few years! Inevitably, in a period of such rapid growth, there are unresolved and contentious issues that are best not revealed to an undergraduate student. Opacity can persist, too, in quite basic areas such as the connection between residue sequence in proteins and their surface functional behaviour. For many years, it was quite widely believed that, once the sequence had been determined, it would provide the key to our understanding of the native structure, and also the key to functionality once the effects of the solvent environment had been taken into account.

*See, for example, 'The Guardian' newspaper, 13th February 1990, p. 3.

When single proteins are used, emulsification and foaming first require rapid adsorption, and then sufficiently rapid intermolecular interaction to provide a coherent and flexible film that will protect the fine droplets, or bubbles, from coalescence. Whilst rapid adsorption at a newly created interface is essential, it is not the sole functional quality that is needed. If it were, then β-casein would be a superb foaming agent. Unfortunately, it forms rather feeble films.

In order to account for the behaviour of the wide range of native and modified proteins (and hydrolysates) that have now been studied, it has become customary to think in terms of solubility, size, hydrophobicity, and structural flexibility. Net charge, too, is considered, and this is reasonable as a low net charge at the interface presumably favours intermolecular association, just as in bulk solution. Solubility, which should relate to composition and sequence, seems also to be important. Effective macromolecular size must matter when the rate of migration to the new interface is a limiting factor. Hydrophobicity and flexibility, however, are more elusive issues. Soon after the concept of surface hydrophobicity became popular, it was shown to correlate well with emulsifying power but not with foaming power[25] for globular proteins. Overall average hydrophobicity, as determined by Bigelow's method,[26] seems to correlate better with foaming power,[27] implying perhaps that there is more denaturation at the air–water than at the oil–water interface. However, a careful comparison of the various experimental methods for assessing surface hydrophobicity has since shown that there is little satisfactory agreement between them,[28] and it has raised serious doubts on the validity of correlations of this type with functionality. The effects of structural flexibility are also not really clear. Adding additional cross-links, to increase structural stability, has been shown[29] to decrease foaming power.

Careful studies of the differences in foaming power between lysozymes from different sources,[30] which differ by only a few residues, convincingly demonstrate that explanations in terms of an average hydrophobicity, or an overall free energy of complete denaturation, are inadequate. Even more striking is the recent study[31] which shows that modifying only a single residue in a protein, and a residue at that which is well removed from the surface in the native structure, can significantly affect foaming power.

The current lack of understanding for systems containing well characterized single proteins is compounded when real foods, containing many biopolymer components, including polysaccharides, and low-molecular-weight surfactants, are considered. A succinct review[32] of the present situation for individual proteins, and for pairs of positively and negatively charged proteins, which give mixed films with enhanced strength, shows just how much more systematic research is needed. One curious feature is that class A gelatin, which has an isoelectric pH near 9, and is therefore positively charged in neutral solution, does not seem to have been considered. In economic terms, class A gelatin must surely rank higher than lysozyme or clupeine, or any protein deliberately modified to raise the isoelectric point.

In view of the overall unsatisfactory situation with regard to protein functionality, an alternative way of looking at the problem is needed, and it might be helpful to direct attention more to the details of structural stability. In order to do so, we must stop thinking in terms of the equilibrium or static structures, which have been derived from X-ray diffraction patterns, and think more in terms of delicately balanced dynamic structures which are perturbed by interaction with ligands, and, I suggest, by surfaces. The stability of the secondary structure is, in a globular protein, greater than that of the tertiary, unless the latter is strongly maintained by sufficient covalent cross-bonds. Moreover, the weakness of the non-covalent constraints close to the surface of the molecule allows quite sizeable local fluctuations in structure to occur.[33] It seems reasonable to expect that these are the first sites for structural reorganization (partial denaturation) on adsorption. The presence of cross-bonds would determine the extent to which further disorganization could be induced. In these limited denaturations, hydrophobic residues from within the molecule would become exposed, even though secondary structures were maintained. Some (all?) of the current experimental techniques for determining hydrophobicity could well disturb the system. Moreover, the free energy of denaturation, as presently measured, need not relate at all to that actually involved in the limited disorganization on adsorption. What is needed, for a start, is information on the proportions of ordered structures that persist during adsorption, and afterwards. It may be possible to base methods on those[12] used for studying the gelation of globular proteins, though it could well be necessary to use radio-labelled proteins, and this might create uncertainties in interpretation, unless one can be certain that the labelling has no significant influence on the surface functionality. Even when a more detailed description of an adsorbed conformation is available, a full understanding of surface and interfacial functionality will emerge only when reliable values for the thickness of the film, and for the distribution of the components within the film, are known.

Ice Cream—Although, as was mentioned at the beginning of this article, milk may seem a quite complicated system to the ordinary chemist, it is now possible to understand much of its colloidal behaviour in terms of fundamental physico-chemical concepts.[34] Our current appreciation for most other food colloids, however, is more rudimentary, as they are even more complex.

Ice-cream is one such product. It is a good example from the teacher's viewpoint: there is a lot of published information, and it has been nicely summarized.[35,36] But there are still a number of intriguing questions to be answered, though I suspect the manufacturers know much more than what has been released already. I shall draw attention to some of the issues by discussing ice-cream making in the light of recently published work on colloids. Only standard, supermarket vanilla ice-cream will be considered. In this way the complications of the natural fat globule membrane in dairy ice-cream will be avoided, as will those arising when the pH is lowered to enhance fruit flavours, or when the texture is changed to give the 'still-quite-soft-from-the-freezer' variety. A student would also have to consider legal and nutritional factors and aspects of processing, but none of these is included here.

Ice-cream is taken to be a dilute, sweetened, oil-in-water emulsion which has been frozen and aerated to form a solid containing some 50 vol% air. In a typical formulation of the supermarket vanilla type, there is *ca.* 9 wt% vegetable fat, 14 wt% added sucrose, 4 wt% glucose syrup solids, 9 wt% skimmed milk powder,

1 wt% whey powder, and 0.5 wt% mixed biopolymer stabilizers and low molecular weight emulsifiers. The balance, some 62.5 wt%, is water, except for very small amounts of colours and flavour.

In essence, the processing of ice-cream involves first blending all the ingredients with heating, to obtain a coarse emulsion, with droplets up to about 15 μm diameter. This blend is then pasteurized (for, say, 20 s at 81 °C) and a fine emulsion is created, whilst it is still hot (70–75 °C), by means of two-stage homogenization. For reasons of microbiological safety, the fine emulsion is cooled rapidly, and then aged at about 4 °C before it is simultaneously aerated and frozen. It leaves the freezer as a semi-solid at about −6 °C, and is fully hardened and stored at about −25 °C, where some 90% of the water is present as ice crystals. These crystals, and any sugar (especially the poorly soluble lactose from the skimmed milk powder) must not be large enough to give a gritty 'mouthfeel'. Typically the crystal size ranges from 25 to 50 μm, with some 8×10^6 crystals per gram of ice-cream.[35] Of similar importance are the sizes of the air bubbles: 60–100 μm is an acceptable size range. The number density of air bubbles is similar to the number density of ice crystals. The correct proportion of air gives a smooth, light product. Too little air gives a soggy feel to the product, whilst too much makes it crumbly. The fat globules are much finer, averaging *ca.* 0.5 μm diameter, and much more numerous (about 2×10^5 globules for every air bubble), so that before aeration the mix has a smooth, creamy texture. The fat is a blend which has a higher solid fraction than milk fat just prior to aeration. Needless to say, the component fats and oils must have bland flavours and odours. An important function of the sugars is to balance the fatty taste with just the right amount of sweetness. The simple sugars lower the freezing point to about −2 °C.

So, to what extent can colloid science account for the nature of the product? Quite a lot is known about the mix, in the state before freezing and aeration. The particle-size distribution in the mix[37] is not unlike that of a cream liqueur,[38] which is a sweetened emulsion with about the same total solids content as ice-cream, stabilized by dairy protein and added emulsifier. As only a very small proportion of the droplets have diameters of 2 μm or more, the mix is stable to creaming and flocculation over at least several weeks. In a refrigerator (+4 °C), the mix eventually deteriorates through microbiological action rather than emulsion instability. Any tendency to creaming is reduced by the stabilizers in the aqueous phase—these are usually blended from guar gum, locust bean gum, alginate, and sodium carboxymethylcellulose—but this cannot be their principal function. The rheology of the emulsion during freezing and when it is consumed, and the slowing of ice crystal growth during storage, are much more important aspects for the stabilizers.

In the creation of the mix, there is more than enough protein available for all the oil–aqueous interface. Indeed, the ratio of protein to fat is about twice that for a cream liqueur, and this is considered to be very high.[39] Only about one half of the protein in a cream liqueur is adsorbed: the remainder has been found in the aqueous solution between the droplets.[38] No information on the protein content of the aqueous phase of ice-cream mix has been published, but it would be reasonable to presume that the depletion by adsorption is not very different from that for a cream liqueur, even though the globules in the latter are protected initially by the natural membrane. Analysis of membrane recovered from ice-cream mix, prior to aeration, has shown that all the available proteins are present.[40] The relative proportions suggest that β-casein is preferentially adsorbed. Electron micrographs have indicated[35] that micellar sub-units and whole micelles are adsorbed in addition to individual molecules. This is to be expected in the turbulent conditions that prevail in homogenization when the pressure suddenly falls.[41] As in a cream liqueur, the interface no doubt also contains added emulsifier (usually commercial glyceryl monostearate, with a monoglyceride content of 40–60%) together with the similar mono- and di-glycerides that are inevitably present at low levels in the fat. The added emulsifier is fat soluble, and will have migrated rapidly to the new interface.

The hydrodynamic thickness of the interfacial layer is 60–90 μm.[37] The arrangement of the various components within this layer is still, however, subject to speculation. Earlier views held that these were in concentric layers, with surfactant next to the oil and proteins on the outside. The proteins, in turn were assumed to be layered, with disordered and denatured proteins inside native proteins. It has more recently been surmized,[42] from the rheological behaviour of adsorbed films at a planar interface, that milk proteins and mono- and di-glycerides form complexes when adsorbed. These would tend to anchor some of the protein next to the fat in the ice-cream mix, and additional protein would be held by protein–protein attractions. The ratio of added emulsifier to fat is about twice as high in ice-cream as it is in a cream liqueur, but in neither case is it known how much emulsifier is adsorbed and whether the boundary next to the fat is entirely uniform with respect to its chemical composition. A mosaic description has been favoured by some authors,[35] and this seems sensible where large protein aggregates (micelles or sub-units) are adsorbed as well as single proteins and emulsifier. In any event the droplet integrity is maintained mainly through steric stabilization. The ionic strength, derived mostly from the minerals in the milk powder and whey, is high enough in the mix to minimize the charge stabilization.

None of the published work deals with the freshly-made mix. No doubt structural rearrangement within the film take place during the cooling and ageing, as with any emulsion, and lead to improvements in film strength. An unusual feature for ice-cream mix, however, is the extent of shrinkage of the droplets. It is estimated that the surface area decreases by some 30%.[35] Part of this occurs during ageing, as the crystal form of the solid fat that develops on rapid cooling slowly adopts a more stable state. Non-adsorbed emulsifier would no doubt be a complication here, by modifying the association of the triglycerides, but the significance of this is not evident from published work. The fatty tails of adsorbed emulsifier would also interact with nearby triglycerides, though some are thought[42] to be involved in the association with protein (casein). What is also obscure is how the large decrease in interfacial area affects the structural organization of the adsorbed film.

The aged mix, as mentioned above, is quite stable when held above the freezing point of the aqueous phase. What is not known, from published accounts, is whether the mix is stable at sub-zero temperatures. A cream liqueur is very stable[43] at temperatures low enough for the fat to become mostly solidified, so that the liqueur resembles ice-cream mix. (The liqueur does not, of course, contain ice-crystals.) Only when the cold liqueur is simultaneously stirred, though not sheared as in aeration, does instability develop. This suggests that instability in ice-cream mix stems from the high shearing action, rather than the low temperature in the freezer, though the loss of water as ice will enhance the effect.

In the freezer, about one half of the water in the mix turns to ice, over some 20 seconds, causing the ionic strength to rise and the pH to change, whilst air is incorporated. It is at this stage that the main textural features of ice-cream are produced, and where defects can arise. An important point to remember is that the foam needs only to have a fairly transient stability, as the bubbles are quite quickly prevented from coalescing by the solid matrix of ice crystals. If the foam is too stable, and survives melt-down, an undesirable 'head' is seen. What is of more importance than long-term stability is the size of the air bubbles and ice crystals.

It is unnecessary for the newly-formed air–water interface to compete for protein with the oil–water interface of the mix, as there is more than enough protein still available in solution. Some destabilization of the emulsion occurs, as clumped droplets[35] can be seen by low-temperature microscopy, and this may be desirable texturally. If carried too far, however, the system in effect is churned, and there is an unacceptable 'buttery' taste, and a rather watery serum on melt-down. In this respect, there is a major difference between ice-cream and whipped cream, since in the latter some liquid fat, coming from shear-induced destabilization, is needed to provide the matrix holding the incorporated air bubbles apart. There are also significant differences between ice-cream and cream, for whipping, even though the sequence of events seem similar.[44] The fat content of cream is very much higher, the fat is mostly liquid, and the globules are larger so that they are destabilized by shear more readily. In addition, the natural membrane protecting the globules in cream is very different from the man-made membrane in ice-cream. Homogenized milk, suitably concentrated, would be much more like ice-cream mix!

In ice-cream, some of the instability to shear is likely to arise from the complex structure of the adsorbed layer. The description of this as a mosaic implies that the layer could have weak and strong regions. In the absence of added surfactant (emulsifier), the adsorbed layer should be more uniform. But can good ice-cream then be made? Certainly the presence of an emulsifier decreases the amount of adsorbed protein, as in a cream liqueur,[38] but in the latter case it is not simply the amount of adsorbed protein that matters: a liqueur made with no emulsifier is less stable than the usual formulation. Rather it is the interaction of the emulsifier with adsorbed protein which underlies the destabilization of the emulsion when it is sheared.

A further cause of instability in the freezer seems to derive from the crystalline fat at the boundary between the globule and the adsorbed layer. If these fat crystals are needle-like, and in a suitable orientation, then the protein layers may be pierced when the solid globules collide, and so globule–globule interaction will be promoted.[36] In the turbulent situation in the freezer, globules may also be damaged by ice-crystals. But only a limited amount of liquid fat emerges, and it is not enough to spread over much of the surface of the air bubbles.

The changes in ionic strength and pH, through the loss of water as ice, are bound to affect the properties of the adsorbed layers at both types of interface in the frozen mix. In particular, the calcium ion concentration is important, through the known effects on calcium-sensitive caseins. It seems surprising that there appear to be no published values of pCa for ice-cream at any stage of its manufacture. In cream liqueurs, ionic calcium has to be controlled by sequestration with citrate,[45] to achieve the necessary emulsion stability, and pCa (and pH) are crucial factors for optimum stability.[43]

No less important than the changes in the emulsion droplets are the changes in the aqueous phase. Only when the temperature falls low enough, to harden the product, is there any chance of completely solidifying the aqueous phase, as a glass, and it is by no means certain that this ever occurs as different authors have given very different values for the temperature needed to bring this about.[36] It is safer, at present, to think in terms of a concentrated, extremely viscous solution in which non-specific interactions between polysaccharide stabilizer and residual proteins occur. Interactions requiring ionic bonding, though, are not likely to be important, in view of the ionic strength. About 10% of the water remains liquid, in hardened ice-cream, and provides the medium through which coarsening of the ice crystals occurs as the temperature slowly rises and falls. The rate at which water molecules diffuse in this highly concentrated solution remains to be established. This is most probably the rate-determining step in the Ostwald ripening of the crystals.

It is often asserted that the initial size of the ice crystals is determined, in part at least, by the stabilizers; but no conclusive evidence is available. The rate of nucleation of ice in sucrose solutions has been shown not to be changed, within experimental error, when the usual stabilizers for ice-cream are present,[46] even at levels much higher than those reached when ice-cream leaves the freezer. It seems that the original population of crystal sizes is determined by the operation of the freezer itself—a very thin layer of mix is frozen extremely quickly, and then fractured by the scraper to give the crystals.

In this account, as in others on ice-cream, the sucrose has been ignored as a feature of the colloidal behaviour. However, sucrose is well known to be important in enhancing the stability of food foams, and it has been shown[47] that, at very high sugar concentrations, this could be due to an increase in the strength of the adsorbed protein film. As sucrose is a non-reducing sugar, it cannot interact covalently with proteins, and so the mode of interaction to cause the improved rheological behaviour is by no means clear. It is tempting to say that sucrose promotes denaturation to achieve the observed effect, but, at a level similar to that used in ice-cream, sucrose appears to protect β-lactoglobulin against thermal denaturation.[48] There seem to have been no detailed studies of protein films adsorbed at the oil–sucrose solution interface. As foams and emulsions are of such widespread occurrence in dessert products, surely it is time that this role of sucrose, and other sugars, was investigated?

References

1. C. B. Anfinsen, *Science*, 1973, **181**, 223.
2. P. I. Rose, in 'Theory of the Photographic Process', ed. T. H. James, Macmillan, New York, 4th edn, 1977, Chap. 2.

3. K. A. Piez, in 'Biochemistry of Collagen', ed. G. N. Ramachandran and A. H. Reddi, Plenum, New York, 1976, p. 1.
4. C. Tkocz and K. Kühn, *Eur. J. Biochem.*, 1969, **7**, 454.
5. W. M. Marrs, Leatherhead Food R. A. Report, 1984, No. 461.
6. P. Johnson and J. C. Metcalfe, *Eur. Polym. J.*, 1967, **3**, 423.
7. R. J. A. Grand and G. Stainsby, *J. Photogr. Sci.*, 1975, **23**, 67.
8. G. Stainsby, in 'Science and Technology of Gelatin', ed. A. G. Ward and A. Courts, Academic Press, London, 1977, Chap. 6.
9. M. Djabourov, *Contemp. Phys.*, 1988, **29**, 273.
10. J. P. Busnel, S. M. Clegg, and E. R. Morris, in 'Gums and Stabilisers for the Food Industry', ed. G. O. Phillips, D. J. Wedlock, and P. A. Williams, IRL Press, Oxford, 1988, Vol. 4, p. 105.
11. J. D. Ferry, *Adv. Protein Chem.*, 1948, **4**, 1.
12. A. H. Clark and C. D. Lee-Tuffnell, in 'Functional Properties of Food Macromolecules', ed. J. R. Mitchell and D. A. Ledward, Elsevier Applied Science, London, 1986, p. 203.
13. V. J. Morris, *Chem. Ind.*, 1985 (March 4), 159.
14. G. Stainsby, *Food Chem.*, 1980, **6**, 3.
15. R. P. Carpenter, R. B. Weddle, and F. W. Wood, U. K. Pat. No. 1,443,513, 1976.
16. S. B. Mohamed and G. Stainsby, *Food Chem.*, 1984, **14**, 1.
17. J. E. McKay, G. Stainsby and E. L. Wilson, *Carbohydr. Polym.*, 1985, **5**, 223.
18. S. Ring and G. Stainsby, in 'Gums and Stabilizers for the Food Industry', ed. G. O. Phillips, D. J. Wedlock, and P. A. Williams, Pergamon, Oxford, 1982, Vol. 1, p. 323.
19. R. K. Richardson, G. Robinson, S. B. Ross-Murphy, and S. Todd, *Polym. Bull.*, 1981, **4**, 541.
20. Y. A. Antonov, V. Ya. Grinberg, N. A. Zhuravskaya, and V. B. Tolstoguzov, *J. Texture Studies*, 1980, **11**, 199.
21. E. Dickinson, G. Stainsby, and L. Wilson, *Colloid Polym. Sci.*, 1985, **263**, 933.
22. 'Food Emulsions and Foams', ed., E. Dickinson, Royal Society of Chemistry, London, 1987.
23. 'Food Colloids', ed., R. D. Bee, P. Richmond, and J. Mingins, Royal Society of Chemistry, Cambridge, 1989.
24. 'Advances in Food Emulsions and Foams', ed., E. Dickinson and G. Stainsby, Elsevier Applied Science, London, 1988.
25. A. Kato, Y. Osako, N. Matsudomi, and K. Kobayashi, *Agric. Biol. Chem.*, 1983, **47**, 33.
26. C. C. Bigelow, *J. Theor. Biol.*, 1967, **16**, 187.
27. A.-A. Townsend and S. Nakai, *J. Food Sci.*, 1983, **48**, 588.
28. M. Shimizu, M. Saito, and K. Yamauchi, *Agric. Biol. Chem.*, 1986, **50**, 791.
29. A. Kato, H. Yamaoka, N. Matsudomi, and K. Kobayashi, *J. Agric. Food Chem.*, 1986, **34**, 370.
30. J. R. Bacon, J. W. Hemmant, N. Lambert, R. Moore, and D. J. Wright, *Food Hydrocolloids*, 1988, **2**, 225.
31. A. Kato and K. Yutani, *Protein Eng.*, 1988, **2**, 153.
32. E. Dickinson, in 'Foams: Physics, Chemistry and Structure', ed. A. J. Wilson, Springer-Verlag, London, 1989, p. 39.
33. J. A. McCammon, *Rep. Prog. Phys.*, 1984, **47**, 1.
34. E. Dickinson and G. Stainsby, 'Colloids in Food', Elsevier Applied Science, London, 1982, p. 411.
35. K. G. Berger, in 'Food Emulsions', ed. S. Friberg, Marcel Dekker, New York, 1976, p. 141.
36. J. K. Madden, in 'Foams: Physics, Chemistry and Structure', ed. A. J. Wilson, Springer-Verlag, London, 1989, p. 185.
37. A. E. Bird and G. Stainsby, *J. Sci. Food Agric.*, 1974, **25**, 1339.
38. E. Dickinson, S. K. Narhan, and G. Stainsby, *J. Food Sci.*, 1989, **54**, 77.
39. E. Dickinson and G. Stainsby, *Food Technol.*, 1987, **41(9)**, 74.
40. I. W. L. Roberts, 'Studies of the fat globule membrane in ice cream', Ph.D. Thesis, University of Leeds, 1978.

41. P. Walstra, in 'Encyclopedia of Emulsion Technology', ed. P. Becher, Marcel Dekker, New York, 1983, Vol. 1, p. 57.
42. G. Doxastakis and P. Sherman, *Colloid Polym. Sci.*, 1986, **264**, 254.
43. E. Dickinson, S. K. Narhan, and G. Stainsby, *Food Hydrocolloids*, 1989, **3**, 85.
44. M. Anderson and B. E. Brooker, in 'Advances in Food Emulsions and Foams', ed. E. Dickinson and G. Stainsby, Elsevier Applied Science, London, 1988, p. 221.
45. D. D. Muir and W. Banks, *J. Food Technol.*, 1986, **21**, 229.
46. A. H. Muhr, J. M. V. Blanshard, and S. J. Sheard, *J. Food Technol.*, 1986, **21**, 587.
47. F. MacRitchie and A. E. Alexander, *J. Colloid Sci.*, 1961, **16**, 57.
48. A. Z. Kafel, 'Thermal denaturation of β-lactoglobulin', Ph.D. Thesis, University of Leeds, 1980.

Appendix: The Honours B.Sc. Degree Scheme in Food Science at the University of Leeds

The scheme extends over four years and covers the disciplines set out in Table 1. In his or her first year, the precise blend of basic science courses depends upon each student's A-level background, but in subsequent years all students follow the same scheme. During the first two years, the basic sciences are taught in the main, single subject, departments such as Chemistry, Microbiology, *etc.* Thereafter, all teaching is in the Procter Department, building on the basic sciences and integrating them to form the coherent discipline of food science. The standard reached in pure chemistry at the end of year 2 is similar to that of an Ordinary Degree in Chemistry.

The first year course in food science is largely an introduction to food technology. Food Science 2 continues with process engineering in greater depth, and also includes short components on legislation, quality, statistics, nutrition, colloid science, texture and rheology. Food Science 3 is the course dealing with the chemistry and biochemistry of all the main components of food, including the small molecules of importance in colour and flavour, the lipids, and the biopolymers. Food analysis forms the main subject of the practical work. In the final year, the major food commodities are considered *in toto*. This involves an appreciation of the chemical, microbiological, and physical changes which take place during the various processing techniques, and on storage, together with analytical, nutritional, and legislative aspects.

The courses in biopolymers, in year 3, are supported by lectures and formal laboratory work in food biochemistry, which cover the structures and actions of the more important enzymes, together with the changes they catalyse in the components, including the biopolymers, in cereals, vegetables, meat, milk, eggs,

Table 1 *Honours degree scheme in food science*

Year 1	Food Science 1	Organic Chemistry	Physics Mathematics Biochemistry Biophysics
Year 2	Food Science 2	Physical Chemistry	General Microbiology
Year 3	Food Science 3	Food Microbiology	
Year 4	Food Science 4		

Table 2 *Sequence of topics for lectures on food proteins*

(I)	Backbone:	(a)	structure and types of folding;
		(b)	methods for assessing helical content.
(II)	Side chains:	(a)	sequence and secondary structure;
		(b)	intra- and inter-chain bonding;
		(c)	tertiary and quaternary structure;
		(d)	thermal and chemical denaturation.
(III)	Ionizing groups:	(a)	water and aqueous buffers;
		(b)	charge profile including (i) Donnan, (ii) electrokinetics, and (iii) solubility and separation.
(IV)	Purity and homogeneity:	(a)	chemical;
		(b)	electrochemical;
		(c)	molecular size and shape.

etc. In addition, there is a short course on the synthetic polymers used for can liners, food packaging, *etc.* The sequence of topics for lectures and numerical example classes on food proteins in the biopolymers course is outlined in Table 2. An analogous sequence is used for the (shorter) course on polysaccharides.

The total load on the student in year 3 is approximately 7 lectures a week, together with 11 hours of practical work and some 4 hours of examples classes (numerical work). In the final year, the formal teaching load is deliberately lighter, so that the student has a proper opportunity for independent study. Short optional courses (including topics such as 'concentrated colloidal dispersions') are provided. The laboratory work in year 4 is based around a team project, and, more importantly, an individual research project, rather than relying on set experiments, as in the earlier years.

Progress through the scheme is assessed primarily by written examinations (4 papers) at the end of each year with examples classes and practical work being continuously assessed during the year. The substantial chemical content of the scheme is recognized by the Royal Society of Chemistry by allowing Leeds Food Science graduates with first or second class honours to become graduates of the Society without further examination.

The major area of employment of B.Sc. food scientists is the food manufacturing industry in the UK, which in recent years has taken over 80% of the new Leeds graduates. The postings are usually to research and development, quality control, production management, or marketing. A steady, but much narrower, stream of graduates enters the public sector.

Steric Interactions between Microemulsion Droplets in a Plastic-Crystalline Phase

By David C. Steytler

AFRC INSTITUTE OF FOOD RESEARCH, NORWICH LABORATORY, COLNEY LANE, NORWICH NR4 7UA

Julian Eastoe, Brian H. Robinson

SCHOOL OF CHEMICAL SCIENCES, UNIVERSITY OF EAST ANGLIA, NORWICH NR4 7TJ

and John C. Dore

PHYSICS LABORATORY, UNIVERSITY OF KENT, CANTERBURY CT2 7NZ

1 Introduction

Steric interactions between monolayers of surfactant are of fundamental importance in colloid science in determining the stability of many surfactant-stabilized dispersions. This is of particular significance for food systems in which most dispersions are sterically stabilized. Direct measurement of steric interactions has presented experimentalists with a formidable challenge which has been met by a variety of delicate mechanical techniques. The force box, developed by Israelachvili, has proved one of the most sensitive methods, and has provided unique information in some elegant compression studies of adsorbed monolayers of macromolecules.

In this paper we present preliminary small-angle neutron scattering (SANS) studies of a model water-in-oil microemulsion in an apolar solvent forming a plastic-crystalline phase. The systematic response of inter-droplet separation to pressure in this system suggests the basis of a new technique for probing steric interactions between interpenetrating surfactant layers in sterically stabilized dispersions.

Microemulsions—Water-in-oil (w/o) microemulsions may be formed spontaneously on mixing appropriate proportions of water, oil, and a surfactant stabilizer.[1] When Aerosol–OT (sodium bis-(2-ethylhexyl)sulphosuccinate) is

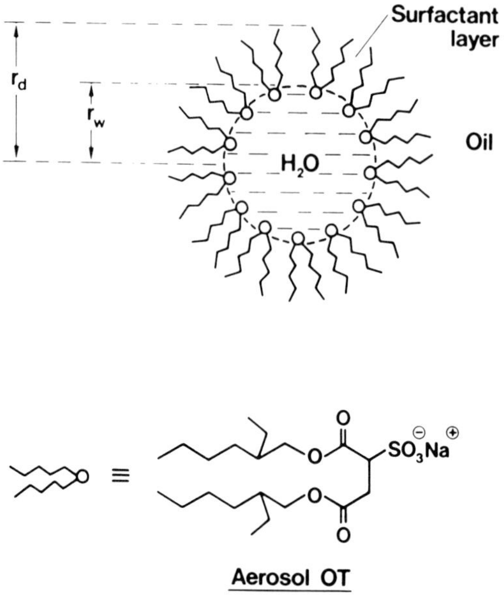

Aerosol OT

Figure 1   *Schematic representation of an AOT-stabilized w/o microemulsion droplet (r_w = water core radius, r_d = overall droplet radius)*

used as surfactant, a system comprising essentially monodisperse spherical water droplets is formed at low volume fractions of the dispersed phase.[2] Unlike emulsions, these dispersions contain nanometre-size droplets, as shown schematically in Figure 1; the systems are both optically transparent and thermodynamically stable. The droplet radius r_w for AOT w/o microemulsions in alkane solvents is readily controlled by the water-to-surfactant ratio w_0 through equation (1) (see below) which can easily be derived from geometrical considerations if all the surfactant is taken to be located at the interface. This linear relationship has been confirmed experimentally using a range of physical methods:[3]

$$(r_w/\text{Å}) = 1.8w_0 \qquad (w_0 = [\text{H}_2\text{O}]/[\text{AOT}]). \qquad (1)$$

The form of the pair potential between microemulsion droplets is of considerable interest as it determines both the equilibrium properties (phase stability) and the dynamic properties (exchange kinetics) in these systems. A schematic representation of the pair potential has recently been proposed[4] to explain the fundamentally different mechanisms of phase separation occurring at the upper and lower temperature phase boundaries. The form of the potential varies with the external parameters (*e.g.* temperature, pressure) reflecting the change in surfactant–oil interactions. Although presenting a barrier to droplet coalescence, and thereby imparting stability, the steric repulsion presented by the fluid surfactant layer is believed to be relatively small, and exchange of the aqueous phase between droplets occurs in approximately 0.1% of the encounters.[5]

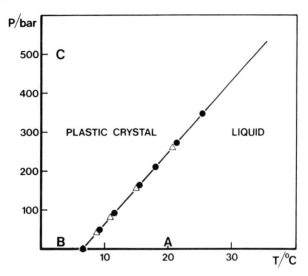

Figure 2 *Phase diagram for the liquid/plastic-crystal phase transition for pure cyclohexane ($\triangle$) and an AOT-stabilized w/o microemulsion in cyclohexane ($\bullet$)*

Plastic Crystals—With most alkane solvents, w/o microemulsions destabilize to form a two-phase system before the freezing temperature (or pressure) of the oil can be reached.[4,6] Although w/o microemulsions have been extensively studied in liquid alkanes, there have been no investigations of the structural changes which occur when the oil component is selectively solidified. (It should be noted also that it is difficult selectively to freeze the aqueous domains.)

Certain symmetrical molecules (*e.g.* cyclohexane) form solid phases at higher temperatures than their asymmetric homologues (*e.g.* n-hexane). These high temperature solid phases exhibit greater plasticity and are commonly referred to as plastic crystals.[7] They are characterized by a high degree of orientational disorder and a correspondingly small entropy of fusion (typically $\Delta S_{fus} < 10\ J\ K^{-1}\ mol^{-1}$). Cyclohexane forms a plastic-crystalline phase in a region of pressure–temperature (p, T) space in which stable microemulsions exist, and so this system was used to examine structural features of microemulsions in solid phases. The liquid/plastic-crystal phase transition for pure cyclohexane is shown in Figure 2.

2 Experimental

Selective freezing of the continuous oil phase in a variety of w/o microemulsions was examined using cyclic alkanes forming plastic-crystalline phases. Microemulsions stabilized by anionic and cationic surfactants were studied. The anionic surfactant was AOT presenting a surfactant layer thickness ($r_d - r_w$) of *ca.* 9 Å.

The cationic surfactant was didodecyldimethyl ammonium bromide (DDAB), and in this case the two longer alkyl chains form a thicker surfactant layer of approximately 17 Å.[8] Experiments were conducted to examine both the macroscopic features and the microscopic structural changes accompanying the liquid/plastic-crystal phase transition of the solvent containing the microemulsion droplets.

Macroscopic Phase Behaviour—A high-pressure optical cell, previously used[4] to study the stability of microemulsions in liquid and near-critical alkanes, was used to locate the liquid/plastic-crystal transition. The phase transition isopleths were recorded by visual observation of the onset of solidification of the cyclohexane component as a function of pressure and temperature. The transition is easily observed as the solid is essentially opaque.

Optical Microscopy—The freezing process was examined by optical microscopy to distinguish any pertinent structural features accompanying the phase transition. A thin film of the microemulsion in the liquid state was gradually cooled in a 2 mm cuvette below the freezing temperature of cyclohexane. The direction of flow of the coolant through the cell holder established a small temperature gradient across the cell such that the freezing front slowly passed through the sample and could be viewed and recorded using a Wild Heerburg M8 optical microscope.

Small-angle Neutron Scattering (SANS)—Measurements were made on the PACE diffractometer at the Orphée reactor at CEN, Saclay, France. Neutron wavelengths of 6.22 and 5.27 Å were used with a sample-to-detector distance of approximately 2 m providing a Q range of 0.0156–0.204 Å^{-1}. The high-pressure cell was similar to that described previously,[4] but necessary modifications to the windows to allow SANS measurements reduced the operating pressure from 1000 to 600 bar.

All measurements in the plastic-crystalline phase were made on systems containing D_2O + H-surfactant + C_6H_{12}, such that there was a single-step change in the contrast profile at the surfactant–water interface for which the contrast is given at the surface of the water core (radius r_w). In order to measure the overall droplet dimension r_d, additional measurements were also made in the liquid phase on systems containing H_2O + H-surfactant + C_6D_{12}.

3   Results

Phase transition measurements have demonstrated that the fusion line is not significantly displaced when a w/o microemulsion is frozen in cyclohexane at low volume fraction of the dispersed phase. Indeed, visually, the phase transition is identical to that of the pure solvent and is completely reversible, indicating that in the solid phase separation of the dispersed aqueous phase does not occur. A typical result for an AOT-stabilized microemulsion is shown in Figure 2. When examined microscopically, the mechanism of freezing of the microemulsion is

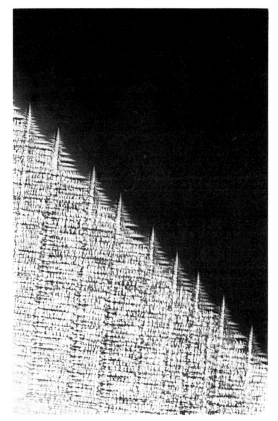

Figure 3   *Dendritic growth of cyclohexane freezing from an AOT-stabilized w/o microemulsion*

characterized by a dendritic growth pattern in which pure solvent is preferentially frozen from the mixture as shown in Figure 3. Previously, dendritic growth has been shown[9] to occur in plastic crystals (*e.g.* cyclohexanol) doped with impurities when the impurity is concentrated in liquid domains between the dendrite branches. We therefore expect the droplets here to be similarly located in small pockets of unfrozen cyclohexane trapped between the branches of the dendrites.

The measured SANS intensity $I(Q)$ from a system of particles in solution contains an intra-particle form factor, $P(Q)$, characterizing the size and shape of individual droplets, and an inter-particle structure factor, $S(Q)$, including structural correlations arising from particle interactions:[10]

$$I(Q) = P(Q)S(Q). \qquad (2)$$

For weakly interacting particles, such as microemulsion droplets, $S(Q)$ approaches unity at low volume fractions ($\phi \to 0$) in the liquid phase. As the droplet

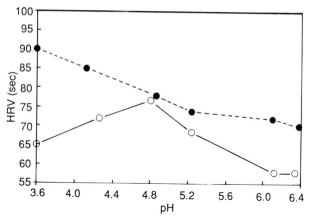

Figure 4   *Dependence of foam stability upon pH for a* 0.2 wt% BSA *pure solution* (○—○) *and a* 0.2 wt% BSA + 0.1 wt% *lysozyme solution* (●—●)

The mechanism of foam stability enhancement at pH 6.0 has already been discussed. The increase in HRV observed at pH values below 4.8 may indicate the positive role of hydrophobic interactions between partially unfolded proteins at acidic pH. Two additional experiments were carried out at pH 4.8 and at pH 3.45, where 0.1 wt% lysozyme was added to BSA solutions of various concentrations. The results obtained[12] were found to verify the trends shown in Figure 4.

4 Conclusions

It is confirmed that foam stability of a heterogeneous protein solution is enhanced when pH conditions favour electrostatic interactions between acidic and basic proteins. In addition, when electrostatic attractions are suppressed by addition of salt, and interactions of hydrophobic nature are expressed, the foam stability of the heterogeneous systems becomes more greatly enhanced.

Acknowledgement. This work was funded as part of the SERC Rolling Programme in Biochemical Engineering at the University of Birmingham.

References

1. T. J. Leeson, Ph.D. Thesis, University of Birmingham, 1989.
2. A. D. Rudin, *J. Inst. Brewing*, 1957, **63**, 506.
3. M. Velissariou, M.Sc. Project Thesis, University of Birmingham, 1988.
4. W. Bumbullis and K. Schügerl, *Eur. J. Appl. Microbiol. Biotechnol.*, 1979, **8**, 17.
5. J. Jollès, J.-P. Périn, and P. Jollès, *Molec. Cell. Biochem.*, 1974, **17**, 39.
6. S. Poole, S. I. West, and C. L. Walters, *J. Sci. Food Agric.*, 1984, **35**, 701.
7. J. R. Mitchell, in 'Developments in Food Proteins', ed. B. J. F. Hudson, Elsevier Applied Science, London, 1986, p. 291.
8. D. C. Clark, J. Mingins, F. E. Sloan, L. J. Smith, and D. R. Wilson, in 'Food Emulsions and Foams', ed. E. Dickinson, Royal Society of Chemistry, London, 1987, p. 110.

9. A.-A. Townsend and S. Nakai, *J. Food Sci.*, 1983, **48**, 588.
10. S. H. Kim and J. E. Kinsella, *J. Food Sci.*, 1987, **52**, 128, 1341.
11. W. Melander and C. Horvath, *Arch. Biochem. Biophys.*, 1977, **183**, 200.
12. M. Velissariou, unpublished data.

Interaction between Polystyrene Latices and Catalase: Primarily Ionic or Hydrophobic?

By J. Stone-Masui*, P. Steenhoudt, L. Marteaux, and W. E. E. Stone†

UNITÉ DE CHIMIE DES INTERFACES, UNIVERSITÉ CATHOLIQUE DE LOUVAIN, PLACE CROIX DU SUD 1, 1348 LOUVAIN-LA-NEUVE, BELGIUM

This contribution represents a preliminary report of our study of the interaction of an enzyme (catalase) with surfaces of variable hydrophobicity. Although numerous studies concerning protein adsorption have been carried out, there is still a need for results on well defined systems.[1-5]

Monodisperse polystyrene latex particles prepared previously by one of us (J.S-M.), and carefully purified by mixed-bed ion exchange resins, were used as the adsorbent. Conductometric titrations coupled with XPS measurements gave the sulphate content of the polymer surfaces (with a 100% variation in the concentration of SO_4^- groups).[6-8] A more hydrophilic system was also used bearing sulphate and OH (phenol) groups (Polylab). The main characteristics of these adsorbents are given in Table 1.

The adsorbate considered here is catalase (Merck) from *Aspergillus niger*. Prior to adsorption, the enzyme was also characterized by XPS. Some of its characteristics[9-11] are reported in Table 2. This enzyme is presently being extensively studied in our laboratory.[12]

Table 1 *Characteristics of the hydrophobic polystyrene latices (PS) and the hydrophilic latex (PS-OH)*

	PS Latices		PS-OH Latex
Particle diameter (μm)	0.220	0.490	0.336
Specific surface area (m^2 g^{-1})	26	11.6	17
Surface charge density (—O—SO$_3^-$) (μC cm^{-2})	3.9^a	8.4^a	3.5^b
Surface area per —O—SO$_3^-$ group (Å^2)	410	190	460

^{a} Determined by conductometric titration, after purification of the latices with mixed-bed ion exchange resins.
^{b} Determined by XPS; 5% of the surface carbons are linked to one OH group (phenol).

*Research Associate FNRS (Belgium).
†Section de Physico-Chimie Minérale (MRAC).

Table 2   *Properties of lyophilized catalase from* Aspergillus niger

Molecular weight (daltons)	350 000
Isoelectric pH	4.4
Sub-units with covalently bound carbohydrate	4
Prosthetic group	protohematin
Stokes radius (nm)	5.8 ± 0.5
Specific volume ($cm^3\,g^{-1}$)	0.73
Optimum pH for catalytic activity	7
Monolayer concentration of enzyme (without conformational deformation) for spheres with hexagonal packing ($mg\,m^{-2}$)	4–5

This study determines, by XPS, the quantity of catalase remaining adsorbed on the particles after centrifugation and separation of the supernatant. Analysis of this supernatant by the colorimetric method of Bradford[13] (the Bio–Rad test) allows a comparison of both methods of adsorption determination. Good agreement is obtained. All adsorption isotherms were determined with an equilibrium time of 24 h at $30 \pm 2\,°C$ and a total latex surface area available for adsorption of $0.2\,m^2$.

For KCl–HCl systems, the results at initial pH values of 3.6, 4.4, and 5.0 are consistent with high affinity adsorption isotherms at all ionic strengths. There is no detectable influence of pH on adsorption. Moreover, adsorption decreases with an increase in ionic strength from 1 to 55 mM KCl, which reflects the importance of ionic interactions between opposite charges, resulting in eventual ion exchange and ion-pair formation even when the net charge of the protein is the same as the one of the adsorbent. As shown by experiments conducted at pH 3.6, more adsorption was observed on increasing the latex particle negative charge. This behaviour is also in agreement with the effective influence of attractive electrostatic interactions in the control of the adsorption of the enzyme.

In an attempt to increase the reproducibility of our results, at first considered to be influenced by pH variations, we decided to fix the pH with appropriate dilutions of a phosphate buffer (20 mM KH_2PO_4, 30 mM Na_2HPO_4) of ionic strength 110 mM. For a low ionic strength of 1 mM, at pH 7, the adsorption isotherm shape was drastically modified and it became the low affinity type. This change can be attributed to protein positive charge shielding by divalent buffer anions. But, as adsorption is still finite, it must be assumed that hydrophobic interaction with the polystyrene surface is effective. Results at ionic strengths of 11 and 55 mM reflect the importance of the kind of small ions. Maintaining the ionic strength by phosphate buffer alone or with added KCl gives very different results. An electrostatic repulsion decrease with increasing ionic strength is observed only for equilibrium catalase concentrations smaller than 600 $\mu g\,ml^{-1}$ in the case of the KCl-containing electrolyte. On the other hand, with phosphate buffer alone, increasing the ionic strength up to 55 mM increases greatly the catalase adsorption. Nevertheless, in none of these experiments is a full monolayer reached (based on the known catalase Stokes radius). The increased adsorption in the presence of the phosphate buffer is attributed to a hydrophobic interaction enhancement. Although the salt concentrations here are much lower than those in the salting out of proteins, Cl^- and HPO_4^{2-} ions influence the adsorption isotherms in a similar manner.

For the more hydrophilic latex PS-OH, even though the sulphate charge is similar to that of the 0.22 μm diameter PS latex, no enzyme adsorption occurs, even at ionic strength 55 mM and pH 7. In this case, the surface–protein hydrophobic interactions are apparently impeded.

We conclude that hydrophobic interactions are the most important driving forces for protein adsorption on PS latices. Ionic charge density and ionic strength also have an effect on the shape of the adsorption isotherm.

References

1. J. D. Andrade, 'Surface and Interfacial Aspects of Biomedical Polymers', Plenum Press, New York, 1985, Vol. 2.
2. E. Dickinson and G. Stainsby, 'Colloids in Food', Applied Science, London, 1982, Chap. 6.
3. W. Norde, *Adv. Colloid Interface Sci.*, 1986, **25**, 267.
4. A. Rosevear, J. F. Kennedy, and J. M. S. Cabzal, 'Immobilised Enzymes and Cells', Adam Hilger, Bristol, 1987.
5. W. Norde and J. Lyklema, *Colloids Surf.*, 1989, **38**, 1.
6. J. Stone-Masui and A. Watillon, *J. Colloid Interface Sci.*, 1975, **52**, 479.
7. J. H. Stone-Masui and W. E. E. Stone, in 'Polymer Colloids II', ed. R. M. Fitch, Plenum Press, New York, 1980, p. 331.
8. W. E. E. Stone and J. H. Stone-Masui, in 'Science and Technology of Polymer Colloids', eds. G. W. Poehlein, R. H. Ottewill, and J. W. Goodwin, M. Nijhoff, Boston, 1983, p. 480.
9. B. P. Wasserman and H. O. Hultin, *Arch. Biochem. Biophys.*, 1981, **212**, 385.
10. K. Kikuchi-Torii, S. Hayashi, H. Nakamoto, and S. Nakamura, *J. Biochem.*, 1982, **92**, 1449.
11. A. A. Mosavi-Movahedi, A. E. Wilkinson, and M. N. Jones, *Int. J. Biol. Macromol.*, 1987, **9**, 327.
12. O. Lardinois and D. Masy, personal communication.
13. S. J. Compton and C. G. Jones, *Anal. Biochem.*, 1985, **151**, 369.

Competitive Adsorption of β-Lactoglobulin in Mixed Protein Emulsions

By Douglas G. Dalgleish

HANNAH RESEARCH INSTITUTE, AYR, SCOTLAND KA6 5HL

Susan E. Euston, Josephine A. Hunt, and Eric Dickinson

PROCTER DEPARTMENT OF FOOD SCIENCE, UNIVERSITY OF LEEDS,
LEEDS LS2 9JT

1 Introduction

It has been established that, for certain proteins, especially the caseins, it is possible for competitive adsorption to occur during the formation of an oil-in-water emulsion containing more than one protein. Also, exchange may in some cases take place between proteins which are adsorbed at an oil–water interface and those which are in the aqueous phase.[1-3] This behaviour of casein-containing emulsions has allowed speculation that interfacial proteins in mixed systems might be under a form of thermodynamic control, such that equilibrium would be established between different adsorbed and free proteins. This, however, has proved not to be the case for systems containing β-lactoglobulin (β-lg) and α-lactalbumin (α-la).[4] In these mixtures, the exchange reactions were found to be much altered from the type which was first seen in the caseins, insofar as additional adsorption of one type of protein on the other, rather than direct replacement, was found to occur. Studies of the electrophoretic mobilities of the emulsion particles[4] and of the surface viscosities of mixed protein films[5] confirm that simple replacement reactions are perhaps likely to be in the minority.

In this paper, we extend the studies of emulsions containing mixtures of α-la and β-lg to other emulsified systems involving β-lg. The purpose of the studies is to extend our knowledge of the competition of proteins at the oil–water interface by investigating a number of binary mixed protein systems having, as their common factor, the presence of β-lg. Thus, we discuss here mixtures of β-lg with the highly surface-active protein β-casein, the moderately surface-active α-la, and the poorly surface-active ovalbumin.

2 Experimental

The proteins β-lg, α-la, and ovalbumin were obtained from Sigma Chemical Co. Additional β-lg and the n-tetradecane were obtained from BDH Chemicals. The β-casein was prepared in the laboratory as described previously.[6]

All experiments were carried out using buffer solutions of 20 mM Tris–HCl, pH 7.0. Oil-in-water emulsions (20 wt% n-tetradecane, 0.5 wt% protein) were prepared using a laboratory-scale single-valve homogenizer[7] at an operating pressure of 300 bar. Particle sizes and distributions were measured using a Coulter counter TAII with a 30 μm orifice.

Mixed emulsions were prepared by dissolving appropriate weights of the two proteins in the required amount of buffer, and then homogenizing with the n-tetradecane. In exchange experiments, only one protein was used in the formation of the emulsion, and the emulsion was then centrifuged at 10 000g for 15 min. The cream was then washed and resuspended in a solution of the second protein, which had been made up to have a concentration the same as that of the first and a volume equal to that of the aqueous phase removed during the centrifugation. The mixture was then stirred for periods of up to 48 hours, with periodic removal of samples.

The protein concentrations in the aqueous phase were determined by centrifugation of the emulsion (10 000g, 15 min), collection of the sub-natant layer, and chromatography of this solution using fast protein liquid chromatography (FPLC) (Pharmacia Ltd.). A Mono-Q column was used to separate the protein components, and the concentrations in the solutions were estimated from the areas of the peaks eluting from the column, compared with those determined for standard amounts of the same proteins.

3 Results

When emulsions were prepared using mixtures of β-lg + α-la, there was no preference for either protein at the interface (Figure 1a), as demonstrated by the straight lines relating interfacial composition and solution composition. Since both proteins show the same affinity for the interface in such mixtures during emulsification, it was expected that they would show similar behaviour in competitive experiments where one protein was used to attempt to displace the other from the oil–water interface. However, this was not found to be the case.[4] When an emulsion formed using α-la was exposed to β-lg, it was evident that displacement of the former by the latter was at best inefficient. As well as displacing a small amount of the adsorbed α-la, an equimolar concentration of β-lg caused additional adsorption since the binding of β-lg was considerably greater than the amount of α-la displaced. At higher concentrations of β-lg, there was an increase in the desorption of α-la. In the reverse experiment, α-la was even less capable of displacing β-lg from the interface.[4]

It was apparent that β-lg was incapable of displacing β-casein from the interface. When an emulsion incorporating β-casein was treated with β-lg, no β-casein appeared in solution; however, some of the β-lg did bind to the emulsion droplets (Figure 2a), so that the protein load increased by about one-third.

D. G. Dalgleish, S. E. Euston, J. A. Hunt, and E. Dickinson   487

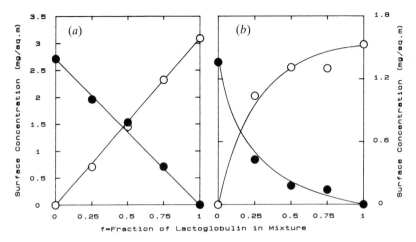

Figure 1   *Proportions of proteins at surfaces of oil-in-water emulsions prepared with mixtures of proteins. The surface concentration is plotted against the weight fraction of β-lg in the emulsifier. (a), α-la + β-lg mixtures: ○, β-lg; ●, α-la. (b), ovalbumin + β-lg mixtures: ○, β-lg; ●, ovalbumin. The lower relative values of surface coverage in (b) appear to be as a result of slightly different emulsification conditions*

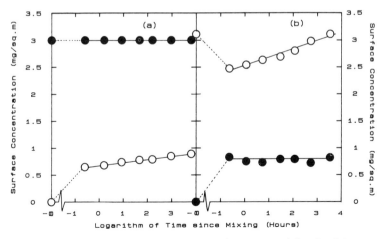

Figure 2   *Protein exchange on mixing washed protein-stabilized oil-in-water emulsions with protein solutions. The surface concentration is plotted against the logarithm of the time that has elapsed since mixing. (a), β-casein emulsion treated with β-lg: ●, β-casein; ○, β-lg. (b), β-lg emulsion treated with β-casein: ○, β-lg; ●, β-casein*

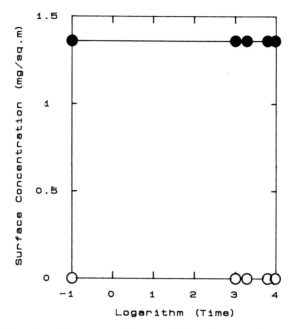

Figure 3 *Protein exchange on mixing washed ovalbumin oil-in-water emulsion with a solution of β-lg. The surface concentration is plotted against the logarithm of the time that has elapsed since mixing:* ●, *ovalbumin;* ○, *β-lg*

Because β-casein is more surface-active than β-lg, its resistance to displacement is not unexpected. However, it was also found that β-casein does not completely displace β-lg from the interface (Figure 2b). Immediately after mixing, β-lg was displaced, and an approximately equimolar amount of β-casein was adsorbed, but, as time passed, the displaced β-lg appeared to re-adsorb along with the β-casein. At no time was there any suggestion that full displacement of the β-lg had occurred, despite the higher surface activity of the β-casein.

Ovalbumin provides a further contrast. It is clear from Figure 1b that ovalbumin and β-lg have very different affinities for the oil–water interface during the emulsion formation, since the surface load *versus* solution composition plots are non-linear, unlike those for the β-lg + α-la mixtures. The interface of newly formed emulsions is dominated by β-lg, so that very little ovalbumin is found on the interfaces, except in emulsions where β-lg is completely absent from the emulsifier. Nevertheless, despite this evidence of much greater affinity of the β-lg, exchange experiments demonstrated that, once a layer of adsorbed ovalbumin had been formed, it could not be displaced by β-lg (Figure 3), even when the emulsion had been in contact with the aqueous protein for 48 hours.

4 Discussion

These results demonstrate the differences in behaviour of different proteins when adsorbed at an oil–water interface. It is clear that (with the exception of the caseins), once a layer of protein has been formed at an interface, it can be very difficult to remove it with another protein. We may contrast the behaviour of ovalbumin in the formation of a mixed emulsion (where it competes very ineffectively with β-lg), with its behaviour once it has formed an interfacial layer (where it is very difficult to displace).

A pair of proteins where competition appears to be partially established is α-la + β-lg. Each of these whey proteins can displace the other from the interface to a limited extent, although rather high concentrations (typically 10×) of the displacing protein are generally required. This also confirms that, once a structured layer of adsorbed protein has been formed, other proteins, even highly surface-active ones (*e.g.* in the β-lg + β-casein mixture), do not find it easy to displace the material which has been adsorbed first.

A feature of many of these experiments is the observation that the presence of a second protein causes the total load of adsorbed protein to increase, as seen in Figures 2a and 2b. This may arise from the formation of multi-layers of protein, with the second binding either on top of or underneath the original layer,[4,7] but it is also possible that the second protein adsorbs in gaps in the original surface layer, or even that it may alter the structure of the first adsorbed protein to allow further adsorption to occur. It is known that the electrophoretic mobility of emulsion particles are altered when exchange reactions are attempted, and that interesting changes in the surface viscosity of protein films also occur.[4,5] Quantitative measurements of compositions of interfacial materials such as those described here should permit us to understand these alterations in physical properties.

Acknowledgement. S. E. Euston and J. A. Hunt acknowledge the receipt of AFRC Cooperative Studentships in conjunction with the Hannah Research Institute.

References

1. D. F. Darling and D. W. Butcher, *J. Dairy Res.*, 1978, **45**, 197.
2. E. W. Robson and D. G. Dalgleish, *J. Food Sci.*, 1987, **52**, 1694.
3. E. Dickinson, S. E. Rolfe, and D. G. Dalgleish, *Food Hydrocolloids*, 1988, **2**, 397.
4. E. Dickinson, S. E. Rolfe, and D. G. Dalgleish, *Food Hydrocolloids*, 1989, **3**, 193.
5. E. Dickinson, S. E. Rolfe, and D. G. Dalgleish, *Int. J. Biol. Macromol.*, 1990, **12**, 189.
6. W. D. Annan and W. Manson, *J. Dairy Res.*, 1969, **36**, 259.
7. E. Dickinson, A. Murray, B. S. Murray, and G. Stainsby, in 'Food Emulsions and Foams', ed. E. Dickinson, Royal Society of Chemistry, London, 1987, p. 86.

Effect of Molecular Weight on the Emulsifying Behaviour of Gum Arabic

By Eric Dickinson, Vanda B. Galazka

PROCTER DEPARTMENT OF FOOD SCIENCE, UNIVERSITY OF LEEDS, LEEDS LS2 9JT

and Douglas M. W. Anderson

CHEMISTRY DEPARTMENT, UNIVERSITY OF EDINBURGH, EDINBURGH EH9 3JJ

1 Introduction

Gum arabic is used to stabilize flavour oil emulsions for the soft drinks industry, but its precise mode of action is far from fully understood. The main reason for this lies with the compositional and structural complexity of the material.

Physico-chemical studies of gum arabic have suggested that it consists of a mixture of highly branched arabinogalactan heteropolymers in combination with a small amount of protein (typically ca. 2 wt%). It is believed that hydroxyproline and serine are involved in covalently linking the carbohydrate to the protein to form an arabinogalactan–protein complex.[1,2] The protein component is mainly associated with a high molecular weight fraction representing only 20–30% of the total gum.[3] The major fraction is of lower molecular mass, containing very little nitrogenous material. It has been demonstrated[4] that the protein-rich fraction adsorbs strongly at the oil–water interface, and is mainly responsible for the emulsifying and stabilizing properties of the natural gum.

When gum arabic is used as a stabilizer for flavour oil emulsions, a stabilizing film is formed at the oil–water interface whose surface viscoelasticity is rather insensitive to dilution of the aqueous phase,[5] thus enabling the emulsion to be stabilized both as a concentrate and as a diluted beverage.

Recent experimentation[6,7] with samples of several different *Acacia* gum species with nitrogen contents in the range 0.1–7.5% (protein contents 0.5–47 wt%) indicates a strong correlation between the proportion of protein in the gum and its surface properties at the oil–water interface.

2 Materials and Methods

The gum samples GAO and GA1 were good quality Sudanese *Acacia senegal* samples. Gum GAO was fractionated on a gel permeation chromatography column packed with Sephacryl S-500 and eluted with 0.3 wt% NaCl solution. Using a molecular weight distribution curve based on a previously established elution profile for GAO, gum fraction GAH was an arbitrary cut intended to separate the 10% of material of highest molecular weight (0.87% N, 5.7 wt% protein). Gum fraction GAL (0.35% N, 2.3 wt% protein) is the material subsequently recovered after removal of GAH from GAO. Gum samples GA2, GA3, and GA4 were three of a series of controlled degradation products derived from the natural gum GA1 to give a range of molecular weights (from 3.1×10^5 down to 2.2×10^5 daltons) each with a nitrogen content of 0.35%. Analytical data for these samples are reported elsewhere.[8] The n-hexadecane was AnalaR grade from Sigma Chemicals. Buffer salts and sodium azide were of reagent grade and used as purchased. Water was double distilled.

Emulsions of n-hexadecane-in-water were prepared using the Leeds one-stage valve mini-homogenizer.[9] To a 1 wt% solution of the gum in phosphate buffer (pH 7) was added an appropriate amount of n-hexadecane to give a premix containing 10 vol% oil. After blending, the coarse premix was homogenized at 300 bar and 25 °C. The resulting emulsion was stored in a water bath at 25 °C; 0.1 wt% sodium azide was added as bactericide. Immediately after emulsion preparation, and at regular intervals after thorough mixing, droplet-size distributions were determined using the Coulter counter model TAII with a 30 μm orifice tube and 0.1 M NaCl as suspending electrolyte.

3 Results and Discussion

Droplet-size distributions for the freshly made n-hexadecane-in-water emulsions are not very different for the three samples GAO, GAH, and GAL. Data based on these distributions show that the high-molecular-weight fraction GAH (0.87% N) gives a coarser emulsion than the residual low molecular weight fraction GAL (0.35% N), with the original gum sample GAO (0.38% N) somewhere in between. Figure 1 shows droplet-size distributions for the same emulsions after storage for 24 hours. Based on the data shown it can be seen that the stabilizing capacities of the gum samples lie in the order: high molecular weight fraction > original gum sample ~ low molecular weight fraction.

Results for interfacial tensions of 10^{-3} wt% gum solutions at the n-hexadecane–water interface after 24 hours show that the high-nitrogen fraction GAH is more surface active than the residual fraction GAL, with the original gum sample GAO lying somewhere in between. Comparison of these results with earlier data[7] on gum samples from different *Acacia* species confirms the good correlation between nitrogen content and limiting long-time surface activity.

Figure 2 shows the droplet-size distribution for emulsions prepared with undegraded (GA1) and degraded samples (GA2, GA3, and GA4) after storage for 5 hours. It is found that there is a general trend of decreasing emulsion stability with increasing degradation (molecular weights: GA1 > GA2 > GA3 > GA4). This result is consistent with the work of Nakamura[10] who found that emulsion stability increases with the weight-average molecular weight of the gum. The results in Figure 2 indicate that when the nitrogen content is kept constant, there is a correlation between the molecular weight of the gum and its stabilizing ability.

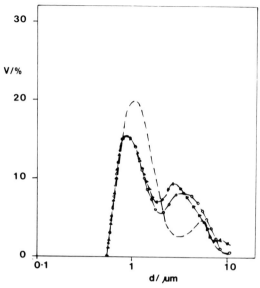

Figure 1 *Droplet-size distributions of n-hexadecane-in-water emulsions prepared with fractionated gum arabic samples after storage for 24 hours. The smoothed percentage differential volume V(d) is plotted against droplet diameter* d: GAO, —○—; GAH, — —; GAL, —▲—

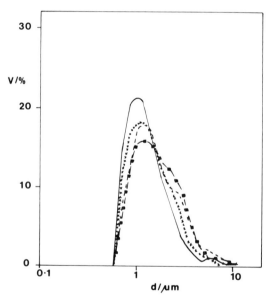

Figure 2 *Droplet-size distributions of emulsions prepared with samples of variously degraded gum arabic after storage for 5 hours. The smoothed percentage differential volume V(d) is plotted against droplet diameter* d: GA1, ——; GA2, × × × ×; GA3, - - -; GA4, —■—

In conclusion, the results presented in this paper are consistent with the idea that it is the protein-rich high-molecular-weight component (the arabino-galactan–protein complex) which provides the functionality of gum arabic as an emulsion stabilizer.

References

1. Y. Akiyama, S. Eda, and K. Kato, *Agric. Biol. Chem.*, 1984, **48**, 235.
2. D. M. W. Anderson, in 'Gums and Stabilisers for the Food Industry', eds. G. O. Phillips, D. J. Wedlock, and P. A. Williams, IRL Press, Oxford, 1988, Vol. 4, p. 31.
3. M.-C. Vandevelde and J.-C. Fenyo, *Carbohydr. Polym.*, 1985, **5**, 251.
4. R. C. Randall, G. O. Phillips, and P. A. Williams, *Food Hydrocolloids*, 1988, **2**, 131.
5. E. Dickinson, D. J. Elverson and B. S. Murray, *Food Hydrocolloids*, 1989, **3**, 101.
6. E. Dickinson, in 'Gums and Stabilisers for the Food Industry', eds. G. O. Phillips, D. J. Wedlock, and P. A. Williams, IRL Press, Oxford, 1988, Vol. 4, p. 249.
7. E. Dickinson, B. S. Murray, G. Stainsby, and D. M. W. Anderson, *Food Hydrocolloids*, 1988, **2**, 477.
8. E. Dickinson, V. B. Galazka, and D. M. W. Anderson, *Carbohydr. Polym.*, in press.
9. E. Dickinson, A. Murray, B. S. Murray, and G. Stainsby, in 'Food Emulsions and Foams', ed. E. Dickinson, Royal Society of Chemistry, London, 1987, p. 86.
10. M. Nakamura, *Yukagaku*, 1986, **35**, 554.

Bridging Flocculation in Emulsions Made with a Mixture of Protein + Polysaccharide

By Eric Dickinson and Vanda B. Galazka

PROCTER DEPARTMENT OF FOOD SCIENCE, UNIVERSITY OF LEEDS,
LEEDS LS2 9JT

1 Introduction

Proteins and polysaccharides have a major role in controlling the stability and rheology of food colloids.[1] In any emulsion, the principal role of the protein is as an emulsifier and stabilizer of emulsion droplets against aggregation and coalescence. High molecular weight polysaccharides are generally added as gelling agents or thickeners in order to modify the rheological behaviour of the aqueous phase, thereby retarding or inhibiting the processes leading to instability (creaming, flocculation, and coalescence). During the preparation of oil-in-water emulsions, colloidal particles may become aggregated by macromolecular bridging which is generally due to there being insufficient emulsifier to completely cover a newly created surface.

Emulsion formation by a binary mixture of hydrocolloids is affected by competition between the components for adsorption at the oil–water interface.[2] Recently, it has been reported that, for high volume fraction oil-in-water emulsions (55 wt% n-tetradecane) made with mixed protein systems (caseinate + gelatin or whey protein + gelatin), it was not possible to produce stable emulsions with certain emulsifier concentrations and compositions.[3,4] Furthermore, light microscopy revealed that these emulsions were flocculated and that the degree of flocculation was dependent upon emulsifier composition.[4] The behaviour of the caseinate + gum arabic system reported in this paper turns out to be qualitatively similar to that of the caseinate + gelatin system,[4] which suggests that bridging flocculation is caused by partial displacement of polysaccharide from the interface by the more surface-active caseinate component during homogenization.

2 Materials and Methods

The food-grade propylene glycol alginate (Kelcoloid LVF) was a gift from Kelco Inc. (San Diego, USA). It was a 'low viscosity' sample with a degree of esterification of 56%. Gum arabic was a commercial sample obtained from Pepsico Inc. (Valhalla, NY). Sodium carboxymethylcellulose (CMC) was obtained from Hercules Inc. (Delaware, USA). The CMC 7MF was a 'medium viscosity' CMC polymer with a degree of substitution of seven carboxymethyl groups for every ten β-glucopyranose residues. Spray-dried sodium caseinate was obtained from the Scottish Milk Marketing Board. The n-hexadecane was AnalaR grade purchased from Sigma Chemicals, and the buffer salts were from BDH Chemicals.

Oil-in-water emulsions (20 vol% n-hexadecane) containing variable concentrations and compositions of protein and polysaccharide in phosphate buffer (pH 7) were prepared at 40 °C using a 'jet homogenizer'[3,5] at an operating pressure of 240 bar. Immediately after preparation, droplet-size distributions were determined using a Type TAII Coulter counter (Coulter Electronics, Luton) with a 30 μm orifice tube.

Evidence for bridging flocculation was examined using the Nomarski differential interference contrast technique.[6] This enabled us to discriminate between air bubbles (low refractive index), which appeared 'flat' on the photomicrographs and n-hexadecane oil droplets (high refractive index), which appeared in relief.

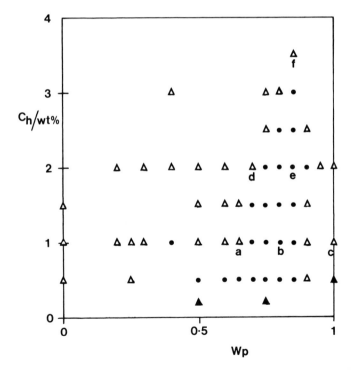

Figure 1 *State stability diagram for caseinate + gum arabic emulsions (20 vol% n-hexadecane) made by jet homogenization. Total protein and polysaccharide content, c_h, is plotted against polysaccharide weight fraction, w_p: △, unflocculated; ●, flocculated; ▲, coalescence. Points (a–f) refer to the pictures in Figure 2*

The degree of flocculation for each emulsion sample was assessed subjectively by giving a value of an aggregation number N_A on a scale from $N_A = 0$ (no aggregation) to $N_A = 7$ (extremely aggregated).

3   Results and Discussion

The degree of flocculation for the caseinate + polysaccharide polymer emulsions was found to depend on emulsifier concentration and composition. Figure 1 shows the state diagram for the caseinate + gum arabic system—it is seen that

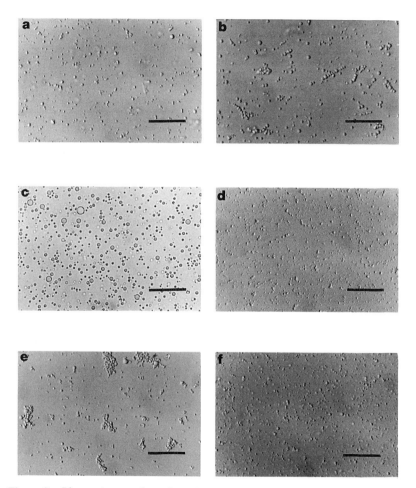

Figure 2   *Photomicrographs of various caseinate + gum arabic emulsions:* (a) $c_h = 1.0$ wt%, $w_p = 0.65$; (b) $c_h = 1.0$ wt%, $w_p = 0.8$; (c) $c_h = 1.0$ wt%, $w_p = 1.0$; (d) $c_h = 2.0$ wt%, $w_p = 0.7$; (e) $c_h = 2.0$ wt%, $w_p = 0.85$; (f) $c_h = 3.5$ wt%, $w_p = 0.85$. *Bars correspond to 100μm*

within a closed boundary in the binary-composition space it is not possible to make an unflocculated emulsion. The extent of flocculation at six points on the state diagram are depicted in Figure 2. Points (a), (b), and (c) of Figure 1 refer to three emulsions of the same total hydrocolloid content (1 wt%) but different compositions. We find that the gum arabic-rich emulsion (a) with $w_p = 0.65$ is essentially unaggregated ($N_A = 1$), but, at the slightly higher gum arabic fraction of $w_p = 0.8$, the resulting emulsion (b) is moderately aggregated ($N_A = 3$). At the gum arabic fraction of $w_p = 1.0$, the emulsion (c) is unflocculated ($N_A = 0$). A similar sequence of changes occurs on increasing the total hydrocolloid content c_h to 2 wt%. Figure 2(d) shows an emulsion with $w_p = 0.7$ which is unaggregated ($N_A = 1$), and an increase of the gum arabic fraction to $w_p = 0.85$ produces a highly aggregated emulsion ($N_A = 6$) (e). When the total hydrocolloid content is high ($c_h = 3.5$ wt%), the resulting emulsion, even at $w_p = 0.85$, (f), is relatively unflocculated ($N_A = 1$).

Results for caseinate + propylene glycol alginate emulsions (not shown) also give a region in the state diagram which is associated with flocculated emulsions, but the region occurs at lower hydrocolloid fractions ($w_p = 0.4$–0.5) than with caseinate + gum arabic. Stable emulsions could not be made with carboxymethyl-cellulose alone, and mixtures of caseinate + CMC ($w_p > 0.1$) produced emulsions which were mainly flocculated.

In the three protein–polysaccharide systems studied, it is inferred that bridging flocculation by caseinate is the cause of droplet aggregation, and that the range of hydrocolloid concentration and composition over which bridging flocculation occurs is related to the competitive adsorption between the milk protein and polysaccharide during homogenization. Droplet-size profiles for the three systems show that the presence of a small concentration of polysaccharide polymer leads to an increase in droplet-size, thus confirming that low concentrations of polysaccharide present during emulsification may have a potentially destabilizing effect on food emulsions.

References

1. E. Dickinson and G. Stainsby, 'Colloids in Food', Applied Science, London, 1982.
2. E. Dickinson, *Food Hydrocolloids*, 1986, **1**, 3.
3. J. Castle, E. Dickinson, A. Murray, and G. Stainsby, in 'Gums and Stabilisers for the Food Industry', ed. G. O. Phillips, D. J. Wedlock, and P. A. Williams, IRL Press, Oxford, 1988, Vol. 4, p. 473.
4. E. Dickinson, F. O. Flint, and J. A. Hunt, *Food Hydrocolloids*, 1989, **3**, 389.
5. I. Burgaud, E. Dickinson, and P. V. Nelson, *Int. J. Food Sci. Technol.*, 1990, **25**, 39.
6. C. F. A. Culling, 'Modern Microscopy', Butterworth, London, 1974.
7. E. Tornberg and N. Ediriweera, in 'Food Emulsions and Foams', ed. E. Dickinson, Royal Society of Chemistry, London, 1987, p. 52.

Emulsifying and Surface Properties of the 11S and 7S Globulins of Soybean

By Eric Dickinson and Yasuki Matsumura

PROCTER DEPARTMENT OF FOOD SCIENCE, UNIVERSITY OF LEEDS,
LEEDS LS2 9JT

1 Introduction

While there have been numerous technical studies involving the use of soy protein isolates as emulsifiers, few fundamental investigations of the emulsifying and surface properties of soy proteins have been reported.[1,2] We report here on the emulsifying and surface properties of the two major soy protein fractions, 11S globulin (3.5×10^5 daltons) and 7S globulin (1.8×10^5 daltons). Information is also presented on the competitive adsorption of the two soy proteins at the emulsion droplet surface.

2 Materials and Methods

Samples of 11S and 7S globulins were prepared and purified according to the methods of Thanh et al.[3,4] from soy bean seeds (Glycine Max, Shiro-tsuru-noko). Oil-in-water emulsions (10 vol% n-tetradecane, 0.5 wt% protein, pH 7.6) were prepared at room temperature using the Leeds one-stage mini-homogenizer[5] at an operating pressure of 300 bar. Two different buffer solutions were used for the aqueous phase: (i) 0.005 M phosphate (*low* ionic strength) and (ii) 0.05 M phosphate + 0.45 M NaCl (*high* ionic strength). Droplet-size distributions were determined using a Coulter counter (model TAII) with a 30 μm orifice tube and 0.18 M NaCl as suspending electrolyte. The emulsion creaming rate was monitored visually by measuring the height of the serum layer as a function of time. Surface viscosity was determined for the individual proteins at the n-tetradecane–water interface (10^{-3} wt% bulk protein concentration, pH 7.6, 25 °C) according to the method of Dickinson et al.[6]

3 Results and Discussion

Figure 1 displays droplet-size distributions for the 11S and 7S globulin emulsions at low ionic strength. The freshly made 11S globulin stabilized emulsion has a narrow distribution with a most-probable droplet size of 1.2 μm, whereas the 7S

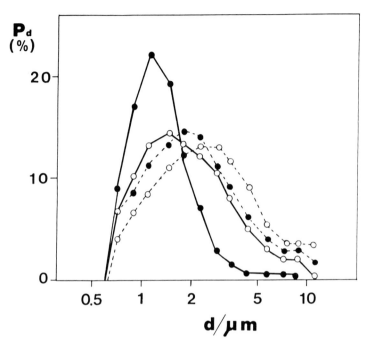

Figure 1   *Droplet-size distributions of emulsions prepared at low ionic strength. The volume-weighted percentage droplet probability P_d is plotted against droplet diameter d: ——, 11S globulin; – – –, 7S globulin; ●, fresh emulsion; ○, emulsion after 1 week*

globulin stabilized emulsion has a broader distribution with a most-probable droplet size of 1.9 µm. This indicates the better emulsifying capacity of 11S globulin as compared with 7S globulin. After one week, both distributions have shifted to larger droplet sizes, but the emulsions can still be regarded as being moderately stable, given the relatively low protein-to-oil ratio. At high ionic strength, however, the emulsifying capacity and emulsion stability are rather poor for both 11S and 7S globulins, as can be seen from the distributions in Figure 2.

The relatively good stability of emulsions made with the individual soy globulins at low ionic strength is confirmed by the creaming data in Figure 3. After one week of storage, the serum layer height was found to represent only *ca*. 10% of the total emulsion height, as compared with *ca*. 50% after only 24 hours at high ionic strength. The poor emulsifying properties at high ionic strength is probably due to flocculation of droplets during and after homogenization, which in turn leads to enhanced coalescence and creaming.

Figure 4 shows time-dependent surface shear viscosities of adsorbed films of the pure proteins at the oil–water interface. What is immediately apparent from Figure 4 is that both soy proteins form highly viscous films at low or high ionic strength. The surface viscosity reached after 48 hours is higher than that for any of

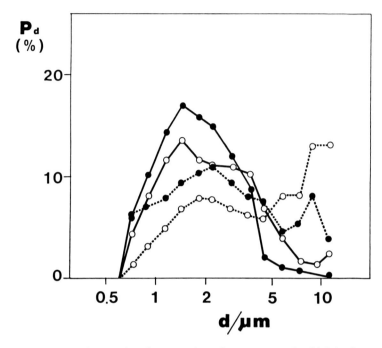

Figure 2   *Droplet-size distributions of emulsions prepared at high ionic strength. The volume-weighted percentage droplet probability P_d is plotted against droplet diameter d: ——, 11S globulin; – – –, 7S globulin; ●, fresh emulsion, ○, emulsion after 1 week*

the milk proteins, even β-lactoglobulin.[7] Such a high surface viscosity is favourable towards the conferring of good long-term stability to emulsions by adsorbed protein films.[8] Like β-lactoglobulin, the surface viscosity of 11S soy globulin is strongly time-dependent, and this could be due to slow formation of —S—S— bonds in the ageing interfacial film. Preliminary experiments involving the displacement of aged 11S globulin from the emulsion droplet surface would seem to indicate that polymerization of the protein at the interface does indeed occur.

Competitive adsorption of the two globulins at the emulsion droplet surface has been investigated. Emulsions were prepared with mixtures of 11S globulin + 7S globulin of differing composition but the same total amount of proteinaceous emulsifier. After centrifugation, the concentrations of the unadsorbed proteins in the serum phase were determined densitometrically by SDS–PAGE, and so the surface concentrations of the adsorbed proteins could be calculated by difference. It was found that, within experimental error, the composition of the adsorbed layer is the same as that of the protein mixture used to make the emulsion. Further sequential exchange experiments between the two proteins at the emulsion droplet surface were performed. An emulsion freshly made with 11S globulin (or 7S globulin) was washed free of unadsorbed protein by centrifugation; the cream was then redispersed and mixed with a solution of 7S globulin (or 11S globulin), and the change in the composition of the protein layer was determined over a

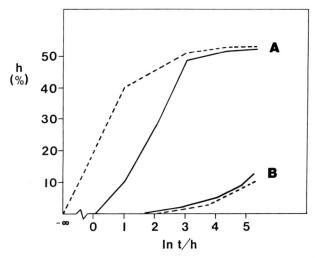

Figure 3 *Stability of n-tetradecane-in-water emulsions (pH 7.6, 0.5 wt% protein) prepared (A) at high ionic strength and (B) at low ionic strength. Serum height h, as a percentage of total emulsion height, is plotted against the logarithm of the storage time* t: ——, 11S globulin; – – –, 7S globulin

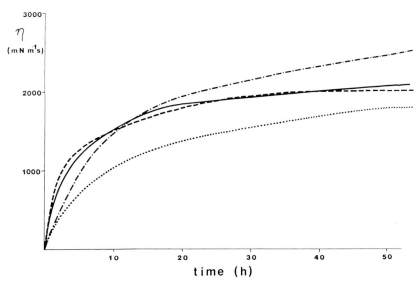

Figure 4 *Surface viscosity of adsorbed films of soy globulins at the n-tetradecane–water interface (pH 7.6, 10^{-3} wt% protein, 25 °C). Surface shear viscosity, η, is plotted against adsorption time:* ——, 11S globulin, high ionic strength (0.5 M); — — —, 7S globulin, high ionic strength (0.5 M); —·—·—, 11S globulin, low ionic strength (0.005 M); – – –, 7S globulin, low ionic strength (0.005 M)

period of 32 hours as just described. It was found that, over this time period, there is almost no replacement (less than 5%) of one globulin by the other at the droplet surface. This lack of interchange between adsorbed protein and protein in solution is similar to that found[9] in emulsion systems containing the two whey proteins, α-lactalbumin + β-lactoglobulin, but is different from the rapid, reversible competitive adsorption which occurs with α_{s1}-casein + β-casein.[10]

References

1. F. Yamauchi, Y. Ogawa, Y. Kamata, and K. Shibasaki, *Agric. Biol. Chem.*, 1982, **46**, 615.
2. H. J. Rivas and P. Sherman, *Colloids Surf.*, 1984, **11**, 155.
3. V. H. Thanh and K. Shibasaki, *J. Agric. Food Chem.*, 1976, **24**, 1117.
4. V. H. Thanh, K. Okubo, and K. Shibasaki, *Plant Physiol.*, 1975, **56**, 19.
5. E. Dickinson, A. Murray, B. S. Murray, and G. Stainsby, in 'Food Emulsions and Foams', ed. E. Dickinson, Royal Society of Chemistry, London, 1987, p. 86.
6. E. Dickinson, B. S. Murray, and G. Stainsby, *J. Colloid Interface Sci.*, 1985, **106**, 259.
7. J. Castle, E. Dickinson, A. Murray, B. S. Murray, and G. Stainsby, in 'Gums and Stabilisers for the Food Industry', ed. G. O. Phillips, D. J. Wedlock, and P. A. Williams, Elsevier Applied Science, London, 1986, Vol. 3, p. 409.
8. E. Dickinson, *Colloids Surf.*, 1989, **42**, 191.
9. E. Dickinson, S. E. Rolfe, and D. G. Dalgleish, *Food Hydrocolloids*, 1989, **3**, 193.
10. E. Dickinson, S. E. Rolfe, and D. G. Dalgleish, *Food Hydrocolloids*, 1988, **2**, 397.

Effect of Ethanol on the Foaming of Food Macromolecules

By Maqsood Ahmed and Eric Dickinson

PROCTER DEPARTMENT OF FOOD SCIENCE, UNIVERSITY OF LEEDS,
LEEDS LS2 9JT

1 Introduction

Processes in which the foaming of liquids is encountered are very numerous industrially. In spite of this, certain aspects of foam stabilization are not completely understood. Foaming in industrial processes is mainly due to the presence of surface active agents and the stability of the foam formed is decisively affected by the nature of the surface active agent. In the foam, when bubbles approach each other, a thin film of the continuous aqueous phase is trapped between them. The behaviour of the thin aqueous film dictates the foam stability. The formation and stability of food foams is determined in large part by the properties of adsorbed films of protein at the air–water interface,[1] and one way of influencing the protein adsorption behaviour is to change the solvent quality of the aqueous continuous phase, *e.g.* by addition of alcohol.

In a recent study of the effect of ethanol content on stability of foam formed from dilute solutions of pure proteins by a shaking technique,[2] there was found to be a very marked ethanol concentration dependence, with maximum foam life times occurring at ethanol contents of 0.2–0.6 wt%. With foam produced from alcoholic drinks, it was also found[3] that optimum foaming exists at a well-defined ethanol concentration.

The work described here compares the effects of ethanol content on the foaming of a protein (β-lactoglobulin), two polysaccharides (gum arabic and propylene glycol alginate (PGA)), and a 1:1 mixture of protein + polysaccharide (β-lactoglobulin + PGA).

2 Materials

The β-lactoglobulin (freeze-dried salt free preparation, batch no. 44064) and ethyl alcohol (99.7 to 100 vol%) were AnalaR grade and were obtained from BDH Chemicals, as were the AnalaR grade buffer salts. Gum arabic was a commercial sample obtained from Pepsico, Inc. (Valhalla, N.Y.). Propylene glycol alginate (Kelcoloid LVF 48379 A) was a gift from Kelco, Inc. (San Diego, USA).

Solutions were made up in phosphate pH 7 buffer of total ionic strength 0.005 M. Care was taken to avoid foaming during solution preparation.

3   Experimental

Foam life times were determined by a shaking method. The advantages of this method over the alternative sparging method for solutions containing a volatile solvent (ethanol) are discussed elsewhere.[2] After thermal equilibration, each test solution was shaken at 25 ± 0.1 °C in a half-filled cylindrical glass bottle (volume 140 ml) with an air-tight plastic lid. The motion was in a vertical plane with a constant frequency of 6 Hz and maximum amplitude of 3 cm. The shaking time was exactly 1 minute, and the foam life time was defined as the time from the cessation of shaking to the complete disappearance of bubbles at the gas–liquid interface. To obtain consistent and reproducible results, it was essential to keep to the same size and shape of bottle, to fill the bottle to exactly the same level, and strictly to control the mechanical shaking conditions as described above. Foam life time was measured with a stopwatch. Quoted values are averages from three runs. Individual values were usually within ±20% of the mean.

Great care was taken in the cleaning of the apparatus and making up of the solutions to avoid adventitious contamination by trace impurities, as foam life time is known to be extremely sensitive to trace contaminations. All glassware was cleaned, first, with Decon 90, and then with concentrated sulphuric acid; after rinsing copiously with double-distilled water, it was dried in an oven.

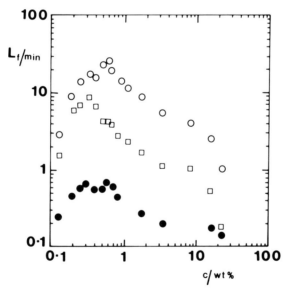

Figure 1   *Foam stability of β-lactoglobulin solutions (0.005 M phosphate buffer, pH 7) shaken at 25 °C. Foam life time L_f is plotted against ethanol concentration c: ○, 1×10^{-4} wt% protein; □, 5×10^{-5} wt% protein; ●, 2×10^{-5} wt% protein*

4 Results and Discussion

The results obtained for the foam life time of foams formed from dilute solutions of food macromolecules containing ethanol at various concentrations are given in Figures 1, 2, 3, and 4 for β-lactoglobulin, gum arabic, propylene glycol alginate, and β-lactoglobulin + propylene glycol alginate, respectively. Irrespective of the macromolecular concentration, the foam life time increases sharply with small additions of alcohol, and then gradually falls off again with further additions of alcohol. The maximum foam life time occurs at ethanol contents of 0.6 wt% for β-lactoglobulin solutions, 3.0 wt% for gum arabic solutions, 0.8 wt% for propylene glycol alginate solutions, and 0.3 wt% for a 1:1 mixture of β-lactoglobulin + propylene glycol alginate.

The foam stability behaviour of aqueous solutions of these food macro-molecules is qualitatively similar to that reported previously[2] for the pure proteins β-casein, gelatin, and lysozyme. As for the existence of a maximum in the foam stability of protein solutions at low alcohol concentration, it would seem that others have been aware of the general phenomenon.[4] The effect of ethanol content on the foaminess of dilute protein solutions has been investigated by various workers[5–8] using a sparging method. We have found,[2] however, that in a sparging system the foaming characteristics of dilute protein solutions are substantially influenced by the gradual stripping of ethanol from the solution by the gas stream, a factor which appears to have been disregarded by these earlier workers.[5–8] Nevertheless, the qualitative trends reported previously[5–8] appear

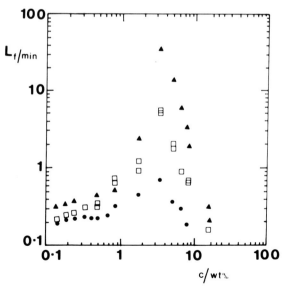

Figure 2 *Foam stability of gum arabic solutions (0.005 M phosphate buffer, pH 7) shaken at 25 °C. Foam life time L_f is plotted against ethanol concentration c:* ▲, 1×10^{-1} *wt% gum arabic;* □, 5×10^{-2} *wt% gum arabic;* ●, 1×10^{-2} *wt% gum arabic*

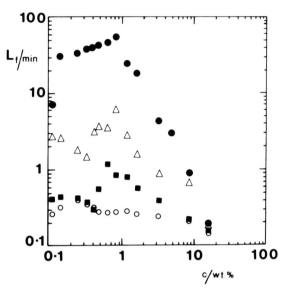

Figure 3   *Foam stability of propylene glycol alginate solutions* (0.005 M *phosphate buffer*, pH 7) *shaken at* 25 °C. *Foam life time* L_f *is plotted against ethanol concentration c:* ●, 5×10^{-3} wt% *propylene glycol alginate;* △, 5×10^{-4} wt% *propylene glycol alginate;* ■, 1×10^{-4} wt% *propylene glycol alginate;* ○, 5×10^{-5} wt% *propylene glycol alginate*

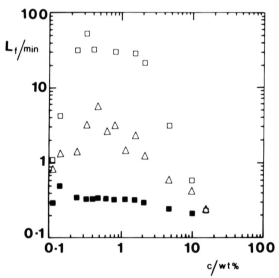

Figure 4   *Foam stability of β-lactoglobulin and propylene glycol alginate mixtures* (0.005 M *phosphate buffer*, pH 7) *shaken at* 25 °C. *Foam life time* L_f *is plotted against ethanol concentration c:* □, 7×10^{-5} wt% (*β-lactoglobulin* + PGA); △, 5×10^{-5} wt% (*β-lactoglobulin* + PGA); ■, 2×10^{-5} wt% (*β-lactoglobulin* + PGA)

similar to those found here. The main result is that foam stabilized by food macromolecules has a maximum stability in the presence of a small amount of alcohol. The reason seems likely to be related to the sharp lowering of the surface tension on addition of a small amount of ethanol to water,[9] the associated formation of smaller gas bubbles,[2] and the accompanying changes in dynamic surface rheological properties of the gas–liquid interface,[10] which may lead to a reduction in the rates of film drainage and bubble coalescence. While small quantities of ethanol enhance foamability, an increasing concentration may lower foam stability by reducing surface viscoelasticity of the adsorbed macromolecular film, as occurs with caseinate at the oil–water interface.[11] In addition, food macromolecules will tend to become more aggregated in the bulk solution in the presence of ethanol, and this will lower their availability for effective adsorption and foam stabilization.[3]

References

1. E. Dickinson, in 'Foams: Physics, Chemistry and Structure', ed. A. J. Wilson, Springer-Verlag, London, 1989, p. 39.
2. M. Ahmed and E. Dickinson, *Colloids Surf.*, 1990, **47**, 353.
3. M. Ahmed and E. Dickinson, *Food Hydrocolloids*, 1990, **4**, 77.
4. A. Prins and K. Van't Riet, *Trends Biotechnol.*, 1987, **5**, 296.
5. W. Bumbullis and K. Schügerl, *Eur. J. Appl. Microbiol. Biotechnol.*, 1979, **8**, 17.
6. K. Kalischewski and K. Schügerl, *Colloid Polym. Sci.*, 1979, **257**, 1099.
7. K. Kalischewski, W. Bumbullis, and K. Schügerl, *Eur. J. Appl. Microbiol. Biotechnol.*, 1979, **7**, 21.
8. A. R. Lawers and R. Ruyssen, in 'Proceeding of 4th International Congr. Surface Active Substances', ed. F. Asinger, Gordon and Breach, London, 1964, Vol. 2, p. 1153.
9. B. Janczuk, T. Bialopiotrowicz, and W. Wojcik, *Colloids Surf.*, 1989, **36**, 391.
10. A. Prins, in 'Advances in Food Emulsions and Foams', eds. E. Dickinson and G. Stainsby, Elsevier Applied Science, London, 1988, p. 91.
11. E. Dickinson and C. M. Woskett, *Food Hydrocolloids*, 1988, **2**, 187.

Gelation in a Synthetic Polypeptide System

By J. C. Horton and A. M. Donald

CAVENDISH LABORATORY, MADINGLEY ROAD, CAMBRIDGE CB3 0HE

1 Introduction

The gelation of biological and quasi-biological systems has been investigated for many years now. The synthetic polypeptide poly(γ-benzyl-α,L-glutamate) (PBLG) is an example of a quasi-biological material which can form gels. It can be regarded as a man-made analogue of a protein, in that it consists of successive peptide links to each of which is attached the same $-(CH_2)_2-$ $(C{=}O)-O-CH_2-C_6H_5$ side-group. In helicoidal solvents it adopts the characteristic α-helix conformation found in many biological materials. This confers rigidity on the molecule which leads to the molecule exhibiting liquid crystallinity. In 1956, Flory[1] proposed a phase diagram for rod-like molecules which experimental workers have subsequently shown to be correct in most of its major features. In particular, Miller and co-workers[2,3] have examined the PBLG + dimethyl formamide (DMF) system in some detail. The work described here involves PBLG in benzyl alcohol (BA)—also a helicoidal solvent. This latter system gels at room temperature rather than below (which is the case for PBLG + DMF).

The phase diagram for rod-like molecules as presented by Flory (see Figure 1) shows three separate regions, two of which correspond to a single phase (an isotropic phase at low concentrations and an anisotropic one at high concentrations), and one of which is biphasic. This biphasic region, consisting of two coexisting phases (in accordance with the usual rules of phase diagrams), can be conveniently divided into three sections: the wide biphasic, the chimney, and the cap. The wide biphasic and the chimney both consist of one isotropic phase and one anisotropic—they differ in the range of concentration which they span (see Figure 1). The cap is different from the other two in that it consists of two anisotropic phases. Flory was uncertain whether it would exist in reality, and he considered it might only be an artefact of his calculations. For some systems (e.g. PCBL* + DMF[4]), there are data suggesting that indeed it does not, and that the phase boundary in that part of the phase diagram is simply a monotonically

---

* PCBL, poly(ε-carbobenzoxy-α,L-lysine), is similar to PBLG but with the side group $-(CH_2)_4-NH-(C{=}O)-O-CH_2-C_6H_5$.

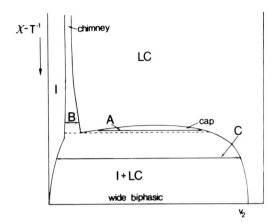

Figure 1 *Flory's phase diagram for rod-like molecules. The axes are solvent quality (χ) and volume fraction of polymer (v_2). (Note: $\chi \sim T^{-1}$, where T is temperature.) The annotation on the diagram is: I, isotropic; LC, liquid crystalline. The tie lines marked A, B, and C, indicated by double-headed arrows, are referred to in the text. The dotted lines divide the biphasic into its three constituent parts: chimney, cap, and wide biphasic*

decreasing function of the volume fraction of polymer, v_2. However, there is evidence that it does occur in the PBLG + DMF[4] and PBLG + BA[5] systems. In particular, Hill and Donald[5] observed two liquid crystalline phases resulting from a phase separation which took place in the hot stage of an optical polarizing microscope, although they were concerned that the small scale of the experiment might have affected its validity. The aim of the present work was to attempt to reproduce the earlier result for PBLG + BA in the bulk, and to relate gelation to the Flory phase diagram.

2 Experimental

Samples of PBLG + BA gels were made up with 99% purity BA to a concentration of 15 vol%. The molecular weight of the PBLG was 2.6×10^5 daltons. Details of the preparation of these stock gels can be found elsewhere.[6]

Samples from the stock gels were placed in small-bore test tubes and immersed in a hot oil bath (with temperature set to between 80 and 110 °C) and the products were examined. The main techniques used for this were X-ray scattering (using a powder camera), differential scanning calorimetry (DSC), and polarized light microscopy.

Gels were also made up from dried BA. This was of two sorts. Initially, BA was dried in the laboratory by standing over K_2CO_3, filtering, and then distilling at reduced pressure (~ 10 mm Hg). This material was subsequently superseded by BA dried by the manufacturers (Aldrich) and supplied under N_2.

3   Results

Three distinct classes of separations were found, each being a separation into two phases. These will be referred to as A, B, and C. Basically, they appear to be the three possible combinations from taking three different phases two at a time, and we have identified them as occurring in the cap, the chimney, and the wide biphasic, respectively. Reasons for this are given below.

In each case, one phase lies above the other in the tube, and some sort of boundary between the two can be seen. One of the phases is isotropic, whilst the other two are anisotropic. In the two separations which involve the isotropic phase, the separation of the two phases is very distinct and resembles a classic water meniscus. In the third separation, in which both the phases are anisotropic, the boundary is not as obvious, but it can be seen clearly when well illuminated. The three phases will now be described.

The isotropic phase is very fluid at elevated temperature. When returned to room temperature it may still be fluid, but it shows white entities visible to the naked eye (as when it occurs in separation C), or it may be a very weak gel (as in separation B). It is notable in that it shows sharp rings in its X-ray spectrum; so far we have observed five.[7] Needle-shaped particles are visible when the isotropic phase is viewed between crossed polars. These disappear when the sample is heated to *ca.* 60 °C. This is the same temperature at which this material shows a strong endotherm in the DSC experiment.

The two anisotropic phases can be distinguished by the naked eye. That from B (which we shall designate LC1) appears clear, but the one from C (LC2) gives a frosted appearance. When viewed between crossed polars, LC2 shows the finger prints characteristic of a cholesteric phase (see Figure 2), but LC1 does not. However, LC1 does show faint lines in the X-ray spectrum, whereas LC2 only shows the diffuse halo of the solvent. Both phases show dips in the DSC trace at *ca.* 60 °C, but not as strong as that displayed by the isotropic phase. It is worth

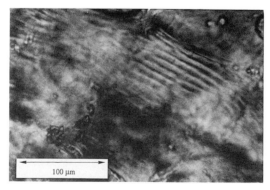

100 µm

Figure 2   *The finger prints of the anisotropic phase LC2 as seen between crossed polars*

noting that the unseparated gel also shows a dip at *ca.* 60 °C in its DSC trace[6] and only a diffuse halo in the powder camera.

The gels prepared from dried BA could not be induced to show a phase separation at any temperature. However, a gel made with commercially dried BA which had been deliberately contaminated with water did separate.

4　Discussion

Phase separation A involves two anisotropic phases, and it therefore can be attributed to the cap. Separations B and C both involve an isotropic phase, and so can be fitted onto Flory's phase diagram in the chimney and the wide biphasic. Matching the anisotropic phases of B and C with those found in A (according to whether or not they produce finger prints) suggests that B occurs in the chimney and C in the wide biphasic.

That such phase separations can be found in a system which subsequently gels at some lower temperature is of considerable interest since phase separation (or, to be more precise, spinodal decomposition) has been proposed as a mechanism for causing gelation in the PBLG + DMF system. In particular, Russo *et al.*[8] have found evidence for spinodal decomposition from light scattering. However, a second conjecture, favoured by several groups, is that the junction zones of the gel are crystalline.[9,10] We believe the two mechanisms can be reconciled by looking at results from a somewhat different field. Phase separation has been invoked by Berghmans and others[11,12] as causing gelation in atactic polystyrene (a-PS) solutions. Essentially, the Berghmans scheme is that, if an a-PS solution is in the two-phase region of its phase diagram, the usual separation into two phases will occur at sufficiently high temperatures, but that at lower temperatures the polymer-rich phase may reach a concentration at which it becomes glassy before it can reach its equilibrium concentration. The glassy particles so formed in this latter case then confer rigidity on the system which is deemed to have gelled. In the results presented here, it has been shown that the isotropic phase, and, to a lesser extent, LC1, shows sharp lines in its *X*-ray spectrum. LC2 and the unseparated gel do not. However, all four of these materials show endotherms (of various strengths) in their DSC traces at *ca.* 60 °C. There is strong evidence from the optical microscope that there are crystals present in the isotropic phase and that they melt at *ca.* 60 °C. It seems probable therefore that the 60 °C transition in the other materials is also caused by the melting of crystals, although, since the signal is not as strong, the crystals may be smaller or less perfect (or both) in these three cases. This may be caused by the high viscosities of the last three materials making it difficult for large crystals to form in them. (This would also be consistent with the observation of faint lines in the *X*-ray spectrum of LC1.)

The use of dry BA in the preparation of a gel seems to suppress the occurrence of phase separations. Russo and Miller[13] have shown that adding a non-solvent (*i.e.* decreasing the overall quality of the solvent) can be shown on the Flory phase diagram as displacing the biphasic to higher temperatures. This means that on the phase diagram of a wet gel there is more likelihood of finding a position which is inside the biphasic (making phase separation possible) but above *ca.* 60 °C where gelation sets in.

5 Conclusions

Three types of phase separation have been seen in the PBLG + BA system. It appears that the mechanism of gelation can be related to these phase separations. At high temperatures, phase separation proceeds in accord with the Flory phase diagram. However, on cooling to room temperature, crystals may form in any of the phases, and, under normal conditions of preparation, this leads to large scale separation being prevented by the crystals. Consequently, the system gels.

Acknowledgement. The authors acknowledge financial support from the Agricultural and Food Research Council.

References

1. P. J. Flory, *Proc. Roy. Soc. (London)*, 1956, **A234**, 73.
2. E. L. Wee and W. G. Miller, *J. Phys. Chem.*, 1971, **75**, 1446.
3. W. G. Miller, C. C. Wu, E. L. Wee, G. L. Santee, J. H. Rai, and K. G. Goebel, *Pure Appl. Chem.*, 1974, **38**, 37.
4. P. S. Russo and W. G. Miller, *Macromolecules*, 1983, **16**, 1690.
5. A. Hill and A. M. Donald, *Liquid Crystals*, 1989, **6**, 93.
6. A. Hill and A. M. Donald, *Polymer*, 1988, **29**, 1426.
7. J. C. Horton and A. M. Donald, in 'Physical Networks', ed. W. Burchard and S. B. Ross-Murphy, Elsevier Applied Science, London, 1990, p. 159.
8. P. S. Russo, P. Magestro, and W. G. Miller, in 'Reversible Polymeric Gels and Related Systems', ed. P. S. Russo, American Chemical Society, Washington, D.C., 1987, p. 152.
9. B. Ginzburg, T. Siromyatnikova, and S. Frenkel, *Polym. Bull.*, 1985, **13**, 139.
10. S. Sasaki, K. Tokuma, and I. Uematsu, *Polym. Bull.*, 1983, **10**, 539.
11. J. Arnauts and H. Berghmans, *Polym. Comm.*, 1987, **28**, 66.
12. R. M. Hikmet, S. Callister, and A. Keller, *Polymer*, 1988, **29**, 1378.
13. P. S. Russo and W. G. Miller, *Macromolecules*, 1984, **17**, 1324.

Electrokinetic Properties of Gels

By T. J. Lewis, J. P. Llewellyn, and M. J. van der Sluijs

INSTITUTE OF MOLECULAR AND BIOMOLECULAR ELECTRONICS, UNIVERSITY OF
WALES, BANGOR, GWYNEDD LL57 1UT

1 Introduction

The equilibrium state of a gel is determined by a balance of forces arising from the rubber-like elasticity of the polymer strands, the polymer–polymer affinity, and the interactions between the electrical charges on the polymers and the ions in the liquid medium. These components, together with any external forces, determine the gel volume, and dramatic changes in this volume may be elicited by small changes in the balance of the forces. Tanaka *et al.* have shown[1] that an electric field of only $1\,\mathrm{V\,cm^{-1}}$, which acts directly on the charges of the system, is sufficient to collapse a polyacrylamide gel. In the present paper, we report further on the electrokinetic effect in gels. Employing alternating electric fields at frequencies in the range 200 Hz–100 kHz, we have examined the induced cyclic dimensional changes of aqueous gels of gelatin and agarose, both of which have a polyelectrolyte structure with counter-ions present in the aqueous medium.

An understanding of the electrokinetic effect in gels is of particular interest for two reasons. The intermediate gel phase is an important component in living systems, but an appreciation of the role played by the interaction of these gels with biologically-generated electric fields is presently lacking. Secondly, the application of the electrokinetic effect will be valuable in providing information on the electrical and viscoelastic properties of bio-molecules, such as proteins, when incorporated into a gel matrix.

2 Experimental Method

To avoid extraneous effects such as unwanted heating and conduction, it is necessary to employ electric fields of magnitude less than $2.5 \times 10^3\,\mathrm{V\,m^{-1}}$ which produce displacements in the gel in the picometre to nanometre range. Fortunately, these very small displacements can be measured by optical interferometry using the method developed by Kwaaitaal[2] and others.

In our initial adaptation of this method, the gel sample, in the form of a circular disc, was placed between thin stainless steel electrodes, one of which was polished and used as the mirror in one arm of a Michelson interferometer. The cyclic gel displacement caused a small modulation of the fringe intensity, which was detected with a photodiode, and measured by a phase sensitive detector. In later modifications of this method, the gel sample, between its stainless steel electrodes, was totally immersed in water on the surface of which was floated a small glass mirror. This mirror faithfully reproduced the motion of the upper electrode on the gel provided the depth of water above the gel was not greater than about 1 mm.

An essential feature of the method is a feed-back loop in the interferometer which prevents any fringe movement due to temperature change, mechanical vibration or acoustic noise. We are able to measure sinusoidal displacements down to 10^{-12} m, and, using a two-channel phase-sensitive detector, both the in- and the out-of-phase components of the electrokinetic motion with respect to the applied field can be found.

3   Results

A typical frequency response of a gelatin gel, which, in the range we have studied, is linear in the electric field, is shown in Figure 1. The peaks, G_1 and G_2, are interpreted as arising from resonances due to standing waves in the gel. They shift to higher frequencies and become broader as the gel thickness is reduced (Figure 2). The product of resonant frequency and thickness is constant for the G_1 peak indicating the same vibrational mode, but for the G_2 peak the product decreases

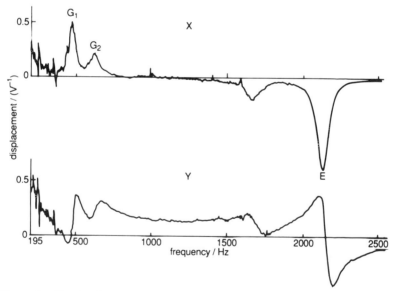

Figure 1   *The in-phase component* X *(top) and out-of-phase component* Y *(bottom) of the displacement of a 2 mm thick, 22 mm diameter, 12 wt% gelatin gel sample subjected to 0.2 V RMS. Peaks G_1, G_2, and E are assigned to gel and electrode resonances, respectively*

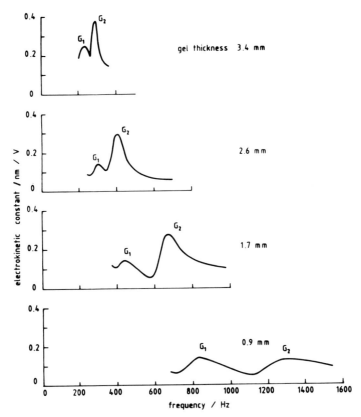

Figure 2   *The effect of gel thickness on the G_1 and G_2 resonances for a* 12 wt%
gelatin gel

systematically as the thickness increases. The reasons for this are not clear at
present.

Since the rigidity modulus of the gel will determine its wave velocity, and
therefore the resonant frequency, the G_1 and G_2 peaks may be expected to
depend on gelatin concentration. The observed dependence for a 2 mm sample is
shown in Figure 3 and indicates a linear relationship. Extrapolation to zero gelatin
concentration produces a resonant frequency of about 250 Hz.

The resonant peak E (Figure 1) can be attributed to the lowest resonant mode
of the stainless steel top electrode on the gels.[3] This resonance is slightly modified
by the presence of the gel, and if the electrode is embedded within the gel it is
suppressed altogether.

The gel samples exhibited a degree of electrokinetic asymmetry. Application of
a steady voltage bias across the cell during measurements reduced the response,
the reduction increasing with bias voltage and being much more pronounced
when the top electrode was biased positively. This asymmetry is likely to be

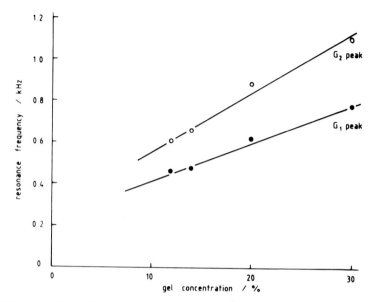

Figure 3   *The effect of gelatin concentration on the G_1 and G_2 resonant frequencies for constant sample thickness* 2 mm

related to the observation that, after a sample was prepared by allowing the gelatin solution to cool between the electrodes, a persistent potential of about 100 mV was established between the electrodes with the upper one generally negative. We believe this to be a manifestation of the Costa Ribeiro effect[4] responsible for charge separation across an advancing solid–liquid interface in a solidifying molten dielectric.

Measurement of the electrokinetic effect in a 2 wt% agarose gel produces the results shown in Figure 4, which are similar in magnitude and with resonances in the same frequency range as were obtained with the gelatin samples. However, the agarose sample was made significantly thicker than the gelatin one, 13 mm as opposed to 2 mm, so that the electrodes could be embedded in it, and this has produced a corresponding reduction in the resonant frequencies. A more detailed investigation of the agarose gels is under way.

4   Discussion

The electrokinetic effect in these gels is quite significant; indeed the response (of the order of 1 nm V^{-1}) is comparable to that of conventional piezo-electric transducers (0.5 nm V^{-1}). There are several factors which lead us to believe that the origin of the effect is at the electrode–gel interfaces.

Firstly, the experiments of Tanaka *et al.*[1] using steady fields comparable in strength to ours showed that gel contraction occurred preferentially at one electrode. With alternating fields, this contraction may be expected to occur

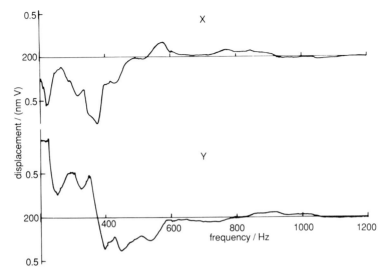

Figure 4   *The in-phase component* X (*top*) *and out-of-phase component* Y
(*bottom*) *of the displacement of a* 13 mm *thick,* 22 mm *diameter,* 2 wt%
agarose gel sample subjected to 2 V RMS

cyclically at each electrode in turn generating propagating waves with radial
symmetry. A mode conversion due to the Poisson ratio would allow this
transverse motion to be converted to a longitudinal one at the top electrode and
mirror.

Secondly, classical theory of piezo-electric transduction shows[5] that generation
is largely confined to regions where the product of the electric field and the
generating coefficient has maximum rate of change. In practice, this region will be
the electrode–transducer interface. In the case of aqueous polyelectrolyte gels,
the metal electrode contacts are likely to produce double-layers of large capacitance. Modulation of the dimensions of this capacitance by the applied field, as
well as the ion migration effects described by Tanaka,[6] could be responsible for
generating the electrokinetic effect. The presence of this large capacitance has led
us to conclude that the Y component of displacement shown in Figures 1 and 4 is in
phase with the field in the interfacial regions of the system.

Thirdly, using the methods given by Mason[7] and others, it is possible to model
the system in terms of the analogue electrical circuit representation of the acoustic
impedances of the gel, the electrodes, and the surrounding media. In this model,
the active generating element is assumed to be a thin layer of gel in contact with
each electrode. Using realistic values of the various acoustic impedances, we have
shown that dissipative effects reduce and broaden higher frequency resonances so
that only resonances at the lowest frequencies are significant. From this analysis
we conclude that the G_1 and G_2 peaks for the gelatin gel (Figure 1) correspond to
wave velocities in the gelatin gel of about 2 m s^{-1}, in keeping with the estimates of

rigidity moduli.[8] We believe that the same electrokinetic mechanisms are responsible for the resonances of the agarose gel, but that a lower dissipation allows more of them to be observed.

Acknowledgement. This work is supported by a research grant from the Science and Engineering Research Council.

References

1.   T. Tanaka, I. Nishio, S. T. Sun, and S. Ueno-Nishio, *Science*, 1982, **218**, 467.
2.   T. Kwaaitaal, *Rev. Sci. Instruments*, 1974, **45**, 39.
3.   A. B. Wood, *Proc. Phys. Soc.*, 1935, **47**, 794.
4.   B. Gross, *Phys. Rev.*, 1954, **94**, 1545.
5.   M. O'Donnell, L. J. Busse, and J. G. Miller, in 'Methods of Experimental Physics', ed. P. D. Edmonds, Academic Press, New York, 1981, Vol. 19, p. 29.
6.   T. Tanaka, *Sci. Am.*, 1981, **244**, 110.
7.   W. P. Mason, 'Electrochemical Transducers and Wave Filters', 2nd edn, Van Nostrand, Toronto, 1948, p. 185.
8.   J. D. Ferry, *J. Am. Chem. Soc.*, 1948, **70**, 2244.

New Improved Method* for Measuring the Surface Dilational Modulus of a Liquid

By J. J. Kokelaar, A. Prins

DEPARTMENT OF FOOD SCIENCE, WAGENINGEN AGRICULTURAL UNIVERSITY,
P.O. BOX 8129, 6700 EV WAGENINGEN, THE NETHERLANDS

and M. de Gee

DEPARTMENT OF MATHEMATICS, WAGENINGEN AGRICULTURAL UNIVERSITY,
THE NETHERLANDS

1 Introduction

The surface dilational modulus E of a liquid is defined according to the equation

$$E = A(\mathrm{d}\gamma/\mathrm{d}A) = \mathrm{d}\gamma/\mathrm{d}\ln A, \tag{1}$$

where A is the surface area of liquid, and γ is the surface tension. In order to measure the surface dilational modulus of a liquid, a sinusoidal change in surface area $\mathrm{d}A$ is applied to the liquid surface, and the resulting change in surface tension $\mathrm{d}\gamma$ is measured by using the Wilhelmy plate technique. The measurement is usually carried out in a Langmuir trough, in which one of the four barriers, which enclose a rectangular part of the surface, is moved. This technique has two disadvantages. The first one is of a fundamental nature: the change in surface area is not completely dilational, it contains also a shear component. The second one is of a practical nature: there is always a certain amount of leakage at the very point where the moving barrier touches the other barriers.

2 The New Method

The advantages of the new method described here are as follows.

(1) The moving barrier is absolutely free of leakage.
(2) The change in surface area is almost pure dilational.
(3) The method can also be applied to liquid–liquid interfaces.
(4) The construction of the apparatus is simple and cheap.

* Patent pending.

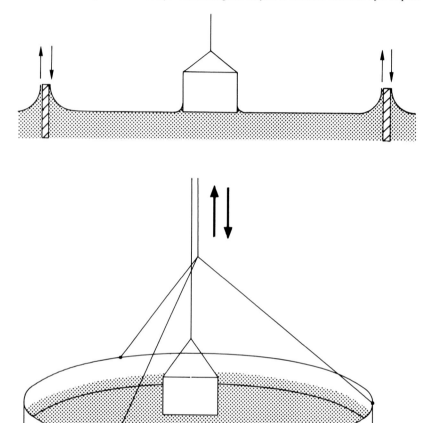

Figure 1   *Cylindrical barrier in a liquid surface. Side elevation (upper) and view from above (lower). Arrows represent directions of motion of the barrier*

The essential part of the new method is a cylindrical ring placed vertically in the liquid surface (Figure 1). When the inner wall of the ring is completely wetted, the contact angle of the liquid meniscus inside the ring is 0°. By simply moving the ring up and down, the surface area inside the ring can be enlarged or diminished. Because of the movement of the ring, there is always a liquid film right above the point where the liquid meniscus meets the inner wall of the ring. This film is being enlarged or diminished (Figure 2).

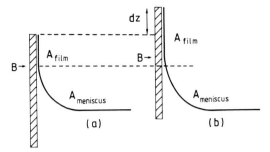

Figure 2 *Surface area of the liquid meniscus inside the ring* (a) *before and* (b) *after a vertical displacement dz of the ring*

3 Calculation of Parameters

In order to calculate properly the surface dilational modulus E, both the surface area of the meniscus A and the change in surface area dA inside the ring have to be known. The quantity A is the surface of the liquid meniscus inside the ring including the film present above point B in Figure 2a.

In a first-order approach, dA is equal to the pre-set vertical displacement of the ring dz multiplied by $2\pi R$, where R is the inner radius of the ring. In this approach, it is assumed that the point where the meniscus meets the inner wall remains on

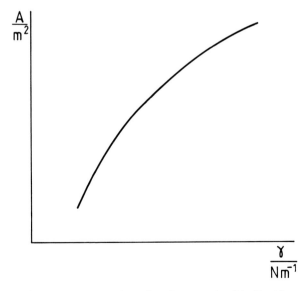

Figure 3 *Schematic representation of surface area* A *of the liquid meniscus inside the ring as a function of surface tension* γ

the same height with respect to the 'zero' level of the meniscus. This is not correct, and so, in a second-order approach, one has to take into account that due to the change in surface tension the shape of the liquid meniscus will change, which implies that the capillary rise (point B) changes as well. Hence, in this approach, the change in surface area consists of two parts: (i) the change in surface area of the adhering film against the ring, dA_{film}; and (ii) the change in surface area of the liquid meniscus up to the point where it meets the wall, $dA_{meniscus}$. We then have

$$dA = dA_{film} + dA_{meniscus}.$$ (2)

Both dA_{film} and $dA_{meniscus}$ are functions of γ and dz, and this relationship can be obtained using a special computer program. This results in a graph of A as a function of γ as shown schematically in Figure 3, which can be used to calculate the true value of the surface dilational modulus of the liquid, as well as the storage modulus, the loss modulus, and the loss angle of the modulus.

4 Conclusion

The new technique described here is an improvement over the usual determination of the surface dilational modulus of a liquid carried out by means of a Langmuir trough. The complete absence of leakage opens up the possibility of investigating the behaviour of surfactants when they are allowed to reach the surface only by diffusion through the liquid. To this end, the surfactant is spread over the surface at the outside of the ring only. The combined recording of the surface tension and the surface dilational modulus of the surface inside the ring gives worthwhile information about the properties of that surfactant.

Application of Fractal Aggregation Theory to the Heat Coagulation of Milk Products

By J. A. Nieuwenhuijse

cc FRIESLAND R&D, P.O. BOX 2206, 8901 MA LEEUWARDEN, THE NETHERLANDS

and P. Walstra

DEPARTMENT OF FOOD SCIENCE, WAGENINGEN AGRICULTURAL UNIVERSITY, P.O. BOX 8129, 6700 EV WAGENINGEN, THE NETHERLANDS

1 Introduction

About 80% of the protein in milk is casein. The casein is present as colloidal particles, the so-called 'casein micelles'. Casein monomers have little secondary structure, and so they are not able to be denatured by heat. Nevertheless, an intensive heat treatment of milk does result in coagulation, principally by aggregation of the casein micelles. Heat coagulation of milk products has been extensively studied, but attention so far has been focused almost exclusively on the effect of compositional and processing factors on the heat coagulation time (HCT), *i.e.* the time needed at a certain high temperature (120–140 °C) to cause visible coagulation in an agitated sample of milk. Here, we present some results of the application of fractal aggregation theory to the heat coagulation of milk products.

2 The Model

An aggregate of fractal (*i.e.* scale invariant) geometry can be described by the equation

$$N_p = (R_a/a)^D \tag{1}$$

where N_p is the number of particles in a growing floc, R_a is the radius of the floc, a is the radius of the primary particle, and D is the fractal dimensionality ($\leqslant 3$ in 3-dimensional space). Fractal geometry applies at scales larger than a, and for a sufficiently large value of N_p. The maximum number of particles in a floc, N_a, is equal to $(R_a/a)^3$. The volume fraction of particles in a growing floc ϕ_a is thus given by[1]

$$\phi_a = N_p/N_a = (R_a/a)^{D-3}. \tag{2}$$

Thus, ϕ_a decreases as R_a increases, and a gel will be formed when ϕ_a becomes equal to the total volume fraction of primary particles, ϕ_t, in the liquid. At that moment, we have $\phi_t = (R_{a,gel}/a)^{D-3}$ and $N_{p,gel} = (R_{a,gel}/a)^D$, and so we have:

$$N_{p,gel} = \phi_t^{D/(D-3)}. \tag{3}$$

For aggregates to be formed, the particles must encounter each other. To calculate the encounter frequency, one may use the Smoluchowski equation for slow coagulation:

$$N_t = \frac{N_0}{1 + (4kTN_0/3\eta)t} \frac{1}{W}, \tag{4}$$

where N_0 and N_t are the initial number of particles and the number at time t, respectively, η is the viscosity of the liquid in which the particles are suspended, kT has its usual meaning, and W is the stability factor. Of course, at the time of formation of a gel, t_{gel}, the quantity $N_{p,gel}$ equals $N_0/N_{t,gel}$; combination of equations (3) and (4), and using ϕ_t rather than N_0, then yields:

$$t_{gel} = \frac{\phi_t^{D/(D-3)} - 1}{\phi_t} \frac{\pi\eta a^3}{kT} W. \tag{5}$$

Representing the results in a log t_{gel} *versus* log ϕ_t plot, and comparing the plots with theoretical plots, yields an average value for W during the process of the coagulation, and one for D at the moment of gelation.

3   Validity of the Model

For the above model to be valid, some conditions must be fulfilled. Since the theory is based only on the dependency of t_{gel} on ϕ_t, W should not depend on the heating time, and so not on ϕ_t. If, for instance, W were to depend on some heat-induced change in the solvent phase, milk with a low ϕ_t would be affected by such a change to a greater extent than milk with a high ϕ_t. In addition, D at gelation should be independent of the heating period needed to induce gelation. Aggregation of particles may be either simple flocculation, leading to a gel, or flocculation followed by fusion, leading to large particles; and, if the ratio of the rate of flocculation over that of fusion changes during heating, D also changes. It is established[2] that heat coagulation of concentrated milk below the optimum pH is simple flocculation, and that, above the optimum pH, fusion predominates until shortly before the HCT. Hence, it appears that, below the optimum pH, the model is applicable, but it is not applicable above the optimum pH.

Preliminary results indicate that the heat coagulation of recombined cream also is simple flocculation, at least at or below the natural pH. This is not too surprising, since aggregation of fat globules predominates the mechanism of heat coagulation in cream, and no significant coalescence of fat globules was found during (indirect UHT) heating.[3]

4   Results

Concentrated skim milk (25 wt% total solids) was made from pre-heated milk (3 min at 120 °C). The protein concentration was varied by means of ultrafiltration (thus without altering the ratio of dissolved salts and lactose to water), *i.e.* retentate or the concentrated milk was diluted with permeate. The pH was altered by adding 1 M HCl, and the total solids content by adding the appropriate amount of demineralized water. The quantity ϕ_t was determined by viscometry, and t_{gel} was defined as the gelation or coagulation time (minus 70 s for 'heating-up') as determined by the so-called subjective heat stability test. Figure 1 is the HCT–pH plot, and Figure 2 is the log t_{gel} *versus* log ϕ_t plot. It can be seen that we have $D \approx 2.3$ and $W \approx 10^5$. The former value is fairly close to that found for other casein gels,[1] but the latter value is about a factor of 10 lower than found in kinetic studies on the heat coagulation of unconcentrated milk.[4]

5   Conclusions

At conditions where simple flocculation predominates, the heat coagulation strongly depends on the initial volume fraction of particles. Fractal aggregation

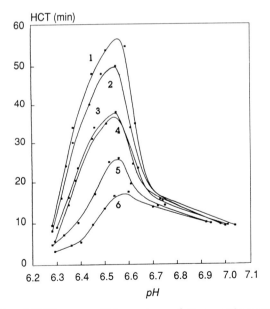

Figure 1   *Plot of HCT versus pH for various relative protein concentrations*: (1) 0.85; (2) 0.91; (3) 1.0 *(retentate + permeate)*; (4) 1.0 *(original concentrate)*; (5) 1.13; (6) 1.27

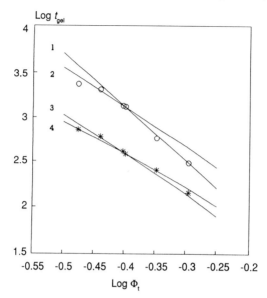

Figure 2   *Plot of logarithm of gelatin time* t_{gel} *versus logarithm of particle volume fraction* ϕ_t: *, *experiment* (pH 6.3); ○, *experiment* (pH 6.4); (1) *theory* (D = 2.5, W = 8.27 × 10⁴); (2) D = 2.3, W = 4.17 × 10⁵; (3) D = 2.3, W = 1.23 × 10⁵; (4) D = 2.1, W = 3.19 × 10⁵

theory describes this dependence rather well. Completely reliable values for D and W cannot be obtained, however, because ϕ_t has been determined at 20 °C while its value at 120 °C is needed in equation (5). In addition, the HCT depends not only on the time needed for a gel to be formed, but also on the strength of the gel; hence it is not an accurate measure of t_{gel} for all kinds of milk products. Other complicating factors are that the flocs are in most cases not very large and probably interpenetrate somewhat when the gel is formed. In addition, W is not likely to be completely independent of the heating time.

References

1. L. G. B. Bremer, T. van Vliet, and P. Walstra, *J. Chem. Soc., Faraday Trans. 1*, 1989, **85**, 3359.
2. J. A. Nieuwenhuijse, A. Sjollema, M. A. J. S. van Boekel, T. van Vliet, and P. Walstra, *Neth. Milk Dairy J.*, to be published.
3. J. P. Melsen and P. Walstra, *Neth. Milk Dairy J.*, 1989, **43**, 63.
4. M. A. J. S. van Boekel, J. A. Nieuwenhuijse, and P. Walstra, *Neth. Milk Dairy J.*, 1989, **43**, 129.

Sedimentation in Aqueous Xanthan + Galactomannan Mixtures

By H. Luyten, T. van Vliet, and W. Kloek

DEPARTMENT OF FOOD SCIENCE, WAGENINGEN AGRICULTURAL UNIVERSITY,
P.O. BOX 8129, 6700 EV WAGENINGEN, THE NETHERLANDS

1 Introduction

The sedimentation of suspended particles in a liquid dispersion is regulated by the rheological properties of the dispersion at low stresses and low shear rates.[1] Here, some experimental results for the viscosity of xanthan + galactomannan mixtures as determined by sedimentation and with a rheometer are compared.

2 Materials and Methods

Sedimentation experiments were performed with glass beads of 0.2–5 mm diameter in a glass tube of 18 mm diameter. Sedimentation viscosities were calculated with the Stokes equation corrected for the wall effect and for non-laminar flow. Apparatus viscosity was measured with a Deer Rheometer PDR 81 equipped with concentric cylinders.

The polysaccharides used were a gift from Suiker Unie (Roosendaal, The Netherlands). The tube and the rheometer were filled with the hot polysaccharide mixture after heating the dispersion to 100 °C. Measurements were performed at 20 °C.

3 Results and Discussion

The apparent viscosity of glycerol + water mixtures, as well as of dispersions of xanthan in 0.1 wt% aqueous NaCl, as determined by sedimentation or in the rheometer, were, at comparable shear stresses, equal within the experimental accuracy. However, for xanthan + galactomannan mixtures, the apparent sedimentation viscosities were considerably higher than those measured in the rheometer at the same stress. Some examples of the results are shown in Figures 1 to 3. Differences in viscosities measured by the two techniques were found to get less when the gel structure in the xanthan + locust bean gum mixtures was disturbed (e.g. by gently shaking) before the measurements were done.

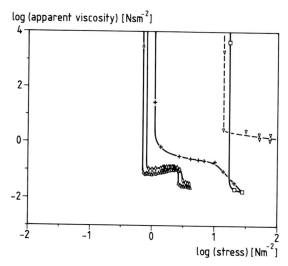

Figure 1   *The logarithm of the apparent viscosity of 0.05 wt% xan-*
than + 0.05 wt% locust bean gum in 0.1 wt% aqueous NaCl as a
function of the logarithm of the applied stress. Results are from sedimen-
tation (– – –) and Deer Rheometer (——) experiments

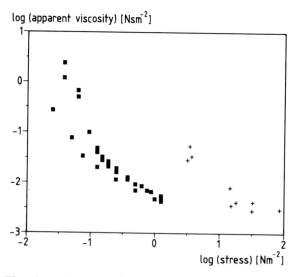

Figure 2   *The logarithm of the apparent viscosity of 0.05 wt% xan-*
than + 0.05 wt% guar gum in 0.1 wt% aqueous NaCl as a function of
the logarithm of the applied stress. Results are from sedimentation (+)
and Deer Rheometer (■) experiments

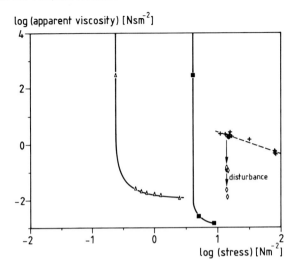

Figure 3 *The logarithm of the apparent viscosity of 0.02 wt% xanthan + 0.02 wt% locust bean gum in 0.1 wt% aqueous NaCl as a function of the logarithm of the applied stress. Results are from sedimentation (– – –) and Deer Rheometer (——) experiments*

A possible explanation for the difference in viscosity could be that the dispersions are inhomogeneous, *e.g.* due to the presence of lumps with a higher polysaccharide concentration in the dispersion. If these lumps are large compared to the size of the glass beads, but small compared to the geometry of the rheometer, the apparent viscosity which is relevant for the sedimentation of particles could indeed be higher than the mean viscosity of a larger volume of material which consists of fairly concentrated lumps in a low viscous liquid.

There are several indications that lumps could be present.

(i) A xanthan + locust bean gum mixture which had been disturbed appeared to be inhomogeneous. On standing, a low-viscosity liquid was expelled. After a very gentle disturbance of the gel, the lumps could be seen to be up to a few centimetres in diameter.

(ii) At first sight, a xanthan + guar gum mixture appeared homogeneous, but with a light microscope some transparent particles were visible. The size of these particles was up to a few tenths of a millimetre.

(iii) The yield stress in a xanthan + locust bean gum mixture was poorly reproducible (Figures 1 and 3); it was found to vary by a factor of 30.

Further research on these inhomogeneities is currently being carried out.

4 Conclusions

The rheological properties of some xanthan + galactomannan mixtures depends on the dimensions of the volume of material which is actually involved in the testing procedure. This may be due to inhomogeneities in the material; these are larger than the volume just around a suspended particle which determines the sedimentation rate, but smaller than the amount of material that is tested in a concentric cylinder rheometer. These inhomogeneities could consist of lumps having a higher polysaccharide concentration.

Acknowledgement. This work was supported by Suiker Unie and by the Dutch Innovation Oriented Program Carbohydrates (IOP-k).

Reference

1. T. van Vliet and P. Walstra, in 'Food Colloids', ed. R. D. Bee, P. Richmond, and J. Mingins, Royal Society of Chemistry, Cambridge, 1989, p. 206.

Molecular Motion in Carbohydrate + Water Mixtures in the Liquid and Glassy States as Studied by Spin-probe ESR

By M. J. G. W. Roozen

DEPARTMENT OF FOOD SCIENCE, WAGENINGEN AGRICULTURAL UNIVERSITY,
P.O. BOX 8129, 6700 EV WAGENINGEN, THE NETHERLANDS

and M. A. Hemminga

DEPARTMENT OF MOLECULAR PHYSICS, WAGENINGEN AGRICULTURAL
UNIVERSITY, P.O. BOX 8128, 6700 ET WAGENINGEN, THE NETHERLANDS

1 Introduction

The glass transition of amorphous carbohydrate + water mixtures has become a topic of increasing interest during the last few years.[1-8] The glass transition is the phenomenon in which a solid amorphous phase exhibits a discontinuous change in the specific heat on changing the temperature.[9] Due to an extremely high viscosity, the translational mobility of molecules at temperatures below the glass transition temperature is virtually zero.[10] Consequently, a food product in the glassy state would not be expected to decrease in quality during storage. It is therefore useful to study the molecular properties of glassy states in more detail. To this end, the techniques of conventional and saturation-transfer electron spin resonance, ESR, have been employed to study the rotational mobility of spin probes above and below the glass transition temperature in several carbohydrate + water mixtures.

2 Theory

To describe the isotropic rotational motion of small spin-probe molecules in liquids, a modified Stokes–Einstein relation has been used:[11-14]

$$\tau_c = (\eta V/k_b T)k + \tau_0, \tag{1}$$

in which τ_c is the rotational mobility of the spin probe (a measure of the time during which the molecule persists in a given orientation), η the Stokes viscosity,

V is the volume of the spin probe, k_b is Boltzmann's constant, and τ_0 is the zero viscosity rotational correlation time (which has often been assumed to be negligible[14-16]). The parameter k depends on specific solvent–probe interactions and on the geometry of the probe molecules. This can be expressed in the form[17]

$$k = S(1 - \theta) + \theta, \qquad (2)$$

in which θ is a geometric function that depends on the shape of the spin-probe molecule. In the mixtures studied here, hydrogen bonding is the most important interaction between the spin probe and the solvent molecules. The parameter S can then be interpreted as a measure of the effectiveness of this bonding. For non-spherical molecules, S can be negative, implying that the molecules displace less solvent than is expected from their geometry. This is the case if a cavity is formed in which the molecule can carry out free rotation; k is then equal to zero, so that S equals $-\theta(1 - \theta)^{-1}$ [see equation (2)].

3   Methods

Carbohydrate + water mixtures were made by mixing sucrose (Merck) or malto-dextrin (Boehringer) and spin-probe (Aldrich) solutions. The following spin probes have been used:

1   2

The probe concentration in the samples was 0.2–0.5 mg ml^{-1}. ESR spectra were recorded on a Bruker 200D ESR spectrometer with nitrogen-flow temperature control. The recording conditions for the ESR spectra and the methods used for estimating τ_c from the spectra have been described elsewhere.[18]

4   Results

To analyse the interactions between the probe and the solvent in sucrose + water mixtures at temperatures above the freezing point, graphs of the values of τ_c *versus* η/T were plotted [see equation (1); data not shown]. To obtain values for k from the slopes of these graphs, the volume of the probe (see Table 1) is

Table 1   *Van der Waals volume* V *and shape parameter*
θ for different spin probes

Probe	θ^a	$V^b/\text{Å}^3$
1	0.05	180
2	0.10	310

[a] C. M. Hu and R. Zwanzig, *J. Chem. Phys.*, 1974, **60**, 4345.
[b] A. Bondi, *J. Chem. Phys.*, 1964, **68**, 441.

considered to be independent of the composition of the solution. As shown in Table 1, the van der Waals incremental volumes as tabulated by Bondi[19] are used.

In Figure 1, the parameter k is given as a function of the sucrose content in sucrose + water mixtures. The value of k at which S becomes negative is indicated with the arrow labelled 1. At concentrations above 40 wt%, a decrease in k is observed; this point is indicated with the arrow labelled 2.

The rotational correlation time for both spin probes after rewarming of a rapidly cooled 20 wt% sucrose + water mixture is shown in Figure 2. Between −70 and −31 °C, log τ_c decreases almost linearly with temperature. At temperatures between −31 and −28 °C (as indicated by the arrows), τ_c starts to decrease much more strongly.

In Figure 3, the rotational mobility for spin probe 1 in malto-hepta-ose + water mixture is shown. In rewarming the rapidly cooled 20 wt% malto-hepta-ose mixture, a strong decrease in τ_c at about −13 °C is observed, and for a 85 wt% mixture at 18 °C.

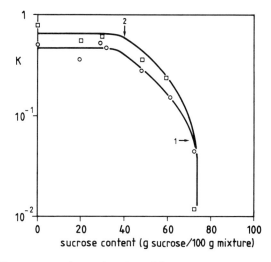

Figure 1   *The parameter* k *as a function of the sucrose content in sucrose + water mixtures*: ○, *spin probe* 1; □, *spin probe* 2

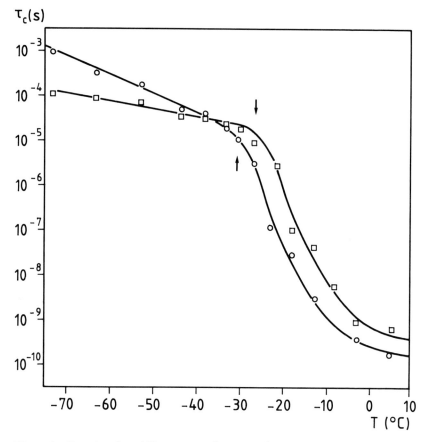

Figure 2   *Rotational mobility τ_c as a function of temperature T in a 20 wt% sucrose + water mixture*: ○, *spin probe* 1; □, *spin probe* 2

5   Discussion

Figure 1 shows that k is constant and close to unity in a sucrose + water mixture between 0 and 40 wt% sucrose: this can be explained by the fact that the probe molecules can form many hydrogen bonds with the solvent. At concentrations above 40 wt% sucrose, k decreases, indicating that fewer hydrogen bonds are formed between the probe molecules and the solvent. This is in agreement with the conclusions of Flink[20] that sucrose + water mixtures at concentrations over 40 wt% sucrose change from a solution of hydrated sucrose molecules to a sucrose–water phase, in which all the water molecules are involved in hydrogen bonds with sucrose. Arrow 1 in Figure 1 shows that k becomes equal to θ (and thus S becomes negative) at about 70 wt% sucrose. A negative value of S indicates that the probe molecules displace less solvent than can be expected from their shape. This means that they must be present in cavities without being part of the lattice.

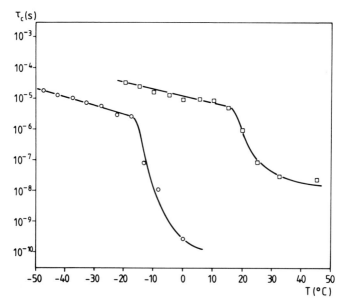

Figure 3   *Rotational mobility τ_c as a function of temperature T in a malto-hepta-ose + water mixture for spin probe* 1: ○, 80 wt% *water*; □, 15 wt% *water*

It is possible that the cavities are not a property of the lattice, but are induced by the probe molecule itself.

In Figure 2, it can be seen that the mobility of spin probes in a rapidly cooled 20 wt% sucrose solution starts (at rewarming) to increase strongly at temperatures between −33 and −28 °C. This agrees with the results of Levine and Slade,[21] who observed the glass transition temperature T_g for this mixture to be at −32 °C by means of differential scanning calorimetry measurements. At temperatures where part of the water is frozen, the spin probes are not present in an ice lattice, but in a concentrated amorphous solution (CAS). The value for η below T_g is *ca.* 10^{12} Pa s, which gives a value of k from equation (1) of 10^{-9}. This very low value of k implies that S is negative, and that the probes are present in cavities. The Stokes viscosity of the CAS decreases from *ca.* 10^{12} Pa s below the glass transition to *ca.* 10^3 Pa s at temperatures 20 °C above this transition due to the following effects.[6] (1) Ice starts to melt at the glass transition temperature, which gives rise to a plasticizing of the CAS.[22] (2) The free volume of the system increases—this is the volume of the mixture which is not occupied by molecules—which permits molecules to diffuse resulting in a dramatic decrease in the Stokes viscosity.[23,24] The relative decrease of τ_c as shown in Figure 2 is, however, much smaller than the decrease of η; this can be explained as follows. When the ice starts to melt due to the glass transition, k will increase strongly, as can be deduced from Figure 1. This effect will reduce the decrease in τ_c due to the glass transition.

As shown in Figure 3, spin probe 1, although it may be present in cavities, senses the glass transition of malto-hepta-ose + water mixtures: for a 20 wt% malto-hepta-ose solution at $-13\,°C$ and for an 85 wt% mixture at $18\,°C$. This is in reasonable agreement with results of Levine and Slade[21] and Orford *et al.*,[8] respectively.

Malto-tri-ose + water and malto-penta-ose + water mixtures show similar behaviour near the glass transition temperatures.

6   Conclusions

Spin probe ESR is a suitable technique to obtain information about the existence of hydrogen bonds and the presence of cavities in highly concentrated carbohydrate systems, and it may also sense the glass transition temperature.

Acknowledgements. We thank P. Walstra and T. van Vliet for fruitful and stimulating discussions. We are grateful to Unilever Research, Colworth House (UK), for financial support.

References

1.   F. Franks, in 'Properties of Water in Foods', ed. D. Simatos and J. Multon, Nijhoff Publishers, Dordrecht, 1985, p. 497.
2.   C. G. Biliaderis, C. M. Page, T. J. Maurice, and B. O. Juliano, *J. Agric. Food Chem.*, 1986, **34**, 16.
3.   H. Levine and L. Slade, in 'Food Structure: Its Creation and Evaluation', ed. J. M. V. Blanshard and J. R. Mitchell, Butterworth, London, 1988, p. 149.
4.   L. Slade and H. Levine, in 'Food Structure: Its Creation and Evaluation', ed. J. M. V. Blanshard and J. R. Mitchell, Butterworth, London, 1988, p. 115.
5.   D. Simatos and M. Karel, in 'Food Preservation and Moisture Control', ed. C. C. Seow, Elsevier Applied Science, London, 1988, p. 1.
6.   D. Simatos, G. Blond, and M. Le Meste, *Cryo-Letters*, 1989, **10**, 77.
7.   G. Blond, *Cryo-Letters*, 1989, **10**, 299.
8.   P. D. Orford, R. Parker, S. G. Ring, and A. C. Smith, *Int. J. Biol. Macromol.*, 1989, **11**, 91.
9.   S. R. Elliot, 'Physics of Amorphous Materials', Longman, London, 1983, Chap. 1.
10. F. Franks, 'Biophysics and Biochemistry at Low Temperatures', Cambridge Univ. Press, Cambridge, 1985, Chap. 3.
11. R. E. D. McClung and D. Kivelson, *J. Chem. Phys.*, 1968, **59**, 3380.
12. D. Kivelson, M. G. Kivelson, and I. Oppenheim, *J. Chem. Phys.*, 1970, **52**, 1811.
13. B. Kowert and D. Kivelson, *J. Chem. Phys.*, 1976, **64**, 5206.
14. J. L. Dote, D. Kivelson, and R. N. Schwartz, *J. Phys. Chem.*, 1981, **85**, 2169.
15. D. Kivelson and P. Madden, *Annu. Rev. Phys. Chem.*, 1980, **31**, 523.
16. D. R. Bauer, J. I. Brauman, and R. Pecora, *J. Chem. Soc.*, 1973, 6840.
17. D. Hoel and D. Kivelson, *J. Chem. Phys.*, 1975, **62**, 1323.
18. M. J. G. W. Roozen and M. A. Hemminga, *J. Phys. Chem.*, 1990, **94**.
19. A. Bondi, *J. Phys. Chem.*, 1964, **68**, 441.
20. J. M. Flink, in 'Physical Properties of Foods', ed. M. Peleg and E. B. Bageley, AVI Publishing Co., Westport, CT, 1983, p. 473.
21. H. Levine and L. Slade, *Cryo-Letters*, 1988, **9**, 21.
22. F. Franks, *Cryo-Letters*, 1986, **7**, 207.
23. M. L. Williams, R. F. Landel, and D. J. Ferry, *J. Am. Chem. Soc.*, 1955, **77**, 3701.
24. T. Soesanto and M. C. Williams, *J. Phys. Chem.*, 1981, **85**, 3338.

Mechanical Properties of Wheat Starch Plasticized with Glucose and Water

By Anne-Louise Ollett, Roger Parker, and Andrew C. Smith

AFRC INSTITUTE OF FOOD RESEARCH, NORWICH LABORATORY, COLNEY LANE, NORWICH NR4 7UA

1 Introduction

The deformation and fracture behaviour of systems containing food biopolymers is important both to food processing operations and to product texture. Much work has concentrated on food systems which contain high proportions of water making them soft and rubbery. In terms of modulus–temperature behaviour,[1,2] these materials are in the 'rubbery plateau' region. Relatively little work exists on systems containing low proportions of water (<25 wt%) corresponding to stiffer materials in the glass transition or 'glassy plateau' regions, which are typically described as leathery, or hard and brittle foods, respectively.

In this study, the flexural behaviour of solid wheat starch is investigated in the vicinity of its glass transition. The effects of different concentrations of water and glucose on the modulus and fracture properties are studied. The morphology of the fracture surfaces is investigated using scanning electron microscopy. This is a brief account; a fuller description without the complementary theoretical calculations is published elsewhere.[3]

2 Experimental

Sample Preparation—The samples were cut from narrow sheets extruded using a Baker–Perkins twin-screw extruder. Commercial starch and glucose monohydrate were pre-mixed to give glucose contents of 0%, 8.4%, 17.4%, 34.5%, and 67.8% (as a weight percentage of the dry solids). This was extruded at 6.0 kg h^{-1} with a water content of 30% (wet weight basis). A temperature of 140 °C in the central sections of the extruder ensured complete melting, and cooling the final sections prevented the melt from boiling and gave a bubble-free extrudate. Controlled drying of the samples resulted in water contents in the range 7–25%. The water content was determined gravimetrically.

Flexural Testing—The force–deflection behaviour of each sample was measured in a three-point bend test using an Instron 1122. Sample dimensions were in accordance with British Standard BS2782 and were typically 3.0 mm thick, 12.0 mm wide, and 100 mm long.

The maximum tensile stress σ and strain ε were calculated using the standard small deformation formulae,[4]

$$\sigma = \frac{3Fl}{2bh^2}, \tag{1}$$

where F is the load, and l, b, and h are the load support span, the sample breadth and the height, respectively, and

$$\varepsilon = \frac{6hY}{l^2}, \tag{2}$$

where Y is the deflection at the load nose. The samples were strained to a maximum of 0.04. The flexural modulus, $E = \sigma/\varepsilon$, was calculated from the initial slope of the stress–strain curve.

Scanning Electron Microscopy—After coating with gold to a depth of 20 nm, samples were viewed in a Philips 501B scanning electron microscope using a 15 kV accelerating voltage.

3 Results

Small Deformation—A summary of the flexural moduli of all the samples is shown in Figure 1. The moduli decrease with increasing water and glucose content. The early work of Tobolsky[1] and Ferry[2] categorized the state of pure amorphous polymers according to their moduli. At low water contents, when the modulus is above 1 GPa, the materials are in the glassy plateau region, *i.e.* below the glass transition temperatures T_g. At higher water contents, the moduli are in the range 50–150 MPa which is indicative of a partially crystalline rubber. Intermediate values correspond to materials in the region of their glass transitions. At all but the highest glucose content the glass transition is spanned by only 2 wt% of water content. This indicates the strong influence of water content upon the glass transition temperature of the mixture.

The composition dependence of T_g can be compared with the predictions of the classic thermodynamic theory of Couchman and Karasz.[5] The glass transition is modelled as an equilibrium second-order phase transition. Whilst the approximations made in the original derivation of the theory[5] made it applicable to compatible polymer blends, more recently the theory has been successfully applied to polymer + diluent mixtures.[6] The glass transition temperature of a multicomponent mixture is given by

$$T_g = \frac{\sum_i w_i \Delta C_{pi} T_{gi}}{\sum_i w_i \Delta C_{pi}}, \tag{3}$$

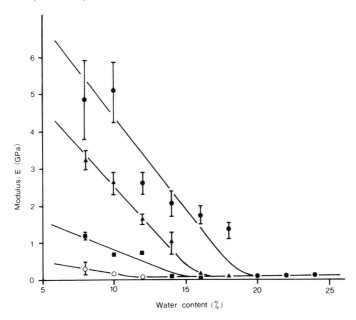

Figure 1   *Flexural modulus* E *versus water content for different glucose contents*:
●, 0 wt%; ▲, 8.4 wt%; ■, 17.4 wt%; ○, 67.8 wt%

where w_j, ΔC_{pi}, and T_{gi} are the mass fraction, heat capacity increment, and glass transition temperature of the ith component, respectively. The heat capacity increment is the increase in heat capacity at the glass transition, and it can be measured using a differential scanning calorimeter. The values of ΔC_p and T_g for starch are not measurable directly. They have been estimated by extrapolation of the values of a homologous series of malto-oligosaccharides.[7] These values together with those for glucose[7] and water[8] are summarized in Table 1. Values of T_g calculated using equation (3) are shown in Figure 2. Water depresses T_g of pure starch (A) most strongly. In the range of water contents studied, 7 to 25 wt%, T_g is depressed over 100 K. The effect is progressively weakened as glucose is substituted for starch. This explains the relatively broad glass transition region in Figure 1 for the 67.8 wt% glucose sample (E in Figure 2). A more detailed comparison can be made by using the experimentally determined compositions at which the

Table 1   *Glass transition temperature* T_g *and heat capacity increment* ΔC_p *of starch, glucose, and water*

Component	T_g/K	$\Delta C_p/J\ K^{-1}\ g^{-1}$
Starch	500	0.47
Glucose	303	0.88
Water	134	1.94

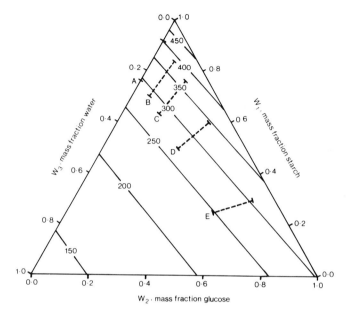

Figure 2 *Theoretical prediction of* T_g *for starch + glucose + water mixtures. Dashed lines show composition of samples studied. Glucose contents: A, 0 wt%; B, 8.4 wt%; C, 17.4 wt%; D, 34.5 wt%; E, 67.8 wt%*

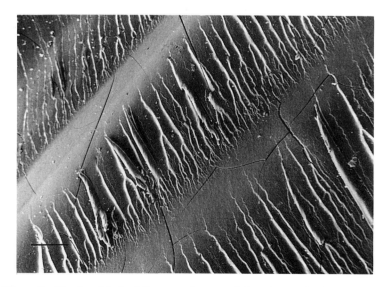

Figure 3 *Bands of forked lines running parallel to the crack direction. Length bar* $= 100\,\mu m$

glass transition occurs at 293 K to predict a theoretical T_g using equation (3). This comparison shows that theory overestimates T_g of our samples by between 11 and 44 K. This is similar to the level of agreement found with calorimetric determinations of T_g.[7,9]

Large Deformations and Fracture—Only samples with flexural moduli above 1 GPa fractured in this study. Samples with lower moduli showed non-Hookean behaviour and remained intact at the 0.04 strain limit. A limited study of yielding was made by repeatedly loading samples to below their fracture stress and measuring any permanent deformation. There was no significant yielding, and it was concluded that only brittle fracture was occurring.

Scanning electron microscopy showed that the same features occurred on the fracture surfaces throughout the range of water contents. Crack growth was initially perpendicular to the sample surface and bent away from this plane as the crack neared the top surface. Bands of forked lines running parallel to the crack direction marked these curved surfaces (Figure 3). These features have also been observed on the fracture surface of synthetic glassy polymers and are described as 'ribs' of 'river markings'.[10]

Acknowledgement.—The authors would like to thank Mrs S. A. Clark for obtaining the scanning electron micrographs.

References

1. A. V. Tobolsky, 'Properties and Structure of Polymers', John Wiley, New York, 1960.
2. J. D. Ferry, 'Viscoelastic Properties of Polymers', 1st edn, John Wiley, New York, 1961.
3. A.-L. Ollett, R. Parker, and A. C. Smith, *J. Mater. Sci.*, 1990.
4. L. E. Nielsen, 'Mechanical Properties of Solid Polymers and Composites', Marcel Dekker, New York, 1974, Vol. 1.
5. P. R. Couchman and F. E. Karasz, *Macromolecules*, 1978, **11**, 117.
6. G. ten Brinke, F. E. Karasz, and T. S. Ellis, *Macromolecules*, 1983, **16**, 244.
7. P. D. Orford, R. Parker, S. G. Ring, and A. C. Smith, *Int. J. Biol. Macromol.*, 1989, **11**, 91.
8. M. Sugisaki, H. Suga, and S. Seki, *Bull. Chem. Soc. Jpn.*, 1968, **41**, 2591.
9. K. J. Zeleznak and R. C. Hoseney, *Cereal Chem.*, 1987, **64**, 121.
10. R. P. Kusy and D. T. Turner, *Polymer*, 1977, **18**, 391.

Structure of Microemulsion-based Organo-Gels

By Peter J. Atkinson, Brian H. Robinson

SCHOOL OF CHEMICAL SCIENCES, UNIVERSITY OF EAST ANGLIA,
NORWICH NR4 7TJ

Malcolm J. Grimson

AFRC INSTITUTE OF FOOD RESEARCH, NORWICH LABORATORY, COLNEY LANE,
NORWICH NR4 7UA

Richard K. Heenan

RUTHERFORD–APPLETON LABORATORY, CHILTON, DIDCOT, OXFORDSHIRE
OX11 0QZ

and Andrew M. Howe

SURFACE SCIENCE GROUP, RESEARCH DIVISION, KODAK LIMITED, HARROW
HA1 4TY

1 Introduction

Water-in-oil microemulsions stabilized by the anionic surfactant Aerosol–OT (AOT) are optically clear one-phase dispersions of discrete spherical water droplets in a continuous oil medium.[1] Enzymes and other macromolecules can be solubilized in such systems without changing the nature of the microemulsion.[2] Gelatin may be dissolved in an n-heptane microemulsion at 50 °C and, above a critical concentration of the biopolymer of a few wt%, the whole system gels on cooling to below 30 °C.[3,4] This is a surprising result considering that >85% of the volume can be hydrocarbon, in which gelatin is insoluble. The organo-gels thus formed are optically clear and stable in contact with a co-existing oil phase. Enzymes solubilized in the organo-gels retain their activity, making the systems suitable for reverse enzyme synthesis of important food materials, such as flavouring esters, in high yield and under very mild conditions.[5]

Small-angle neutron scattering (SANS) and electrical conductivity measurements have been used to investigate the structure of these novel organo-gels. The results demonstrate that there are significant changes in the microstructure of the aqueous domains on gelation of the microemulsion.[6] Phase stability studies show that gelatin modifies the microemulsion phase boundaries with respect to temperature, and that a similar melting temperature (*ca.* 30 °C) is observed for all gelled samples.

2   Experimental

Microemulsion-based organo-gels are made by first dissolving gelatin powder (Sigma, Bloom 300, acid-hydrolysed pig-skin gelatin) in water at 50 °C. The required quantity of AOT in n-heptane is then added to the aqueous solution and the mixture shaken vigorously. After allowing the mixture to cool, with occasional shaking to maintain a single phase, a gel is formed below 30 °C. A typical composition would be 0.1 mol dm^{-3} AOT, w_0 (= [H$_2$O]/[AOT]) = 60, 3.5 wt% gelatin in heptane.

SANS spectra were recorded on the LOQ instrument of ISIS, the spallation neutron source at the SERC Rutherford–Appleton Laboratory. Electrical conductivity was measured with a Philips conductivity bridge operating at 1 kHz. All measurements were made at 25 °C.

3   Results and Discussion

SANS allows exploration of the size range 1–100 nm, and so is well suited to microemulsion-based systems. The spectra are plotted as intensity $I(Q)$ (measured in cm^{-1} per steradian) against wave-vector Q (measured in Å^{-1}). The

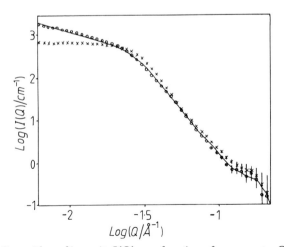

Figure 1   *Logarithm of intensity* I(Q) *as a function of wave-vector* Q *for* 10 vol% D$_2$O, 0.1 mol dm^{-3} AOT, 3.5 wt% *gelatin, n-heptane organo-gel* (○). *Solid line is least squares fit with rods of length* >700 Å *and radius* 100 Å, *plus* 72% *of scatter from gelatin-free microemulsion of the same composition* (x)

wave-vector depends on the scattering angle θ and the neutron wavelength λ through the relationship $Q = (4\pi/\lambda) \sin (\theta/2)$. Small Q equates to large distances in real space and *vice versa*; the available Q-range on LOQ is 0.006–0.22 Å^{-1}.

Figure 1 shows SANS spectra (in log–log form) for an organo-gel with 3.5 wt% gelatin and for the parent (non-polymeric) microemulsion. Addition of gelatin enhances strongly the scattering at low Q, suggesting that larger scale structures are present in the organo-gel than in the microemulsion. In the low-Q region, a slope of -1 is found, which is the same as that observed for aqueous gelatin gels, and indicates the presence of long rod-like structures. Below 30 °C, aqueous gelatin prefers to adopt a rod-like helical conformation with radius 0.7 nm and length up to 300 nm.[7] At high Q, scattering from the interface between the aqueous and oil domains dominates. The slope of -4 indicates a smooth, sharp interface. The high-Q scattering is unchanged on gelation, suggesting that the interfacial area does not change on addition of gelatin. The data cannot be fitted as rods alone in the mid-Q region. However, by including a contribution to the scatter from the parent microemulsion droplets, very good fits may be obtained as shown in Figure 1. The fit to the SANS data for the gel in Figure 1 gives a rod length of >700 Å, with a rod radius of 100 Å in the presence of polydisperse droplets of number-mean radius 56 Å.

Compared to the parent microemulsions, the electrical conductivity of organo-gels is very high, and increases with disperse phase volume fraction (constant water: AOT: gelatin ratio) as shown in Figure 2. Transport of the conducting ions probably occurs through the water channels, which are stabilized by the percolating network (of gelatin helices) that must form for the system to gel.

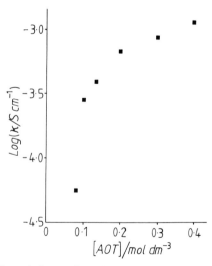

Figure 2 *Logarithm of electrical conductivity* K *versus* AOT *concentration for dilution of a* 40 vol% H_2O, 14 wt% *gelatin,* 0.4 mol dm^{-3} AOT *organo-gel with n-heptane*

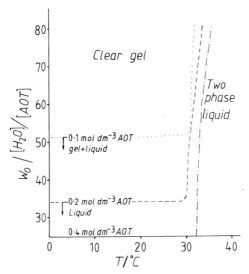

Figure 3   *Phase stability map of* w_0 *versus temperature* T *for organo-gels with the same* [AOT]: *gelatin ratio* (3.5 wt% *gelatin in* 0.1 mol dm^{-3} AOT *samples*) *showing the organo-gel phase boundaries*

Phase stability studies show that the gelatin modifies microemulsion phase behaviour considerably (Figure 3). In particular, the single-phase gel region formed at high w_0 is stable over a wide range of temperature. On warming, the organo-gels are stable until the temperature reaches *ca.* 30 °C, when melting occurs. This observation supports the suggestion that gelatin helices dominate the network part of the structure, since the organo-gel melting temperature corresponds to the gelatin helix–coil transition temperature. The low w_0 transition is of the sol–gel type: as the volume fraction of the dispersed phase is increased, the w_0 value at which the whole system may be gelled decreases since there is sufficient material for percolation and gelation to occur.

Thus, we propose that the organo-gel structure comprises a network of gelatin/water rods in equilibrium with microemulsion-like droplets, all stabilized by a monolayer of AOT. This model explains the observed properties of the gels with respect to electrical conductivity, phase stability, and neutron scattering.

References

1. B. H. Robinson, *Nature (London)*, 1987, **320**, 309.
2. P. L. Luisi and L. Magid, *Crit. Rev. Biochem.*, 1986, **20**, 409.
3. G. Haering and P. L. Luisi, *J. Phys. Chem.*, 1986, **90**, 5892.
4. C. Quellet and H. F. Eicke, *Chimia*, 1986, **40**, 233.
5. G. D. Rees, T. R. Jenta, M. da G. Nascimento, and B. H. Robinson, *Biochim. Biophys. Acta*, 1990.
6. P. J. Atkinson, M. J. Grimson, R. K. Heenan, A. M. Howe, and B. H. Robinson, *J. Chem. Soc., Chem. Commun.*, 1989, 1307.
7. D. R. Eyre, *Science*, 1986, **207**, 1317.

Factors affecting the Formation and Stabilization of Gas Cells in Bread Doughs

By J. Castle and S. S. Sahi

FLOUR MILLING AND BAKING RESEARCH ASSOCIATION, CHORLEYWOOD,
HERTFORDSHIRE WD3 5SH

1 Introduction

When the Chorleywood Bread Process was first introduced the need for a fundamental understanding of the process then became apparent, but we still have little understanding of the mechanisms involved in bread production at a molecular level. Wheat flour dough is a complex dispersion of air cells, lipid droplets, and starch granules in a continuous hydrated gluten matrix. A uniform distribution of bubble nuclei are created during mixing.[1] Carbon dioxide diffuses into the bubble nuclei in proof to form gas cells and to create a semi-solid foam, and on baking this becomes a sponge-like structure. The bubble wall material must be present in the aqueous phase of the dough, and be able to adsorb rapidly at the gas–liquid interface. The bubble walls must have considerable visco-elasticity as the bubbles expand to occupy >90% of the total volume.

We report measurements on aqueous extracts from flours of different baking qualities. Their interfacial properties were determined using a Langmuir trough and a surface rheometer. The response of films to compression and expansion corresponds to the forces the bubbles experience during proof and in the oven, and their response to shear reflects forces during mixing.

2 Experimental

Aqueous extracts (0.05–25 wt%) were prepared using double-distilled deionized water. Flour + water mixtures were gently stirred, left for 20 minutes, and then the solids were removed by centrifugation for 30 minutes at 1100g. Several flours were examined: CWRS (Canadian Western Red Spring), a blend of flours of excellent baking quality, and UK cultivars giving both good and poor bread-making performances (HG1 and HG2).

The effect of extract concentration on surface pressure was measured using a Wilhelmy plate after 30 minutes (γ(water) = 70 mN m^{-1}). The results in Figure 1 show that only a small proportion of the extract is needed to generate surface pressure values in the range 15–20 mN m^{-1}, as is observed in adsorption studies of

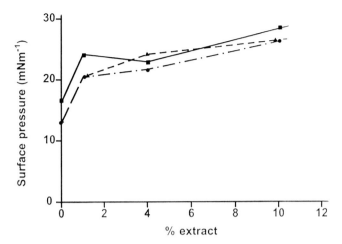

Figure 1 *Plot of surface pressure against concentration of aqueous extract for three flours*: ■, CWRS; ▲, HG1; ●, HG2

pure protein solutions.[2] Increasing the concentration has little additional effect; and there is only a slight difference between the flours.

The response of the films to compression and expansion was measured on films formed from 25 wt% aqueous extracts spread onto water at the maximum available area (0.056 m^2). Compression–expansion cycles were studied to determine the surface pressure above which an irreversible loss in molecular area

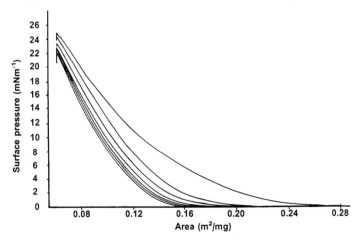

Figure 2 *Generalized form of the compression cycles for a spread film of an aqueous extract held for 20 minute periods at 15 mN m^{-1}. Surface pressure is plotted against area. Expansion cycles are not shown for clarity*

occurs, when the molecules in the film desorb into the sub-phase. This was found to occur at surface pressures greater than 10 mN m^{-1}, and films were therefore held at 15 mN m^{-1} for a period of 20 minutes before expanding and recompressing. The shape of the compression curve is similar for the three different flours (Figure 2), but the magnitude of displacement of the curves after consecutive compression cycles varies.

The loss of area decreases after each compression; subsequent compressions have a diminishing effect on desorption. The irreversible loss in area at $\Pi = 10$ mN m^{-1} has been used to calculate the rates of desorption from the compression curves. The rate constants decrease in the order HG2 > HG1 > CWRS suggesting that flours of poor baking quality lose film material into the sub-phase at greater rates than flours of good baking quality.

The response of the films to shear was measured at the surface of a 10 wt% aqueous extract. Both surface viscosity and surface elasticity data were obtained over a 90 minute period. The surface viscosity decreased with time due to pentosan degradation by enzymic action (this effect has also been observed in bulk solution). However, the surface elasticity increased over the same time period to give films of greater mechanical strength.

3 Conclusions

Small amounts of flour rapidly reduce the surface tension of water, increasing concentrations showing little effect on the final value attained. The films possess the ability to withstand compression/expansion cycles although the curves become displaced to lower molecular areas. This may be due to desorption of the molecules, or the molecules may 'buckle' during compression, but not unfold during the experiment. Interfacial films develop elastic properties on ageing, whereas the film viscosity decreases rapidly with time. There are differences between the three flours of different baking qualities although these are not as large as might be expected.

Acknowledgement. We acknowledge the financial support of the Chief Scientist's Group of the Ministry of Agriculture, Fisheries, and Food. The results are the property of the Ministry and are Crown Copyright.

References

1. J. C. Baker and M. D. Mize, *Cereal Chem.*, 1941, **18**, 19.
2. D. E. Graham and M. C. Phillips, *J. Colloid Interface Sci.*, 1979, **70**, 427.

Galactomannans as Emulsifiers

By Dov Reichman

ADUMIM CHEMICALS LTD., 90610 MISHOR ADUMIM, ISRAEL

and Nissim Garti

CASALI INSTITUTE OF APPLIED CHEMISTRY, SCHOOL OF APPLIED SCIENCE AND
TECHNOLOGY, THE HEBREW UNIVERSITY OF JERUSALEM, 91904 JERUSALEM,
ISRAEL

1 Introduction

Polysaccharides, and other hydrocolloids, are widely used in food products to
modify texture, to control water mobility, and to improve moisture retention.
The contribution of polysaccharides to the stability of food emulsions has been
traditionally related primarily to their rheological effects on the continuous
phase,[1] and there is limited information in the literature about their interfacial
properties.[2] Among hydrocolloids, food proteins are known to be useful emulsifiers, although flocculation often takes place. It has been shown[3] that when the
protein load is insufficient to give complete coverage during homogenization,
flocculation by macromolecular bridging may occur. In this type of flocculation,
macromolecular material on one droplet becomes shared, through partial adsorption, with the initially uncoated surface of another.[4]

Polysaccharides are known to have less surface activity in comparison to
proteins.[5] This inferiority is related to their low flexibility and the monotonic
repetition of the monomer units in the backbone.[6] Gum arabic is an exception. It
is known in the literature as an emulsifier and stabilizer, and its emulsifying
properties have been investigated as a function of protein content and protein/
peptide ratio.[7]

In latex dispersions containing emulsifiers and gums, it has been shown[8] that
the gums affect dispersion stability even at low levels. Such stabilization cannot be
explained by viscosity effects alone. In oil-in-water emulsions containing both
emulsifiers and gums, it has been proposed[9] that the gums adsorb on to the
surfactant, forming a combined structure of a primary surfactant layer covered by
an adsorbed polymer layer.

It has been demonstrated in our previous work[10] that galactomannans,
although highly hydrophilic in nature, can adsorb at interfaces, reducing surface

tension to levels whereby they can be categorized as surfactants. Moreover, they can interact synergistically with food emulsifiers. In the present study, we report on the preparation of non-viscous oil-in-water emulsions with galactomannans acting as the sole emulsifier without additional surfactant.

2 Materials and Methods

Locust bean gum (LBG) and guar gum were obtained from Sigma Chemicals and were used without further treatment. Guar gum was dissolved while stirring in distilled water at room temperature. The LBG solution was prepared by mixing the gum at room temperature for 30 minutes, heating the solution to 80 °C for 15 minutes, and then cooling to room temperature. Sodium azide (0.05 wt%) was added as a preservative. The hydrocarbons n-tetradecane and toluene were selected as the oil phase solvents because of their purity (Sigma Chemicals). Soya oil (supplied by Tet-Bet, Israel) was used only for comparison studies since it was found to contain up to 0.8 wt% monoglycerides. Treatment with Florisil 100–200 mesh[11] did not reduce the monoglyceride level significantly.

Oil-in-water emulsions were prepared by using a Silverson homogenizer (Silverson Machines Ltd., UK) for pre-emulsification followed by shearing the emulsion twice through a microfluidizer (5 bar) (Microfluidics Corporation, Newton, MA). In order to prevent air incorporation in the Silverson, low-speed homogenization was applied for 10 minutes just to enable homogeneous mixing. A Coulter Counter (model TA II, Coulter Electronics Ltd., UK) was used to evaluate the droplet-size distribution. Microscopic examination was carried out on a Nikon Optiphot (AFX-πA, Nikon, Japan) equipped with a Nikon camera (FX 35 WA). Emulsion viscosity was measured by a Brookfield cone-and-plate viscometer (model DV-II). Measurements were made at 30 ± 0.1 °C using a Brookfield EX-100 constant temperature bath.

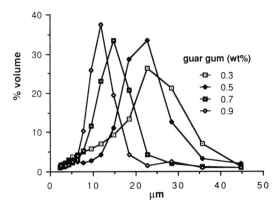

Figure 1   *Droplet-size distribution of 5 wt% n-tetradecane emulsions with various concentrations. Percentage volume is plotted against droplet diameter (μm)*

3 Results and Discussion

Droplet-size distributions of emulsions containing 5 wt% n-tetradecane, at different levels of guar, are recorded in Figure 1. The Coulter counter measurements were taken 24 hours after emulsion preparation. The results indicate that, as the guar gum content increases, the droplet-size distribution becomes narrower and smaller. The difference in the viscosity between the emulsions of high and low guar content is less than 50 mPa s. Emulsion stability toward coalescence has been checked by determining the droplet-size distribution one week and one month after preparation. Emulsions prepared with 0.3 wt% guar exhibit low stability toward coalescence. The droplet-size distribution indicates formation of large droplets with a wide distribution after just one week of incubation (Figure 2a). The emulsions showed complete oil separation after one month of storage. At 0.9 wt% guar (Figure 2b), however, good stability is achieved. The rate of coalescence is relatively small, and changes seem to take place more than a week

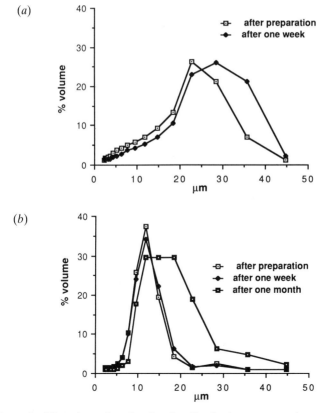

Figure 2 *Time-dependent droplet-size distribution, presented as percentage volume, plotted against droplet diameter, for emulsions prepared with 5 wt% n-tetradecane: (a) 0.3 wt% guar gum, (b) 0.9 wt% guar gum*

(a)

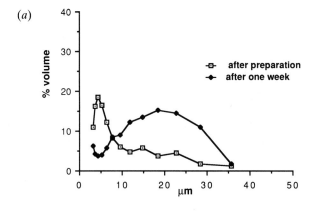

(b)

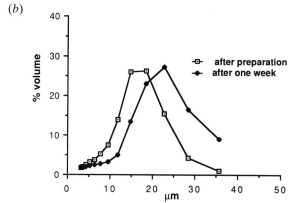

Figure 3   *Time-dependent droplet-size distribution of emulsion of* 5 wt% *n-tetradecane stabilized by* (a) 0.1 wt% *LBG and* (b) 0.5 wt% *LBG*

after preparation. A similar trend has been found with 5 wt% n-tetradecane emulsions stabilized by 0.1 and 0.5 wt% LBG (Figure 3). These results demonstrate that oil-in-water emulsions can be created by using galactomannans as emulsifiers without additional surfactant.

When changing the n-tetradecane content from 1 to 20 wt%, keeping the guar content at 0.9 wt%, the most probable droplet size varies from 3.5 to 20 μm, and the size distribution becomes wider (Figure 4). At 10 wt% and above, the stability decreases significantly.

The tendency of the emulsion to flocculate or to coalesce was detected under the light microscope. Most emulsions showed strong flocculation tendency which seems to be affected mainly by the hydrocolloid/oil ratio. A similar tendency was found in protein-stabilized emulsions.[3] Emulsions prepared with low LBG concentration (<0.5 wt%) showed significant flocculation, and seemed creamy in appearance, probably due to the gelling properties of the gum. At 0.7 and 0.9 wt%, no flocculation or gelling effect was detected. The differences in

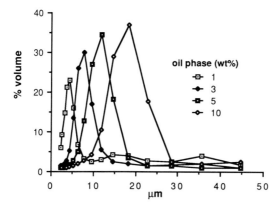

Figure 4 *Droplet-size distribution of emulsions prepared with various n-tetradecane levels* (1–10 wt%) *and stabilized with* 0.9 wt% *guar gum*

flocculation between guar- and LBG-stabilized emulsions can be explained in terms of more effective droplet coverage which is achieved by the unsubstituted LBG regions. The emulsion stabilized with guar gum, which had good droplet coverage (0.9 wt% guar, 0.1 wt% n-tetradecane) did not show any flocculation.

Throughout the entire study efforts were made to prevent the build-up of viscosity of the continuous aqueous phase by the hydrocolloids in order to avoid complicating effects on the stability. The viscosity results for the guar-stabilized emulsions as a function of n-tetradecane content are presented in Table 1.

Where there is sufficient coverage of the hydrocolloid on the oil droplets, the emulsion is fairly stable. Although the viscosity increases with increasing hydrocolloid content, the change on going from stable to unstable emulsions is relatively small. When reducing the oil content at constant hydrocolloid content, the viscosity decreases but the coverage efficiency increases, and, as a result, good stability to coalescence is achieved (Figure 5). It was demonstrated that, even at low viscosity, stability towards coalescence and flocculation can be obtained.

Microscope examination under polarized light of most of the emulsions prepared with LBG or guar showed liquid crystals arranged in Maltese crosses around the oil droplets (Figure 6a,b). It has been suggested[12] that polysaccharides are anticipated to form liquid crystalline phases. Hydroxypropyl cellulose was reported[13] to form liquid crystalline structures in aqueous solutions and cholesteric mesophases in some organic solvents.[14] No reports are available, to the best of our knowledge, on the liquid crystalline organization of hydrocolloids

Table 1 *Viscosity of oil-in-water emulsions at* 30 °C *prepared with various n-tetradecane levels and* 0.9 wt% *guar gum*

Oil content/wt%	1	3	5	10
Viscosity/mPa s	28	40	84	100

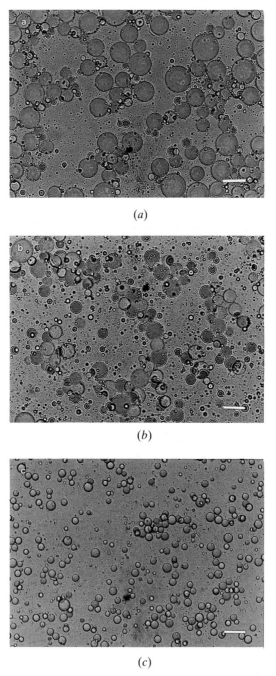

(*a*)

(*b*)

(*c*)

Figure 5 *Photomicrographs of n-tetradecane emulsions stabilized by guar gum*
(bar corresponds to 20 μm): (a) 0.6 wt% guar, 10 wt% oil; (b) 0.9 wt%
guar, 10 wt% oil; (c) 0.9 wt% guar, 5 wt% oil

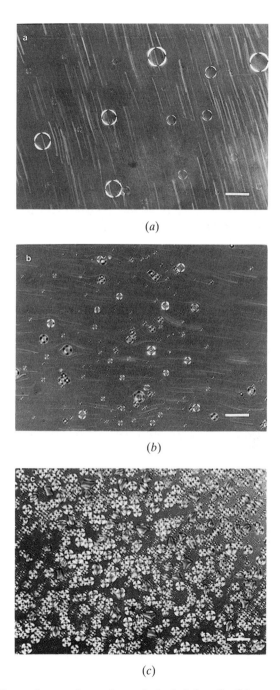

Figure 6 *Photomicrographs under polarized light of oil-in-water emulsions stabilized by galactomannans (bar corresponds to 20 μm): (a) 0.9 wt% LBG, 5 wt% n-tetradecane; (b) 0.9 wt% guar, 5 wt% toluene; (c) 0.9 wt% guar, 5 wt% soya oil*

in emulsions. Moreover no indications exist on the liquid crystalline formation of guar or LBG, neither in the aqueous phase, nor in the oil.

The amount of liquid crystals was determined qualitatively and found to change when replacing the nature of the oil from n-tetradecane to soybean and toluene.[15] Moreover, the tendency towards coverage by liquid crystals seems to increase when using LBG instead of guar, suggesting that the smooth regions of the mannose units in LBG (unsubstituted with galactose) allow a higher degree of order as in the gel structures formed in aqueous solutions.

The fact that the hydrocolloids adsorb at the oil–water interface as lamellar liquid crystals leads to improved stability toward coalescence. It clearly indicates that the stability is due to steric and mechanical effects arising from the interfacial orientation of the galactomannans at the interface rather than by their presence in the continuous aqueous phase.

In conclusion, the present study stresses the ability of locust bean gum and guar gum to adsorb on oil droplets, to form layers of lamellar liquid crystals, and to enhance emulsion stability toward coalescence and flocculation.

References

1. A. Gayot, M. Carpentier, and A. Meybeck, *J. Pharm. Belg.*, 1986, **41**, 299.
2. J. N. BeMiller, in 'Gums and Stabilisers for the Food Industry', ed. G. O. Phillips, D. J. Wedlock, and P. A. Williams, IRL Press, Oxford, 1988, Vol. 4, p. 3.
3. E. Dickinson, F. O. Flint, and J. A. Hunt, *Food Hydrocolloids*, 1989, **3**, 389.
4. T. F. Tadros and B. Vincent, in 'Encyclopedia of Emulsion Technology', ed. P. Becher, Marcel Dekker, New York, 1983, Vol. 1, p. 129.
5. E. Dickinson and G. Stainsby, in 'Advances in Food Emulsions and Foams', ed. E. Dickinson and G. Stainsby, Elsevier Applied Science, London, 1988, p. 1.
6. G. Stainsby, in 'Functional Properties of Food Macromolecules', ed. J. R. Mitchell and D. A. Ledward, Elsevier Applied Science, London, 1986, p. 315.
7. E. Dickinson, B. S. Murray, G. Stainsby, and D. M. W. Anderson, *Food Hydrocolloids*, 1988, **2**, 477.
8. B. Bergenstahl, S. Fogler, and P. Stenius, in 'Gums and Stabilisers for the Food Industry', ed. G. O. Phillips, D. J. Wedlock, and P. A. Williams, Elsevier Applied Science, London, 1986, Vol. 3, p. 285.
9. B. Bergenstahl, in 'Gums and Stabilisers for the Food Industry', ed. G. O. Phillips, D. J. Wedlock, and P. A. Williams, IRL Press, Oxford, 1988, Vol. 4, p. 363.
10. D. Reichman and N. Garti, in 'Gums and Stabilisers for the Food Industry', ed. G. O. Phillips, D. J. Wedlock, and P. A. Williams, IRL Press, Oxford, 1990, Vol. 5, p. 441.
11. A. G. Gaonkar, *J. Am. Oil Chem. Soc.*, 1989, **66**, 1090.
12. J. M. V. Blanshard, *Prog. Food Nutr. Sci.*, 1982, **6**, 3.
13. R. S. Webowyj and D. G. Gray, *Mol. Cryst. Liq. Cryst.*, 1978, **34**, 97.
14. T. Tsutsui and R. Tanaka, *Polym. J.*, 1980, **12**, 473.
15. D. Reichman and N. Garti, in preparation.

Computer Simulation of Macromolecular Adsorption

By Eric Dickinson and Stephen R. Euston

PROCTER DEPARTMENT OF FOOD SCIENCE, UNIVERSITY OF LEEDS,
LEEDS LS2 9JT

1 Introduction

Linear polymer chain models are limited in their application to many biological systems. It is often the case that biopolymers do not conform to the idealized representation of a linear, flexible chain (*e.g.* globular proteins have a complex, compact tertiary structure held together by covalent linkages). It has been proposed[1] that such molecules are better represented as soft, or deformable, particles when adsorbed at an interface. In this text we describe two models[2-4] in order to compare and contrast the application of the linear chain model and the soft particle model to adsorbed macromolecules. In the first model, we look particularly at the competitive adsorption of linear polymer chains and small displacer molecules. The second model attempts to treat the unfolding of a compact macromolecule at a hard planar wall.

2 Flexible Linear Chain Model

A Monte Carlo model was set up consisting of 50-segment linear lattice chains (defined on a three-dimensional tetrahedral lattice) competing with two-segment displacer molecules for space at an interface.[2,3] Interactions in the system were controlled by several parameters: polymer segment adsorption energy ($E_p^{ads} = -0.75$ kT); displacer adsorption energy ($E_d^{ads} = -3.0$ kT); polymer–polymer interaction ($E_{pp} = 0.0$ kT); polymer–displacer interaction ($E_{pd} = +2.0$ kT); and displacer–displacer interaction ($E_{dd} = +1.5$ kT). Formation of polymer–displacer aggregates and micelle-type displacer aggregates were prevented by setting the polymer–displacer and displacer–displacer interactions as repulsive.

Initially, polymer molecules were allowed to adsorb at a plane interface, and to form an equilibrium layer with respect to the formation of train, loop, and tail segments. When an equilibrium had been achieved, displacer molecules were defined at random on the lattice, and allowed to adsorb and compete with the polymer for interfacial lattice sites. Depending on the displacer concentration, the relative adsorption energies of polymer and displacer segments, and any displacer–displacer or polymer–displacer interactions, the polymer segments may be either totally displaced, partially displaced, or not displaced at all. Simulation runs of 5–8×10^7 Monte Carlo steps were required to achieve an equilibrium polymer layer at the interface. Further runs of 6–10×10^7 Monto Carlo steps were needed to reach a new equilibrium after addition of displacer molecules. Configurations were sampled every 5×10^3 Monte Carlo steps after equilibrium had been achieved. Figure 1 shows a situation where the displacer concentration is

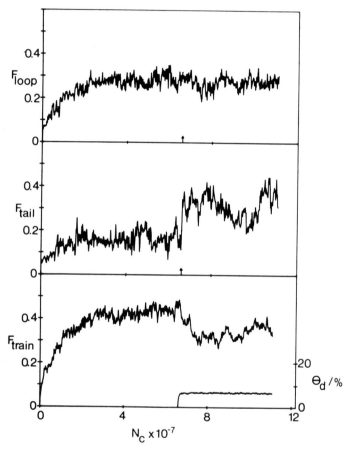

Figure 1   *Fractions of polymer segments existing as trains (F_{train}), loops (F_{loop}), and tails (F_{tail}) over a typical simulation run. The fractions of the total number of segments are plotted as a function of the number of Monte Carlo steps (N_c). The displacer surface coverage (θ_d) is also plotted. The displacer was added at the point marked ($\uparrow$). Other parameters are: total displacer concentration $= 0.49\%$; $E_p^{ads} = -0.75\,kT$; $E_d^{ads} = -3.0\,kT$; $E_{pp} = 0.0\,kT$; $E_{dd} = 1.5\,kT$; $E_{pd} = 2.0\,kT$*

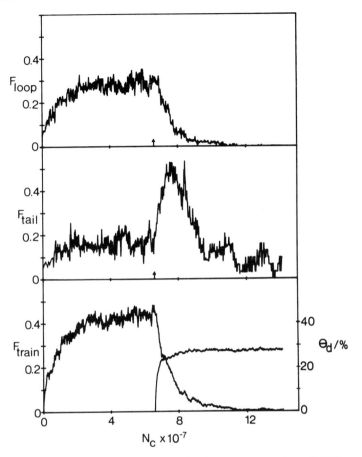

Figure 2 *As Figure 1, except that displacer concentration* = 2.93%

insufficient to achieve total displacement, but partial displacement of polymer segments occurs. This manifests itself as a change in the equilibrium fraction of polymer segments in trains, loops, and tails. At a high displacer concentration (see Figure 2), there is an almost total displacement of polymer segments from the surface. Figure 3 presents two-dimensional projections of the three-dimensional polymer configurations at different displacer concentrations. In these figures the horizontal lines mark the position of the interface. In a system where no displacer is present (Figure 3a), polymer chains adopt the usual train, loop, tail conformation. At low displacer concentrations (Figures 3b and 3c), the polymer molecule is incompletely displaced, but a reduction in the number of train segments, and the formation of longer, dangling tails, is clearly evident. At a high enough displacer concentration, the polymer molecule can become completely detached from the surface (see Figure 3d).

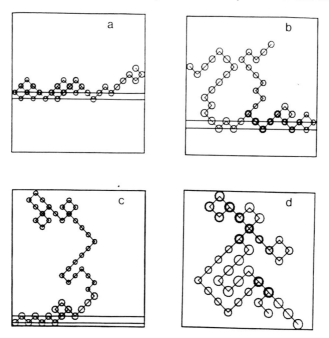

Figure 3   *Two-dimensional projections of individual chains at different displacer concentrations*: (a) 0%; (b) 0.49%; (c) 0.89%; (d) 3.9%

3   Soft Particle Model for Adsorbed Compact Macromolecules

A large proportion of biological macromolecules adopt a complex, compact particle-like structure both in the bulk phase and when adsorbed at surfaces. Following an idea proposed by de Feijter and Benjamins,[1] we have extended a model which investigated the bulk phase properties of deformable particles[5,6] in order to account for their behaviour when adsorbed at a hard wall.[4]

A Monte Carlo simulation of deformable particles was set up, which represented the molecules as cyclic lattice chains of 100 segments, adsorbed at a hard wall on a two-dimensional square lattice. The deformability and structure of the particles were controlled by an interaction between non-connected segments on the chain (E_1). Each segment was also allowed to adsorb at a hard wall with an adsorption energy E_s. The simulation cell had dimensions of 120 lattice sites in the x-direction, and 40 in the y-direction, with hard walls defined at $y = 0$ and $y = 41$, and periodic boundary conditions in the x-direction. Simulation runs of $5–10 \times 10^6$ Monte Carlo steps were required to achieve equilibrium, after which runs were carried out for a similar number of steps with sampling of configurations every 3×10^3 Monte Carlo steps.

Figure 4 shows instantaneous configurations for an isolated inflated deformable particle (repulsive $E_1 = 0.1\,kT$), at different adsorption energies E_s. These

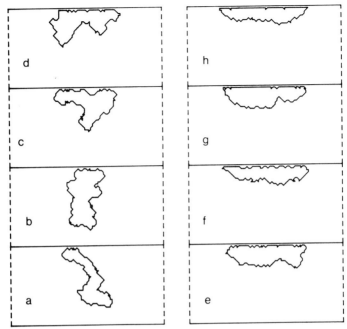

Figure 4 *Configurations of an isolated* 100-*segment deformable particle adsorbed at a hard wall with* $E_1 = 0.1\,kT$ *at different adsorption energies* E_s: (a) 0.0 kT; (b) −0.4 kT; (c) −0.8 kT; (d) −1.2 kT; (e) −1.6 kT; (f) −2.0 kT; (g) −3.0 kT; (h) −4.0 kT

configurations show that the presence of a repulsive E_1 interaction (which mimics a finite surface tension between internal and external phases) has the effect of preventing the particle from forming a completely flattened configuration at the surface. This is because the formation of a highly flattened configuration requires a significant loss of entropy that may not be compensated for by change in potential energy on the formation of a few extra surface–segment contacts. By introducing an attractive E_1 interaction (Figures 5 and 6), we can model the unfolding of a highly compact particle at the surface. In Figure 5, where $E_1 = -0.05\,kT$, at low E_s the particle adsorbs in a highly folded configuration, and only unfolds significantly as the surface becomes more strongly adsorbing. In Figure 6, with E_1 increased to $-0.1\,kT$, the process of unfolding becomes more difficult, with the particle maintaining its compact structure at the highest E_s values used.

4 Conclusions

The two models presented in this work clearly show the differences between a model for flexible linear chains, and one more suited to simulating compact

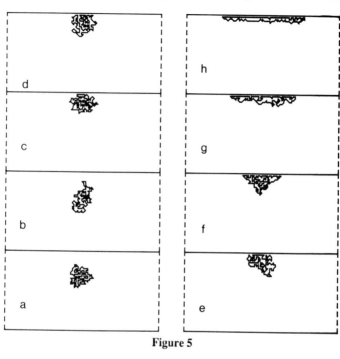

Figure 5

Figure 6

globular proteins. Whereas linear chains readily adopt flattened conformations when adsorbed at interfaces, the process of unfolding is much more complicated for compact folded macromolecules. Figures 4 to 6 clearly show how the degree of unfolding of a compact molecule is a strong function of the balance between segment–segment and segment–surface interactions.

References

1. J. A. de Feijter and J. Benjamins, *J. Colloid Interface Sci.*, 1982, **66**, 865.
2. E. Dickinson and S. R. Euston, *Molec. Phys.*, 1989, **68**, 407.
3. E. Dickinson, S. R. Euston, and C. M. Woskett, Proceedings of the 10th Scandinavian Symposium on Surface Chemistry, Turku, Finland, 1989.
4. E. Dickinson and S. R. Euston, *J. Chem. Soc., Faraday Trans.*, 1990, **86**, 805.
5. E. Dickinson, *Phys. Rev. Lett.*, 1984, **53**, 728.
6. E. Dickinson and S. R. Euston, *Molec. Phys.*, 1989, **66**, 865.

Figure 5 *Configurations of an isolated 100-segment deformable particle adsorbed at a hard wall with* $E_1 = -0.05$ kT *at different adsorption energies* E_s: (a) 0.0 kT; (b) -0.8 kT; (c) -1.2 kT; (d) -1.6 kT; (e) -2.0 kT; (f) -2.5 kT; (g) -3.0 kT; (h) -4.0 kT

Figure 6 *Configurations of an isolated 100-segment deformable particle adsorbed at a hard wall with* $E_1 = -0.1$ kT *at different adsorption energies* E_s: (a) 0.0 kT; (b) -0.8 kT; (c) -1.2 kT; (d) -1.6 kT; (e) -2.0 kT; (f) -2.5 kT; (g) -3.0 kT; (h) -4.0 kT

Models of Protein Adsorption at Gas–Liquid Interfaces

By Roger Douillard, Jacques Lefebvre, and Vinh Tran

INSTITUT NATIONAL DE LA RECHERCHE AGRONOMIQUE, LA GÉRAUDIÈRE,
B.P. 527, 44026 NANTES CEDEX 03, FRANCE

1 Introduction

Our understanding of the mechanisms of protein adsorption at the air–buffer interface is presently poor due to the difficulties in interpreting the concentration and pressure isotherms. A recent approach by Guzman *et al.*[1] throws some light on the significance of the concentration isotherms. They should be visualized as resulting from the formation of two adsorption layers: the first one is actually located at the interface, and the second one results from adsorption of protein onto the first layer. This model gives a very good fit to experimental data for β-casein, bovine serum albumin, and lysozyme. Unfortunately, it relies on the hypothesis that each layer is in equilibrium with the protein solution just below the interface. This hypothesis is not realistic since, at equilibrium, the first layer is in equilibrium with the second one which itself is in equilibrium with the solution below the interface. In addition, the model does not take into account the conformational changes that are known to be induced by protein adsorption.[2,3]

In this communication, we present a model of protein adsorption where the two layers at the interface are in equilibrium with each other. Moreover, protein in the first layer may occur in two configurations, as suggested by Graham and Phillips.[4] To calculate the surface pressure, we use the equation derived by Frisch and Simha[5] for flexible polymers with only part of their segments on the interface. This model fits well the experimental concentration and pressure adsorption isotherms of bovine serum albumin and lysozyme given in the literature.[6]

2 Adsorption Isotherms

Concentration Isotherm—Protein in the solution, just below the interface, is at concentration C_s. Γ_2 is the concentration per unit area of the interface in the second layer, Γ_{11} and Γ_{12} are respectively the concentrations per unit area of the protein in its less and more unfolded conformations in the first layer. The kinetics of protein adsorption are thus depicted by the set of equilibria:

$$C_s \underset{k_{d2}}{\overset{k_{a2}}{\rightleftharpoons}} \Gamma_2 \underset{k_{d11}}{\overset{k_{a11}}{\rightleftharpoons}} \Gamma_{11} \underset{k_{d12}}{\overset{k_{a12}}{\rightleftharpoons}} \Gamma_{12}. \tag{1}$$

In equation (1), k_a and k_d are adsorption and desorption (or conformation change) rate constants. Taking $\hat{a}_{11}$ and $\hat{a}_{12}$ as the areas per unit mass of protein under the less and more unfolded conformations, the rates of change of the various surface concentrations are given by:

$$\frac{d\Gamma_{12}}{dt} = k_{a12}\Gamma_{11}(1 - \hat{a}_{11}\Gamma_{11} - \hat{a}_{12}\Gamma_{12}) - k_{d12}\Gamma_{12}, \tag{2}$$

$$\frac{d\Gamma_{11}}{dt} = k_{a11}\Gamma_2(1 - \hat{a}_{11}\Gamma_{11} - \hat{a}_{12}\Gamma_{12}) - k_{d11}\Gamma_{11} - \frac{d\Gamma_{12}}{dt}, \tag{3}$$

$$\frac{d\Gamma_2}{dt} = k_{a2}C_s(\Gamma_{11} + \Gamma_{12}) - k_{d2}\Gamma_2 - \frac{d\Gamma_{11}}{dt} - \frac{d\Gamma_{12}}{dt}. \tag{4}$$

We assume, in equation (4), that the rate of adsorption of protein into the layer 2 is proportional to the number of protein molecules adsorbed in layer 1.

At equilibrium, the rates of change of Γ_{11}, Γ_{12}, and Γ_2 vanish, and the following isotherm results from equations (2) to (4):

$$\Gamma_{11}(1 - \hat{a}_{11}\Gamma_{11} - \hat{a}_{12}\Gamma_{12}) = K_{12}\Gamma_{12}, \tag{5}$$

$$\Gamma_2(1 - \hat{a}_{11}\Gamma_{11} - \hat{a}_{12}\Gamma_{12}) = K_{11}\Gamma_{11}, \tag{6}$$

$$C_s(\Gamma_{11} + \Gamma_{12}) = K_2\Gamma_2, \tag{7}$$

where $K_{11} = k_{d11}/k_{a11}$, $K_{12} = k_{d12}/k_{a12}$, and $K_2 = k_{d2}/k_{a2}$. We define as an intermediary variable:

$$X = \Gamma_{12}/\Gamma_{11}. \tag{8}$$

We can write, by combining (5) to (7):

$$X^2 + X - K_{11}K_2/K_{12}C_s = 0. \tag{9}$$

Thus, solving equation (9), and using the resulting value of X, we get, from equations (5), (8) and (7), the concentrations Γ_{11}, Γ_{12}, and Γ_2:

$$X = \frac{-1 + \sqrt{1 + 4K_{11}K_2/K_{12}C_s}}{2}, \tag{10}$$

$$\Gamma_{11} = \frac{1 - XK_{12}}{\hat{a}_{11} + X\hat{a}_{12}}, \tag{11}$$

$$\Gamma_{12} = X\Gamma_{11}, \tag{12}$$

$$\Gamma_2 = C_s\Gamma_{11}(1 + X)/K_2. \tag{13}$$

The concentration isotherm is therefore given by:

$$\Gamma_t = \frac{(1 - XK_{12})}{\hat{a}_{11} + X\hat{a}_{12}}(1 + X)(1 + C_s/K_2).$$ (14)

Pressure Isotherm—According to statistical thermodynamics theories, the surface pressure of an adsorbed flexible polymer is a function of the number of its segments occupying sites in the interface. Only the molecules of the first layer are involved. Following Frisch and Simha,[5] surface pressure Π is given by:

$$\frac{\Pi}{\Pi_0} = \frac{s(v - 1)}{2v(1 - 1/t)} \ln\left[1 - \frac{2}{s}\left(1 - \frac{1}{t}\right)\frac{pA_0}{A}\right] - \ln\left(1 - \frac{pA_0}{A}\right)$$ (15)

with $\Pi_0 = kT/A_0$, where k is the Boltzmann constant, T is absolute temperature, A_0 the actual area of a surface site, s is the surface coordination number (which should normally vary between 2 and 6), v is the number of segments of the polymer molecule exactly located in the interface, t is the total number of segments of the adsorbed molecule, $p = v/t$, and A is the apparent area per polymer segment,

$$A = M/N\Gamma t,$$ (16)

where M is the molecular weight of the polymer, and N is the Avogadro number.

For the case of a protein adsorbing in two conformations with concentrations Γ_{11} and Γ_{12}, each conformation should be considered as a population. The variables of equation (15), which are statistical means, become:

$$\Gamma = \Gamma_{11} + \Gamma_{12},$$ (17)

$$v = (v_{11}\Gamma_{11} + v_{12}\Gamma_{12})/\Gamma,$$ (18)

$$s = (v_{11}s_{11}\Gamma_{11} + v_{12}s_{12}\Gamma_{12})/(v_{11}\Gamma_{11} + v_{12}\Gamma_{12}).$$ (19)

The Calculations—Experimental data used was that obtained by Graham and Phillips[6] for adsorption of purified bovine serum albumin and lysozyme. Models were fitted to those data by the method of the complex[7,8] using software developed by Tran and Buléon.[9]

3   Results and Discussion

Concentration Isotherms—It can be seen from the form of equations (14) and (10) that the isotherm does not pass through the origin of the axes. In fact, the concentration in the adsorption layers is zero below a limiting value C_{s1} of C_s:

$$C_{s1} = K_{11}K_{12}K_2/(1 + K_{12}).$$ (20)

The concentrations of the three populations of adsorbed molecules, Γ_{11}, Γ_{12}, and Γ_2, can be observed in Figures 1A and 2A. It is clearly visible how each

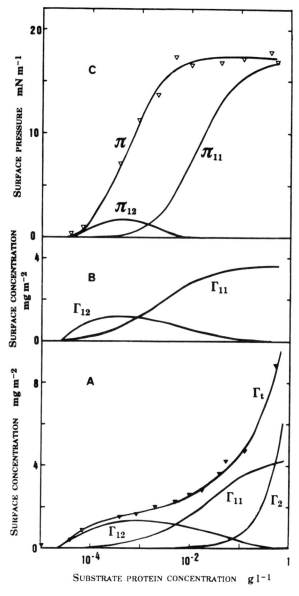

Figure 1 *Adsorption isotherms for bovine serum albumin. Surface concentration*
(▼) and surface pressure (▽) data are from Graham and Phillips.[6]
Surface concentrations calculated from the model are shown in (A)
when surface concentration data are used, and in (B) *when surface*
pressure data are used. Surface pressures Π_{11}, Π_{12}, *and* Π *calculated for*
populations Γ_{11}, Γ_{12}, *and* $\Gamma_{11} + \Gamma_{12}$ *are shown in* (C)

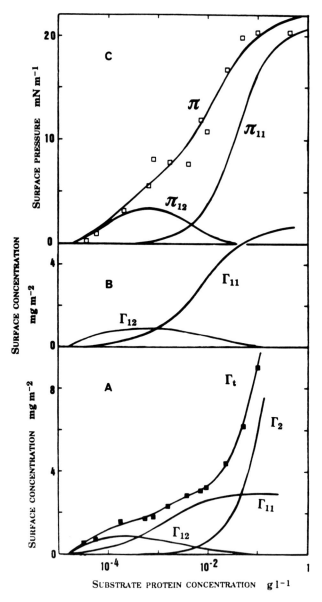

Figure 2 *Adsorption isotherms for lysozyme. Surface concentration (■) and surface pressure (□) data are from Graham and Phillips.[6] Surface concentrations calculated from the model are shown in (A) when surface concentration data are used, and in (B) when surface pressure data are used. Surface pressures Π_{11}, Π_{12}, and Π calculated for populations Γ_{11}, Γ_{12}, and $\Gamma_{11} + \Gamma_{12}$ are shown in (C)*

population contributes to the very good fit between the model and the experimental data.

Pressure Isotherms—Using equations (15) to (19), we obtained the very good agreement seen in Figures 1C and 2C. Surprisingly, each concentration Γ_{11} or Γ_{12} alone would be responsible of a very low pressure. This observation reinforces, in equation (15), the non-linear dependence of surface pressure on surface concentration.

The areas per adsorbed site (A_0) are given in Table 1. These values are very close to the 14 Å² or 15 Å² previously reported for the cross section of one amino-acid residue of a protein adsorbed at the air–water interface.[1,4,10] We may thus conclude that the segments of the proteins considered in the model must be identified with amino-acid residues. This interpretation is consistent with the number of adsorbed segments for each protein; ν_{11} and ν_{12} account indeed for 20–72% of the total number of amino acids per molecule. These values seem consistent with previous interpretations in which the protein molecules adopt conformations with trains of amino acids in the interface and loops or tails in the bulk.[4] The values of ν_{11} and ν_{12} which we have computed allow quantification of the proportions of train or loop and tail conformations in each molecule.

Evaluation of the Consistency of the Model—Regarding the two sets of values of the parameters given by the fitting of our model to the concentration and pressure isotherms (see Table 1), we observe reasonably good agreement. This is also apparent when comparing the dependence of the populations Γ_{11} and Γ_{12} on the concentration in the bulk, as determined from concentration or pressure isotherms (Figure 1A and 1B, Figure 2A and 2B).

Only the area per protein unit mass ($\hat{a}$) may reasonably be compared with previous data since the equilibrium constants derived have different meanings

Table 1 *Parameters used to fit the model to the concentration and pressure isotherms*

	BSA		Lysozyme	
	Conc.	*Press.*	*Conc.*	*Press.*
$\hat{a}_{11}$/m² kg⁻¹	2.27×10^5	2.78×10^5	3.37×10^5	1.68×10^5
$\hat{a}_{12}$/m² kg⁻¹	5.42×10^5	4.79×10^5	6.68×10^5	8.57×10^5
K_{11}	1.02×10^{-3}		2.52×10^{-3}	
K_{12}	4.31×10^{-2}	1.20×10^{-1}	1.51×10^{-1}	9.32×10^{-2}
K_2/kg m⁻³	5.10×10^{-1}		4.73×10^{-2}	
$K_{11}K_2$/kg m⁻³	5.2×10^{-4}	2.18×10^{-4}	1.19×10^{-4}	1.61×10^{-4}
s_{11}		5.92		5.42
s_{12}		2.52		5.95
A_0/Å²		15.3		12.5
ν_{11}		160		25.8
ν_{12}		310		92.9
C_{s1}/kg m⁻³	2.15×10^{-5}	2.34×10^{-5}	1.56×10^{-5}	1.37×10^{-5}
t		581		129

according to the model used. The maximum value of $\hat{a}$ should be of the order of $10^6 \, \mathrm{m}^2 \, \mathrm{kg}^{-1}$. Values previously computed for BSA and lysozyme[3] are of the order of 0.3–$0.4 \times 10^6 \, \mathrm{m}^2 \, \mathrm{kg}^{-1}$. All the values we calculate seem to be of the right order of magnitude. Moreover, $\hat{a}_{12}$ is found to be significantly greater than $\hat{a}_{11}$, a fact which is consistent with unfolded and folded conformations, respectively.

Concerning the values of C_{s1} for the two proteins, it should be noted that they are very similar when calculated from the concentration or pressure isotherms. There is less than 13% difference (Table 1).

The s_{11} values are surprisingly large. This would seem to mean that, for both proteins, an appreciable flexibility or close packing of the polypeptide chain occurs in the less extended conformation of the first layer. The value of s_{12} is still close to 6 in the extended conformation of lysozyme, while it is much smaller with BSA. As far as these values have any true significance, we may conclude that the extended conformations of the two proteins must have very dissimilar characteristics.

References

1. F. MacRitchie, *Adv. Protein Chem.*, 1978, **32**, 283.
2. F. MacRitchie, *Adv. Colloid Interface Sci.*, 1986, **25**, 341.
3. R. Z. Guzman, R. G. Carbonell, and P. K. Kilpatrick, *J. Colloid Interface Sci.*, 1986, **114**, 536.
4. D. E. Graham and M. C. Phillips, *J. Colloid Interface Sci.*, 1979, **70**, 427.
5. H. L. Frisch and R. Simha, *J. Chem. Phys.*, 1957, **27**, 702.
6. D. E. Graham and M. C. Phillips, *J. Colloid Interface Sci.*, 1979, **70**, 415.
7. M. J. Box, *Comput. J.*, 1965, **8**, 42.
8. M. J. Box, *Comput. J.*, 1966, **9**, 67.
9. V. Tran and A. Buléon, *J. Appl. Cryst.*, 1987, **20**, 430.
10. P. Joos, Proceedings of the 5th International Congress on Surface Activity, Barcelona, 1969, Vol. 2, p. 513.

Interactions between Whey Proteins and Lipids in Emulsions

By B. Closs, M. Le Meste, J.-L. Courthaudon, B. Colas, and D. Lorient

ENS.BANA, CAMPUS UNIVERSITAIRE, 21000 DIJON, FRANCE

1 Introduction

Many of the oil-in-water emulsions produced by the food industry are stabilized by milk proteins. However, the mechanism for emulsion stabilization remains only partially understood. Among the factors which are presumed to be involved in the mechanism of emulsion stabilization by proteins, we have studied: (1) the affinity of the milk proteins for the lipids, and (2) the contribution of the flexibility of the protein side chains. We base our approach on mobility measurements using the electron spin resonance method (ESR). Interactions between proteins and lipids are revealed by a change in the mobility of lipid molecules dispersed in solution or incorporated in oil droplets. Two paramagnetic fatty acids homologues were selected.

In the present work, we have measured (i) the relative affinities of the whey protein components for the paramagnetic fatty acids, (ii) the amount and nature of the proteins adsorbed on the oil droplets, and (iii) the emulsifying properties of these proteins.

2 Materials and Methods

Protein Preparation—Whey proteins were prepared in the laboratory from fresh skimmed milk.[1] Samples of α-lactalbumin and β-lactoglobulin were purified from the whey protein preparation by ion exchange chromatography on DEAE Trisacryl M using Tris–HCl buffer (0.02 M), pH 7.6, 0–0.3 M NaCl. Slab polyacrylamide electrophoresis confirmed the purity of the samples.

Emulsion Preparation—One volume of soya oil was added to three volumes of a 2.5 wt% protein solution at pH 7. The mixture was homogenized with a Polytron PCU-2 at 12 000 rpm for 2 min.

ESR Analysis—In dilute solution, and in the absence of any interaction, the ESR spectrum of lipids is characterized by fast molecular reorientation (Figure 1a). On

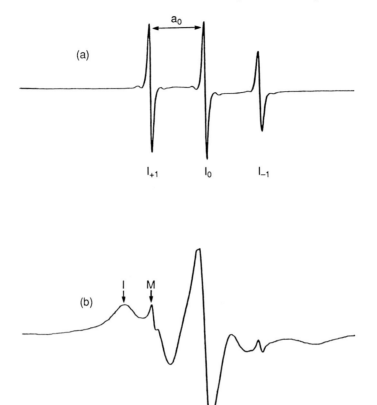

Figure 1 *ESR spectra of 5SA in* (a) *buffer and* (b) *whey protein solution*

the other hand, a powder spectrum is obtained when the nitroxide labels are randomly oriented and have very slow motions (Figure 1b). Intermediate spectra, or a superposition of both previous kinds, are also often obtained. Changes in the relative contributions of both previous types of spectra are reflected in modifications in the line amplitude ratio $R = I/M$ (Figure 1), where I corresponds to the amplitude for radicals with slow motions and M to that for radicals with rapid motions. This ratio R was considered as representing, qualitatively, the affinity of the protein for the fatty acids. The molecular ratio of paramagnetic fatty acid to protein was 0.5. ESR experiments were performed at room temperature ($20 \pm 1\,°C$) with a Varian E9 spectrometer. The 5- and 16-doxyl stearic acids (5SA, 16SA) were obtained from Aldrich.

3 Results and Discussion

Fatty-acid–Protein Interaction—The mobility of the polar end of the fatty acid is reflected in the behaviour of the 5SA nitroxide. The 16SA nitroxide indicates the

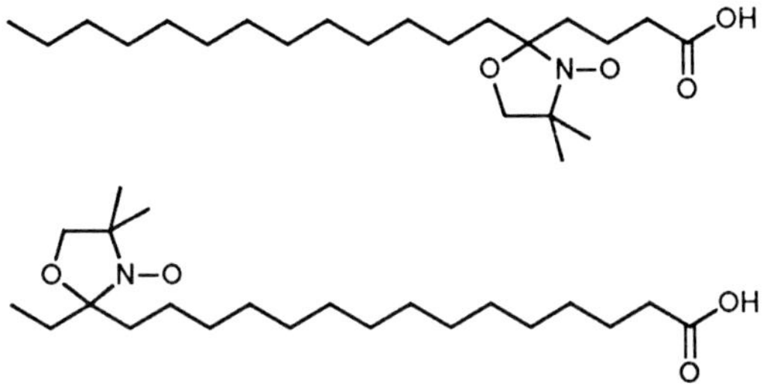

Figure 2   *Chemical structures of nitroxide radicals*

behaviour of the apolar chain of the fatty acids (Figure 2). The spectra obtained with 5SA in buffer alone or in the presence of whey proteins are shown in Figure 1. The spectrum of 5SA or 16SA $(0.25\,\text{mg ml}^{-1})$ in buffer reveals that the nitroxides have rapid motions and are well dispersed. On adding whey proteins to this solution, a high proportion of 5SA molecules were found to have lost mobility. The resulting ESR spectrum is composed of two components reflecting two populations of radicals: one with slow motions (population I), and the other one with much more rapid motions (population M), the mobility being similar to that in the buffer solution without protein. The comparison with simulated spectra allows us to estimate that more than 95% of the radicals have slow motions. They are assumed therefore to be interacting with the proteins. With 16SA, all the fatty acids remained mobile. The proteins can be arranged in order of increasing values of R, *i.e.* of increasing affinity for the 5SA: α-lactalbumin $< \beta$-lactoglobulin = whey protein mixture. The high affinity of β-lactoglobulin for the fatty acids was confirmed at pH 5; at this pH, β-lactoglobulin is protected by the fatty acids against aggregation, whereas α-lactalbumin is not.

The nitroxide fatty acids were also used as probes dispersed in the oil, in order to study the interactions between proteins and lipids in food emulsions. Upon

Table 1   *Interactions between whey proteins and 5- and 16-doxyl stearic acids (5SA; 16SA) in solutions and in emulsions (I/M = relative amplitudes of the immobilized and mobile radical spectra, D_{rot} = reorientational frequency of the nitroxide radicals, a_0 = polarity of the medium)*

Sample	I/M	D_{rot}/s^{-1}	a_0/G
5SA Nitroxide			
Buffer	0	1.0×10^9	16.4
Whey proteins	1.0	0.5×10^9	15.7
Emulsion (buffer)	0	1.0×10^8	14.5
Emulsion + whey proteins	0.8	7.0×10^7	14.5
16SA Nitroxide			
Buffer	0	1.6×10^9	15.7
Whey proteins	0	0.9×10^9	15.5
Emulsion (buffer)	0	1.6×10^8	14.5
Emulsion + whey proteins	0	2.3×10^6	14.5

incorporation into the oil droplets, the mobility of the nitroxide was reduced by a factor of *ca.* 10, independent of the position of the nitroxide on the chain of the fatty acid. The apolar nature of the environment of the nitroxides was confirmed by the a_0 value. In the presence of whey proteins, strongly immobilized 5SA molecules were observed. Conversely, 16SA molecules were less affected: that is, no population with slow motions was apparent. The differences in behaviour between β-lactoglobulin and α-lactalbumin, already observed in solution, were confirmed in emulsions containing the 5SA nitroxides. In the emulsion, the mobility of the amino-acid side chains of the proteins is less than in solution.[2] The polarity of the environment of these residues is reduced.

Modifications of the whey proteins have been performed in order to investigate the effect of changes in protein structure on interaction with lipids. (I) With trypsic hydrolysis of the β-lactoglobulin, despite the resulting higher accessibility of the protein residues, the fatty-acid–protein interactions were inhibited by the progress of the hydrolysis. (II) With disulphide bridge reduction and carboxy-methylation with iodoacetamide, the fatty-acid–protein interactions were significantly enhanced.

Adsorption of Proteins on the Oil Droplets—Despite its lower surface hydrophobicity, measured by phase partioning,[3] β-lactoglobulin was found to be adsorbed preferentially on the oil droplets. This was observed with the whey proteins, or with mixtures of β-lactoglobulin and α-lactalbumin. These results confirm the higher affinity of β-lactoglobulin for the lipids revealed by ESR measurements. Dickinson *et al.*[4] observed that it was more difficult to displace β-lactoglobulin than α-lactalbumin from the oil–water interface. However, Euston[5] showed, by measuring the interfacial concentration of proteins, no preference for either α-lactalbumin or β-lactoglobulin at the n-tetradecane–water interface of freshly made emulsions. Reduced proteins were found to be adsorbed preferentially relatively to the control proteins.[1]

Emulsifying Properties of Whey Proteins—The proteins β-lactoglobulin and α-lactalbumin have similar emulsfying activity (IAE). However, the stabilizing effect of β-lactoglobulin is higher (Table 2). The reduction of the disulphide bridges decreases the emulsifying properties of both proteins. On the other hand, limited trypsic hydrolysis improves these properties for β-lactoglobulin.

4 Conclusions

From our results it appears that in emulsions stabilized by whey proteins: (1) milk proteins and lipids interact preferentially through their polar groups; (2) the polar and very accessible amino-acid residues of the protein participate in the interaction; (3) the milk proteins can be depicted as adsorbing on the fat droplets without affecting the major features of the lipid organization; and (4) the affinity of β-lactoglobulin for fatty acids is higher than the affinity of α-lactalbumin. No close correlation was found between the affinity of the proteins for the lipids and the emulsifying properties of the proteins.

Table 2 *Surface properties of whey proteins. Influence of reduction and carboxymethylation of SH groups and proteolysis by trypsin (IAE = emulsifying activity, Dh = degree of hydrolysis)*

	IAE (m² g⁻¹)	Emulsion stability; separated oil (vol%)[a]	Amount of adsorbed protein (mg protein/ml oil)[b]
β-lactoglobulin			
native	690 ± 30	7 ± 1	
reduced	387 ± 20	60 ± 7	1.65
Dh = 0.5%	780 ± 25	16 ± 2	
Dh = 5%	820 ± 35	29 ± 3	
α-lactalbumin			
native	670 ± 35	40 ± 5	
reduced	417 ± 18	55 ± 5	0.57
Dh = 0.5%	280 ± 10	56 ± 5	
Dh = 5%		92 ± 8	

[a] After centrifugation of the emulsion (1000 g, 100 min).
[b] Measurements carried out by SDS–PAGE on lipid phase of the emulsion after centrifugation (1000 g, 40 min).

References

1. B. Closs, Ph.D. Thesis, University of Burgundy, 1990.
2. M. Le Meste, B. Colas, B. Closs, and J.-L. Courthaudon, in 'Macromolecular Interactions and Food Colloid Stability', eds. N. Parris and R. Barford, ACS Symposium Series, Washington, D.C., 1990, in press.
3. M. Shimizu, M. Saito, and Y. Yamauchi, *Agric. Biol. Chem.*, 1986, **49**, 189.
4. E. Dickinson, S. E. Rolfe, and D. G. Dalgleish, *Food Hydrocolloids*, 1989, **3**, 193.
5. S. E. Euston, 'Competitive adsorption of milk proteins at oil–water interfaces', Ph.D. Thesis, University of Leeds, 1989.